Algorithmic Learning in a Random World

Vladimir Vovk • Alexander Gammerman •
Glenn Shafer

Algorithmic Learning in a Random World

Second Edition

Springer

Vladimir Vovk
Royal Holloway
University of London
London, UK

Alexander Gammerman
Royal Holloway
University of London
London, UK

Glenn Shafer
Rutgers University
Newark, NJ, USA

ISBN 978-3-031-06651-1 ISBN 978-3-031-06649-8 (eBook)
https://doi.org/10.1007/978-3-031-06649-8

This Springer imprint is published by the registered company Springer Nature Switzerland AG
The registered company address is: Gewerbestrasse 11, 6330 Cham, Switzerland

Contents

Part I Set Prediction

Preface to the Second Edition

The second edition contains three new chapters, and the old chapters have been revised, updated, and, in many cases, expanded. The approximate correspondence between the chapters in the first and second editions is summarized in Table 1. These are the main changes in the second edition:

- The structure of the book has become clearer. Now it is divided into four parts, the first two dealing with prediction under the assumption of randomness, the third part devoted to testing the assumption of randomness, and the fourth devoted to prediction under generalized randomness.
- A new chapter, Chap. 7, has been added in Part II. It treats a new subject, conformal predictive distributions and their application to decision-making.
- The part of the book devoted to testing (Part III) has been greatly expanded. It has two new chapters, Chaps. 9 and 10.
- Sections 3.2–3.5 of the first edition have been removed, since they are based on criteria of efficiency for conformal prediction different from the ones that we recommend in the second edition. This material has been replaced by a thorough analysis of various criteria of efficiency (Sect. 3.1 in the second edition).

Table 1 The chapters in the second edition (from 1 to 13) and their counterparts in the first edition (given in parentheses, from 1 to 10); "corr. to" stands for "roughly corresponds to"

1 (corr. to 1)		
Part I		
2 (corr. to 2)	3 (corr. to 3)	4 (corr. to 4)
Part II		
5 (corr. to 5)	6 (corr. to 6)	7 (new)
Part III		
8 (corr. to 7)	9 (new)	10 (new)
Part IV		
11 (corr. to 8)	12 (corr. to 9)	13 (corr. to 10)

- We have simplified the definition of validity of Venn prediction (Chap. 6 of the second edition, corresponding to Chap. 6 of the first edition). The price to pay is that our new notion of validity is much more basic; however, it is also more accessible to readers who are not familiar with game-theoretic probability [317, 320].

The development of the theory of conformal prediction has been active, especially over the last several years. The goal of covering all new developments would have been too ambitious, and in the new edition, we concentrated on the work (often our own) that is most closely connected with the material of the first edition. Some of the new developments are briefly described in a new section, Sect. 13.5.

In the new edition, all misprints and errors in the first edition, known to us today, have been corrected. We are grateful to Stijn Vanderlooy, Fabio Stella, David Lindsay, Misha Dashevsky, and Dima Devetyarov for pointing out some of them. Discussions and correspondence with numerous colleagues have led to several clarifications and are gratefully appreciated; those colleagues include Dmitry Adamskiy, Ernst Ahlberg, Anastasios Angelopoulos, Alexander Balinsky, Rina Foygel Barber, Stephen Bates, David Bell, Tony Bellotti, Claus Bendtsen, Henrik Boström, Brian Burford, Evgeny Burnaev, Emannuel Candés, Lars Carlsson, Giovanni Cherubin, Alexey Chervonenkis, Dave Cohen, Nicolo Colombo, David Cooper, A. Philip Dawid, Wang Di, Thomas Dietterich, David L. Dowe, Martin Eklund, Paul Embrechts, Ola Engkvist, Valentina Fedorova, Matteo Fontana, Patrizio Giovannotti, Leo Gordon, Danqiao Guo, Yuri Gurevich, Jan Hannig, Glenn Hawe, Emily Hector, Alan Hutchinson, Drago Indjic, Ian Jacobs, Ulf Johansson, Wouter Koolen, Antonis Lambrou, Martin Larsson, Steffen Lauritzen, Rikard Laxhammar, Jing Lei, Lihua Lei, Pitt Lim, Xiaohui Liu, Philip M. Long, Zhiyuan Luo, Valery Manokhin, Ryan Martin, Lutz Mattner, Sally McClean, Robert C. Merton, Ilya Muchnik, Khuong Nguyen, Ulf Norinder, Ilia Nouretdinov, Nell Painter, Nicola Paoletti, Harris Papadopoulos, Dusko Pavlovic, Ivan Petej, Aaditya Ramdas, Daljit Rehal, Yaniv Romano, Johannes Ruf, Alessio Sancetta, Teddy Seidenfeld, Jieli Shen, Evgueni Smirnov, James Smith, Ryan Tibshirani, Paolo Toccaceli, Alexandre Tsybakov, Vladimir Vapnik, Jesus Vega, Denis Volkhonskiy, George Vostrov, Vladimir V'yugin, Ruodu Wang, Larry Wasserman, Chris Watkins, Peter Westfall, Bob Williamson, Minge Xie, Meng Yang, Fedor Zhdanov, Chenzhe Zhou, and Andrej Zukov Gregoric. Thanks to the members of the TEX-LATEX Stack Exchange community, first of all David Carlisle, Phelype Oleinik, and Werner for help with LATEX. In our computational experiments, we have used `scikit-learn` [274].

We thank the following companies and funding bodies for generous financial support: AFOSR, Amazon Web Services, AstraZeneca, BBSRC, BMBF, Centrica, Cyprus Research Promotion Foundation, EPSRC, EU Horizon 2020 Research and Innovation programme, Hewlett-Packard, Leverhulme Magna Carta Doctoral

Centre at Royal Holloway, University of London, Mitie, MRC, National Natural Science Foundation of China, Natural Sciences and Engineering Research Council of Canada, NHS England, Royal Society, Stena Line, US NSF, VLA, and Yandex.

London, UK — Vladimir Vovk
London, UK — Alexander Gammerman
Newark, NJ, USA — Glenn Shafer
July 2022

Preface to the First Edition

This book is about prediction algorithms that learn. The predictions these algorithms make are often imperfect, but they improve over time, and they are *hedged*: they incorporate a valid indication of their own accuracy and reliability. In most of the book we make the standard assumption of randomness: the examples the algorithm sees are drawn from some probability distribution, independently of one another. The main novelty of the book is that our algorithms learn and predict simultaneously, continually improving their performance as they make each new prediction and find out how accurate it is. It might seem surprising that this should be novel, but most existing algorithms for hedged prediction first learn from a training dataset and then predict without ever learning again. The few algorithms that do learn and predict simultaneously do not produce hedged predictions. In later chapters we relax the assumption of randomness to the assumption that the data come from an online compression model. We have written the book for researchers in and users of the theory of prediction under randomness, but it may also be useful to those in other disciplines who are interested in prediction under uncertainty.

This book has its roots in a series of discussions at Royal Holloway, University of London, in the summer of 1996, involving AG, Vladimir Vapnik and VV. Vapnik, who was then based at AT&T Laboratories in New Jersey, was visiting the Department of Computer Science at Royal Holloway for a couple of months as a part-time professor. VV had just joined the department, after a year at the Center for Advanced Study in Behavioral Sciences at Stanford. AG had become the head of department in 1995 and invited both Vapnik and VV to join the department as part of his programme (enthusiastically supported by Norman Gowar, the college principal) of creating a machine learning centre at Royal Holloway. The discussions were mainly concerned with Vapnik's work on support vector machines, and it was then that it was realized that the number of support vectors used by such a machine could serve as a basis for hedged prediction.

Our subsequent work on this idea involved several doctoral students at Royal Holloway. Ilia Nouretdinov has made several valuable theoretical contributions. Our other students working on this topic included Craig Saunders, Tom Melluish, Kostas Proedrou, Harris Papadopoulos, David Surkov, Leo Gordon, Tony Bellotti,

Daniil Ryabko, and David Lindsay. The contribution of Yura Kalnishkan and Misha Vyugin to this book was less direct, mainly through their work on predictive complexity, but still important. Thank you all.

GS joined the project only in the autumn of 2003, although he had earlier helped develop some of its key ideas through his work with VV on game-theoretic probability; see their joint book—*Probability and Finance: It's Only a Game!*—published in 2001.

Steffen Lauritzen introduced both GS and VV to repetitive structures. In VV's case, the occasion was a pleasant symposium organized and hosted by Lauritzen in Aalborg in June 1994. We have also had helpful conversations with Masafumi Akahira, Satoshi Aoki, Peter Bramley, John Campbell, Alexey Chervonenkis, Philip Dawid, José Gonzáles, Thore Graepel, Gregory Gutin, David Hand, Fumiyasu Komaki, Leonid Levin, Xiao Hui Liu, George Loizou, Zhiyuan Luo, Per Martin-Löf, Sally McClean, Boris Mirkin, Fionn Murtagh, John Shawe-Taylor, Sasha Shen', Akimichi Takemura, Kei Takeuchi, Roger Thatcher, Vladimir V'yugin, David Waltz, and Chris Watkins.

Many ideas in this book have their origin in our attempts to understand the mathematical and philosophical legacy of Andrei Nikolaevich Kolmogorov. Kolmogorov's algorithmic notions of complexity and randomness (especially as developed by Martin-Löf and Levin) have been for us the main source of intuition, although they almost disappeared in the final version. VV is grateful to Andrei Nikolaevich for steering him in the direction of compression modelling and for his insistence on its independent value.

We thank the following bodies for generous financial support: EPSRC through grants GR/L35812, GR/M14937, GR/M16856, and GR/R46670; BBSRC through grant 111/BIO14428; MRC through grant S505/65; the Royal Society; the European Commission through grant IST-1999-10226; NSF through grant 5-26830.

University of London (VV, AG, and GS)
Rutgers University (GS)
July 2004

Vladimir Vovk
Alexander Gammerman
Glenn Shafer

Notation and Abbreviations

Sets, Bags, and Sequences

$\emptyset$	The empty set
$\mathbb{N}$	The positive integer numbers, $\{1, 2, \dots\}$
$\mathbb{N}_0$	The nonnegative integer numbers, $\{0, 1, \dots\}$
$\mathbb{Z}$	The integer numbers
$\mathbb{Q}$	The rational numbers
$\mathbb{R}$	The real numbers
$\overline{\mathbb{R}}$	The extended real numbers, $\mathbb{R} \cup \{-\infty, \infty\}$
$\{z_1, \dots, z_n\}$	Set (each element enters only once)
$\wr z_1, \dots, z_n \int$	Bag (can contain more than one copy of the same element); Sect. 2.2.1
$(z_1, \dots, z_n)$	Sequence (the parentheses and commas may be omitted)
$\Box$	The empty sequence
$[a_1, \dots, a_n]$	The set of all infinite continuations of a finite sequence $a_1, \dots, a_n$
$\lvert A \rvert$	The size of a set or bag A
Z^n	The set of all sequences of elements of Z of length n
$Z^{(n)}$	The set of all bags of elements of Z of size n
Z^*	The set of all finite sequences of elements of Z
$Z^{(*)}$	The set of all bags (always finite) of elements of Z
Z^∞	The set of all infinite sequences of elements of Z
2^Z	The set of all subsets of a set Z
Y^X	The set of all functions of the type $X \to Y$

Stochastics

$\mathbb{P}$	Probability
$\mathbb{E}$	Expectation

$\mathbf{P}(Z)$	The set of all probability distributions on Z (measurable space)
$\mathbf{B}_\delta$	The Bernoulli distribution on $\{0, 1\}$ with parameter δ: $\mathbf{B}_\delta\{1\} = \delta$ and $\mathbf{B}_\delta\{0\} = 1 - \delta$
$\mathrm{Ber}(\pi)$	The power distribution $\mathbf{B}_\pi^\infty$ on $\{0, 1\}^\infty$
$\mathbf{U}$	The uniform distribution on $[0, 1]$
$\mathbf{N}_{\mu,\sigma^2}$	The normal distribution on $\mathbb{R}$ with mean μ and variance σ^2 (often written as $\mathbf{N}(\mu, \sigma^2)$)
Φ	The distribution function of the standard normal distribution $\mathbf{N}_{0,1}$
ϕ	The density function of the standard normal distribution $\mathbf{N}_{0,1}$
$\mathbf{t}_{\delta,n}$	The $(1 - \delta)$-quantile of the t-distribution: $\mathbb{P}\{\xi \geq \mathbf{t}_{\delta,n}\} = \delta$, where ξ has Student's t-distribution with n degrees of freedom
$\mathbf{z}_\delta$	The $(1 - \delta)$-quantile of the standard normal distribution: $\mathbb{P}\{\xi \geq \mathbf{z}_\delta\} = \delta$, where ξ has the normal distribution $\mathbf{N}_{0,1}$
$Q_\mathbf{X}$	Marginal distribution of Q on X (Sect. 3.1.4)
$Q_{\mathbf{Y}\mid\mathbf{X}}$	Conditional Q-distribution on $\mathbf{Y}$ given $x \in \mathbf{X}$ (Sect. 3.1.5)
$\Omega \hookrightarrow Z$	Markov kernel from Ω to Z (Sect. A.4)
Pf^{-1}	The image of P under a mapping f (Sect. A.1.1)
ρ	Variation distance between probability distributions (Sect. 7.6)
$\stackrel{\text{law}}{\longrightarrow}$	Convergence in law (Sect. A.1.2)

Machine Learning

$\mathbf{X}$	Object space (Sect. 2.1.1)
$\mathbf{Y}$	Label space, $\lvert\mathbf{Y}\rvert > 1$ (Sect. 2.1.1)
$\mathbf{Z}$	The example space ($\mathbf{Z} = \mathbf{X} \times \mathbf{Y}$, Sect. 2.1.1)
$\mathbf{H}$	Feature space (with the feature mapping $F : \mathbf{X} \to \mathbf{H}$; Sect. 2.3.4)

Programming

$+=$	is used in the sense of Python: $a \mathrel{+}= b$ is equivalent to $a := a + b$
$-=$	$a \mathrel{-}= b$ is equivalent to $a := a - b$

Confidence Prediction

ϵ	Significance level
Γ_n^ϵ	The prediction set at trial n (Sect. 2.1.3)
err_n^ϵ	The indicator of error at trial n (Eq. (2.8))
Err_n^ϵ	The cumulative number of errors up to trial n (Eq. (2.9))

OE_n^ϵ	The cumulative observed excess up to trial n (Protocol 3.1)
OF_n^ϵ	The cumulative observed fuzziness up to trial n (Protocol 3.1)
τ_n	The nth random number used by a randomized confidence predictor (Sect. 2.1.4)

Other Notations

1_A	The indicator function of a set or property A, i.e., $1_A = 1$ if A holds and $1_A = 0$ if not
$f\|_A$	The restriction of a function or Markov kernel f to a subset A of its domain
$\operatorname{diam} A$	The diameter (largest distance between points) of A
$\operatorname{co} A$	The convex hull of a set A in a linear space
$u \cdot v$	The scalar product of vectors u and v
I_n	The identity $n \times n$ matrix (n is omitted if clear from the context)
X'	Matrix X transposed
X^{-1}	The inverse of matrix X
$\operatorname{rank} X$	The rank of matrix X
$u \vee v$	The maximum of u and v, also denoted $\max(u, v)$
$u \wedge v$	The minimum of u and v, also denoted $\min(u, v)$
u^+	$u \vee 0$
u^-	$(-u) \vee 0$
$F(t-)$	The limit of $F(u)$ as u approaches t from below
$F(t+)$	The limit of $F(u)$ as u approaches t from above
$f_n = O(g_n)$	$\limsup_{n\to\infty}(f_n/g_n) < \infty$ (used for $f_n, g_n > 0$)
$f_n = \Theta(g_n)$	$f_n = O(g_n)$ and $g_n = O(f_n)$

Abbreviations

AA	Aggregating Algorithm
ACP	Adjusted conditional probability (idealized conformity measure)
ASP	Adjusted signed predictability (idealized conformity measure)
a.s.	Almost surely (i.e., with probability one)
BKJ	Bayes–Kelly–Jeffreys (martingale)
BRR	Bayesian ridge regression
CCP	Cross-conformal predictor
CCPS	Cross-conformal predictive system
CLS	Conformalized least squares
CP	Conditional probability (idealized conformity measure)
CPD	Conformal predictive distribution
CPS	Conformal predictive system

CRR	Conformalized ridge regression
CSD	Cumulative sum diagram
CTM	Conformal test mmartingale
DIR	Direct isotonic regression
DPS	Deterministic predictive system
FOCVP	Fully object-vonditional Venn pPredictor
GCM	Greatest convex minorant
ICP	Inductive conformal predictor
ICPS	Inductive conformal predictive system
IID	Independent and identically distributed
KRRPM	Kernel ridge regression prediction machine
LCCT	Label-conditional conformal transducer
LSPM	Least squares prediction machine
MCP	Mondrian conformal predictor
MCT	Mondrian conformal transducer
NNR	Nearest neighbours regression
OCM	Online compression model
OOS	One-off structure
PAVA	Pair-adjacent violators algorithm
PDMS	Predictive decision-making system
RPD	Randomized predictive distribution
RPS	Randomized predictive system
SP	Signed predictability (idealized conformity measure)
SVM	Support vector machine
USPS	For the USPS dataset, see Sect. B.1
w.r.	with respect

Chapter 1
Introduction

Abstract In this introductory chapter, we sketch the existing work in machine learning on which we build and then outline the contents of the book.

Keywords Machine learning · Conformal prediction · Venn prediction · Assumption of randomness

1.1 Machine Learning

The rapid development of computer technology during the last several decades has made it possible to solve ever more difficult problems in a wide variety of fields. The development of software has been essential to this progress. The painstaking programming in machine code or assembly language that was once required to solve even simple problems has been replaced by programming in high-level object-oriented languages. We are concerned with the next natural step is this progression—the development of programs that can *learn*, i.e., automatically improve their performance with experience.

The need for programs that can learn was already recognized by Alan Turing [358] in 1950, who argued that it may be too ambitious to write from scratch programs for tasks that even humans must learn to perform. Consider, for example, the problem of recognizing hand-written digits. We are not born able to perform this task, but we learn to do it quite robustly. Even when the hand-written digit is represented as a grey-scale matrix, as in Fig. 1.1, we can recognize it easily, and our ability to do so scarcely diminishes when it is slightly rotated or otherwise perturbed. We do not know how to write instructions for a computer that will produce equally robust performance.

The essential difference between a program that implements instructions for a particular task and a program that learns is adaptability. A single learning program may be able to learn a wide variety of tasks: recognizing hand-written digits and faces, diagnosing patients in a hospital, estimating house prices, etc.

Recognition, diagnosis, and estimation can all be thought of as special cases of prediction. A person or a computer is given certain information and asked to

V. Vovk et al., *Algorithmic Learning in a Random World*,
https://doi.org/10.1007/978-3-031-06649-8_1

Fig. 1.1 A hand-written digit

predict the answer to a question. A broad discussion of learning would go beyond prediction to consider the problems faced by a robot, who needs to act as well as predict. The literature on machine learning, has emphasized prediction, however, and the research reported in this book is in that tradition. We are mostly interested in algorithms that learn to predict well.

1.1.1 Learning Under Randomness

One learns from experience. This is as true for a computer as it is for a human being. In order for there to be something to learn there must be some stability in the environment; it must be governed by constant, or evolving only slowly, laws. And when we learn to predict well, we may claim to have learned something about that environment.

The traditional way of making the idea of a stable environment precise is to assume that it generates a sequence of examples randomly from some fixed probability distribution, say Q, on a fixed space of possible examples, say $\mathbf{Z}$. These mathematical objects, $\mathbf{Z}$ and Q, describe the environment.

The environment can be very complex; $\mathbf{Z}$ can be large and structured in a complex way. This is illustrated by the USPS dataset from which Fig. 1.1 is drawn (see Sect. B.1 in Appendix B). Here an example is any 16×16 image with 31 shades of grey, together with the digit the image represents (an integer between 0 to 9). So there are $31^{16\times 16} \times 10$ (this is approximately 10^{382}) possible examples in the space $\mathbf{Z}$.

In most of this book, we assume that each example consists of an *object* and its *label*. In the USPS dataset, for example, an object is a grey-scale matrix like the one in Fig. 1.1, and its label is the integer between 0 and 9 represented by the grey-scale matrix.

In the problem of recognizing hand-written digits and other typical machine-learning problems, it is the space of objects, the space of possible grey-scale images, that is large. The space of labels is either a small finite set (in what is called *classification problems*) or the set of real numbers (*regression problems*).

When we say that the examples are chosen randomly from Q, we mean that they are independent in the sense of probability theory and all have the distribution Q. They are independent and identically distributed (IID). We call this the *randomness assumption* (or, sometimes, statistical randomness assumption, when we want to emphasize that it is not related to Kolmogorov's algorithmic randomness).

Of course, not all work in machine learning is concerned with learning under randomness. In learning with expert advice, for example, randomness is replaced by a game-theoretic set-up [45, 381]; here a typical result is that the learner can predict almost as well as the best strategy in a pool of possible strategies. In reinforcement learning, which is concerned with rational decision-making in a dynamic environment [351], the standard assumption is Markovian. In this book, we will consider extensions of learning under randomness in Parts III and IV.

1.1.2 Learning Under Unconstrained Randomness

Sometimes we make the randomness assumption without assuming anything more about the environment: we know the space of examples $\mathbf{Z}$, we know that examples are drawn independently from the same distribution, and this is all we know. We know nothing at the outset about the probability distribution Q from which each example is drawn. In this case, we say we are *learning under unconstrained randomness*. Most of the work in this book, like much other work in machine learning, is concerned with learning under unconstrained randomness.

The strength of modern machine-learning methods often lies in their ability to make hedged predictions under unconstrained randomness in a *high-dimensional* environment, where examples have a very large (or infinite) number of components. We already mentioned the USPS dataset, where each example consists of 257 components (16×16 pixels and the label). In machine learning, this number is now considered small, and the problem of learning from the USPS dataset is sometimes regarded as a toy problem.

1.2 A Shortcoming of Statistical Learning Theory

Machine learning has made significant strides in its study of learning under unconstrained randomness. We now have a wide range of algorithms that often work very well in practice: neural networks, decision trees, nearest neighbours algorithms, and naive Bayes methods have been used for decades; newer algorithms include support vector machines, random forests, and boosting, and in recent years neural networks have again become extremely popular in the form of deep learning.

From a theoretical point of view, traditional machine learning's most significant contributions to learning under unconstrained randomness fall under the umbrella of *statistical learning theory*. This theory, which began with the discovery of VC dimension by Vapnik and Chervonenkis in the late 1960s and was partially rediscovered independently by Valiant [359] in 1984, has produced both deep mathematical results and learning algorithms that work well in practice (see Vapnik [366] for a review).

Given a "training" set of examples, a prediction algorithm of machine learning produces what we call a *prediction rule*—a function mapping the objects into the labels. This is illustrated in Fig. 1.4 (where "prediction rule" is referred to as "general rule"). Usually, the value taken by a prediction rule on a new object is a *simple prediction*—a guess that is not accompanied by any statement concerning how accurate it is likely to be. Statistical learning theory does guarantee, however, that, under some conditions on the prediction algorithm, as the training set becomes bigger and bigger these predictions will become more and more accurate with greater and greater probability: they are *probably approximately correct*.

How probably and how approximately? This question has not been answered as well as we might like by statistical learning theory. This is because the theoretical results that might be thought to answer it, the bounds that demonstrate arbitrarily good accuracy with sufficiently large sizes of the training set, are usually too loose to tell us anything interesting for training sets that we actually have. This happens in spite of the empirical fact that the predictions often perform very well in practice. Consider, for example, the problem of recognizing hand-written digits, which we have already discussed. Here we are interested in giving an upper bound on the probability that our learning algorithm fails to choose the right digit; we might like this probability to be less than 0.05, for example, so that we can be 95% confident that the prediction is correct. Unfortunately, typical upper bounds on the probability of error provided by the theory, even for relatively clean datasets such as the USPS dataset we have discussed, are greater than 1; bounds less than 1 can usually be achieved only for very straightforward problems or with very large datasets. This is true even for newer results in which the bound on the accuracy depends on the training set (cf. Sect. 13.2.3).

1.2.1 The Hold-Out Estimate of Confidence

Fortunately, there are less theoretical and more effective ways of estimating the confidence we should have in predictions output by machine-learning algorithms. One of the most effective is the oldest and most naive: the "hold-out" estimate. In order to compute this estimate, we split the available examples into two parts, a training set and a hold-out sample, playing the role of a test set. We apply the algorithm to the training set in order to find a prediction rule, and then we apply this prediction rule to the hold-out sample. The observed rate of errors on the hold-out sample tells us how confident we should be in the prediction rule when we apply it to new examples (for details, see Sect. 1.2.1).

1.2.2 The Contribution of This Book

When we use a hold-out sample to obtain a bound on the probability of error, or when we use an error bound from statistical learning theory, we are *hedging* the prediction—we are adding to it a statement about how strongly we believe it. In this book, we develop a different way of producing hedged predictions. Aside from the

Fig. 1.2 In the problem of digit recognition, we would like to attach lower confidence to the prediction for the image in the middle than to the predictions for the images on the left and the right

elegance of our new methods, at least in comparison with the procedure that relies on a hold-out sample, the methods we develop have several important advantages.

As mentioned in the preface to the first edition, we do not always have the rigid separation between learning and prediction, which is the feature of the traditional approaches that makes hedged prediction feasible. In our basic learning protocol learning and prediction are blended, yet our predictions are hedged.

Second, the hedged predictions produced by our new algorithms are much more confident and accurate. We have, of course, a different notion of a hedged prediction, so the comparison can be only informal; but the difference is so big that there is little doubt that the improvement is real from the practical point of view.

A third advantage of our methods is that the confidence with which the label of a new object is predicted is always tailored not only to the previously seen examples but also to that object (see Fig. 1.2).

1.3 The Online Framework

The new methods presented in this book are quite general; they can be tried out, at least, in almost any problem of learning under randomness. The framework in which we introduce, and then study, these methods is somewhat unusual, however. Most previous theoretical work in machine learning has been in an *offline* framework: one uses a batch of old examples to produce a prediction rule, which is then applied to new examples. We begin instead with a framework that is *online*: one makes predictions sequentially, basing each new prediction on all the previous examples instead of repeatedly using a rule constructed from a fixed batch of examples.

1.3.1 Online Learning

Our basic framework is *online* because we assume that the examples are presented one by one. Each time, we observe the object and predict the label. Then we observe the label and go on to the next example. We start by observing the first object x_1 and predicting its label y_1. Then we observe y_1 and the second object x_2, and predict its label y_2. And so on. At the nth step, we have observed the previous examples

$$(x_1, y_1), \ldots, (x_{n-1}, y_{n-1}) \tag{1.1}$$

and the new object x_n, and our task is to predict y_n. The quality of our predictions should improve as we accumulate more and more old examples. This is the sense in which we are learning.

1.3.2 Online/Offline Compromises

The methods we present in this book are most naturally described and are most amenable to mathematical analysis in the online framework. However, they also extend to relaxations of the online protocol that make it close to the offline setting, and this is important, because most practical problems have at least some offline aspects. If we are concerned with recognizing hand-written zip codes, for example, we cannot always rely on a human teacher to tell us the correct interpretation of each hand-written zip code; why not use such an *ideal teacher* directly for prediction? The relaxation of the online protocol considered in Sect. 3.3 includes "slow teachers", who provide the feedback with a delay, and "lazy teachers", who provide feedback only occasionally. In the example of zip codes recognition, this relaxation allows us to replace constant supervision by using returned letters for teaching or by occasional lessons.

1.3.3 One-Off and Offline Learning

We will sometimes discuss prediction algorithms in the simplest possible setting, which we will refer to as *one-off learning*. A one-off predictor is given a training set (1.1) and a test object x_n; its task is to predict the label y_n of x_n. It is applied only once.

A one-off predictor can be embedded into both online and offline frameworks (see Fig. 1.3). In the online framework, new examples arrive sequentially, their labels are predicted given their objects, and they are added to the training set, as described earlier.

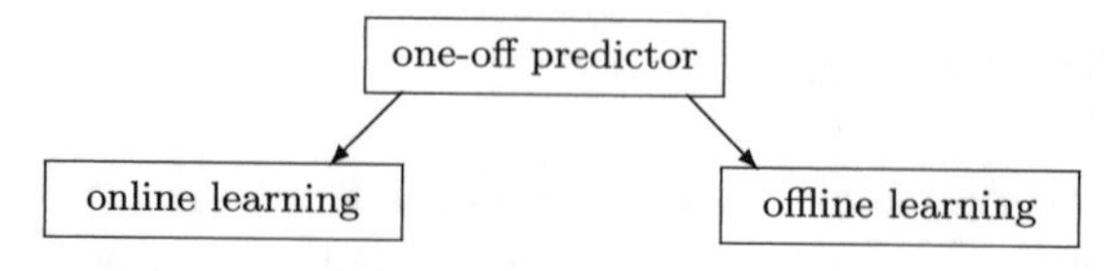

Fig. 1.3 One-off, online, and offline learning: two ways of embedding one-off predictors into a bigger learning protocol

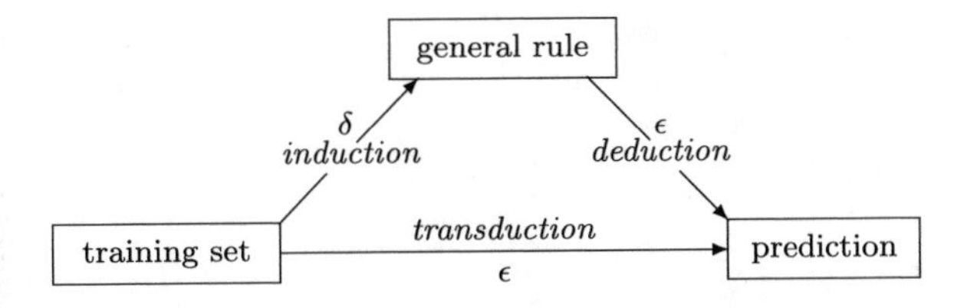

Fig. 1.4 Inductive and transductive (or eductive) prediction

In the offline framework (and sometimes in the one-off framework), there is a shift in our notation. While in the one-off and online settings we usually write $z_1, \dots, z_{n-1}$ for the old examples and z_n for the new example (especially in the online setting), now that we are also interested in the case where the same rule may be applied to many new examples, we usually write $z_1, \dots, z_l$ for the old examples and $z_{l+1}, \dots, z_{l+k}$ for the new examples. The one-off predictor is applied independently k times to the old examples $z_1, \dots, z_l$ and the new object x_i, $i = l + 1, \dots, l + k$, to predict its label y_i.

1.3.4 Induction, Transduction, and the Online Framework

Vapnik's [366] distinction between induction and transduction, as applied to the problem of prediction, is depicted in Fig. 1.4. In *inductive prediction* we first move from examples in hand to some more or less general rule, which we might call a prediction or decision rule, a model, or a theory; this is the *inductive step*. When presented with a new object, or a test set of new objects, we derive a prediction from the general rule; this is the *deductive step*. In *transductive prediction*, we take a shortcut, moving from the old examples directly to the prediction(s) about the new object(s).

In philosophy transduction had been known as eduction since its introduction by W. E. Johnson in 1924 in Part III [182, Chap. IV] of his Magnum Opus *Logic* [180–182]. He points out that the notion (unlike the term) of eduction is well-known, being "the kind of inference which Mill speaks of as from particulars to particulars" (Johnson prefers "inference from instances to instances", which in our terminology becomes "inference from examples to examples"). In the same book Johnson introduced the notion of exchangeability and used it as an assumption for eduction. We will continue the discussion of exchangeability in Sect. 2.9.1.

Typical examples of the inductive step are estimating parameters in statistics and finding a "concept" (to use Valiant's [359] terminology) in statistical learning theory. Examples of transductive prediction are estimation of future observations in statistics (see, e.g., [64, Sect. 7.5]) and nearest neighbours algorithms in machine learning.

In machine learning, transduction is usually discussed in the case where the test set contains a significant number of new objects, whereas in this book we are often

William Ernest Johnson (1858–1931).

The photograph is in public domain

interested in the case of a single new object. When this is not clear from the context, we will refer to this case as *single-object transduction*.

In the case of simple predictions, the distinction between induction and single-object transduction is less than crisp. A method for doing single-object transduction is a method for predicting the label y of a new object x from a training set $x_1, y_1, \dots, x_l, y_l$ and x itself. Such a method gives a prediction for any object that might be presented as x, and so it defines, at least implicitly, a rule, which might be extracted from $x_1, y_1, \dots, x_l, y_l$ (induction), stored, and then subsequently applied to x to predict y (deduction). So any real distinction is really at a practical and computational level: do we extract and store the general rule or not?

For hedged predictions the difference between transduction and induction goes deeper. We will typically want different notions of hedged prediction in the two frameworks. Mathematical results about induction typically involve two parameters, often denoted ϵ (the desired accuracy of the prediction rule) and δ (the probability we are willing to tolerate of failing to achieve the accuracy of ϵ), whereas results about transduction involve only one parameter, which we will denote ϵ (in this book, the probability of error we are willing to tolerate); see Fig. 1.4. A detailed discussion can be found in Chap. 13, which also contains a historical perspective on the three main approaches to hedged prediction (inductive, transductive, and Bayesian).

When we work in the online protocol, we would want to use a general rule extracted from $x_1, y_1, \dots, x_{n-1}, y_{n-1}$ only once, to predict y_n from x_n. After observing x_n and then y_n, we have a larger dataset, $x_1, y_1, \dots, x_n, y_n$, and we can use it to extract a new, possibly improved, general rule before trying to predict y_{n+1} from x_{n+1}. So from a purely conceptual point of view, induction seems silly in the online framework; it is more natural to say that we are doing transduction, even in cases where the general rule is easy to extract. As a practical matter, however, the computational cost of a transductive method may be high, and in this case, it may be sensible to compromise with the inductive approach. After accumulating a certain number of examples, we might extract a general rule and use it for a while, only updating it as frequently as is practical.

1.4 Conformal Prediction

Most of this book is devoted to a particular method that we call "conformal prediction". When we use this method, we predict that a new object will have a label that makes it similar to the old examples in some specified way, and we use the degree to which the specified type of similarity holds within the old examples to estimate our confidence in the prediction. Our conformal predictors are, in other words, "confidence predictors".

We need not explain here exactly how conformal prediction works. This is the topic of the next chapter. But we will explain informally what a confidence predictor aims to do and what it means for it to be valid and efficient.

1.4.1 Nested Prediction Sets

Suppose we want to pinpoint a target that lies somewhere within a rectangular field. This could be an online prediction problem; for each example, we predict the coordinates $y_n \in [a_1, a_2] \times [b_1, b_2]$ of the target from a set of measurements x_n.

We can hardly hope to predict the coordinates y_n exactly. But we can hope to have a method that gives a subset Γ_n of $[a_1, a_2] \times [b_1, b_2]$ where we can be confident y_n lies. Intuitively, the size of the *prediction set* Γ_n should depend on how great a probability of error we want to allow, and in order to get a clear picture, we should specify several such probabilities. We might, for example, specify the probabilities 1%, 5%, and 20%, corresponding to *confidence levels* 99%, 95%, and 80%. When the probability of the prediction set failing to include y_n is only 1%, we declare 99% confidence in the set (highly confident prediction). When it is 5%, we declare 95% confidence (confident prediction). When it is 20%, we declare 80% confidence (casual prediction). We might also want a 100% confidence set, but in practice this might be the whole field assumed at the outset to contain the target.

Figure 1.5 shows how such a family of prediction sets might look. The casual prediction pinpoints the target quite well, but we know that this kind of prediction can be wrong 20% of the time. The confident prediction is much bigger. If we want

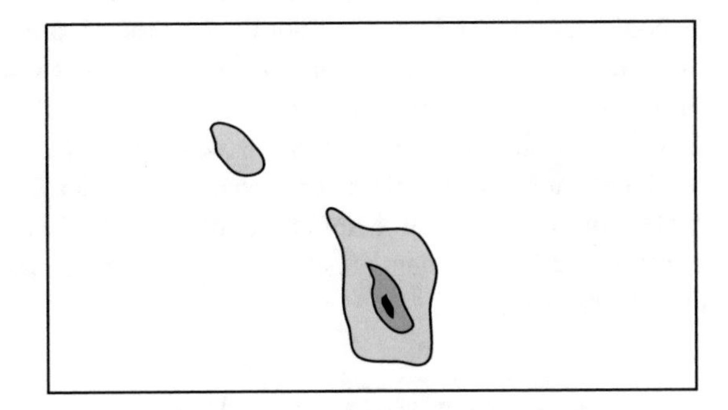

Fig. 1.5 An example of a nested family of prediction sets (casual prediction in black, confident prediction in dark grey, and highly confident prediction in light grey)

to be highly confident (make a mistake only for each 100th example, on average), we must accept an even lower accuracy; there is even a completely different location that we cannot rule out at this level of confidence. In principle, a confidence predictor outputs prediction sets for all confidence levels, and these sets are nested, as in Fig. 1.5.

There are two important desiderata for confidence predictors:

- They should be *valid*, in the sense that in the long run the frequency[1] of error does not exceed ϵ at each chosen confidence level $1-\epsilon$.
- They should be *efficient*, in the sense that the prediction sets they output are as small as possible.

We would also like the predictor to be as conditional as possible—e.g., we want it to take full account of how difficult the particular example is.

1.4.2 Validity

Our conformal predictors are always valid (although we have different kinds of validity for different varieties of conformal predictors; here we concentrate on the basic variety, full conformal predictors). Figure 1.6 confirms empirically the validity for one particular conformal predictor that we study in Chap. 3. The solid, dash-dot, and dotted lines show the cumulative number of errors for the confidence levels 99%, 95%, and 80%, respectively; the value of each graph at n is the total number of errors made over the first n examples. As expected, the number of errors made grows linearly, and the slope is approximately 20% for the confidence level 80%, 5% for the confidence level 95%, and 1% for the confidence level 99%. Some details of our computational experiments will be described in Appendix B (Sect. B.5).

As we will see in Chap. 2, a precise discussion of the validity of conformal predictors actually requires that we distinguish two kinds of validity: conservative and exact. In general, a conformal predictor is conservatively valid: the probability it makes an error when it outputs a prediction set at a confidence level $1-\epsilon$ is no greater than ϵ, and there is little dependence between errors it makes when predicting successive examples (at successive *trials*, or *steps*, as we will say). This implies, by the law of large numbers, that the long-run frequency of errors at confidence level $1-\epsilon$ is about ϵ or less. In practice, the conservativeness is often not great, especially when n is large, and so we get empirical results like those in Fig. 1.6, where the long-run frequency of errors is close to ϵ. From a theoretical point of view, however, we must introduce a small element of deliberate randomization into the prediction process in order to get exact validity, where the probability of a prediction set at a confidence level $1-\epsilon$ being in error is exactly ϵ, errors are made independently at different trials, and the long-run frequency of errors converges to ϵ.

[1] By “frequency” we usually mean “relative frequency”.

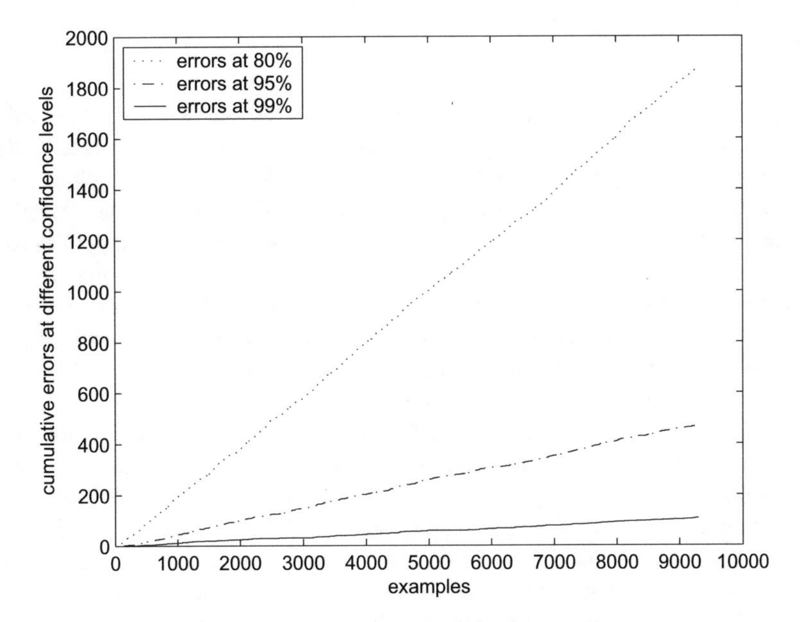

Fig. 1.6 Online performance of a conformal predictor ("the 1-nearest neighbour conformal predictor", described in Sect. 3.2.1) on the USPS dataset (9298 hand-written digits, randomly permuted) for the confidence levels 80, 95, and 99%. The figures in this book are not too much affected by statistical variation (due to the random choice of the permutation of the dataset)

1.4.3 Efficiency

Machine learning has been mainly concerned with two types of problems:

- Classification, where the label space $\mathbf{Y}$ is a small finite set (often binary).
- Regression, where the label space is the real line (or, sometimes, its simple subset, such as the positive real numbers).

In regression problems, the prediction set output by a conformal predictor is often an interval of values, and a natural measure of efficiency of such a prediction is the length of the interval, at different confidence levels. In classification problems, a natural measure of efficiency is the size of the prediction set—the number of labels in it, perhaps apart from the true label (which will be in the prediction set with high probability for a valid predictor). Measuring efficiency in the case of classification will be one of the major topics of Chap. 3.

As we will see in Chaps. 2 and 3, there are many conformal predictors for any particular online prediction problem, whether it is a classification problem or a regression problem. Indeed, we can construct a conformal predictor from any method for scoring the similarity (conformity, as we call it) of a new example to old ones. All these conformal predictors, it turns out, are valid. Which is most efficient—which produces the smallest prediction sets in practice—will depend on details of the environment that we may not know in advance.

1.4.4 Conditionality

The goal of conditionality can be explained with a simple example discussed by David Cox [62]. Suppose there are two categories of objects, "easy" (easy to predict) and "hard" (hard to predict). We can tell which objects belong to which category, and the two categories occur with equal probability; about 50% of the objects we encounter are easy, and 50% hard. We have a prediction method that applies to all objects, hard and easy, and has error rate 5%. We do not know what the error rate is for hard objects, but perhaps it is 8%, and we get an overall error rate of 5% only because the rate for easy objects is 2%. In this situation, we may feel uncomfortable, when we encounter a hard object, about appealing to the average error rate of 5% and saying that we are 95% confident of our prediction.

Whenever there are features of objects that we know make the prediction easier or harder, we would like to take these features into account—to condition on them. This is done by conformal predictors almost automatically: they are designed for specific applications so that their predictions take fullest possible account of the individual object to be predicted. What is not achieved automatically is the validity separately for hard and easy objects. It is possible, for example, that if a figure such as Fig. 1.6 were constructed for easy objects only, or for hard objects only, the slopes of the cumulative error lines would be different. We would get the correct slope if we average the slope for easy objects and the slope for hard objects, but we would ideally like to have the "conditional validity": validity for both categories of objects. As we show in Sect. 4.6, this can be achieved by modifying the definition of conformal predictors. In fact, the conditional validity is handled by a general theory that also applies when we segregate examples not by their difficulty but by their time of arrival, as when we are using an inductive rule that we update only at specified intervals.

1.4.5 Flexibility of Conformal Predictors

A useful feature of our method is that a conformal predictor can be built on top of almost any machine-learning algorithm. The latter, which we call the *underlying algorithm*, may produce hedged predictions, simple predictions, or simple predictions complemented by ad hoc measures of confidence; our experience is that it is always possible to transform it into a conformal predictor that inherits its predictive performance but is, of course, valid, just like any other conformal predictor. In this book we explain how to build conformal predictors using such methods as nearest neighbours, support vector machines, bootstrap, boosting, neural networks, decision trees, random forests, ridge regression, logistic regression, and any Bayesian algorithm (see Sects. 2.3, 2.4, 3.2, and 4.3).

1.5 Probabilistic Prediction Under Unconstrained Randomness

There are many ways to do classification and regression under unconstrained randomness and for high-dimensional examples. Conformal predictors, for example, combine good theoretical properties with high accuracy in practical problems. It is true that the environment has to be benign, in some sense, for any learning method to be successful, but there are no obvious insurmountable barriers for classification and regression. The situation changes if we move to the harder problem of *probabilistic prediction*: that of guessing the probability distribution for the new object's label. Features of data that can reasonably be expected in typical machine-learning applications become such barriers.

For simplicity, we will assume in this section, apart from the last short subsection, that the label is binary, 0 or 1. In this case the probabilistic prediction for the label of the new object boils down to one number, the predicted probability that the label is 1.

The problem of probabilistic prediction is discussed in Chaps. 5–7 and 12–13. Probabilistic prediction is impossible in an important sense, but there are also senses in which it is possible. So this book gives more than one answer to the question "Is probabilistic prediction possible?" We start with a "yes" answer.

1.5.1 *Universally Consistent Probabilistic Predictor*

Stone [347] showed that a nearest neighbours probabilistic predictor (whose probabilistic prediction is the fraction of objects classified as 1 among the k nearest neighbours of the new object, with a suitably chosen k) is *universally consistent*, in the sense that the difference between the probabilistic prediction and the true conditional probability given the object that the label is 1 converges to zero in probability. The only essential assumption is randomness;[2] there are no restrictive regularity conditions.

Stone's actual result was more general, and it has been further extended in different directions. We will state similar results in Chaps. 6 and 7 in the context of conformal prediction.

[2] The other assumption made by Stone was that the objects were coming from a Euclidean space; since "Euclidean" is equivalent to "standard Borel" in the context of existence of a universally consistent probabilistic predictor, this assumption is very weak. See Sect. A.1.1 for further information about standard Borel spaces.

1.5.2 Probabilistic Prediction Using a Finite Dataset

The main obstacle in applying Stone's theorem is that the convergence it asserts is not uniform. The situation that we typically encounter in practice is that we are given a set of examples and a new object, and we would like to estimate the probability that the label of the new object is 1. It is well understood that in this situation the applicability of Stone's theorem is very limited (see, e.g., [82, Sect. 7.1]). In Chap. 5 we give another, more direct, formalization of this observation.

We say that a dataset consisting of old examples and one new object is *diverse* if no object in it is repeated (in particular, the new object is different from all old objects). The main result of Sect. 5.3 asserts that any nontrivial (not empty and not containing 0 and 1) valid prediction interval for the conditional probability given the new object that the new label is 1 is inadmissible if the dataset is diverse and randomness is the only assumption.

The assumption that the dataset is diverse is related to the assumption of a high-dimensional environment. If the objects are, for example, complex images, we will not expect precise repetitions among them.

1.5.3 Venn Prediction

The results of Chap. 5 show that it is impossible to estimate the true conditional probabilities under the conditions stated; that chapter also contains a result (Theorem 5.5) that it is impossible to find conditional probabilities that are as good (in the sense of the algorithmic theory of randomness) as the true probabilities. If, however, we are prepared to settle for less and only want probabilities that are "well calibrated" (in other words, have a frequentist justification), a modification of conformal predictors which we call Venn predictors will achieve this goal, in a strong non-asymptotic sense. This is the subject of Chap. 6.

1.5.4 Conformal Predictive Distributions

Venn predictors only work in the case of classification. It turns out that in the case of regression conformal prediction is capable of outputting predictive distribution functions, which we call conformal predictive distributions. This is the topic of Chap. 7. It is interesting that while our approach to probabilistic prediction in the case of classification requires a nontrivial modification of conformal prediction, no such modifications are needed for regression. An application of conformal predictive distributions discussed in Chap. 7 is to decision making: practically meaningful loss functions can be combined with conformal predictive distributions to arrive at optimal decisions.

1.6 Beyond Randomness

In this book we also consider testing the assumption of randomness and alternatives to this assumption. The most radical alternative is introduced in Chaps. 11 and 12 under the name of "online compression modelling".

1.6.1 Testing Randomness

This is the topic of Part III. We start in Chap. 8 by adapting the mathematical apparatus developed in the previous chapters to testing the assumption of randomness. The usual statistical approach to testing (sometimes called the "Neyman–Pearson–Wald" theory) is essentially offline: in the original Neyman–Pearson approach (see, e.g., [222]), the sample size is chosen a priori, and in Wald's [419] sequential analysis, the sample size is data-dependent but still at some point a categorical decision on whether the null hypothesis is rejected or not is taken (with probability one). Our approach is online: we constantly update the strength of evidence against the null hypothesis of randomness. Finding evidence against the null hypothesis involves gambling against it, and the strength of evidence equals the gambler's current capital. For further details and the history of this approach to testing, see [316, 317, 320]. The main mathematical finding of Sect. 8.1 is that there exists a wide family of "exchangeability martingales", which can be successfully applied to detecting lack of randomness. In the rest of Chap. 8 we discuss different strategies for deciding when a machine-learning algorithm needs to be retrained because of the change in the distribution of the data.

In Chap. 9 we explore the efficiency of our approach to testing randomness, which we call *conformal testing*. (It turns out surprisingly efficient, although we only consider the binary case.) And in Chap. 10 we adapt the ideas of conformal testing to protecting machine-learning algorithms against changes in the distribution of the data.

1.6.2 Online Compression Models

As we will see in Chap. 11, the idea of conformal prediction generalizes from learning under randomness, where examples are IID, to "online compression models". In an online compression model, it is assumed that the data can be summarized in a way that can be updated as new examples come in, and the only known probabilities are backward probabilities—probabilities, given a summary, for the previous summary and the last example summarized.

Online compression models derive from the work of Andrei Kolmogorov. They open a new direction for broadening the applicability of machine-learning methods, giving a new meaning to the familiar idea that learning can be understood as information compression.

In Chap. 11 we consider in detail two important online compression models (exchangeability and Gaussian) and their variants. In Chap. 12 we extend the idea of Venn prediction to online compression modelling and introduce another interesting model of this kind, the hypergraphical model, which can be used in causal inference.

1.7 Context

Each chapter of this book ends with a section entitled "Context". These sections are set in a small font and may use mathematical notions and results not introduced elsewhere in the book. Almost all of them are preceded by a section (also set in a small font) containing longer proofs.

Turing suggested the idea of machine learning in his paper published in *Mind* as an approach to solving his famous "imitation game" [358, Sect. 1].

The terms "transduction" and "transductive inference" were introduced in machine learning by Vapnik [365, Sect. 6.1] in 1995. His setting of transduction, without using this terminology, was introduced in his 1971 booklet [363] and then repeated in [370] and [364]. See [366, Comments and bibliographical remarks, Chap. 8] and [367, Chap. 3] for further historical information. For a very readable review of Johnson's eduction, see Dale [68].

An alternative term for our "prediction rule" that is popular in machine learning is "prediction model". We do not use "prediction model" since the word "model" is already overloaded in this book.

An empirical study of various bounds on prediction accuracy is reported in [211]. It found the hold-out estimate to be a top performer.

Part I
Set Prediction

Chapter 2
Conformal Prediction: General Case and Regression

Abstract In this chapter we formally introduce conformal predictors. After giving the necessary definitions, we will prove that when a conformal predictor is used in the online mode, its output is valid, not only in the asymptotic sense that the sets it predicts for any fixed confidence level $1 - \epsilon$ will be wrong with frequency at most ϵ (approaching ϵ in the case of smoothed conformal predictors) in the long run, but also in a much more precise sense: the error probability of a smoothed conformal predictor is ϵ at every trial and errors happen independently at different trials. In Sect. 2.6 we will see that conformal prediction is indispensable for achieving this kind of validity. The basic procedure of conformal prediction might look computationally inefficient when the label set is large, but in Sect. 2.3 we show that in the case of, e.g., least squares regression (where the label space $\mathbb{R}$ is uncountable) there are ways of making conformal predictors much more efficient. In addition to validity, we also discuss the efficiency of conformal predictors.

Keywords Conformal prediction · Underlying algorithm · Conformalized ridge regression · Conformalized nearest neighbours regression

2.1 Confidence Predictors

The conformal predictors we define in this chapter are confidence predictors—they make a range of successively more specific predictions with successively less confidence. In this section we define precisely what we mean by a confidence predictor and its validity.

2.1.1 Assumptions

We assume that Reality outputs successive pairs

$$(x_1, y_1), (x_2, y_2), \dots, \tag{2.1}$$

V. Vovk et al., *Algorithmic Learning in a Random World*,
https://doi.org/10.1007/978-3-031-06649-8_2

called *examples*. Each example (x_i, y_i) consists of an *object* x_i and its *label* y_i. The objects are elements of a measurable space $\mathbf{X}$ called the *object space*, and the labels are elements of a measurable space $\mathbf{Y}$ called the *label space*. We assume that $\mathbf{X}$ is nonempty and that $\mathbf{Y}$ contains at least two essentially different elements.[1] When we need a more compact notation, we write z_i for (x_i, y_i). We set $\mathbf{Z} := \mathbf{X} \times \mathbf{Y}$ and call $\mathbf{Z}$ the *example space*. Thus the infinite data sequence (2.1) is an element of the measurable space $\mathbf{Z}^\infty$.

When we say that the objects are *absent*, we mean that $|\mathbf{X}| = 1$. In this case x_i do not carry any information and do not need to be mentioned; we will then identify $\mathbf{Y}$ and $\mathbf{Z}$ and refer to the examples as *observations*. (The distinction between examples and observations will not always be clear-cut.)

Our standard assumption is that Reality chooses the examples independently from some probability distribution Q on $\mathbf{Z}$—i.e., that the infinite sequence $z_1, z_2, \dots$ is drawn from the *power probability measure* (or *power distribution*) Q^∞ in $\mathbf{Z}^\infty$. Most of the results of this book hold under this *randomness assumption*, or *randomness model*, but usually we need only the slightly weaker assumption that the infinite data sequence (2.1) is drawn from a distribution P on $\mathbf{Z}^\infty$ that is *exchangeable*. The statement that P is exchangeable means that for every positive integer n, every permutation π of $\{1, \dots, n\}$, and every measurable set $E \subseteq \mathbf{Z}^n$,

$$P\left\{(z_1, z_2, \dots) \in \mathbf{Z}^\infty : (z_1, \dots, z_n) \in E\right\}$$
$$= P\left\{(z_1, z_2, \dots) \in \mathbf{Z}^\infty : (z_{\pi(1)}, \dots, z_{\pi(n)}) \in E\right\} . \tag{2.2}$$

(In other words, the distribution of the examples does not depend on their order.) Every power distribution is exchangeable, and under a natural regularity condition ($\mathbf{Z}$ is a standard Borel space), any exchangeable distribution on $\mathbf{Z}^\infty$ is a mixture of power distributions; for details, see Sect. A.5. In our mathematical results, we usually use the randomness assumption or the exchangeability assumption depending on which one leads to a stronger statement.

2.1.2 Simple Predictors and Confidence Predictors

We assume that at the nth trial Reality first announces the object x_n and only later announces the label y_n. If we simply want to predict y_n, then we need a function

$$D : \mathbf{Z}^* \times \mathbf{X} \to \mathbf{Y} . \tag{2.3}$$

[1] Formally, the σ-algebra on $\mathbf{Y}$ is assumed to be different from $\{\emptyset, \mathbf{Y}\}$. It is convenient to assume that for each pair of distinct elements of $\mathbf{Y}$ there is a measurable set containing only one of them; we will do this without loss of generality, and then our assumption about $\mathbf{Y}$ is that $|\mathbf{Y}| > 1$.

We call such a function a *simple predictor*, always assuming it is measurable. For any sequence of old examples, say $x_1, y_1, \ldots, x_{n-1}, y_{n-1} \in \mathbf{Z}^*$, and any new object, say $x_n \in \mathbf{X}$, it gives $D(x_1, y_1, \ldots, x_{n-1}, y_{n-1}, x_n) \in \mathbf{Y}$ as its prediction for the new label y_n.

As we explained in Sect. 1.4, however, we have a more complicated notion of prediction. Instead of merely choosing a single element of $\mathbf{Y}$ as our prediction for y_n, we want to give a range of more or less precise predictions, each labelled with a degree of confidence. We want to give subsets of $\mathbf{Y}$ large enough that we can be confident that y_n will fall in them, while also giving smaller subsets in which we are less confident. An algorithm that predicts in this sense requires an additional input $\epsilon \in (0, 1)$, which we call the *significance level*; the complementary value $1 - \epsilon$ is called the *confidence level*. Given all these inputs, say $x_1, y_1, \ldots, x_{n-1}, y_{n-1}, x_n, \epsilon$, an algorithm of the type that interests us, say Γ, outputs a subset

$$\Gamma_n^\epsilon = \Gamma^\epsilon(x_1, y_1, \ldots, x_{n-1}, y_{n-1}, x_n)$$

of $\mathbf{Y}$. (We position ϵ as a superscript instead of placing it with the other arguments.) We require this subset to shrink as ϵ is increased: Γ must satisfy

$$\Gamma^{\epsilon_1}(x_1, y_1, \ldots, x_{n-1}, y_{n-1}, x_n) \subseteq \Gamma^{\epsilon_2}(x_1, y_1, \ldots, x_{n-1}, y_{n-1}, x_n) \tag{2.4}$$

whenever $\epsilon_1 \ge \epsilon_2$. Intuitively, once we observe the *incomplete data sequence*

$$x_1, y_1, \ldots, x_{n-1}, y_{n-1}, x_n \tag{2.5}$$

and choose the significance level ϵ, Γ predicts that

$$y_n \in \Gamma^\epsilon(x_1, y_1, \ldots, x_{n-1}, y_{n-1}, x_n) , \tag{2.6}$$

and the smaller ϵ is the more confident the prediction is. According to condition (2.4), we are more confident in less specific predictions.

Formally, we call a measurable function

$$\Gamma : \mathbf{Z}^* \times \mathbf{X} \times (0, 1) \to 2^{\mathbf{Y}} \tag{2.7}$$

($2^{\mathbf{Y}}$ is the set of all subsets of $\mathbf{Y}$) that satisfies (2.4), for all significance levels $\epsilon_1 \ge \epsilon_2$, all positive integers n, and all incomplete data sequences (2.5), a *confidence predictor* (or *deterministic confidence predictor*). The requirement that Γ be measurable means that for each n the set of sequences $\epsilon, x_1, y_1, \ldots, x_n, y_n$ satisfying (2.6) is a measurable subset of $(0, 1) \times (\mathbf{X} \times \mathbf{Y})^n$.

2.1.3 Validity

Let us begin with an intuitive explanation of the notions of exact and conservative validity. For each significance level ϵ, we want to have confidence $1-\epsilon$ in our prediction (2.6) about y_n. This means that the probability of the prediction being in error—the probability of the event (2.6) not happening—should be ϵ. Moreover, since we are making a whole sequence of predictions, first for y_1, then for y_2, and so on, we would like these error events to be independent. If these conditions are met no matter what exchangeable probability distribution P governs the sequence of examples, then we say that the confidence predictor Γ is "exactly valid", or, more briefly, "exact". This means that making errors with Γ is like getting heads when making independent tosses of a biased coin whose probability of heads is always ϵ. If the probabilities for errors are allowed to be even less than this, then we say that the confidence predictor Γ is "conservatively valid" or, more briefly, "conservative".

When we make independent tosses of a biased coin whose probability of heads is always ϵ, the frequency of heads will converge to ϵ with probability one—this is the strong law of large numbers. So the frequency of errors at significance level ϵ for an exactly valid confidence predictor converges to ϵ with probability one. As we will see, confidence predictors sometimes have this asymptotic property even when they are not exactly valid. So we give the property a name of its own; we call a confidence predictor that has it "asymptotically exact". Similarly, we call a confidence predictor for which the frequency of errors is asymptotically no more than ϵ (for which the upper limit of the frequency of errors is at most ϵ) with probability one "asymptotically conservative".

In order to restate these definitions in a way sufficiently precise to exclude possible misunderstandings, we now introduce a formal notation for the errors Γ makes when it processes the data sequence

$$\omega = (x_1, y_1, x_2, y_2, \dots)$$

at significance level ϵ. Whether Γ makes an error on the nth trial can be represented by a number that is 1 in the case of an error and 0 in the case of no error:

$$\begin{aligned} \mathrm{err}_n^\epsilon(\Gamma, \omega) &:= 1_{y_n \notin \Gamma^\epsilon(x_1, y_1, \dots, x_{n-1}, y_{n-1}, x_n)} \\ &:= \begin{cases} 1 & \text{if } y_n \notin \Gamma^\epsilon(x_1, y_1, \dots, x_{n-1}, y_{n-1}, x_n) \\ 0 & \text{otherwise}, \end{cases} \end{aligned} \tag{2.8}$$

and the number of errors during the first n trials is

$$\mathrm{Err}_n^\epsilon(\Gamma, \omega) := \sum_{i=1}^{n} \mathrm{err}_i^\epsilon(\Gamma, \omega) . \tag{2.9}$$

Protocol 2.1 Online confidence prediction

1: $\mathrm{Err}_0^\epsilon := 0$ for all $\epsilon \in (0, 1)$
2: **for** $n = 1, 2, \ldots$
3: Reality outputs $x_n \in \mathbf{X}$
4: Predictor outputs $\Gamma_n^\epsilon \subseteq \mathbf{Y}$ for all $\epsilon \in (0, 1)$
5: Reality outputs $y_n \in \mathbf{Y}$
6: $\mathrm{err}_n^\epsilon := 1_{y_n \notin \Gamma_n^\epsilon}$ for all $\epsilon \in (0, 1)$
7: $\mathrm{Err}_n^\epsilon := \mathrm{Err}_{n-1}^\epsilon + \mathrm{err}_n^\epsilon$ for all $\epsilon \in (0, 1)$.

The online learning protocol with the error counts err_n^ϵ and Err_n^ϵ is shown as Protocol 2.1. This is a game protocol, and we can consider strategies for both players, Reality and Predictor. We have assumed that Reality's strategy is randomized; her moves are generated from an exchangeable probability distribution P on $\mathbf{Z}^\infty$. A confidence predictor Γ is, by definition, a measurable strategy for Predictor.

If ω is drawn from an exchangeable probability distribution P, the number $\mathrm{err}_n^\epsilon(\Gamma, \omega)$ is the realized value of a random variable, which we designate $\mathrm{err}_n^\epsilon(\Gamma)$. Formally, the confidence predictor Γ is *exactly valid* if for each ϵ,

$$\mathrm{err}_1^\epsilon(\Gamma), \mathrm{err}_2^\epsilon(\Gamma), \ldots \tag{2.10}$$

is a sequence of independent Bernoulli random variables with parameter ϵ—i.e., if it is a sequence of independent random variables each of which has probability ϵ of being one and probability $1 - \epsilon$ of being zero—no matter what exchangeable distribution P we draw ω from. Unfortunately, the notion of exact validity is vacuous for confidence predictors: see Theorem 2.2 below.

The notion of conservative validity is more complex; now we only require that $\mathrm{err}_n^\epsilon(\Gamma)$ be *dominated* by a sequence of independent Bernoulli random variables with parameter ϵ. Let $\mathbf{U}$ be the uniform probability measure on $[0, 1]$. For each probability measure P on $\mathbf{Z}^\infty$, the probability space

$$(\mathbf{Z}^\infty \times [0, 1], P \times \mathbf{U}) \tag{2.11}$$

is the product measurable space $\mathbf{Z}^\infty \times [0, 1]$ equipped with the product probability measure $P \times \mathbf{U}$. Formally, the confidence predictor Γ is *conservatively valid* if, for any exchangeable probability distribution P on $\mathbf{Z}^\infty$, there exists a family

$$\xi_n^{(\epsilon)}, \quad \epsilon \in (0, 1), \quad n = 1, 2, \ldots, \tag{2.12}$$

of $\{0, 1\}$-valued random variables on the product probability space $(\mathbf{Z}^\infty \times [0, 1], P \times \mathbf{U})$ such that:

- for all n and ϵ, $\xi_n^{(\epsilon)}$ depends only on the first n examples;
- for any fixed ϵ, $\xi_1^{(\epsilon)}, \xi_2^{(\epsilon)}, \ldots$ is a sequence of independent Bernoulli random variables with parameter ϵ on the probability space $(\mathbf{Z}^\infty \times [0, 1], P \times \mathbf{U})$;
- for all n and ϵ, $\mathrm{err}_n^\epsilon(\Gamma) \le \xi_n^{(\epsilon)}$.

Intuitively, the auxiliary probability space $([0, 1], \mathbf{U})$ represents extra randomization; the random variables $\text{err}_n^\epsilon(\Gamma)$ are extended to the product probability space $\mathbf{Z}^\infty \times \mathbf{U}$ in a natural way, by ignoring the extra randomization.

To conclude, we define asymptotic validity. The confidence predictor Γ is *asymptotically exact* if, for any exchangeable probability distribution P on $\mathbf{Z}^\infty$ generating examples $z_1, z_2, \ldots$ and any significance level ϵ,

$$\lim_{n\to\infty} \frac{\text{Err}_n^\epsilon(\Gamma)}{n} = \epsilon \tag{2.13}$$

with probability one. It is *asymptotically conservative* if, for any exchangeable probability distribution P on $\mathbf{Z}^\infty$ and any significance level ϵ,

$$\limsup_{n\to\infty} \frac{\text{Err}_n^\epsilon(\Gamma)}{n} \le \epsilon \tag{2.14}$$

with probability one.

Proposition 2.1 *An exact confidence predictor is asymptotically exact. A conservative confidence predictor is asymptotically conservative.*

This proposition is an immediate consequence of the law of large numbers.

2.1.4 Randomized Confidence Predictors

We will also be interested in randomized confidence predictors, which depend, additionally, on elements of an auxiliary probability space. The main advantage of randomization in this context is that, as we will see, there are many randomized confidence predictors that are exactly valid. Formally, we define a *randomized confidence predictor* to be a measurable function

$$\Gamma : (\mathbf{X} \times [0, 1] \times \mathbf{Y})^* \times (\mathbf{X} \times [0, 1]) \times (0, 1) \to 2^{\mathbf{Y}} \tag{2.15}$$

which, for all significance levels $\epsilon_1 \ge \epsilon_2$, all positive integer n, and all incomplete data sequences

$$x_1, \tau_1, y_1, \ldots, x_{n-1}, \tau_{n-1}, y_{n-1}, x_n, \tau_n , \tag{2.16}$$

where $x_i \in \mathbf{X}$, $\tau_i \in [0, 1]$, and $y_i \in \mathbf{Y}$ for all i, satisfies

$$\begin{aligned} &\Gamma^{\epsilon_1}(x_1, \tau_1, y_1, \ldots, x_{n-1}, \tau_{n-1}, y_{n-1}, x_n, \tau_n) \\ &\qquad \subseteq \Gamma^{\epsilon_2}(x_1, \tau_1, y_1, \ldots, x_{n-1}, \tau_{n-1}, y_{n-1}, x_n, \tau_n) . \end{aligned} \tag{2.17}$$

We will always assume that $\tau_1, \tau_2, \ldots$ are random variables that are independent (between themselves and of anything else) and distributed uniformly on [0, 1]; this is what one expects to get from a random number generator. (Usually there is no dependence of Γ on $\tau_1, \ldots, \tau_{n-1}$ in its argument (2.16).)

We define $\mathrm{err}_n^\epsilon(\Gamma, \omega)$ and $\mathrm{Err}_n^\epsilon(\Gamma, \omega)$ as before, only now they also depend on the τ's. In other words, $\mathrm{err}_n^\epsilon(\Gamma, \omega)$ is defined by (2.8) with x_i now being *extended objects* $x_i \in \mathbf{X} \times [0, 1]$; $\mathrm{Err}_n^\epsilon(\Gamma, \omega)$ is defined, as before, by (2.9).

In general, many definitions for randomized confidence predictors are special cases of the corresponding definitions for deterministic confidence predictors: the latter should just be applied to the *extended object space* $\mathbf{X} \times [0, 1]$ (whose elements consist of both the object and the random number to be used at a given trial).

If Γ is a randomized confidence predictor, P is an exchangeable distribution on $\mathbf{Z}^\infty$, and $n \in \mathbb{N}$,

$$\mathrm{err}_n^\epsilon(\Gamma, (x_1, \tau_1, y_1, x_2, \tau_2, y_2, \ldots)) \text{ and } \mathrm{Err}_n^\epsilon(\Gamma, (x_1, \tau_1, y_1, x_2, \tau_2, y_2, \ldots)) ,$$

where $(x_1, y_1), (x_2, y_2), \ldots$ are drawn from P and $\tau_1, \tau_2, \ldots$ are drawn independently from $\mathbf{U}^\infty$, the uniform distribution on $[0, 1]^\infty$, are random variables, which will be denoted $\mathrm{err}_n^\epsilon(\Gamma)$ and $\mathrm{Err}_n^\epsilon(\Gamma)$, respectively. We say that Γ is *exactly valid* if for each $\epsilon \in (0, 1)$, (2.10) is a sequence of independent Bernoulli random variables with parameter ϵ.

We are not really interested in the notion of conservative validity for randomized confidence predictors.

2.1.5 Confidence Predictors Over a Finite Horizon

A trivial modification of online prediction is were Reality generates only N examples $z_1, \ldots, z_N$. We have the same protocol, Protocol 2.1, except that n now goes from 1 to N. We will refer to it as the *online protocol with horizon* N.

This modification, however, is important, because it allows us to relax the assumption about the data-generating mechanism and still have valid confidence predictors. We say that a probability measure P on $\mathbf{Z}^N$ is *exchangeable* if we have the following simplified version of (2.2): for every permutation π of $\{1, \ldots, N\}$ and every measurable $E \subseteq \mathbf{Z}^N$,

$$P(E) = P\left\{(z_1, \ldots, z_N) \in \mathbf{Z}^N : (z_{\pi(1)}, \ldots, z_{\pi(N)}) \in E\right\} .$$

The definitions of conservative and exact validity directly carry over to the online protocol with horizon N and exchangeable data-generating measures on $\mathbf{Z}^N$. Achieving validity in this sense (e.g., by the method of conformal prediction, as discussed in the next section) is valuable since the assumption that $z_1, \dots, z_N$ are exchangeable is much weaker than the assumption that the sequence can be extended to a longer exchangeable sequence $z_1, \dots, z_{N+1}$, let alone the assumption that it can be extended to an infinite exchangeable sequence $z_1, z_2, \dots$. (See Sect. 2.9.1 below.)

The requirement of exchangeability of examples $z_1, \dots, z_N$ in a benchmark dataset (such as the USPS dataset) is often violated (see, e.g., Sect. 8.1.2), but it is easy to enforce by randomly permuting it. This way we will enforce the exchangeability of the examples, whereas there is no easy way to enforce their randomness.

2.1.6 One-Off and Offline Confidence Predictors

Let us start from a terminological remark. In the offline mode of prediction we are given two collections of examples, which we usually refer to as the *training set* and *test set* (as we already did in Chap. 1), in order to use familiar terms. In the one-off mode, the test set consists of only one example. Formally the training and test sets, however, are not sets but sequences. The order of examples in them often does not matter, but even in such cases, they are still not sets since they may contain the same element more than once (and so they are bags in the terminology that will be introduced in Sect. 2.2.1).

In the one-off mode we are given a training set $z_1, \dots, z_l$ of examples $z_i = (x_i, y_i)$ and a *test object* $x_{l+1} \in \mathbf{X}$, and the problem is to predict the label y_{l+1} of x. A *one-off confidence predictor* is a measurable function

$$\Gamma : \mathbf{Z}^l \times \mathbf{X} \times (0, 1) \to 2^{\mathbf{Y}}$$

(cf. (2.7)) that always satisfies (2.4), where $n := l + 1$. Similarly, a *one-off randomized confidence predictor* is a measurable function

$$\Gamma : (\mathbf{X} \times [0, 1] \times \mathbf{Y})^l \times (\mathbf{X} \times [0, 1]) \times (0, 1) \to 2^{\mathbf{Y}}$$

(cf. (2.15)) that always satisfies (2.17) with $n := l + 1$.

Let us say that a one-off confidence predictor is *conservatively valid under randomness* if its probability of error at any significance level $\epsilon \in (0, 1)$ is at most ϵ under any power distribution Q^{l+1} on the examples. Being *conservatively valid under exchangeability* is defined in the same way except for replacing the power distributions with the exchangeable distributions on $\mathbf{Z}^{l+1}$. Finally, in being *conservatively valid under infinite exchangeability* the assumption of exchangeability of $z_1, \dots, z_{l+1}$ is replaced by the assumption that those examples are the first $l + 1$ in an infinite exchangeable sequence $z_1, z_2, \dots$. All these notions of validity are weak in that they do not involve any claim of independence of errors (similar to what we required for conservative or exact validity of online predictors).

The task of designing valid one-off confidence predictors can be set under various assumptions, such as randomness, infinite exchangeability (where the available examples are the first $l+1$ of an infinite exchangeable sequence), and the exchangeability of only the available examples $z_1, \ldots, z_l, (x_{l+1}, y_{l+1})$. In the next section we will see that even the last version is feasible.

For one-off confidence predictors, both randomized and deterministic, we have the useful notion of *exact validity under randomness*, i.e., their probability of error at any significance level $\epsilon \in (0,1)$ being exactly ϵ under any power distribution Q^{l+1} on the examples. In the same way we define *exact validity under infinite exchangeability* and *exact validity under exchangeability*. These notions, however, are only useful for randomized predictors.

Theorem 2.2 *No one-off confidence predictor is exactly valid under randomness.*

See Sect. 2.8.1 for a proof.

One-off confidence predictors can be used in the offline setting as well, where, along with the training set $z_1, \ldots, z_l$, we also have a test set $z_{l+1}, \ldots, z_{l+k}$, where $z_i = (x_i, y_i)$. For each test object x_i, the prediction set for y_i is $\Gamma^\epsilon(z_1, \ldots, z_l, x_i)$.

2.2 Conformal Predictors

We start by defining the concept of a nonconformity measure. Intuitively, this is a way of measuring how strange an example is in a bag of examples. There are many different nonconformity measures, and each one defines a conformal predictor and a smoothed conformal predictor.

2.2.1 Bags

The order in which examples appear often does not make any difference, and in order to formalize this point, we need the concept of a bag (also called a multiset). A *bag* of size $n \in \mathbb{N}$ is a collection of n elements some of which may be identical; a bag resembles a set in that the order of its elements is irrelevant, but it differs from a set in that repetition is allowed. To identify a bag, we must say what elements it contains and how many times each of these elements is repeated. We write $\Lbag z_1, \ldots, z_n \Rbag$ for the bag consisting of elements $z_1, \ldots, z_n$, some of which may be identical with each other. The empty bag $\Lbag\Rbag$ contains no elements. We are only interested in finite bags.

We write $\mathbf{Z}^{(n)}$ for the set of all bags of size n of elements of a measurable space $\mathbf{Z}$. The set $\mathbf{Z}^{(n)}$ is itself a measurable space. It can be defined formally as the power space $\mathbf{Z}^n$ with a non-standard σ-algebra, consisting of measurable subsets of $\mathbf{Z}^n$ that contain all permutations of their elements. We write $\mathbf{Z}^{(*)}$ for the set of all bags of elements of $\mathbf{Z}$ (the union of all the $\mathbf{Z}^{(n)}$).

Operations such as union ($\cup$), intersection ($\cap$), and difference ($\setminus$) are defined for bags in a natural way. For example, if $A, B \in Z^{(*)}$ are two bags, the number of times their union $A \cup B$ contains any $z \in Z$ is the number of times A contains z plus the number of times B contains z.

2.2.2 *Nonconformity and Conformity*

As we have already said, a nonconformity measure is a way of scoring how strange an example is as an element of a bag. Formally, a *nonconformity measure* is a measurable mapping

$$A : \mathbf{Z}^{(*)} \times \mathbf{Z} \to \overline{\mathbb{R}} ; \tag{2.18}$$

to each possible bag of examples and each possible example, in or outside the bag, A assigns a numerical score indicating how different the new example is from the old ones. In our definitions we will ignore the values $A(\sigma, z)$ for examples $z \notin \sigma$.

It is easy to invent nonconformity measures, especially when we already have methods for prediction at hand:

- If the examples are merely numbers ($\mathbf{Z} = \mathbb{R}$), then we might define the nonconformity of an example as the absolute value of its difference from the average of all examples in the bag. Alternatively, we could use instead the absolute value of its difference from the median of all examples.
- In a regression problem where examples are pairs of numbers, say $z_i = (x_i, y_i)$, we might define the nonconformity of an example (x, y) as the absolute value of the difference between y and the predicted value $\hat{y}$ calculated from x and all examples in the bag.

We will discuss the latter way of detecting nonconformity further at the end of this section. But whether a particular function on $\mathbf{Z}^{(*)} \times \mathbf{Z}$ is an appropriate way of measuring nonconformity will always be open to discussion, and we do not need to enter into this discussion at this point. In our general theory, we will call *any* measurable function on $\mathbf{Z}^{(*)} \times \mathbf{Z}$ taking values in the extended real line a nonconformity measure.

Given a nonconformity measure A, a sequence $z_1, \dots, z_n$ of examples, and an example z, we can score how different z is from the bag $\Lbag z_1, \dots, z_n \Rbag$; namely, $A(\Lbag z_1, \dots, z_n \Rbag, z)$ is called the *nonconformity score* for z. We will sometimes refer to $\Lbag z_1, \dots, z_n \Rbag$ as the *comparison bag* in this context.

Of course, instead of looking at functions that we feel measure nonconformity, we could look at functions that we feel measure conformity. We call such a function, say B, a conformity measure, and we can use it to define *conformity scores* $B(\Lbag z_1, \dots, z_n \Rbag, z)$. Formally, a *conformity measure* is a measurable function of the type $\mathbf{Z}^{(*)} \times \mathbf{Z} \to \overline{\mathbb{R}}$ (so there is no difference between conformity measures and nonconformity measures as mathematical objects). If we begin in this way, then nonconformity appears as a derivative idea. Given a conformity measure B we can define a nonconformity measure A using any strictly decreasing transformation, say $A := -B$ or perhaps (if B takes only positive values) $A := 1/B$. Given our goal, prediction, beginning with conformity might seem the more natural approach. As we will explain shortly, our strategy for prediction is to predict that a new label will be among the labels that best make a new example conform with old examples, and it is more natural to emphasize the labels that we include in the prediction (the most conforming ones) rather than the labels that we exclude (the most nonconforming ones). But in practice, it is often more natural to begin with nonconformity measures. For example, when we compare a new example with an average of old examples, we will usually first define a distance between the two rather than devise a way to measure their closeness. For this reason, we often (but far from always) emphasize nonconformity rather than conformity. Doing so is consistent with tradition in mathematical statistics, where test statistics are usually defined so as to measure discrepancy rather than agreement.

2.2.3 *p-Values*

Given a nonconformity measure A and a bag $\Lbag z_1, \dots, z_n \Rbag$, we can compute the nonconformity score

$$\alpha_i := A\left(\Lbag z_1, \dots, z_n \Rbag, z_i\right) \tag{2.19}$$

for each example z_i in the bag. Because a nonconformity measure A may be scaled however we like, the numerical value of α_i does not, by itself, tell us how unusual A finds z_i to be. For that, we need a comparison of α_i to the other α_j. A convenient way of making this comparison is to compute the fraction

$$\frac{\left|\{j = 1, \dots, n : \alpha_j \geq \alpha_i\}\right|}{n}. \tag{2.20}$$

This fraction, which lies between $1/n$ and 1, is called the *p-value* for z_i. It is the fraction of the examples in the bag as nonconforming as z_i. If it is small (close to its lower bound $1/n$ for a large n), then z_i is very nonconforming (an outlier). If it is large (close to its upper bound 1), then z_i is very conforming.

If we begin with a conformity measure rather than a nonconformity measure, then we can define the p-value for z_i by

$$\frac{\left|\{j = 1, \dots, n : \beta_j \leq \beta_i\}\right|}{n},$$

where the β_j are the conformity scores. This gives the same result as we would obtain from (2.20) using a nonconformity measure A obtained from B by means of a strictly decreasing transformation.

2.2.4 Definition of Conformal Predictors

Every nonconformity measure determines a confidence predictor. Given a new object x_n and a level of significance, this predictor provides a prediction set that should contain the object's label y_n. We obtain the set by supposing that y_n will have a value that makes (x_n, y_n) conform with the previous examples. The level of significance determines the amount of conformity (as measured by the p-value) that we require.

Formally, the *conformal predictor determined by a nonconformity measure* A is the confidence predictor Γ obtained by setting

$$\Gamma^{\epsilon}(x_1, y_1, \ldots, x_{n-1}, y_{n-1}, x_n) \tag{2.21}$$

equal to the set of all labels $y \in \mathbf{Y}$ such that

$$\frac{|\{i = 1, \ldots, n : \alpha_i \geq \alpha_n\}|}{n} > \epsilon , \tag{2.22}$$

where

$$\begin{aligned} \alpha_i &:= A\big(\sigma, (x_i, y_i)\big), \quad i = 1, \ldots, n-1, \\ \alpha_n &:= A\big(\sigma, (x_n, y)\big), \\ \sigma &:= \wr (x_1, y_1), \ldots, (x_{n-1}, y_{n-1}), (x_n, y) \int . \end{aligned} \tag{2.23}$$

We will sometimes refer to σ defined in this way as the *augmented training set*. In general, a *conformal predictor* is a conformal predictor determined by some nonconformity measure.

The left-hand side of (2.22) is the p-value of (x_n, y) in the bag consisting of it and the old examples (cf. (2.20)). So our prediction with significance level ϵ (or confidence level $1 - \epsilon$) is that the value of y_n will make (x_n, y_n) have a p-value greater than ϵ when it is bagged with the old examples. We are 98% confident, for example, that we will get a value for y_n that gives (x_n, y_n) a p-value greater than 0.02. In other words, we are 98% confident that

$$\frac{\text{number of examples that conform worse or the same as } (x_n, y_n)}{n}$$

will exceed 0.02.

If A is a conformity measure, the conformal predictor determined by A is defined by (2.22) with "$\geq$" replaced by "$\leq$".

2.2.5 *Validity*

Proposition 2.3 *All conformal predictors are conservative.*

It follows by Proposition 2.1 that a conformal predictor is asymptotically conservative. Of course, more can be said. Using the law of the iterated logarithm instead of the law of large numbers, we can strengthen (2.14) to

$$\limsup_{n\to\infty} \frac{\mathrm{Err}_n^\epsilon(\Gamma) - n\epsilon}{\sqrt{2\epsilon(1-\epsilon)n \ln\ln n}} \leq 1 .$$

We will also state two finite-sample implications of Proposition 2.3: Hoeffding's inequality (see Sect. A.6.3) implies that, for any positive integer N and any constant $\delta > 0$,

$$P\left\{\omega : \mathrm{Err}_N^\epsilon(\Gamma, \omega) \geq N(\epsilon + \delta)\right\} \leq \mathrm{e}^{-2N\delta^2} ;$$

the central limit theorem implies that, for any constant c,

$$\limsup_{N\to\infty} P\left\{\omega : \mathrm{Err}_N^\epsilon(\Gamma, \omega) \geq N\epsilon + c\sqrt{N}\right\} \leq \Phi\left(-c/\sqrt{\epsilon(1-\epsilon)}\right) ,$$

where Φ is the standard normal distribution function. For a graphical illustration of asymptotic conservativeness, see Fig. 1.6.

2.2.6 *Smoothed Conformal Predictors*

In this section we introduce a modification of conformal predictors which will allow us to simplify and strengthen Proposition 2.3. The *smoothed conformal predictor determined by the nonconformity measure* A is the following randomized confidence predictor Γ: the set

$$\Gamma^\epsilon(x_1, \tau_1, y_1, \ldots, x_{n-1}, \tau_{n-1}, y_{n-1}, x_n, \tau_n)$$

consists of the $y \in \mathbf{Y}$ satisfying

$$\frac{|\{i = 1, \ldots, n : \alpha_i > \alpha_n\}| + \tau_n |\{i = 1, \ldots, n : \alpha_i = \alpha_n\}|}{n} > \epsilon , \tag{2.24}$$

where α_i are defined, as before, by (2.23). The left-hand side of (2.24) is called the *smoothed p-value*.

The main difference of (2.24) from (2.22) is that in the former we treat the borderline cases $\alpha_i = \alpha_n$ more carefully. Instead of increasing the p-value by $1/n$ for each $\alpha_i = \alpha_n$, we increase it by a random amount between 0 and $1/n$.

When n is not too small, it is typical for almost all $\alpha_1, \dots, \alpha_n$ to be different, and then there is very little difference between conformal predictors and smoothed conformal predictors.

Proposition 2.4 *Any smoothed conformal predictor is exactly valid.*

This proposition will be proved in Chap. 11 (as a special case of Theorem 11.1). It immediately implies Proposition 2.3: if a smoothed conformal predictor Γ and a conformal predictor $\bar{\Gamma}$ are constructed from the same nonconformity measure, the latter's errors $\overline{\mathrm{err}}_n^\epsilon$ never exceed the former's errors err_n^ϵ, $\overline{\mathrm{err}}_n^\epsilon \le \mathrm{err}_n^\epsilon$; besides, the random numbers τ_n in the definition of conformal predictors can be merged into one random number

$$\sum_{n=1}^{\infty} \tau_n 2^{-n} \sim \mathbf{U} \tag{2.25}$$

required in the definition of conservative validity. It also implies

Corollary 2.5 *Every smoothed conformal predictor is asymptotically exact.*

2.2.7 Finite-Horizon Conformal Prediction

The definition of conformal predictors and smoothed conformal predictors also works for the online protocol with horizon N, discussed in Sect. 2.1.5; we just apply the recipe (2.21)–(2.23) (with (2.24) in place of (2.22) for smoothed conformal predictors) for $n = 1, \dots, N$.

Propositions 2.3 and 2.4 continue to hold in the online protocol with horizon N under the assumption, discussed in Sect. 2.1.5, that $z_1, \dots, z_N$ are exchangeable (rather than being the first N examples of an infinite exchangeable sequence). When talking about conservative or exact validity in the online protocol with horizon N, we will mean the definitions given before with n ranging over $1, \dots, N$ and with $z_1, \dots, z_N$ assumed exchangeable.

Proposition 2.6 *In the online protocol with a finite horizon, any conformal predictor is conservatively valid, and any smoothed conformal predictor is exactly valid.*

Proof This is a special case of Theorem 11.2 in Chap. 11.

2.2.8 *One-Off and Offline Conformal Predictors*

The notion of conformal predictor simplifies in the one-off and offline modes, but the price to pay is a weaker notion of validity. In the one-off mode, given a training set $z_1, \ldots, z_l$, $z_i = (x_i, y_i)$, we are predicting the label of a test object $x \in \mathbf{X}$. The *one-off conformal predictor* outputs the prediction set

$$\Gamma^\epsilon(x_1, y_1, \ldots, x_l, y_l, x) := \left\{ y \in \mathbf{Y} : \frac{|\{j = 1, \ldots, l+1 : \alpha_j \geq \alpha_{l+1}\}|}{l+1} > \epsilon \right\}, \tag{2.26}$$

where the nonconformity scores $\alpha_1, \ldots, \alpha_{l+1}$ are computed from a nonconformity measure A:

$$\begin{aligned} \alpha_j &:= A\left(\lbag z_1, \ldots, z_l, (x, y) \rbag, z_j\right), \quad j = 1, \ldots, l, \\ \alpha_{l+1} &:= A\left(\lbag z_1, \ldots, z_l, (x, y) \rbag, (x, y)\right). \end{aligned} \tag{2.27}$$

It is a one-off confidence predictor. We will refer to $n := l + 1$ as its *size*.

According to Proposition 2.4, one-off conformal predictors are conservatively valid under infinite exchangeability and, therefore, under randomness. They are even conservatively valid under exchangeability. As we pointed out in Sect. 2.1.6, our notions of validity in the one-off mode do not involve any claim of independence of errors. An advantage of these weaker notions of validity is that they continue to hold for the following generalized notion of conformal predictor.

The definition of nonconformity scores α_i requires that they should not depend on the order of the examples in the comparison bag. Let us now drop this requirement. A *nonconformity statistic* is defined to be a measurable mapping $S : \mathbf{Z}^* \to \mathbb{R}$; this is a standard notion of a test statistic in the context of permutation tests. Now we can modify (2.26) and (2.27) as

$$\Gamma^\epsilon(z_1, \ldots, z_l, x) := \left\{ y \in \mathbf{Y} : \frac{|\{\pi : S(z_{\pi(1)}, \ldots, z_{\pi(l)}, z_{\pi(l+1)}) \geq S(z_1, \ldots, z_l, z_{l+1})\}|}{(l+1)!} > \epsilon \right\}, \tag{2.28}$$

where $z_{l+1} := (x, y)$ and π ranges over all $(l+1)!$ permutations of the set $\{1, \ldots, l+1\}$. Notice that (2.26) is a special case of (2.28) corresponding to nonconformity statistics that do not depend on the ordering of their arguments apart from the last one:

$$S(z_1, \ldots, z_n) := A(\lbag z_1, \ldots, z_n \rbag, z_n)$$

for some nonconformity measure A. We will say that (2.28) is a *weak conformal predictor*. As in the case of conformal predictors, we will also use conformity statistics, in a similar sense to conformity measures.

In the offline mode of prediction, we are given a training set $z_1, \dots, z_l$ of examples $z_i = (x_i, y_i)$ and the problem is to predict the labels $y_i, i = l+1, \dots, l+k$, of the test examples $z_{l+1}, \dots, z_{l+k}$, given their objects. In this case we apply the one-off conformal predictor to the training set $z_1, \dots, z_l$ and the test object x_i for each $i = l + 1, \dots, l + k$ separately.

2.2.9 General Schemes for Defining Nonconformity

There are many different ways of defining nonconformity measures, but here we will only explain the most basic approaches, which in the following sections will be illustrated in the case of regression, $\mathbf{Y} = \mathbb{R}$. As already mentioned, our nonconformity measures are often based on an underlying algorithm, which is a standard prediction algorithm outputting point predictions (i.e., they are simple predictors in the terminology of Sect. 2.1.2) or probabilistic predictions (this will be relevant in Part II) without any guarantees of validity. In Sects. 2.3 and 2.4 we use ridge regression and nearest neighbours as underlying algorithms, respectively, and in Sect. 2.9.5 we give references to more interesting and sophisticated examples, such as the lasso [224]. We will give more examples in later chapters: in Sect. 3.2 for the case of classification and in Sects. 4.2–4.3 for the cases of both classification and regression. An unusual example will be given in Sect. 4.3.8, where a conformal predictor will be constructed without using an underlying simple predictor.

Conformity to a Bag

Suppose we are given a bag $\Lbag z_1, \dots, z_n \Rbag$ and we want to estimate the nonconformity of each example z_i as an element of the bag, as in (2.19). (It is clear that the values (2.19) are all we need to know about the nonconformity measure, and we will often define nonconformity measures by specifying the nonconformity scores (2.19).) There is a natural solution if we are given a simple predictor (2.3) whose output does not depend on the order in which the old examples are presented. The simple predictor D then defines a prediction rule $D_{\Lbag z_1,\dots,z_n \Rbag} : \mathbf{X} \to \mathbf{Y}$ by the formula

$$D_{\Lbag z_1,\dots,z_n \Rbag}(x) := D(z_1, \dots, z_n, x) .$$

A natural measure of nonconformity of z_i is the deviation of the predicted label

$$\hat{y}_i := D_{\Lbag z_1,\dots,z_n \Rbag}(x_i) \tag{2.29}$$

from the true label y_i. In this way any simple predictor, combined with a suitable measure of deviation of $\hat{y}_i$ from y_i, leads to a nonconformity measure and, therefore, to a conformal predictor.

The simplest way of measuring the deviation of $\hat{y}_i$ from y_i is to take the absolute value of their difference as α_i,

$$\alpha_i := \left| y_i - \hat{y}_i \right| . \tag{2.30}$$

Another approach is to take $\alpha_i := \left| y_i - \hat{y}_{(i)} \right|$, where $\hat{y}_{(i)}$ is the *deleted prediction*

$$\hat{y}_{(i)} := D_{\lbag z_1,\dots,z_{i-1},z_{i+1},\dots,z_n \rbag}(x_i)$$

computed by applying to x_i the prediction rule found from the dataset with the example z_i deleted. The rationale behind this deletion is that z_i, even if it is an outlier, can influence the prediction rule $D_{\lbag z_1,\dots,z_n \rbag}$ so heavily that $\hat{y}_i$ will become close to y_i, even though y_i can be very far from $\hat{y}_{(i)}$. In the rest of this subsection we will only discuss the *ordinary* predictions (2.29), but the modifications to the deleted predictions will be obvious.

More generally, the prediction rule $D_{\lbag z_1,\dots,z_n \rbag}$ may map $\mathbf{X}$ to some *prediction space* $\hat{\mathbf{Y}}$, a measurable space not necessarily coinciding with $\mathbf{Y}$. For example, the elements of $\hat{\mathbf{Y}}$ may be probability distributions on $\mathbf{Y}$, or simple predictions complemented by some measure of their accuracy (such as an estimated variance or, for multidimensional labels, covariance matrix). An *invariant predictor* is a function D that maps each bag $\lbag z_1, \dots, z_n \rbag$ of each size n to a prediction rule $D_{\lbag z_1,\dots,z_n \rbag} : \mathbf{X} \to \hat{\mathbf{Y}}$ and such that the function

$$(\lbag z_1, \dots, z_n \rbag, x) \mapsto D_{\lbag z_1,\dots,z_n \rbag}(x)$$

of the type $\mathbf{Z}^{(n)} \times \mathbf{X} \to \hat{\mathbf{Y}}$ is measurable for all n. A *discrepancy measure* is a measurable function $\Delta : \mathbf{Y} \times \hat{\mathbf{Y}} \to \overline{\mathbb{R}}$. Given an invariant predictor D and a discrepancy measure Δ, we define a function A as follows: for any $((x_1, y_1), \dots, (x_n, y_n)) \in \mathbf{Z}^*$, the nonconformity (or conformity) scores

$$\alpha_i = A(\lbag (x_1, y_1), \dots, (x_n, y_n) \rbag, (x_i, y_i))$$

are defined by the formula

$$\alpha_i := \Delta \left(y_i, D_{\lbag (x_1,y_1),\dots,(x_n,y_n) \rbag}(x_i) \right) . \tag{2.31}$$

This way we obtain a nonconformity measure.

A simple special case of (2.31) is

$$\alpha_i := \frac{|y_i - \hat{y}_i|}{\hat{\sigma}_i}, \tag{2.32}$$

where $\hat{\sigma}_i$ is an estimate of the accuracy of $\hat{y}_i$ computed from $\wr z_1, \dots, z_n \int$ and x_i. It corresponds to an invariant predictor that outputs pairs $(\hat{y}, \hat{\sigma})$ as its predictions.

Conformity to a Property

So far we talked about an important but not the only aspect of conformity, where $A(\sigma, z)$ measures how well z conforms to the bag σ. A standard feature of formalizations of this aspect, such as (2.30) or (2.32), is the presence of the operation of taking the absolute value.

Another important case is where $A(\sigma, z)$ measures how well z conforms to a property that elements of σ may satisfy to different degrees. An example is the modification

$$\alpha_i := y_i - \hat{y}_i \tag{2.33}$$

of (2.30). Defined in this way, α_i show how well z_i conforms to the property of having a large label. The modification

$$\alpha_i := \hat{y}_i - y_i \tag{2.34}$$

of (2.33) shows how well z_i conforms to the property of having a small label. The asymmetric versions (2.33) and (2.34) will often be used later in this chapter and in the rest of the book. The asymmetric versions are still special cases of the general scheme (2.31).

2.2.10 Deleted Conformity Measures

When defining a specific conformity measure A in the following chapters, we will often find it convenient to do it in terms of its *deleted version*

$$A^{\text{del}}(\sigma, z) := A(\sigma \cup z, z) . \tag{2.35}$$

Of course, the conformity measure can be restored given its deleted version as

$$A(\sigma, z) = A^{\text{del}}(\sigma \setminus \{z\}, z)$$

(more, precisely, its values on the relevant part of its domain, $\{(\sigma, z) : z \in \sigma\}$, can be restored).

2.3 Conformalized Ridge Regression

In this section we will implement conformal prediction based on (2.33), (2.34), or their combination concentrating on the case where the underlying simple predictor is ridge regression. In the next section we will adapt the resulting prediction algorithm to nearest neighbours regression.

Ridge regression and nearest neighbours are two of the most standard regression algorithms. In the case of regression there is an obvious difficulty in implementing the idea of conformal prediction: it appears that to form a prediction set (2.21) we need to examine each potential classification y (cf. (2.23)). We will see, however, that in some important cases there is a feasible way to compute (2.21); in particular, this is the case for ridge regression and nearest neighbours regression. (We will make little effort to optimize the computational resources required, so "feasible" essentially means "avoiding examining infinitely many cases" here. Much faster algorithms will be constructed in Sect. 4.2.)

2.3.1 *Least Squares and Ridge Regression*

Ridge regression and its special case, least squares, are among the most widely used regression algorithms. Least squares is the classical algorithm (going back to Gauss and Legendre), and ridge regression is its modification proposed in the 1960s. In this subsection we will describe least squares and ridge regression as simple predictors; for further details, see, e.g., [252].

Suppose $\mathbf{X} = \mathbb{R}^p$ (objects are vectors consisting of p attributes), $\mathbf{Y} = \mathbb{R}$ (we are dealing with the problem of regression), and we are given a training set $z_1, \dots, z_n$. To approximate the data, the ridge regression procedure recommends calculating the value $w \in \mathbb{R}^p$ where

$$a \|w\|^2 + \sum_{i=1}^{n} (y_i - w \cdot x_i)^2 \to \min \tag{2.36}$$

is attained; a is a nonnegative constant called the *ridge parameter*. The ridge regression prediction $\hat{y}$ for the label y of an object x is then $\hat{y} := w \cdot x$. Least squares is the special case corresponding to $a = 0$.

We can naturally represent the ridge regression procedure in a matrix form. Let Y_n be the (column-) vector

$$Y_n := (y_1, \dots, y_n)' \tag{2.37}$$

of the labels and X_n be the $n \times p$ matrix (*data matrix*) formed from the objects

$$X_n := (x_1, x_2, \dots, x_n)' . \tag{2.38}$$

Now we can represent the ridge regression procedure (2.36) as

$$a \|w\|^2 + \|Y_n - X_n w\|^2 \to \min , \tag{2.39}$$

or

$$Y_n' Y_n - 2w' X_n' Y_n + w'(X_n' X_n + a I_p) w \to \min ,$$

I_p being the identity $p \times p$ matrix. Taking the derivative in w we obtain

$$2(X_n' X_n + a I_p) w - 2X_n' Y_n = 0 ,$$

or

$$w = (X_n' X_n + a I_p)^{-1} X_n' Y_n . \tag{2.40}$$

Standard statistical textbooks mainly discuss the case $a = 0$ (least squares; in this case the usual assumption is that the matrix $X_n' X_n$ is invertible). It is easy to see, however, that not only least squares is a special case of ridge regression, but ridge regression can be reduced to least squares as well: the solution of (2.39) for a general $a \geq 0$ can be found as the solution to the least squares problem

$$\left\| \overline{Y} - \overline{X} w \right\|^2 \to \min ,$$

where $\overline{Y}$ is Y_n extended by adding p 0s on top and $\overline{X}$ is X_n extended by adding the $p \times p$ matrix $\sqrt{a} I_p$ on top.

2.3.2 *Basic CRR*

A natural conformal predictor based on ridge regression as underlying algorithm uses (2.30) as nonconformity measure, with $\hat{y}$ computed from ridge regression, and indeed this is the approach that we followed in the first edition [402, Sect. 2.3]. However, a much simpler conformal predictor can be obtained as the combination of the conformal predictor determined by (2.33) as nonconformity measure (and producing upper bounds on the true labels as its predictions) and the conformal predictor determined by (2.34) as nonconformity measure (and producing lower bounds on the true labels as its predictions). This is the strategy that we will follow.

Therefore, we start from the one-sided version of (2.31), with $\Delta(y, \hat{y}) := y - \hat{y}$, considered as nonconformity measure; we will refer to it as the *upper CRR*. The predictions will usually be prediction rays bounded above, i.e., prediction sets of the form $(-\infty, u]$ for some $u \in \mathbb{R}$. (And some applications only require this kind of predictions.)

Now the nonconformity scores α_i are simply the *residuals* $e_i := y_i - \hat{y}_i$, also known as ordinary residuals. Two slightly more sophisticated approaches will be considered in the following subsection.

From (2.40) we can see that the ridge regression prediction for an object x is

$$x'w = x'(X_n'X_n + aI_p)^{-1}X_n'Y_n ;$$

therefore, the predictions $\hat{y}_i$ for the objects x_i are given by

$$\hat{Y}_n := (\hat{y}_1, \ldots, \hat{y}_n)' = X_n(X_n'X_n + aI_p)^{-1}X_n'Y_n .$$

The matrix

$$H_n := X_n(X_n'X_n + aI_p)^{-1}X_n' \tag{2.41}$$

is called the *hat matrix* (since it transforms the y_i into the hatted form $\hat{y}_i$) and plays an important role in the standard regression theory. This matrix, as well as $I_n - H_n$, is symmetric and idempotent when $a = 0$ (remember that a symmetric matrix M is *idempotent* if $MM = M$). Therefore, the vector of nonconformity scores $(\alpha_1, \ldots, \alpha_n)'$ can be written in the form

$$Y_n - H_nY_n = (I_n - H_n)Y_n .$$

Now suppose that we know the incomplete data sequence (2.5), we are given a significance level ϵ, and we want to compute the prediction set output by the conformal predictor determined by the conformity scores $\alpha_i = e_i$. Let y be a possible label for x_n and

$$Y^y := (y_1, \ldots, y_{n-1}, y)' = (y_1, \ldots, y_{n-1}, 0)' + y(0, \ldots, 0, 1)'$$

be the vector of labels. The vector of nonconformity scores is $(I_n - H_n)Y^y = A + yB$, where

$$A := (I_n - H_n)(y_1, \ldots, y_{n-1}, 0)'$$
$$B := (I_n - H_n)(0, \ldots, 0, 1)' .$$

The components of A and B, respectively, will be denoted by $a_1, \ldots, a_n$ and $b_1, \ldots, b_n$.

If we define

$$S_i := \{y : a_i + b_iy \geq a_n + b_ny\} , \tag{2.42}$$

the definition of the p-values can be rewritten as

Algorithm 2.2 Upper conformalized ridge regression (simplified)

Input: a training set $(x_i, y_i) \in \mathbb{R}^p \times \mathbb{R}$, $i = 1, \dots, n-1$, and a test object $x_n \in \mathbb{R}^p$.
1: Set $C := I_n - X_n(X_n' X_n + aI_p)^{-1} X_n'$, X_n being defined by (2.38)
2: set $A = (a_1, \dots, a_n)' := C(y_1, \dots, y_{n-1}, 0)'$
3: set $B = (b_1, \dots, b_n)' := C(0, \dots, 0, 1)'$
4: **for** $i = 1, \dots, n-1$
5: **if** $b_n - b_i > 0$ set $t_i := (a_i - a_n)/(b_n - b_i)$
6: **else** set $t_i := \infty$
7: sort $t_1, \dots, t_{n-1}$ in the ascending order obtaining $t_{(1)} \le \dots \le t_{(n-1)}$
8: output $\left(-\infty, t_{(\lceil (1-\epsilon)n \rceil)}\right]$ as prediction set.

$$p^y := \frac{|\{i = 1, \dots, n : y \in S_i\}|}{n} ;$$

remember that the prediction set includes the labels y with $p^y > \epsilon$.

The case where $b_n \le b_i$ for $i < n$ is clearly anomalous: it means that the predicted label $\hat{y}_i$ is at least as sensitive as $\hat{y}_n$ to changes in the postulated label y of x_n. (And Lemmas 2.16 and 2.17 below will show that this case does not happen from some n on almost surely under natural assumptions.) Our implementation of the conformal predictor will simply set $S_i := \mathbb{R}$ in this case. This convention makes our conformal predictor slightly conservative, but it is natural: we do not want to use rays S_i pointing in the wrong direction for making our prediction set narrower.

Let us now find the upper CRR prediction set under the assumption that $b_n > b_i$ for all $i < n$. In this case each set (2.42) is

$$S_i = (-\infty, t_i], \qquad \text{where } t_i := \frac{a_i - a_n}{b_n - b_i},$$

except for $S_n := \mathbb{R}$; notice that only $t_1, \dots, t_{n-1}$ are defined. The p-value p^y for any potential label y of x_n is

$$p^y = \frac{|\{i = 1, \dots, n : y \in S_i\}|}{n} = \frac{|\{i = 1, \dots, n-1 : t_i \ge y\}| + 1}{n} .$$

Therefore, the upper CRR prediction set at significance level ϵ is the ray

$$(-\infty, t_{(k_n)}] , \tag{2.43}$$

where $k_n := \lceil (1-\epsilon)n \rceil$ and $t_{(k)}$ stands, as usual, for the kth order statistic (defined in Algorithm 2.2) of $t_1, \dots, t_{n-1}$.

The procedure is summarised as Algorithm 2.2. The algorithm is given a significance level ϵ and outputs the corresponding prediction ray.

The prediction set for the lower CRR is very similar; the main change is that (2.43) becomes

Algorithm 2.3 Lower CRR (simplified)

1: **for** $i = 1, \dots, n-1$
2: **if** $b_n - b_i > 0$ set $t_i := (a_i - a_n)/(b_n - b_i)$
3: **else** set $t_i := -\infty$
4: sort $t_1, \dots, t_{n-1}$ in the ascending order obtaining $t_{(1)} \le \dots \le t_{(n-1)}$
5: output $\left[t_{(\lfloor \epsilon n \rfloor)}, \infty\right)$ as prediction set.

Algorithm 2.4 CRR (simplified)

1: **for** $i = 1, \dots, n-1$
2: **if** $b_n - b_i > 0$ set $u_i := l_i := (a_i - a_n)/(b_n - b_i)$
3: **else** set $l_i := -\infty$ and $u_i := \infty$
4: sort $u_1, \dots, u_{n-1}$ in the ascending order obtaining $u_{(1)} \le \dots \le u_{(n-1)}$
5: sort $l_1, \dots, l_{n-1}$ in the ascending order obtaining $l_{(1)} \le \dots \le l_{(n-1)}$
6: output $\left[l_{(\lfloor (\epsilon/2)n \rfloor)}, u_{(\lceil (1-\epsilon/2)n \rceil)}\right]$ as prediction set.

$$[t_{(\lfloor \epsilon n \rfloor)}, \infty) . \tag{2.44}$$

See Algorithm 2.3, in which we drop the first few lines of Algorithm 2.2 introducing the notation.

If we need a bounded prediction interval, we can combine upper and lower CRR and use the conformity scores

$$\alpha_i := \left|\{j = 1, \dots, n : e_j \ge e_i\}\right| \wedge \left|\{j = 1, \dots, n : e_j \le e_i\}\right| ,$$

where $e_i = y_i - \hat{y}_i$ are the residuals, as before. Notice that the corresponding prediction sets Γ^ϵ at significance level ϵ will satisfy

$$\Gamma^\epsilon \subseteq \Gamma^{\epsilon/2,\text{upper}} \cap \Gamma^{\epsilon/2,\text{lower}} , \tag{2.45}$$

where the superscripts “upper” and “lower” indicate the prediction sets output by the upper and lower CRR, respectively. If $\epsilon < 1/2$ and all $n - 1$ residuals for the training set are different, the “$\subseteq$” in (2.45) becomes “$=$”.

The prediction set for the full CRR defined by the right-hand side of (2.45) is the combination of (2.43) and (2.44):

$$\Gamma^\epsilon = \left[t_{(\lfloor (\epsilon/2)n \rfloor)}, t_{(\lceil (1-\epsilon/2)n \rceil)}\right] .$$

The procedure is shown as Algorithm 2.4. Of course, sorting needs to be done just once: since the sequences $u_1, \dots, u_{n-1}$ and $l_1, \dots, l_{n-1}$ are so similar, the sorting operation in line 5 becomes trivial once the sorting operation in line 4 has been carried out.

Let us suppose that the number p of attributes is constant. The computation time of Algorithms 2.2–2.4 is $O(n \log n)$. Indeed,

$$A = (y_1, \dots, y_{n-1}, 0)' - X_n \left[(X_n' X_n + a I_n)^{-1} X_n' (y_1, \dots, y_{n-1}, 0)' \right]$$

and

$$B = (0, \dots, 0, 1)' - X_n \left[(X_n' X_n + a I_n)^{-1} X_n' (0, \dots, 0, 1)' \right]$$

can be computed in time $O(n)$, and the sorting can be done in time $O(n \log n)$ (see, e.g., [55, Part II]).

2.3.3 Two Modifications

The algorithms developed in the previous subsection can be easily modified to allow two alternative ways of computing nonconformity scores. To simplify formulas, we assume $a = 0$ in this subsection (i.e., we will consider least squares). *Conformalized least squares* (CLS) algorithms are CRR algorithms with the ridge parameter a set to 0.

First we consider the case where α_i is defined to be the *deleted residual* $e_{(i)} := y_i - \hat{y}_{(i)}$, where $\hat{y}_{(i)}$ is the least squares prediction for y_i computed from x_i based on the training set $x_1, y_1, \dots, x_{i-1}, y_{i-1}, x_{i+1}, y_{i+1}, \dots, x_n, y_n$. It is well known in statistics that to compute the deleted residuals $e_{(i)}$ we do not need to perform n regressions; they can be easily computed from the usual residuals $e_i = y_i - \hat{y}_i$ by the formula

$$e_{(i)} = \frac{e_i}{1 - h_i}, \tag{2.46}$$

where h_i is the ith diagonal element of the hat matrix H. (For a proof, see [252, Appendix C.7].)

Deleted CLS are defined in the same way as CLS except that the nonconformity scores $\alpha_i := e_i$ are replaced by the deleted versions $\alpha_i := e_{(i)}$. Algorithms 2.2–2.4 will implement the deleted versions if A and B are redefined as follows:

$$a_i := \frac{a_i}{1 - h_i}, \quad b_i := \frac{b_i}{1 - h_i}, \qquad i = 1, \dots, n .$$

It is clear that each

$$h_i = x_i' (X_n' X_n)^{-1} x_i$$

can be computed from $(X_n' X_n)^{-1}$ in time $O(1)$ (again assuming that the number p of attributes is constant), and so the deleted CLS can also be implemented in time $O(n \log n)$.

Algorithm 2.5 Studentized CLS (simplified once again)

1: Define a and b by (2.48)
2: **for** $i = 1, \ldots, n-1$ set $t_i := (a_i - a_n)/(b_n - b_i)$
3: sort $t_1, \ldots, t_{n-1}$ in the ascending order obtaining $t_{(1)} \le \cdots \le t_{(n-1)}$
4: output $\left[t_{(\lfloor(\epsilon/2)n\rfloor)}, t_{(\lceil(1-\epsilon/2)n\rceil)}\right]$ as prediction set.

Another natural modification of CLS is half-way between the CLS and the deleted CLS: the nonconformity scores are taken to be

$$\alpha_i := \frac{e_i}{\sqrt{1-h_i}} \tag{2.47}$$

in the case of the upper CLS and its appropriate modifications for the CLS and lower CLS. We will explain the motivation behind this choice momentarily, but first describe how to implement the *studentized CLS* determined by these nonconformity scores. The implementation is just Algorithms 2.2–2.4 with A and B redefined as

$$a_i := \frac{a_i}{\sqrt{1-h_i}}, \quad b_i := \frac{b_i}{\sqrt{1-h_i}}, \qquad i = 1, \ldots, n\,. \tag{2.48}$$

The computation time is again $O(n \log n)$.

An advantage of the studentized version is that the checks in line 5 in Algorithm 2.2 and line 2 in Algorithm 2.3 become superfluous, since $b_n > b_i$ provided all a_i and b_i are defined (i.e., the denominators in (2.48) are non-zero). This follows from Proposition 7.8 in Chap. 7. Correspondingly, Algorithm 2.4 simplifies to Algorithm 2.5.

Now we explain the standard motivation for studentized CLS. (Remember that this motivation has no bearing on the validity of the constructed conformal predictor; in particular, the smoothed version of this confidence predictor will make errors independently with probability ϵ at each significance level ϵ regardless of whether the assumption of normal noise we are about to make is satisfied or not; cf. Sect. 13.4.) Imagine that the labels y_i are generated from the deterministic objects x_i in the following way:

$$y_i = w \cdot x_i + \xi_i\,, \tag{2.49}$$

where ξ_i are independent normal random variables with the mean 0 and same variance σ^2 (random noise). Set $\xi := (\xi_1, \ldots, \xi_n)'$. Since the vector of residuals is $e = (I_n - H_n)Y_n$ (see above), we obtain

$$e = (I_n - H_n)(X_n w + \xi) = (I_n - H_n)\xi \tag{2.50}$$

for any fixed w (the true parameters); therefore, the covariance matrix of the residuals is

$$\text{var}(e) = \text{var}((I_n - H_n)\xi) = (I_n - H_n)\text{var}(\xi)(I_n - H_n)' = \sigma^2(I_n - H_n) ,$$

since $\text{var}(\xi) = \sigma^2 I_n$ and $I_n - H_n$ is symmetric and idempotent. We can see that the variance of e_i is $(1 - h_i)\sigma^2$, and the scaling of the residuals e_i by dividing by $\sqrt{1 - h_i}$ will equalize their variances (and even their distributions, since by (2.50), e_i are normally distributed).

Notice that according to our motivational model (2.49) the level of noise ξ_i does not depend on the observed object x_i (the variance of ξ_i remains the same, σ^2). Even in this case, it may be useful to scale residuals. If we suspect that noise can be different in different parts of the object space, heavier scaling may become necessary for satisfactory prediction.

2.3.4 Dual Form Ridge Regression

Least squares and ridge regression procedures can only deal with situations where the number of parameters p is relatively small since they involve inverting a $p \times p$ matrix. They are carried over to high-dimensional problems using the so-called "kernel trick", introduced in machine learning by Vapnik (see, e.g., [365, 366]). (The formulas arrived at by using the kernel trick coincide with those obtained by means of Gaussian processes and reproducing kernel Hilbert spaces; see Sect. 2.9.6.)

We first state the ridge regression procedure in the "dual form". The traditional statistical approach to dualization is to use the easy-to-check matrix equality

$$X_n(X_n'X_n + aI_p)^{-1} = (X_nX_n' + aI_n)^{-1}X_n , \tag{2.51}$$

which can be equivalently rewritten as

$$(X_n'X_n + aI_p)^{-1}X_n' = X_n'(X_nX_n' + aI_n)^{-1} .$$

(To see that (2.51) is true, multiply it by $(X_nX_n' + aI_n)$ on the left and $(X_n'X_n + aI_p)$ on the right.) Using (2.51), we can rewrite the ridge regression prediction for an object x based on examples $(x_1, y_1, \dots, x_n, y_n)$ as

$$\hat{y} = Y_n'X_n(X_n'X_n + aI_p)^{-1}x = Y_n'(X_nX_n' + aI_n)^{-1}X_nx , \tag{2.52}$$

where the $n \times p$ matrix X_n and vector Y_n are defined as before. The crucial property of the representation $\hat{y} = Y_n'(X_nX_n' + aI_n)^{-1}X_nx$ is that it depends on the objects $x_1, \dots, x_n, x$ only via the scalar products between them. In particular, if the object space $\mathbf{X}$ is mapped into another Euclidean space (called the *feature space*) $\mathbf{H}$, $F : \mathbf{X} \to \mathbf{H}$ (assumed measurable), and ridge regression is performed in the feature space, the prediction (2.52) can be written in the form

$$\hat{y} = Y_n'(K_n + aI_n)^{-1}k_n , \tag{2.53}$$

where K_n is the matrix with elements $(K_n)_{i,j} := \mathcal{K}(x_i, x_j)$, k_n is the vector with elements $(k_n)_i := \mathcal{K}(x, x_i)$, and $\mathcal{K}$ is the *kernel*, defined by

$$\mathcal{K}(x^{(1)}, x^{(2)}) := F(x^{(1)}) \cdot F(x^{(2)}) \tag{2.54}$$

for all $x^{(1)}, x^{(2)} \in \mathbf{X}$. The prediction rule (2.53) is also known as *kernel ridge regression*. The hat matrix in the dual representation is

$$H_n = X_n(X_n'X_n + aI_p)^{-1}X_n' = (X_nX_n' + aI_n)^{-1}X_nX_n' ,$$

and if ridge regression is carried out in the feature space, this becomes

$$H_n = (K_n + aI_n)^{-1}K_n . \tag{2.55}$$

Now it is easy to represent the CRR algorithms in the kernel form: the only difference from Algorithms 2.2–2.4 is that now C is defined as $I_n - H_n$ with H_n given by (2.55).

The computation time of the kernel form of the CRR algorithms is $O(n^2)$ at the nth step of the online protocol. This can be seen from the well-known (see, e.g., [155, (8)]) and easy-to-check formula

$$\begin{pmatrix} K & k \\ k' & \kappa \end{pmatrix}^{-1} = \begin{pmatrix} K^{-1} + dK^{-1}kk'K^{-1} & -dK^{-1}k \\ -dk'K^{-1} & d \end{pmatrix} , \tag{2.56}$$

where K is a square matrix, k a vector, κ a number, and

$$d := \frac{1}{\kappa - k'K^{-1}k} \tag{2.57}$$

(formula (2.56) assumes that the denominator in (2.57) is different from 0 so that d is defined and, of course, that K^{-1} is defined). Indeed, by this formula $(K_n + aI_n)^{-1}$ can be updated from the previous trial of the online learning protocol in time $O(n^2)$, and both

$$A = (y_1, \dots, y_{n-1}, 0)' - (K_n + aI_n)^{-1}\left[K_n(y_1, \dots, y_{n-1}, 0)'\right]$$

and

$$B = (0, \dots, 0, 1)' - (K_n + aI_n)^{-1}\left[K_n(0, \dots, 0, 1)'\right]$$

can be computed in time $O(n^2)$. We mentioned some conditions on the validity of formula (2.56), but those conditions are satisfied in our context, at least when $a > 0$:

the theorem on normal correlation (see, e.g., [329, Theorem 2.13.2]) implies that d is well-defined (and positive) whenever the matrix $\left(\begin{smallmatrix} K & k \\ k' & \kappa \end{smallmatrix}\right)$ is positive definite; the latter condition is satisfied when the formula is used for updating $(K_n + aI_n)^{-1}$, $a > 0$, to $(K_{n+1} + aI_{n+1})^{-1}$ (in which case $K = K_n + aI_n$, $k = k_n$, and $\kappa = \mathcal{K}(x_n, x_n) + a$).

The above construction also works in the case where $\mathbf{H}$ is an arbitrary Hilbert space. This is especially obvious in the case of a finite horizon, where only the linear span of the observed objects matters.

The essence of the kernel trick is that one does not need to consider the feature space in explicit form [366, Sect. 10.5.2]. It is clear that any kernel (2.54) is symmetric,

$$\mathcal{K}(x^{(1)}, x^{(2)}) = \mathcal{K}(x^{(2)}, x^{(1)}), \quad \forall x^{(1)}, x^{(2)} \in \mathbf{X},$$

and nonnegative definite,

$$\sum_{i=1}^{m} \sum_{j=1}^{m} \mathcal{K}(x^{(i)}, x^{(j)}) a_i a_j \geq 0 \tag{2.58}$$

for all $x^{(1)}, \ldots, x^{(m)} \in \mathbf{X}$ and all $a_1, \ldots, a_m \in \mathbb{R}$. (To see that (2.58) is true, notice that

$$\begin{aligned} &\left\| a_1 F(x^{(1)}) + \cdots + a_m F(x^{(m)}) \right\|^2 \\ &= \left(a_1 F(x^{(1)}) + \cdots + a_m F(x^{(m)}) \right) \cdot \left(a_1 F(x^{(1)}) + \cdots + a_m F(x^{(m)}) \right) \geq 0 .) \end{aligned}$$

It turns out that the opposite statement is also true if we ignore the requirement that the feature mapping be measurable: any function $\mathcal{K} : \mathbf{X}^2 \to \mathbb{R}$ that is symmetric and nonnegative definite can be represented in the form (2.54) (see, e.g., [341, Theorem 4.16]). In interesting cases, the feature mapping F to a Hilbert space $\mathbf{H}$ can be chosen measurable.

2.4 Conformalized Nearest Neighbours Regression

Least squares and ridge regression are just two of the standard regression algorithms; conformal predictors can be implemented in a feasible way for nonconformity measures based on many other regression algorithms. Such an implementation is especially simple in the case of the nearest neighbours algorithm.

The idea of the k-nearest neighbours algorithms, where $k \in \mathbb{N}$ is the number of "neighbours" taken into account, is as follows. Suppose the object space $\mathbf{X}$ is a metric space (for example, the usual Euclidean distance is often used if $\mathbf{X} = \mathbb{R}^p$). To give a prediction for a new object x, find the k objects $x_{i_1}, \ldots, x_{i_k}$ among the known examples that are nearest to x in the sense of the chosen metric (assuming, for simplicity, that there are no ties). In the problem of classification, the predicted label $\hat{y}$ of x is obtained by "voting": it is defined to be the most frequent label among $y_{i_1}, \ldots, y_{i_k}$. In regression, we can take, e.g., the mean or the median of $y_{i_1}, \ldots, y_{i_k}$.

We will only consider the version of the k-nearest neighbours regression (k-NNR) where the prediction $\hat{y}$ for a new object x based on the training set (x_i, y_i), $i = 1, \ldots, n$, is defined to be the arithmetic mean of the labels of the k nearest neighbours of x among $x_1, \ldots, x_n$. It will be easy to see that the more robust procedure where arithmetic mean is replaced by median also leads to a feasible conformal predictor.

Let us consider, in the context of (2.33), the nonconformity scores $\alpha_i := y_i - \hat{y}_{(i)}$, where $\hat{y}_{(i)}$ is the k-NNR prediction for x_i based on the training set (x_j, y_j), $j = 1, \ldots, i-1, i+1, \ldots, n$. The conformal predictor determined by this nonconformity measure (*k-NNR conformal predictor*) is implemented by the upper CRR algorithm (Algorithm 2.2) with the only modification that a_i and b_i are now defined as follows (we assume that $n > k$ and that all distances between the objects are different):

- a_n is the minus arithmetic mean of the labels of x_n's k nearest neighbours and $b_n = 1$;
- if $i < n$ and x_n is among the k nearest neighbours of x_i, a_i is x_i's label minus the arithmetic mean of the labels of those nearest neighbours with x_n's label set to 0, and $b_i = -1/k$;
- if $i < n$ and x_n is not among the k nearest neighbours of x_i, a_i is x_i's label minus the arithmetic mean of the labels of x_i's k nearest neighbours, and $b_i = 0$.

The nonconformity score for the ith example is $\alpha_i = a_i + b_i y$, where y is the label for x_n that is being tried.

Plugging the values a_i and b_i as defined in the previous paragraph into Algorithms 2.2, 2.3, and 2.4, we obtain *upper conformalized nearest neighbours*, *lower conformalized nearest neighbours*, and *conformalized nearest neighbours*, respectively. In none of the NNR-based algorithms there is need to check $b_n - b_i > 0$ (see, e.g., line 5 in Algorithm 2.2), since now we always have $b_n - b_i = 1 + 1/k$.

Instead of measuring distance in the original example space $\mathbf{X}$ we can measure it in the feature space, which corresponds to using the function

$$\begin{aligned}\rho(x^{(1)}, x^{(2)}) &:= \left\| F(x^{(1)}) - F(x^{(2)}) \right\|^2 \\ &= \left(F(x^{(1)}) - F(x^{(2)})\right) \cdot \left(F(x^{(1)}) - F(x^{(2)})\right) \\ &= \mathcal{K}(x^{(1)}, x^{(1)}) + \mathcal{K}(x^{(2)}, x^{(2)}) - 2\mathcal{K}(x^{(1)}, x^{(2)})\end{aligned}$$

as the distance (remember that in the case of k-NNR conformal prediction we are only interested in the distance up to a monotonic transformation).

2.5 Efficiency of Conformalized Ridge Regression

In this section we describe and start implementing in a special case what we will call the *Burnaev–Wasserman programme* (we learned the idea of this research programme from Evgeny Burnaev and Larry Wasserman). The specific conformal predictors constructed so far were obtained by "conformalizing" various underlying algorithms, which are typically chosen from among standard prediction algorithms used in machine learning. This will remain our main strategy in the rest of the book. The Burnaev–Wasserman programme concerns the efficiency of this strategy.

Let us start from conformalizing Bayesian prediction algorithms. Suppose we know, or postulate, the true probability distribution generating the data, and it is a power distribution. With a known data-generating distribution, we can often find the optimal prediction algorithm, the *Bayes algorithm*, under this distribution and for natural definitions of optimality. In this chapter we are interested in Bayes algorithms outputting prediction sets, which we will call *Bayesian predictions* (an example will be given in Sect. 2.5.2). The Bayesian predictions satisfy strong properties of validity, but they are guaranteed only under the known or postulated probability distribution.

A *set predictor* is a measurable function

$$\Gamma : \mathbf{Z}^* \times \mathbf{X} \to 2^{\mathbf{Y}} . \tag{2.59}$$

It is required to be measurable in the same sense as the confidence predictor (2.7) in Sect. 2.1.2; therefore, a set predictor is a confidence predictor that does not depend on ϵ.

To obtain a set predictor that is valid under the exchangeability model, we can run the conformal predictor on top of the Bayes algorithm, "conformalizing" the latter. Namely, we design a nonconformity measure based on the Bayes algorithm (or directly on the data-generating distribution) and then feed it into the recipe for conformal prediction (or one of its modifications discussed in Chap. 4).

The conformalizing procedure improves the guaranteed property of validity in an important respect: now we have validity under the exchangeability model. This may be important when the data-generating distribution is postulated rather than known, which is usually the case in practice. The main question asked in the Burnaev–Wasserman programme is whether we have to pay in terms of efficiency, under the postulated data-generating distribution, for the gain in validity. If the price is low, conformalizing is akin to buying a cheap insurance policy to protect us against the postulated data-generating distribution being wrong. We will discuss the insurance metaphor in a different but similar context in Sect. 10.1. If the price of protection turns out to be high, this might mean that the Bayesian assumption is too fragile.

Our discussion so far is also applicable to narrow parametric models in place of Bayesian models, first of all Gaussian models. There may be nearly optimal prediction algorithms under such models, and then the task is to explore whether we lose much in efficiency, under those models, when conformalizing those nearly optimal algorithms.

2.5.1 Hard and Soft Models

In our informal discussions in the framework of the Burnaev–Wasserman programme, we will sometimes distinguish between hard and soft models. Despite some common features, this distinction is very different from that between null and alternative hypotheses in the Neyman–Pearson theory of hypothesis testing. In particular, we are often interested in the case where the hard model (almost invariably the exchangeability or randomness model) is more general than the soft model (which may be a Bayesian model).

The *hard models* are the usual statistical models: our working hypothesis is that the dataset was generated by one of the probability distributions in the model. In particular, the validity of our confidence predictors is allowed to depend on the hard model.

In addition to the accepted hard model, one often has other a priori information about the data-generating distribution: e.g., only a few parameters might provide the bulk of the information relevant to prediction. Whereas we might hesitate to include such a priori information in the hard model explicitly, since it might destroy the validity of our confidence predictor if this information happened to be far from the truth, we might still be able to use such information, comprising our *soft model*, in designing accurate confidence predictors.

In this section our hard model will be the exchangeability model, and our soft model will be Bayesian. Bayesian confidence predictors are only valid under the soft model, and therefore, we will be interested in the corresponding conformal predictors. Whether the soft model is true or not affects only the efficiency, but not validity, of the conformal predictors.

Separation of the available information about the data-generating distribution into the hard model and soft model increases robustness of confidence predictors with respect to modelling errors. If such an error occurs in the soft model, the validity of predictions is not affected. At worst the predictions will become useless, but they will not become misleading (with high probability under any distribution in the hard model).

2.5.2 Bayesian Ridge Regression

In this subsection we give only a brief summary of Bayesian ridge regression; see Sect. 13.4.1 for details. We are interested in the case where the number $n-1$ of training examples is large, whereas the number p of attributes is fixed.

We assume that, given the objects, the labels $y_1, \ldots, y_n$ are generated by the rule (2.49), where w is a random vector now distributed as $\mathbf{N}(0, (\sigma^2/a)I)$ ($I := I_p$ being the unit $p \times p$ matrix), each ξ_i is distributed as $\mathbf{N}(0, \sigma^2)$, the random elements $w, \xi_1, \ldots, \xi_n$ are independent (given the objects), and σ and a are given positive numbers.

The conditional distribution for the label y_n of the test object x_n given the training set (x_i, y_i), $i = 1, \ldots, n-1$, and x_n is

$$\mathbf{N}\left(\hat{y}_n, (1 + g_n)\sigma^2\right)$$

(see (13.27)), where

$$\begin{aligned} \hat{y}_n &:= x_n'(X'X + aI)^{-1}X'Y \,, \\ g_n &:= x_n'(X'X + aI)^{-1}x_n \,, \end{aligned} \tag{2.60}$$

$X = X_{n-1}$ is the data matrix for the training set (the $(n-1) \times p$ matrix whose ith row is x_i', $i = 1, \ldots, n-1$), and $Y = Y_{n-1}$ is the vector $(y_1, \ldots, y_{n-1})'$ of the training labels. Therefore, the *Bayesian prediction interval* is

$$(B_*, B^*) := \left(\hat{y}_n - \sqrt{1 + g_n}\sigma \mathbf{z}_{\epsilon/2}, \hat{y}_n + \sqrt{1 + g_n}\sigma \mathbf{z}_{\epsilon/2}\right) , \tag{2.61}$$

where ϵ is the significance level (the permitted probability of error, so that $1 - \epsilon$ is the required coverage probability) and $\mathbf{z}_{\epsilon/2}$ is the $(1 - \epsilon/2)$-quantile of the standard normal distribution $\mathbf{N}_{0,1}$.

The prediction interval (2.61) enjoys several desiderata: it is unconditionally valid, in the sense that its error probability is equal to the given significance level ϵ; it is also valid conditionally on the training set and the test object x_n; finally, this prediction interval is the shortest possible conditionally valid interval. We will refer to the class of algorithms producing prediction intervals (2.61) (and depending on the parameters σ and a) as *Bayesian ridge regression* (BRR).

2.5.3 Efficiency of CRR

In the rest of this section, the hard model is, as already mentioned, the exchangeability model, and the soft model assumes that the objects x_i are IID while y_i are generated from (2.49), $w \sim \mathbf{N}(0, (\sigma^2/a)I)$, and $\xi_i \sim \mathbf{N}(0, \sigma^2)$, all independent. (It is slightly more natural to consider the exchangeability model rather than the randomness model as our hard model since a typical element of the soft model is an exchangeable but not power distribution.) In this subsection we will see that under this soft model complemented by natural (and standard) assumptions CRR is asymptotically close to BRR (2.61) (optimized under the soft model), and therefore is approximately conditionally valid and efficient. On the other hand, we have the unconditional validity of CRR under the exchangeability assumption, regardless of whether (2.49) holds.

We assume an infinite sequence of examples $(x_1, y_1), (x_2, y_2), \dots$ but consider only the first n of them and let $n \to \infty$. We make both the randomness assumption about the objects $x_1, x_2, \dots$ (the objects are generated independently from the same distribution) and the assumption (2.49); however, we can relax the assumption that w is distributed as $\mathbf{N}(0, (\sigma^2/a)I)$. These are all the assumptions used in our main result:

(A1) The random objects $x_i \in \mathbb{R}^p$, $i = 1, 2, \dots$, are IID.
(A2) The second-moment matrix $\mathbb{E}(x_1 x_1')$ of x_1 exists and is non-singular.
(A3) The random vector $w \in \mathbb{R}^p$ is independent of $x_1, x_2, \dots$.
(A4) The labels $y_1, y_2, \dots$ are generated by $y_i = w \cdot x_i + \xi_i$, where ξ_i are Gaussian noise variables distributed as $\mathbf{N}(0, \sigma^2)$ and independent between themselves, of the objects x_i, and of w.

Notice that the assumptions imply that the random examples (x_i, y_i), $i = 1, 2, \dots$, are IID given w. It will be clear from the proof that the assumptions can be relaxed further (but we have tried to make them as simple as possible).

Theorem 2.7 *Under the assumptions (A1)–(A4), the prediction sets output by CRR are intervals from some n on almost surely, and the differences between the upper and lower end-points of the prediction intervals for BRR and CRR are asymptotically Gaussian:*

$$\frac{\sqrt{n}}{\sigma}(B^* - C^*) \xrightarrow{\text{law}} \mathbf{N}\left(0, \frac{(\epsilon/2)(1-\epsilon/2)}{\phi^2(\mathbf{z}_{\epsilon/2})} - \mu'\Sigma^{-1}\mu\right), \tag{2.62}$$

$$\frac{\sqrt{n}}{\sigma}(B_* - C_*) \xrightarrow{\text{law}} \mathbf{N}\left(0, \frac{(\epsilon/2)(1-\epsilon/2)}{\phi^2(\mathbf{z}_{\epsilon/2})} - \mu'\Sigma^{-1}\mu\right), \tag{2.63}$$

where ϕ is the density of the standard normal distribution $\mathbf{N}_{0,1}$, $\mu := \mathbb{E}(x_1)$ is the expectation of x_1, and $\Sigma := \mathbb{E}(x_1 x_1')$ is the second-moment matrix of x_1.

The theorem will be proved in Sect. 2.8.2, and in the rest of this subsection we will discuss it. We can see from (2.62) and (2.63) that the symmetric difference between the prediction intervals output by BRR and CRR shrinks to 0 as $O(n^{-1/2})$ in Lebesgue measure with high probability.

Let us first see what the typical values of the standard deviation (the square root of the variance) in (2.62) and (2.63) are. The second term in the variance does not affect it significantly since $0 \le \mu'\Sigma^{-1}\mu \le 1$. Indeed, denoting the covariance matrix of x_1 by C and using the Sherman–Morrison formula (see Remark 2.8 below), we have:

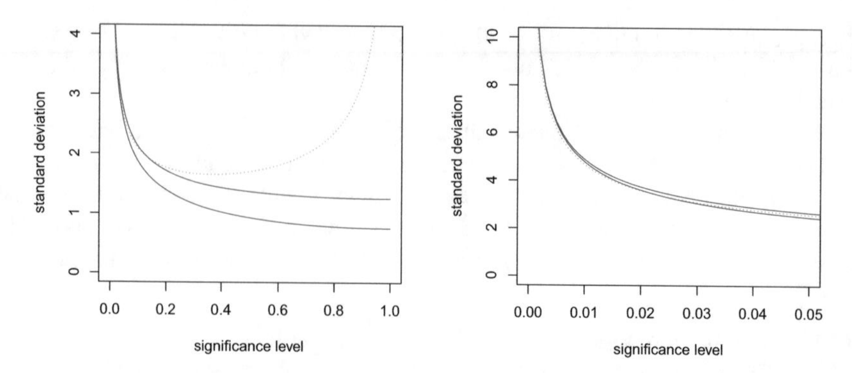

Fig. 2.1 The limits for the standard deviation in Theorem 2.7 as a function of the significance level $\epsilon \in (0, 1)$ (left) and $\epsilon \in (0, 0.05]$ (right) shown as solid blue lines; the asymptotic expression in (2.65) is shown as a dotted red line. In all cases $\sigma = 1$

$$\mu' \Sigma^{-1} \mu = \mu'(C + \mu\mu')^{-1}\mu = \mu' \left(C^{-1} - \frac{C^{-1}\mu\mu' C^{-1}}{1 + \mu' C^{-1}\mu} \right) \mu$$

$$= \mu' C^{-1}\mu - \frac{(\mu' C^{-1}\mu)^2}{1 + \mu' C^{-1}\mu} = \frac{\mu' C^{-1}\mu}{1 + \mu' C^{-1}\mu} \in [0, 1] \qquad (2.64)$$

(we write $[0, 1]$ rather than $(0, 1)$ because C is permitted to be singular.[2]) The first term, on the other hand, can affect the variance more significantly, and the significant dependence of the variance on ϵ is natural: the accuracy of the conformal predictor degrades for small ϵ since, instead of using all data for estimating the end-points of the prediction interval, it relies, under the exchangeability model, on the scarcer information provided by examples in the tails of the distribution generating the labels. Figure 2.1 illustrates the dependence of the standard deviation of the asymptotic distribution on ϵ. The upper solid line in it corresponds to $\mu' \Sigma^{-1} \mu = 0$, and the lower solid line corresponds to $\mu' \Sigma^{-1} \mu = 1$. The possible values for the standard deviation lie between these two lines. The asymptotic behaviour of the standard deviation as $\epsilon \to 0$ is given by

$$\sqrt{\epsilon(1 - \epsilon/2)\pi \exp(\mathbf{z}^2_{\epsilon/2}) - \theta} \sim (-\epsilon \ln \epsilon)^{-1/2} \qquad (2.65)$$

uniformly in $\theta \in [0, 1]$.

The assumptions (A1)–(A4) do not involve a, and Theorem 2.7 continues to hold if we set $a := 0$. Theorem 2.7 can thus also be considered as an efficiency result about conformalizing the standard non-Bayesian least squares procedure; this procedure outputs precisely (B_*, B^*) with $a := 0$ as its prediction intervals

[2] If C is singular, we can apply (2.64) to $\Sigma_\epsilon := \Sigma + \epsilon I$ and $C_\epsilon := C + \epsilon I$, where $\epsilon > 0$, in place of Σ and C, respectively, and let $\epsilon \to 0$.

(see, e.g., [313, Sect. 5.3.1]). The least squares procedure has guaranteed coverage probability under weaker assumptions than BRR (not requiring assumptions about w); however, its validity is not conditional, similarly to CRR.

Remark 2.8 The Sherman–Morrison formula (see, e.g., [155, (3)]) is

$$(K + uv')^{-1} = K^{-1} - \frac{K^{-1}uv'K^{-1}}{1 + v'K^{-1}u}, \tag{2.66}$$

where K is a square matrix and u and v are vectors. It can be checked easily by multiplying the right-hand side by $K + uv'$ on the left and simplifying.

2.6 Are There Other Ways to Achieve Validity?

In this section we will see that conformal predictors are essentially the only confidence predictors in very natural classes that satisfy our non-asymptotic properties of validity.

Let us say that a confidence predictor is *invariant* if $\Gamma^\epsilon(z_1, \ldots, z_{n-1}, x_n)$ does not depend on the order in which $z_1, \ldots, z_{n-1}$ are listed. Since we assume exchangeability, the invariant confidence predictors constitute a natural class (see, e.g., the description of the "sufficiency principle" in [64]; later in this book, however, we will also study confidence predictors that are not invariant, such as inductive conformal predictors in Chap. 4).

If Γ_1 and Γ_2 are (deterministic) confidence predictors, we will say that Γ_1 is *at least as good as* Γ_2 if, for any $\epsilon \in (0, 1)$, any n, and and $x_1, y_1, x_2, y_2, \ldots$,

$$\Gamma_1^\epsilon(x_1, y_1, \ldots, x_{n-1}, y_{n-1}, x_n) \subseteq \Gamma_2^\epsilon(x_1, y_1, \ldots, x_{n-1}, y_{n-1}, x_n) .$$

We will also express this relation by saying that Γ_1 *dominates* Γ_2.

We have given these definitions in the context of online predictors, but they are also applicable to one-off confidence predictors: just remove the quantifier over n. As a gentle introduction, we start from a result about one-off predictors.

Proposition 2.9 *Any one-off confidence predictor that is conservatively valid under exchangeability is dominated by a weak conformal predictor. If, additionally, it is invariant, then it is dominated by a conformal predictor.*

Proof Let n be the size of a conservatively valid weak conformal predictor Γ. Define $S(z_1, \ldots, z_n)$ as the infimum of ϵ such that Γ makes an error when fed with $z_1, \ldots, z_{n-1}$ as the training set and z_n's object as the test object. Let us check that the weak conformal predictor $\tilde{\Gamma}$ determined by S as conformity statistic is at least as good as Γ.

For any sequence $(z_1, \ldots, z_n) \in \mathbf{Z}^n$, let P be the uniform probability measure on all permutations of $(z_1, \ldots, z_n)$, and let $\epsilon \in (0, 1)$. Since P is exchangeable, Γ's property of validity implies that the fraction of the permutations at which $S \le \epsilon$ is at most ϵ (this is true with any $\epsilon' > \epsilon$ in place of ϵ, and so it is true for ϵ as well). Therefore, $\tilde{\Gamma}$ is at least as good as Γ.

The following result asserts that, in the online mode of prediction, invariant conservatively valid confidence predictors are conformal predictors or can be improved to become conformal predictors.

Theorem 2.10 *Suppose* $\mathbf{Z}$ *is a standard Borel space. Let* Γ *be an invariant conservatively valid confidence predictor. Then there is a conformal predictor that is at least as good as* Γ*.*

Theorem 2.10 will be proved in Sect. 2.8.3, and the proof will be based on the same idea as the proof of Proposition 2.9. It is interesting that the proof will not use the fact that the random variables $\xi_1^{(\epsilon)}, \xi_2^{(\epsilon)}, \ldots$ from the definition of Γ's conservative validity are independent. The independence then appears automatically as property of the errors of (or, as explained in the next section, of the p-values produced by) the resulting conformal predictor.

2.7 Conformal Transducers

There are two convenient ways to represent a conformal predictor: as a confidence predictor and as a "confidence transducer". So far we have been using the first way; the goal of this section is to introduce the second way, which is simpler mathematically, and to discuss connections between the two. This will be particularly needed in Chap. 7 and Part III.

2.7.1 *Definitions and Properties of Validity*

A *randomized confidence transducer* is a function f mapping the nonempty sequences in $(\mathbf{X} \times [0, 1] \times \mathbf{Y})^*$ to $[0, 1]$. It is called "transducer" because it can be regarded as mapping each input sequence $(x_1, \tau_1, y_1, x_2, \tau_2, y_2, \ldots)$ in $(\mathbf{X} \times [0, 1] \times \mathbf{Y})^\infty$ into the output sequence $(p_1, p_2, \ldots)$ of *p-values* defined by $p_n := f(x_1, \tau_1, y_1, \ldots, x_n, \tau_n, y_n)$, $n = 1, 2, \ldots$; we often drop "confidence" in "confidence transducer". We say that f is an *exactly valid randomized transducer* (or just *exact randomized transducer*) if the output p-values $p_1 p_2 \ldots$ are always distributed according to the uniform distribution $\mathbf{U}^\infty$ on $[0, 1]^\infty$, provided the input examples $z_n = (x_n, y_n)$, $n = 1, 2, \ldots$, are generated by an exchangeable probability distribution on $\mathbf{Z}^\infty$ and the numbers $\tau_1, \tau_2, \ldots$ are generated independently from the uniform distribution $\mathbf{U}$ on $[0, 1]$.

We can extract exact randomized transducers from nonconformity measures: given a nonconformity measure A, for each nonempty sequence

$$(x_1, \tau_1, y_1, \dots, x_n, \tau_n, y_n) \in (\mathbf{X} \times [0,1] \times \mathbf{Y})^*$$

define

$$f(x_1, \tau_1, y_1, \dots, x_n, \tau_n, y_n) := \frac{|\{i : \alpha_i > \alpha_n\}| + \tau_n \, |\{i : \alpha_i = \alpha_n\}|}{n}, \tag{2.67}$$

where $i = 1, \dots, n$ and α_i, $i = 1, 2, \dots$, are computed from $z_i = (x_i, y_i)$ using A by the usual (cf. (2.23)) formula

$$\alpha_i := A(\langle z_1, \dots, z_n \rangle, z_i) . \tag{2.68}$$

Each randomized transducer f that can be obtained in this way will be called a *smoothed conformal transducer*.

Proposition 2.11 *Each smoothed conformal transducer is an exact randomized transducer.*

This proposition is a special case of Theorem 11.1, which will be proved in Chap. 11 (see Sects. 11.2.3 and 11.5.1).

In a similar way we can define (deterministic) *conformal transducers* f: given a nonconformity measure A, for each nonempty sequence $(z_1, \dots, z_n) \in \mathbf{Z}^*$ set

$$f(z_1, \dots, z_n) := \frac{|\{i : \alpha_i \geq \alpha_n\}|}{n},$$

where α_i are computed by (2.68) as before. In general, a (deterministic) *transducer* is a function f of the type $\mathbf{Z}^* \setminus \{\Box\} \to [0, 1]$, $\Box$ being the empty sequence; as before, we associate with f a mapping from $z_1 z_2 \dots$ to the *p-values* $p_1 p_2 \dots$ ($p_n := f(z_1, \dots, z_n)$). We say that f is a *conservatively valid transducer* (or *conservative transducer*) if there exists a sequence ξ_n, $n = 1, 2, \dots$, of $[0, 1]$-valued random variables on the extended probability space (2.11) such that:

- for each n, ξ_n only depends on the first n examples;
- for each n, $p_n \leq \xi_n$;
- the sequence $\xi_1, \xi_2, \dots$ is distributed as $\mathbf{U}^\infty$, provided the examples are generated from an exchangeable distribution P on $\mathbf{Z}^\infty$.

The following implication of Proposition 2.11 is obvious:

Corollary 2.12 *Each conformal transducer is conservative.*

We can fruitfully discuss confidence transducers even in the case of general example spaces $\mathbf{Z}$, not necessarily products $\mathbf{X} \times \mathbf{Y}$ of object and label spaces. But in the latter case we can associate a confidence predictor $\Gamma = f'$ with each confidence transducer f defining (2.21) as

$$\{y \in \mathbf{Y} : f(x_1, y_1, \dots, x_{n-1}, y_{n-1}, x_n, y) > \epsilon\} . \tag{2.69}$$

Vice versa, with any confidence predictor Γ we can associate the confidence transducer $f = \Gamma'$ defined by

$$f(x_1, y_1, \dots, x_n, y_n) := \sup \left\{\epsilon : y_n \in \Gamma^{\epsilon}(x_1, y_1, \dots, x_{n-1}, y_{n-1}, x_n)\right\} . \tag{2.70}$$

Letting x_i in (2.69) and (2.70) range over the extended object space $\mathbf{X} \times [0, 1]$, we obtain the definition of the randomized confidence predictor f' associated with a randomized confidence transducer f and the definition of the randomized confidence transducer Γ' associated with a randomized confidence predictor Γ.

We will see in the next subsection that the expositions of the theory of hedged prediction in terms of conformal transducers and conformal predictors are essentially equivalent. But first we slightly strengthen Proposition 2.4.

A randomized confidence predictor is *strongly exact* if, for any exchangeable probability distribution on $\mathbf{Z}^{\infty}$ and any sequence $(\epsilon_1, \epsilon_2, \dots) \in (0, 1)^{\infty}$ of significance levels, the sequence of random variables $\mathrm{err}_n^{\epsilon_n}(\Gamma)$, $n = 1, 2, \dots$, is distributed as the product $\mathbf{B}_{\epsilon_1} \times \mathbf{B}_{\epsilon_2} \times \dots$ of Bernoulli distributions with parameters $\epsilon_1, \epsilon_2, \dots$. It can be defined in a similar way what it means for a confidence predictor Γ to be strongly conservative.

Theorem 11.1 will also imply the following proposition.

Proposition 2.13 *Any smoothed conformal predictor is strongly exact.*

2.7.2 *Normalized Confidence Predictors and Confidence Transducers*

To obtain full equivalence between confidence transducers and confidence predictors, a further natural restriction has to be imposed on the latter: they will be required to be "normalized". This is a mild restriction since each confidence predictor can be normalized in such a way that its quality does not suffer.

Formally, the *normal form* Γ_{norm} of a confidence predictor Γ is defined by

$$\Gamma^{\epsilon}_{\mathrm{norm}}(x_1, y_1, \dots, x_n) := \bigcup_{\epsilon' > \epsilon} \Gamma^{\epsilon'}(x_1, y_1, \dots, x_n) .$$

We say that Γ is *normalized* if $\Gamma_{\mathrm{norm}} = \Gamma$. These definitions are also applicable to randomized confidence predictors, in which case x_i range over the extended object space $\mathbf{X} \times [0, 1]$. The following proposition lists some basic properties of the operation $_{\mathrm{norm}}$ and normalized confidence predictors.

Proposition 2.14 *All conformal predictors and smoothed conformal predictors are normalized. For any confidence predictor Γ (randomized or deterministic):*

1. Γ_{norm} *is* at least as good as Γ, *in the sense that*

$$\Gamma^{\epsilon}_{\text{norm}}(x_1, y_1, \ldots, x_n) \subseteq \Gamma^{\epsilon}(x_1, y_1, \ldots, x_n)$$

   for all $x_1, y_1, \ldots, x_n$ *and* ϵ*;*
2. $(\Gamma_{\text{norm}})_{\text{norm}} = \Gamma_{\text{norm}}$*;*
3. Γ *is normalized if and only if the set*

$$\left\{\epsilon : y_n \in \Gamma^{\epsilon}(x_1, y_1, \ldots, x_n)\right\}$$

   is open in $(0, 1)$ *for all* $x_1, y_1, \ldots, x_n$*;*
4. *If* Γ *is exact (resp. conservative, resp. strongly exact), then* Γ_{norm} *is exact (resp. conservative, resp. strongly exact).*

Proof All (smoothed) conformal predictors are normalized because all inequalities involving ϵ in (2.22) and (2.24) are strict. Properties 1 and 2 are obvious. Property 3 follow from the following restatement of the definition of Γ_{norm}:

$$y_n \in \Gamma^{\epsilon}_{\text{norm}}(x_1, y_1, \ldots, x_n)$$

if and only if

$$\exists \epsilon' > \epsilon : y_n \in \Gamma^{\epsilon'}(x_1, y_1, \ldots, x_n) .$$

Property 4 follows from

$$\text{err}^{\epsilon}_n(\Gamma_{\text{norm}}) = \inf_{\epsilon' > \epsilon} \text{err}^{\epsilon'}_n(\Gamma) .$$

□

The next proposition asserts the equivalence of confidence transducers and normalized confidence predictors (in particular, the equivalence of conformal transducers and conformal predictors). Remember that f' is the confidence predictor associated with a confidence transducer f, and Γ' is the confidence transducer associated with a confidence predictor Γ.

Proposition 2.15 *For each confidence transducer* f*, the confidence predictor* f' *is normalized and* $f'' = f$*. For each normalized confidence predictor* Γ*,* $\Gamma'' = \Gamma$*. If* $\Gamma = f'$ *is the normalized confidence predictor associated with a confidence transducer* $f = \Gamma'$*,*

$$\text{err}^{\epsilon}_n(\Gamma, (z_1, z_2, \ldots)) = 1_{f(z_1, \ldots, z_n) \leq \epsilon} \tag{2.71}$$

for any data sequence $z_1, z_2, \ldots$*. If a randomized confidence transducer* f *is exact,* f' *is strongly exact. If a randomized confidence predictor* Γ *is strongly exact,* Γ'

is exact. If f is a (smoothed) conformal transducer, f' is a (smoothed) conformal predictor. If Γ is a (smoothed) conformal predictor, Γ' is a (smoothed) conformal transducer.

Proof Most of the statements in the proposition are obvious. Equality (2.71) follows from the definition of f'. This equality implies that the confidence transducer Γ' is exact for any strongly exact randomized confidence predictor Γ. Indeed, the p-values p_n output by Γ' have the uniform distribution $\mathbf{U}$ on $[0, 1]$, and it remains to apply the following simple fact: if a sequence $p_1, p_2, \ldots$ of random variables distributed as $\mathbf{U}$ is such that, for any sequence $(\epsilon_1, \epsilon_2, \ldots) \in (0, 1)^\infty$, the random variables $1_{p_n \le \epsilon_n}$, $n = 1, 2, \ldots$, are independent, then the random variables p_n themselves are independent (see, e.g., [329], the theorem in Sect. 2.5, p. 215).

2.8 Proofs

2.8.1 *Proof of Theorem 2.2*

We may assume that the examples $z_1, \ldots, z_{l+1}$ are generated from a power distribution Q^{l+1} such that the probability distribution Q on $\mathbf{Z}$ is concentrated on the set $\{(x, y^{(1)}), (x, y^{(2)})\} \subseteq \mathbf{Z}$, for some arbitrarily fixed $x \in \mathbf{X}$ and distinct $y^{(1)}, y^{(2)} \in \mathbf{Y}$ (remember that we assumed $|\mathbf{X}| \ge 1$ and $|\mathbf{Y}| > 1$). Therefore, we assume, without loss of generality, that $\mathbf{Z} = \{0, 1\}$. Set $n := l + 1$ and suppose that $\mathrm{err}_n^\epsilon(\Gamma) = 1$ with probability ϵ for all $\epsilon \in (0, 1)$.

For each k let $f(k)$ be the probability that $\mathrm{err}_n^\epsilon(\Gamma, (z_1, \ldots, z_n)) = 1$ where $(z_1, \ldots, z_n) \in \{0, 1\}^n$ is drawn from the uniform distribution on the set of all binary sequences of length n with k 1s. Since the expected value of $f(k)$ is ϵ w.r.t. to any binomial distribution on $\{0, 1, \ldots, n\}$, the standard completeness result (see, e.g., [222, Example 4.3.1]) implies that $f(k) = \epsilon$ for all $k = 0, 1, \ldots, n$. Therefore, $\epsilon\binom{n}{k}$ is an integer for all k and ϵ, which cannot be true.

2.8.2 *Proof of Theorem 2.7*

For concreteness, we concentrate on the convergence (2.62) for the upper end-points of the conformal and Bayesian prediction intervals, which we rewrite as

$$\sqrt{n}(B^* - C^*) \xrightarrow{\text{law}} \mathbf{N}\left(0, \frac{\alpha(1-\alpha)}{f^2(\zeta_\alpha)} - \sigma^2 \mu' \Sigma^{-1} \mu\right), \qquad (2.72)$$

where $\alpha := 1 - \epsilon/2$, f is the probability density of $\mathbf{N}(0, \sigma^2)$, and $\zeta_\alpha := \mathbf{z}_{\epsilon/2}\sigma$ is the α-quantile of $\mathbf{N}(0, \sigma^2)$. We split the proof into a series of steps.

Regularizing the Rays in Upper CRR

The upper CRR looks difficult to analyze in general, since the sets (2.42) may be rays pointing in the opposite directions. Fortunately, the awkward case $b_n \le b_i$ $(i < n)$ will be excluded for large

n under our assumptions (see Lemma 2.17 below). The following lemma gives a simple sufficient condition for its absence.

Lemma 2.16 *Suppose that, for each $c \in \mathbb{R}^p \setminus \{0\}$,*

$$(c \cdot x_n)^2 < \sum_{i=1}^{n-1} (c \cdot x_i)^2 + a \|c\|^2 , \tag{2.73}$$

where $\|\cdot\|$ stands for the Euclidean norm. Then $b_n > b_i$ for all $i = 1, \ldots, n-1$.

Intuitively, in the case of a small a, (2.73) being violated for some $c \neq 0$ means that all $x_1, \ldots, x_{n-1}$ lie approximately in the same hyperplane, and x_n is well outside it. The condition (2.73) holding for all $c \in \mathbb{R}^p \setminus \{0\}$ can be expressed by saying that the matrix $\sum_{i=1}^{n-1} x_i x_i' - x_n x_n' + aI$ is positive definite.

Proof First we assume $a = 0$ (so that ridge regression becomes least squares); an extension to $a \geq 0$ will be easy. In this case H_n is the projection matrix onto the column space $\mathcal{C} \subseteq \mathbb{R}^n$ of the overall data matrix X_n and $I_n - H_n$ is the projection matrix onto the orthogonal complement $\mathcal{C}^\perp$ of $\mathcal{C}$. We can have $b_n \leq b_i$ for $i < n$ (or even $b_n^2 \leq b_1^2 + \cdots + b_{n-1}^2$) only if the angle between $\mathcal{C}^\perp$ and the hyperplane $\mathbb{R}^{n-1} \times \{0\}$ is 45° or less; in other words, if the angle between $\mathcal{C}$ and that hyperplane is 45° or more; in other words, if there is an element $(c \cdot x_1, \ldots, c \cdot x_n)'$ of $\mathcal{C}$ such that its last coordinate is $c \cdot x_n = 1$ and its projection $(c \cdot x_1, \ldots, c \cdot x_{n-1})'$ onto the other coordinates has length at most 1.

To reduce the case $a > 0$ to $a = 0$ add the p dummy objects $\sqrt{a} e_i \in \mathbb{R}^p$, $i = 1, \ldots, p$, labelled by 0 at the beginning of the training set; here $e_1, \ldots, e_p$ is the standard basis of $\mathbb{R}^p$.

Lemma 2.17 *The case $b_n \leq b_i$ for $i < n$ is excluded from some n on almost surely under (A1)–(A4).*

Proof We will check that (2.73) holds from some n on. Let us set, without loss of generality, $a := 0$. Let $\Sigma_l := \frac{1}{l} \sum_{i=1}^{l} x_i x_i'$. Since $\lim_{l \to \infty} \Sigma_l = \Sigma$ a.s.,

$$|\lambda_{\min}(\Sigma_l) - \lambda_{\min}(\Sigma)| \to 0 \quad (l \to \infty) \quad \text{a.s.} ,$$

where $\lambda_{\min}(\cdot)$ is the smallest eigenvalue of the given matrix. Since $\|x_n\|^2 / n \to 0$ a.s.,

$$\frac{1}{n-1} \sum_{i=1}^{n-1} (c \cdot x_i)^2 = c' \Sigma_{n-1} c \geq \lambda_{\min}(\Sigma_{n-1}) \|c\|^2$$

$$> \frac{1}{2} \lambda_{\min}(\Sigma) \|c\|^2 > \frac{\|c\|^2 \|x_n\|^2}{n-1} \geq \frac{(c \cdot x_n)^2}{n-1}$$

for all $c \neq 0$ from some n on.

Proof Proper

As before, X stands for the data matrix X_{n-1} based on the first $n-1$ examples. A simple but tedious computation (whose details can be found in [40, Appendix A]) gives

$$t_i = \frac{a_i - a_n}{b_n - b_i} = \hat{y}_n + (y_i - \hat{y}_i) \frac{1 + g_n}{1 + g_i} , \tag{2.74}$$

where $g_i := x_i'(X'X + aI)^{-1}x_n$ (cf. (2.60)). The term $\hat{y}_n$ in (2.74) is the centre of the Bayesian prediction interval (2.61); it does not depend on i. We can see that

$$B^* - C^* = \sqrt{1 + g_n}\zeta_\alpha - (1 + g_n)V_{(k_n)} \tag{2.75}$$

(cf. (2.61)), where $\zeta_\alpha := \mathbf{z}_{\epsilon/2}\sigma$ (as defined after (2.72)), $k_n := \lceil \alpha n \rceil$ (cf. (2.43)), and $V_{(k_n)}$ is the k_nth order statistic in the series

$$V_i := \frac{r_i}{1 + g_i} \tag{2.76}$$

of residuals $r_i := y_i - \hat{y}_i$ adjusted by dividing by $1 + g_i$. The behaviour of the order statistics of residuals is well studied: see, e.g., the theorem in [43]. The presence of $1 + g_i$ complicates the situation, and so we first show that g_i is small with high probability.

Lemma 2.18 *Let $\eta_1, \eta_2, \ldots$ be a sequence of IID random variables with a finite second moment. Then* $\max_{i=1,\ldots,n} |\eta_i| = o(n^{1/2})$ *in probability (and even almost surely) as $n \to \infty$.*

Proof By the strong law of large numbers the sequence $\frac{1}{n}\sum_{i=1}^{n} \eta_i^2$ converges a.s. as $n \to \infty$, and so $\eta_n^2/n \to 0$ a.s. This implies that $\max_{i=1,\ldots,n} |\eta_i| = o(n^{1/2})$ a.s.

Corollary 2.19 *Under the conditions of the theorem,* $\max_{i=1,\ldots,n} |g_i| = o(n^{-1/2})$ *in probability.*

Proof Similarly to the proof of Lemma 2.17, we have, for almost all sequences $x_1, x_2, \ldots$,

$$\max_{i=1,\ldots,n} |g_i| \le \frac{\|x_n\| \max_{i=1,\ldots,n} \|x_i\|}{\lambda_{\min}(X'X + aI)} < 2\frac{\|x_n\| \max_{i=1,\ldots,n} \|x_i\|}{(n-1)\lambda_{\min}(\Sigma)}$$

from some n on. It remains to combine this with Lemma 2.18 and the fact that, by Assumption (A1), $\|x_n\|$ is uniformly bounded by a constant with high probability.

Corollary 2.20 *Under the conditions of the theorem,* $n^{1/2}\left(r_{(k_n)} - V_{(k_n)}\right) \to 0$ *in probability.*

Proof Suppose that, on the contrary, there are $\epsilon > 0$ and $\delta > 0$ such that $n^{1/2}\left|r_{(k_n)} - V_{(k_n)}\right| > \epsilon$ with probability at least δ for infinitely many n. Fix such ϵ and δ. Suppose, for concreteness, that, with probability at least δ for infinitely many n, we have $n^{1/2}\left(r_{(k_n)} - V_{(k_n)}\right) > \epsilon$, i.e., $V_{(k_n)} < r_{(k_n)} - \epsilon n^{-1/2}$ (the δ may have to be halved for that). The last inequality implies that $V_i < r_{(k_n)} - \epsilon n^{-1/2}$ for at least k_n values of i. By the definition (2.76) of V_i this in turn implies that $r_i < r_{(k_n)} - \epsilon n^{-1/2} + g_i r_{(k_n)}$ for at least k_n values of i. By Corollary 2.19, however, the last addend is less than $\epsilon n^{-1/2}$ with probability at least $1 - \delta$ from some n on (the fact that $r_{(k_n)}$ is bounded with high probability follows, e.g., from Lemma 2.21 below). This implies $r_{(k_n)} < r_{(k_n)}$ with positive probability from some n on, and this contradiction completes the proof.

The last (and most important) component of the proof is the following version of the theorem in [43], itself a version of the famous Bahadur representation theorem [13]. Details of the proof (under our assumptions) are spelled out in [40, Appendix B].

Lemma 2.21 ([43], Theorem) *Under the conditions of Theorem 2.7,*

$$n^{1/2}\left|\left(r_{(k_n)} - \zeta_\alpha\right) - \frac{\alpha - F_n(\zeta_\alpha)}{f(\zeta_\alpha)} + \mu'(\hat{w}_n - w)\right| \to 0 \quad a.s., \tag{2.77}$$

where F_n is the empirical distribution function of the noise $\xi_1, \ldots, \xi_{n-1}$ and $\hat{w}_n := (X'X + aI)^{-1}X'Y$ is the ridge regression estimate of w.

By (2.75), Corollary 2.19, and Slutsky's lemma (see, e.g., [361, Lemma 2.8]), it suffices to prove (2.72) with the left-hand side replaced by $n^{1/2}(V_{(k_n)} - \zeta_\alpha)$. Moreover, by Corollary 2.20 and Slutsky's lemma, it suffices to prove (2.72) with the left-hand side replaced by $n^{1/2}(r_{(k_n)} - \zeta_\alpha)$; this is what we will do.

Lemma 2.21 holds in the situation where w is a constant vector (the conditions of Theorem 2.7 allow the distribution of w to be degenerate). Let R be a Borel set in $(\mathbb{R}^p)^\infty$ such that (2.77) holds for all $(x_1, x_2, \dots) \in R$, where the "a.s." is now interpreted as "for almost all sequences $(\xi_1, \xi_2, \dots)$". By Lebesgue's dominated convergence theorem, it suffices to prove (2.72) with the left-hand side replaced by $n^{1/2}(r_{(k_n)} - \zeta_\alpha)$ for a fixed w and a fixed sequence $(x_1, x_2, \dots) \in R$. Therefore, we fix w and $(x_1, x_2, \dots) \in R$; the only remaining source of randomness is $(\xi_1, \xi_2, \dots)$. Finally, by the definition of the set R, it suffices to prove (2.72) with the left-hand side replaced by

$$n^{1/2}\frac{\alpha - F_n(\zeta_\alpha)}{f(\zeta_\alpha)} - n^{1/2}\mu'(\hat{w}_n - w) . \tag{2.78}$$

Without loss of generality we will assume that $\frac{1}{n}X_n'X_n \to \Sigma$ as $n \to \infty$ (this extra assumption about R will ensure that Lindeberg's condition is satisfied below).

Since $\mathbb{E}(\alpha - F_n(\zeta_\alpha)) = 0$ and

$$\operatorname{var}(\alpha - F_n(\zeta_\alpha)) = \frac{F(\zeta_\alpha)(1 - F(\zeta_\alpha))}{n-1} = \frac{\alpha(1-\alpha)}{n-1} ,$$

where F is the distribution function of $\mathbf{N}(0, \sigma^2)$, we have

$$n^{1/2}\frac{\alpha - F_n(\zeta_\alpha)}{f(\zeta_\alpha)} \xrightarrow{\text{law}} \mathbf{N}\left(0, \frac{\alpha(1-\alpha)}{f^2(\zeta_\alpha)}\right) \quad (n \to \infty)$$

by the central limit theorem (in its simplest form).

Since $\hat{w}_n = (X'X + aI)^{-1}X'Y$ is the ridge regression estimate,

$$\mathbb{E}(\hat{w}_n - w) = -a(X'X + aI)^{-1}w =: \Delta_n,$$
$$\operatorname{var}(\hat{w}_n) = \sigma^2(X'X + aI)^{-1}X'X(X'X + aI)^{-1} =: \Omega_n .$$

Furthermore, as $n \to \infty$,

$$n^{1/2}\Delta_n = -n^{-1/2}a\left(\frac{X'X}{n} + \frac{aI}{n}\right)^{-1} w \sim -n^{-1/2}a\Sigma^{-1}w \to 0 ,$$
$$n\Omega_n = \sigma^2\left(\frac{X'X}{n} + \frac{aI}{n}\right)^{-1}\frac{X'X}{n}\left(\frac{X'X}{n} + \frac{aI}{n}\right)^{-1} \to \sigma^2\Sigma^{-1} .$$

This gives

$$n^{1/2}\mu'(\hat{w}_n - w) \xrightarrow{\text{law}} \mathbf{N}\left(0, \sigma^2\mu'\Sigma^{-1}\mu\right) \quad (n \to \infty)$$

(the asymptotic, and even exact, normality is obvious from the formula for $\hat{w}_n$).

Let us now calculate the covariance between the two addends in (2.78):

$$\begin{aligned}
&\operatorname{cov}\left(n^{1/2}\frac{\alpha - F_n(\zeta_\alpha)}{f(\zeta_\alpha)}, -n^{1/2}\mu'(\hat{w}_n - w)\right) \\
&= \frac{n}{f(\zeta_\alpha)}\operatorname{cov}\left(F_n(\zeta_\alpha) - \alpha, \mu'(\hat{w}_n - w)\right) \\
&= \frac{n}{(n-1)f(\zeta_\alpha)}\sum_{i=1}^{n-1}\operatorname{cov}\left(1_{\xi_i \le \zeta_\alpha} - \alpha, \mu'(\hat{w}_n - w)\right) \\
&= \frac{n}{(n-1)f(\zeta_\alpha)}\sum_{i=1}^{n-1}\mathbb{E}\Big(\left(1_{\xi_i \le \zeta_\alpha} - \alpha\right)\mu'(X'X + aI)^{-1}X'\xi\Big),
\end{aligned}$$

where $\xi = (\xi_1, \dots, \xi_{n-1})'$ and the last equality uses the decomposition $\hat{w}_n - w = \Delta_n + (X'X + aI)^{-1}X'\xi$ with the second addend having zero expected value. Since

$$\mathbb{E}1_{\xi_i \le \zeta_\alpha}\mu'(X'X + aI)^{-1}X'\xi = \sum_{j=1}^{n-1}\mathbb{E}1_{\xi_i \le \zeta_\alpha}A_j\xi_j = \mu_\alpha A_i \ ,$$

where

$$A_j := \mu'(X'X + aI)^{-1}x_j, \quad j = 1, \dots, n-1 \ ,$$

$$\mu_\alpha := \mathbb{E}1_{\xi_i \le \zeta_\alpha}\xi_i = \int_{-\infty}^{\zeta_\alpha} x f(x)\mathrm{d}x \ ,$$

and an easy computation gives $\mu_\alpha = -\sigma^2 f(\zeta_\alpha)$, we have

$$\begin{aligned}
\operatorname{cov}\left(n^{1/2}\frac{\alpha - F_n(\zeta_\alpha)}{f(\zeta_\alpha)}, -n^{1/2}\mu'(\hat{w}_n - w)\right) &= \frac{n}{(n-1)f(\zeta_\alpha)}\sum_{i=1}^{n-1}\mu_\alpha A_i \\
= -\sigma^2\frac{n}{(n-1)}\sum_{i=1}^{n-1}A_i = -\sigma^2\mu'\left(\frac{1}{n}X'X + \frac{a}{n}I\right)^{-1}\bar{x} &\to -\sigma^2\mu'\Sigma^{-1}\mu
\end{aligned}$$

as $n \to \infty$, where $\bar{x}$ is the arithmetic mean of $x_1, \dots, x_{n-1}$. Finally, this implies that (2.78) converges in law to

$$\mathbf{N}\left(0, \frac{\alpha(1-\alpha)}{f^2(\zeta_\alpha)} + \sigma^2\mu'\Sigma^{-1}\mu - 2\sigma^2\mu'\Sigma^{-1}\mu\right) = \mathbf{N}\left(0, \frac{\alpha(1-\alpha)}{f^2(\zeta_\alpha)} - \sigma^2\mu'\Sigma^{-1}\mu\right) ;$$

the asymptotic normality of (2.78) follows from the central limit theorem with Lindeberg's condition, which holds since (2.78) is a linear combination of the noise random variables $\xi_1, \dots, \xi_{n-1}$ with coefficients whose maximum is $o(1)$ as $n \to \infty$ (this uses the assumption $\frac{1}{n}X_n'X_n \to \Sigma$ made earlier).

2.8.3 Proof of Theorem 2.10

Let $\xi_n^{(\epsilon)}$ be the random variables from the definition of the conservative validity of Γ, $\epsilon \in (0,1)$, $n \in \mathbb{N}$, and Q be a probability distribution on $\mathbf{Z}$. For each sequence of examples $(z_1, \dots, z_n) \in \mathbf{Z}^n$ let $f(z_1, \dots, z_n)$ be the following version of the conditional probability under $Q^\infty \times \mathbf{U}$ that $\xi_n^{(\epsilon)} = 1$ given $z_1, \dots, z_n$:

$$f(z_1, \dots, z_n) := \int \xi_n^\epsilon((z_1, \dots, z_n, \dots), \tau)\mathbf{U}(\mathrm{d}\tau) .$$

For each bag $B \in \mathbf{Z}^{(n)}$, let $f(B)$ be the arithmetic mean of $f(z_1, \dots, z_n)$ over all $n!$ orderings of B. We know that the expected value of $f(B)$ is ϵ under any Q^n, and this, by the completeness of the statistic that maps data sequences $(z_1, \dots, z_n)$ to bags $\{z_1, \dots, z_n\}$ (see [222, Example 4.3.4]; since $\mathbf{Z}$ is standard Borel, it can as well be taken to be $\mathbb{R}$), implies that $f(B) = \epsilon$ for almost all (under any Q^∞) bags B.

Let us show that we can omit "almost". If there is a bag for which $f(B) \neq \epsilon$, defining Q to be the uniform probability distribution on the distinct elements of this bag leads to a contradiction: not Q^∞-almost all bags B satisfy $f(B) = \epsilon$.

Define $S(B, \epsilon)$ as the bag of elements z of B such that Γ makes an error at significance level ϵ at trial n when fed with the elements of B ordered in such a way that the nth example is z (since Γ is invariant, whether an error is made depends only on which element is last, not on the ordering of the first $n-1$ elements). It is clear that

$$\epsilon_1 \leq \epsilon_2 \Longrightarrow S(B, \epsilon_1) \subseteq S(B, \epsilon_2)$$

(as Γ^ϵ is a nested family) and

$$\frac{|S(B, \epsilon)|}{n} \leq \epsilon$$

(as $f(B) = \epsilon$). Therefore, the conformal predictor determined by the conformity measure

$$A(B, z) := \inf\{\epsilon : z \in S(B, \epsilon)\}$$

is at least as good as Γ.

2.9 Context

2.9.1 Exchangeability vs Randomness

In this book we establish various notions of validity for conformal prediction under two closely related assumptions, randomness and exchangeability. The relation becomes very simple in the case of infinite sequences, according to de Finetti's theorem (see Sect. A.5.1 in Appendix A): in wide generality, an exchangeable probability measure is a mixture of power probability measures Q^∞. For a finite horizon, the relation becomes less close.

According to Savage [304, Sect. 3.7] and Hewitt and Savage [158, Sect. 1], Haag (1924, [145]) seems to have been the first to discuss the concept of exchangeability; he also hinted at de Finetti's theorem but did not state it explicitly. De Finetti's theorem in the binary case was obtained by

Bruno de Finetti (1906–1985).

Used with permission from the Archives of the Mathematisches Forschungsinstitut Oberwolfach. © Institute of Mathematical Statistics

him in 1930 [75] and independently by Khinchin in 1932 [190]; the extension to the general case (stated as the result about real-valued random variables) is due to de Finetti [76], and an abstract statement first appeared in [158].

In response to Hewitt and Savage's statement about Haag's role, Good [136, Chap. 3] drew attention to W. E. Johnson's role. In the final section, "Appendix on eduction", of his *Logic* [182], Johnson introduced exchangeability under the name of *permutability postulate* and used it in eduction, as we already mentioned in Chap. 1 (Sect. 1.3.4). See [2] for more information about Johnson.

Dale [68] reviews the early work on exchangeability, first of all, Johnson's, Haag's, and de Finetti's. He believes that "It would be unfair to single out any one of these three authors, at the expense of the others, as the father of exchangeability."

Relation between randomness and exchangeability for finite sequences has been explored by Diaconis and Freedman [83]. While an exchangeable distribution P on the examples $z_1, \ldots, z_N$ is not necessarily close in the variation distance to a mixture of power distributions Q^N even for large N, the situation changes if P can be extended to an exchangeable distribution on $z_1, \ldots, z_N, \ldots, z_{N'}$ for $N' \gg N$: there is a mixture of power distributions at a variation distance of $N(N-1)/N'$, which can be made small by increasing N'. They give examples showing that this rate cannot be improved.

In the binary case, the difference between exchangeability and randomness for finite sequences will be discussed in Sects. 9.1.1 and 9.5.1. For a further discussion of exchangeability, see Sect. 11.6.3.

2.9.2 Conformal Prediction

Conformal predictors were first described in [400] (June 1999) and [303]. The independence of errors in the online mode was proved in [383].

The idea of a rudimentary conformal predictor based on Vapnik's support vector machine was described in [126] (January 1997) and originated at the discussions (mentioned in the preface to the first edition) between Gammerman, Vapnik, and Vovk in the summer of 1996. Having had worked for a long time on the algorithmic theory of randomness (the paper [396] being most relevant), Vovk realized that the fact that a small number of support vectors translates into confident predictions (cf. (13.6)) can be used for making hedged predictions. Let us consider the problem of classification ($|\mathbf{Y}|$ is finite and small). From the point of view of the algorithmic theory of randomness, we can make a confident prediction for the label y_n of the new object x_n given a training set $z_1, \ldots, z_{n-1}$ if the algorithmic randomness deficiency is small for only one possible extension $(z_1, \ldots, z_{n-1}, (x_n, y))$, $y \in \mathbf{Y}$; we can then output the corresponding y as a confident prediction. (In the case of regression, $\mathbf{Y} = \mathbb{R}$, a confident prediction for x_n is possible if the algorithmic randomness deficiency is small for a narrow range of $y \in \mathbf{Y}$.)

Alexey Chervonenkis (1938–2014).

Used with permission of his son Mikhail Chervonenkis

Alexey Chervonenkis understood that a small number of support vectors translates into confident predictions already in June 1966, as we learned in 2004 from his recollections, later published in [407, Chap. 1]. See [407, Chap. 11] for a discussion of two approaches to machine learning born during the summer of 1966; in June Chervonenkis thought of using support vectors for expressing confidence (albeit formalizing it only as the inequality (13.6)), and in July he and Vapnik introduced the growth function and started developing VC theory.

Remark 2.22 The reader who is not familiar with the algorithmic theory of randomness (which is not used in this book outside the end-of-chapter remarks) can consult [205, 232, 238, 323, 415]. In the literature on algorithmic randomness the word "algorithmic" is often omitted, but we will always keep it, to avoid confusion with several other, unrelated, notions of randomness used in this book. (In particular, there are no obvious connections between algorithmic randomness and the assumption of statistical randomness.) The algorithmic notion of randomness formalizes the intuitive notion of typicalness: an object $\omega \in \Omega$ is regarded as typical of a probability distribution P on Ω (we will also say "under P" or "w.r. to P") if there is no reason to be surprised when learning that ω was drawn randomly from P. Using the notion of a universal Turing machine, it is possible to introduce the notion of *algorithmic randomness deficiency*, formalizing the degree of deviation from typicalness.

There are two very different approaches to defining algorithmic randomness deficiency: Martin-Löf's [238] and Levin's [229, 230] (the latter simplified by Gács [123] and in [396]). Kolmogorov's [204] original definition is a special case of Martin-Löf's, but becomes a special case of Levin's (as simplified in [396]) if the plain Kolmogorov complexity in it is replaced by prefix complexity. Martin-Löf's definition is more intuitive, being a universal version of the standard statistical notion of p-value, but Levin's definition often leads to more elegant mathematical results.

The paper [126] was based on Levin's definition, which made it difficult to understand. Conformal predictors, which appeared in [400] and [303], were the result of replacing Levin's definition of algorithmic randomness by Martin-Löf's definition in [126]. See [3, Sect. 7] for further connections between conformal prediction and the algorithmic theory of randomness.

After the notion of conformal predictor crystallized, the connection with the algorithmic theory of randomness started to disappear; in particular, in order to obtain the strongest possible results, we replaced the algorithmic notion of randomness with statistical tests. As we said earlier, in this book we hardly ever mention algorithmic theory of randomness outside the end-of-chapter remarks. This evolution does not look surprising: e.g., we argued in [395] that the algorithmic notions of randomness and complexity are powerful sources of intuition, but for stating mathematical results in their strongest and most elegant form it is often necessary to "translate" them into a non-algorithmic form.

For the information on the many precursors of conformal prediction, see Sect. 13.3; Kei Takeuchi's definition is especially close to ours.

2.9.3 Two Equivalent Definitions of Nonconformity Measures

The first edition of this book [402] used the same definition of a nonconformity measure A, but it was applied to the definition of p-values and prediction sets in a different way. In particular, the values $A(\sigma, z)$ for examples $z \notin \sigma$ were also essential. Namely, we had the same definition via (2.22) of the prediction sets (2.21), but the formula (2.23) for computing nonconformity scores was

$$\begin{aligned}\alpha_i &:= A_n\big(\wr (x_1, y_1), \dots, (x_{i-1}, y_{i-1}), (x_{i+1}, y_{i+1}), \dots, (x_{n-1}, y_{n-1}),\\ &\qquad (x_n, y)\int, (x_i, y_i)\big), \qquad i = 1, \dots, n-1,\\ \alpha_n &:= A_n\big(\wr (x_1, y_1), \dots, (x_{n-1}, y_{n-1})\int, (x_n, y)\big)\end{aligned} \tag{2.79}$$

(we used the notation A_n for the restriction $A|_{\mathbf{Z}^n}$ of A to $\mathbf{Z}^n$). Of course, using such *deleted* nonconformity scores does not change the notion of a conformal predictor: they will lead to the same conformal prediction as our regular nonconformity scores computed from a conformity measure A if we compute the deleted nonconformity scores from the deleted nonconformity measure (2.35).

In Chap. 11 we will generalize conformal prediction replacing the exchangeability model by more general online compression models, which we already mentioned in Sect. 1.6.2. The equivalence between using regular and deleted nonconformity scores will hold for all online compression models used in this book, but in the first edition (Sect. 8.6) we considered a model, the Markov model, for which the equivalence is lost. The main definition of conformal prediction in the Markov model given in the first edition is in terms of deleted nonconformity scores, and it becomes impossible for regular nonconformity scores. However, that definition loses its efficiency in the case of asymmetric Markov chains: see Sect. 8.8 (subsection "Kolmogorov's modelling vs standard statistical modelling"). Alternative approaches to conformal prediction for Markov chains that do not use deleted nonconformity scores are discussed in Sect. 7.3 (subsection "Markov sequences") and Sect. 8.8 (the end of subsection "Kolmogorov's modelling vs standard statistical modelling") of the first edition.

The main advantages of our current approach (exemplified by (2.23)) are that it makes most of our formulas simpler (compare, e.g., (2.23) and (2.79)) and that it greatly simplifies the definition of "one-off structures" in Chap. 11 (Sect. 11.2.5). Some formulas, however, become more complicated. In such cases, we may use the deleted definition (2.35).

2.9.4 The Two Meanings of Conformity in Conformal Prediction

In the first edition [402] of this book we only used the first notion of conformity discussed in Sect. 2.2.9, conformity to a bag. The analogue of conformalized ridge regression for the symmetric nonconformity measure (2.30) in place of the asymmetric conformity measures (2.33) and (2.34) was called the ridge regression confidence machine. The asymmetric approach, which we interpret in terms of conformity to a property, was proposed in [40] to facilitate theoretical analysis, but it also simplifies prediction algorithms. Asymmetric definitions such as (2.33) are widely used in conformal predictive distributions, which will be the topic of Chap. 7.

2.9.5 Examples of Nonconformity Measures

The main examples of nonconformity measures that we considered in this chapter (concentrating on the case of regression) are provided by the methods of nearest neighbours, least squares, and ridge regression. The origins of the nearest neighbours method are lost in the mists of time [275], but early modern references include Fix and Hodges's 1951 paper [113] (reprinted in 1989 with a commentary [333]) and [59] (with the method being used by Peter Hart and Charles Cole in 1966 [58]).

The least squares procedure was invented independently by Gauss and Legendre and first published by Legendre [221] in 1805 (for details, see, e.g., Plackett [279] and Stigler [342, 343]). The term "hat matrix" was introduced by John W. Tukey (see [163]). The ridge regression procedure was first described in detail by Arthur E. Hoerl and Robert W. Kennard [168, 169] in 1970. The idea came from Hoerl's [165] ridge analysis, a method of examining high-dimensional quadratic response surfaces (for details, see Hoerl [167]). The link between ridge analysis and ridge regression is provided by Hoerl's 1962 paper [166].

Deleted residuals are also known as PRESS and predicted residuals, and the nonconformity scores (2.47) differ from "internally studentized residuals" only by a factor that does not affect the conformal predictor's output. We did not consider "externally studentized residuals"; for details and history, see, e.g., [54, Sect. 2.2.1].

The CRR was introduced in [40], as we already mentioned, whereas its predecessor, ridge regression confidence machine (RRCM), was developed by Ilia Nouretdinov and published as [264]. For the pseudocode of various implementations of the RRCM, see the first edition of this book (Sect. 2.3) and Appendix B of the arXiv version of [405].

The conformalized version of the lasso was derived by Lei [224]. This result is much more sophisticated than our simple procedure, Algorithm 2.4, based on ridge regression.

An important advance is Romano et al.'s [293] conformalized quantile regression, which employs classical quantile regression to develop a conformal predictor that is adaptive to heteroscedasticity. Another approach to making conformal prediction adaptive to heteroscedasticity is Chernozhukov et al.'s [49] distributional conformal prediction.

2.9.6 Kernel Methods

Kernel methods are a powerful tool for modifying nonconformity measures and making them more versatile. They have their origins in the Hilbert–Schmidt theory of integral equations (see [249]). The fundamental fact that each symmetric nonnegative definite function has representation (2.54) can be proved by many different methods: see, e.g., Mercer [249] (that paper, however, proves a slightly different result, "Mercer's theorem", about continuous kernels and an integral analogue of condition (2.58)) and Aronszajn [7] (Aronszajn's proof is based on Moore's idea; it is reproduced in Wahba [416]). For more recent expositions, see, e.g., [67, 308, 322, 341]. In the first edition [402, p. 37] we gave a simple and standard probabilistic proof.

There are several approaches to kernel ridge regression; the three main ones appear to be the following:

- the approach adopted in this book: the objects are mapped to an arbitrary (not necessarily functional or separable) Hilbert space, and the prediction rule is chosen from among the continuous linear functionals on that space; the main Eq. (2.53) can be obtained, for example, using the Lagrange method analogously to Vapnik's [366] derivation of SVM (see [302]); the approach based on the equality (2.51) is standard in statistics;
- the approach based on functional Hilbert spaces with bounded evaluation functionals (called *reproducing kernel Hilbert spaces*; the prediction rule is chosen from among the elements of such a space; see, e.g., Wahba [416]);
- the approach based on Gaussian processes: one assumes that the labels y_i are obtained from zero-mean normal random variables with covariances $\operatorname{cov}(y_i, y_j)$ defined in terms of $\mathcal{K}(x_i, x_j)$; (2.53) can then be obtained as the expected value of the x's label. In geostatistics this approach is known as kriging; for further details, see Cressie [66] and Sect. 13.7.4.

An especially important class of kernels is formed by the *universal kernels* $\mathcal{K}$ defined to be continuous kernels such that the finite linear combinations $\sum_{i=1}^{n} c_i K(\cdot, x_i)$ are dense (in the maximum norm) in the set of all continuous functions on any compact subset of the object space $\mathbf{X}$, which is assumed to be a Hausdorff topological space. For details, see [250]. In the case of a compact $\mathbf{X}$ the definition goes back to Steinwart [340]; see also [341, Sect. 4.6]. If we consider the right-hand side of (2.53) as function of the test object x (remember that k_n is defined via x), this function can approximate any continuous function of x on any compact, provided $\mathcal{K}$ is a universal kernel. This makes the location of the prediction interval produced by CRR based on a universal kernel fully adaptive; its length, however, will be less adaptive. We will further discuss universal kernels in the context of conformal prediction, namely conformal predictive distributions, in Chap. 7.

Formula (2.56), which we used for the fast updating of the inverse matrix in the kernel CRR, may have been first explicitly given by Banachiewicz [16, 17]; further references and history can be found in [155]. There are similar updating formulas (going back to Gauss [128] and also reviewed in [155]) that could be used in the case of CRR, but the need for speeding up computations is less pressing for CRR since the matrix to be inverted is of size $p \times p$, which we assumed constant.

2.9.7 Burnaev–Wasserman Programme

In Sect. 2.5 we follow Burnaev and Vovk [40]. We state the main result asymptotically, although non-asymptotic results are definitely feasible. In fact, the main interest of strong parametric assumptions as soft models is in the possibility of non-asymptotic results and, even more importantly, in practical efficiency (non-asymptotic, of course).

Whereas in the case of the Burnaev–Wasserman programme asymptotic results serve as a means to express interesting phenomena in a simple way and are by no means indispensable, some of the results in the following chapters are infinitary in a very essential way, similarly to Stone's [347] result discussed in Sect. 1.5.1. Such intrinsically infinitary results include the existence of universal prediction algorithms in the context of classification (Sect. 3.2 of the first edition [402] of this book), probabilistic classification (Sect. 6.3), and probabilistic regression (Sect. 7.6).

2.9.8 Completeness Results

A version (more sophisticated but less precise, involving arbitrary constants) of Proposition 2.9 and Theorem 2.10 was stated and proved in [265]. An analogous result was stated by Takeuchi for his version of conformal predictors.

For a generalization of Theorem 2.10 to online compression models (introduced in Sect. 1.6.2 and defined in Part IV) different from exchangeability, see [405, Theorem 2].

Our results in Sect. 2.6 can be interpreted in terms of complete classes of decision procedures, in the statistical terminology [222, 420], or Pareto frontiers, in the economic and engineering terminology. Strictly speaking, the definitions of both complete classes and Pareto frontiers also need an irreflexive relation of strict domination, alongside the reflexive relation of domination. To simplify exposition, we did not introduce the former, but we can still roughly summarize Sect. 2.6 as the conformal predictors forming a complete class of valid confidence predictors.

Chapter 3
Conformal Prediction: Classification and General Case

Abstract In this chapter we mainly concentrate on classification, where the label space $\mathbf{Y}$ is finite (and equipped with the discrete σ-algebra), after discussing regression in the previous one. Our first topic is criteria of efficiency of conformal predictors (Sect. 3.1); they will be applied in the next chapter (Sect. 4.3.8) to designing new conformal predictors. We give two more examples of nonconformity measures, specific to the case of classification, and illustrate one of the criteria on one of those measures (Sect. 3.2). Finally, we consider the case of "weak teachers", which are allowed to provide the true label with a delay or not to provide it at all, in Sect. 3.3.

Keywords Conformal classification · Criteria of efficiency of conformal prediction · Weak teachers

3.1 Criteria of Efficiency for Conformal Prediction

In this section we will discuss suitable ways to measure the performance of a conformal predictor on a test set. For concreteness we consider the offline mode of prediction. Suppose we are given a test set $z_{l+1}, \ldots, z_{l+k}$ and would like to use it to measure the efficiency of the predictions derived from the training set $z_1, \ldots, z_l$. (Informally, by the efficiency of conformal predictors we mean that the prediction sets they output tend to be small, and by the efficiency of conformal transducers we mean that the p-values they output tend to be small.) For each test example $z_i = (x_i, y_i)$, $i = l + 1, \ldots, l + k$, we have a nested family $(\Gamma_i^\epsilon : \epsilon \in (0, 1))$ of subsets of $\mathbf{Y}$, where $\Gamma_i^\epsilon := \Gamma^\epsilon(z_1, \ldots, z_l, x_i)$, and a system of p-values $(p_i^y : y \in \mathbf{Y})$, where $p_i^y := p^y(z_1, \ldots, z_l, x_i)$.

We will discuss ten criteria of efficiency for such a family or a system, but some of them will depend, additionally, on the actual (or *observed*) label y_i of the test example. Five of them will be applicable to the prediction sets Γ_i^ϵ and so dependent on the significance level ϵ, and the other five will be applicable to systems of p-values $(p_i^y : y \in \mathbf{Y})$ and so independent of ϵ (ϵ-*free*). We start from the *prior* criteria, which do not depend on the observed test labels.

V. Vovk et al., *Algorithmic Learning in a Random World*,
https://doi.org/10.1007/978-3-031-06649-8_3

3.1.1 Basic Criteria

The simplest criteria of efficiency are:

- The *S criterion* (with "S" standing for "sum") measures efficiency by the average sum

$$\frac{1}{k}\sum_{i=l+1}^{l+k}\sum_{y} p_i^y \tag{3.1}$$

  of the p-values; small values are preferable for this criterion. It is ϵ-free.
- The *N criterion* uses the average size

$$\frac{1}{k}\sum_{i=l+1}^{l+k} |\Gamma_i^\epsilon| \tag{3.2}$$

  of the prediction sets ("N" stands for "number": the size of a prediction set is the number of labels in it). Small values are preferable. Under this criterion the efficiency is a function of the significance level ϵ.

Both these criteria are prior. We could generalize the S criterion by replacing the p-values p_i^y in (3.1) by $\phi(p_i^y)$ for a smooth strictly increasing function ϕ, but we will stick to the simplest case (the general case will be briefly discussed in Remark 3.15).

3.1.2 Other Prior Criteria

A disadvantage of the basic criteria is that they look too stringent. Even for a very efficient conformal transducer, we cannot expect all the p-values p^y to be small: the p-value corresponding to the true label will not be small with high probability; and even for a very efficient conformal predictor we cannot expect the size of its prediction set to be zero: with high probability it will contain the true label. The other prior criteria are less stringent. The ones that do not depend on the significance level are:

- The *U criterion* (with "U" standing for "unconfidence") uses the average unconfidence

$$\frac{1}{k}\sum_{i=l+1}^{l+k} \min_{y}\max_{y'\neq y} p_i^{y'} \tag{3.3}$$

  over the test set, where the *unconfidence* for a test object x_i is the second largest p-value $\min_y \max_{y'\neq y} p_i^{y'}$; small values of (3.3) are preferable.

- The *F criterion* uses the average fuzziness

$$\frac{1}{k} \sum_{i=l+1}^{l+k} \left(\sum_y p_i^y - \max_y p_i^y \right) , \tag{3.4}$$

where the *fuzziness* for a test object x_i is defined as the sum of all p-values apart from a largest one, i.e., as $\sum_y p_i^y - \max_y p_i^y$; smaller values of (3.4) are preferable.

Their counterparts depending on the significance level are:

- The *M criterion* uses the percentage of objects x_i in the test set for which the prediction set Γ_i^ϵ at significance level ϵ is *multiple*, i.e., contains more than one label. Smaller values are preferable. As a formula, the criterion prefers smaller

$$\frac{1}{k} \sum_{i=l+1}^{l+k} 1_{|\Gamma_i^\epsilon|>1} . \tag{3.5}$$

When the percentage (3.5) of multiple predictions is the same for two conformal predictors (which is a common situation: the percentage can well be zero when the data is clean and ϵ is not too demanding), the M criterion compares the percentages

$$\frac{1}{k} \sum_{i=l+1}^{l+k} 1_{\Gamma_i^\epsilon=\emptyset} \tag{3.6}$$

of empty predictions, and larger values are preferable. (Intuitively, an empty prediction is a warning that the test object is unusual, and since such a warning presents useful information and the probability of a warning is guaranteed not to exceed ϵ, we want to be warned as often as possible.) This is a widely used criterion; in particular, it was used in the first edition of this book [402] and papers preceding it.
- The *E criterion* (where "E" stands for "excess") uses the average (over the test set, as usual) amount by which the size of the prediction set exceeds 1. In other words, the criterion gives the average number of excess labels in the prediction sets as compared with the ideal situation of one-element prediction sets. Smaller values are preferable for this criterion. As a formula, the criterion prefers smaller

$$\frac{1}{k} \sum_{i=l+1}^{l+k} \left(|\Gamma_i^\epsilon| - 1\right)^+ ,$$

where $t^+ := \max(t, 0)$. When these averages coincide for two conformal predictors, we compare the percentages (3.6) of empty predictions; larger values are preferable.

3.1.3 Observed Criteria

The prior criteria discussed in the previous subsection treat the largest p-value, or prediction sets of size 1, in a special way. The corresponding criteria of this subsection attempt to achieve the same goal by using the observed label.

These are the observed counterparts of the non-basic prior ϵ-free criteria:

- The *OU* ("observed unconfidence") *criterion* uses the average observed unconfidence

$$\frac{1}{k}\sum_{i=l+1}^{l+k} \max_{y \neq y_i} p_i^y$$

  over the test set, where the *observed unconfidence* for a test example (x_i, y_i) is the largest p-value p_i^y for the *false labels* $y \neq y_i$. Smaller values are preferable for this test.
- The *OF* ("observed fuzziness") *criterion* uses the average sum of the p-values for the false labels, i.e.,

$$\frac{1}{k}\sum_{i=l+1}^{l+k} \sum_{y \neq y_i} p_i^y ; \tag{3.7}$$

  smaller values are preferable.

The counterparts of the last group depending on the significance level ϵ are:

- The *OM criterion* uses the percentage of observed multiple predictions

$$\frac{1}{k}\sum_{i=l+1}^{l+k} 1_{\Gamma_i^\epsilon \setminus \{y_i\} \neq \emptyset}$$

  in the test set, where an *observed multiple* prediction is defined to be a prediction set including a false label. Smaller values are preferable.
- The *OE criterion* (OE standing for "observed excess") uses the average number

$$\frac{1}{k}\sum_{i=l+1}^{l+k} \left|\Gamma_i^\epsilon \setminus \{y_i\}\right|$$

Table 3.1 The ten criteria studied in this section: the two basic ones in the upper section; the four other prior ones in the middle section; and the four observed ones in the lower section

ϵ-free	ϵ-dependent
S (sum of p-values)	*N (number of labels)*
U (unconfidence)	M (multiple)
F (fuzziness)	E (excess)
OU (observed unconfidence)	OM (observed multiple)
OF (observed fuzziness)	*OE (observed excess)*

of false labels included in the prediction sets at significance level ϵ; smaller values are preferable.

The ten criteria used in this section are listed in Table 3.1. Half of the criteria depend on the significance level ϵ, and the other half are the respective ϵ-free versions.

In the case of binary classification problems, $|\mathbf{Y}| = 2$, the number of different criteria of efficiency in Table 3.1 reduces to six: the criteria not separated by a vertical or horizontal line (namely, U and F, OU and OF, M and E, and OM and OE) coincide.

3.1.4 *Idealised Setting*

Which are the best criteria of efficiency? To evaluate different criteria, we will consider the limiting case of infinitely large training and test sets. A reasonable criterion should give reasonable results in this simple case as well.

To formalise the intuition of an infinite training set, we assume that the prediction algorithm is directly given the data-generating probability distribution Q on $\mathbf{Z}$ instead of being given a training set. Instead of conformity measures we will use *idealised conformity measures*: functions $A(Q, z)$ of $Q \in \mathbf{P}(\mathbf{Z})$ (where $\mathbf{P}(\mathbf{Z})$ is the set of all probability measures on $\mathbf{Z}$) and $z \in \mathbf{Z}$, measurable in z. We will fix the data-generating distribution Q for the rest of this section, and so write the corresponding conformity scores as $A(z)$. The *idealised conformal predictor* determined by A outputs the following prediction set $\Gamma^\epsilon(x)$ for each object $x \in \mathbf{X}$ and each significance level $\epsilon \in (0, 1)$. For each potential label $y \in \mathbf{Y}$ for x define the corresponding *p-value* as

$$p^y = p(x, y) = p_A(x, y) = p_A(x, y, \tau) := Q\{z \in \mathbf{Z} : A(z) < A(x, y)\} \\ + \tau Q\{z \in \mathbf{Z} : A(z) = A(x, y)\} \quad (3.8)$$

(we often omit pairs of parentheses in expressions such as $A((x, y))$ and $Q(\{\dots\})$ when there is no danger of ambiguity), where τ is a random number distributed uniformly on $[0, 1]$. (The same random number τ is used in (3.8) for all (x, y).) The prediction set is

$$\Gamma^{\epsilon}(x) = \Gamma^{\epsilon}_{A}(x) = \Gamma^{\epsilon}_{A}(x, \tau) := \{y \in \mathbf{Y} : p_A(x, y, \tau) > \epsilon\} . \tag{3.9}$$

The *idealised conformal transducer* determined by A outputs for each object $x \in \mathbf{X}$ the system of p-values $(p^y : y \in \mathbf{Y})$ defined by (3.8); in the idealised case we will usually use the alternative notation $p(x, y)$ for p^y.

We could have used the *idealised conformity order* when defining the p-values (3.8): $z \preceq z'$ is defined to mean $A(z) \leq A(z')$. Let us say that two idealised conformity measures are *equivalent* if they lead to the same idealised conformity order; in other words, A and B are equivalent if, for all $z, z' \in \mathbf{Z}$, $A(z) \leq A(z') \Leftrightarrow B(z) \leq B(z')$.

The standard properties of validity for conformal transducers and predictors mentioned in the previous section simplify in this idealised case as follows:

- If (x, y) is generated from Q and $\tau \in [0, 1]$ is generated from the uniform distribution independently of (x, y), $p(x, y)$ is distributed uniformly on $[0, 1]$.
- Therefore, at each significance level ϵ the idealised conformal predictor makes an error with probability ϵ.

The test set being infinite is formalised by replacing the use of a test set in the criteria of efficiency by averaging with respect to the data-generating probability distribution Q. In the case of the top two and bottom two criteria in Table 3.1 (the ones set in italics) this is done as follows. An idealised conformity measure A is:

- *S-optimal* if, for any idealised conformity measure B,

$$\mathbb{E}_{x,\tau} \sum_{y \in \mathbf{Y}} p_A(x, y) \leq \mathbb{E}_{x,\tau} \sum_{y \in \mathbf{Y}} p_B(x, y) , \tag{3.10}$$

  where the notation $\mathbb{E}_{x,\tau}$ refers to the expected value when x and τ are independent, $x \sim Q_{\mathbf{X}}$, and $\tau \sim \mathbf{U}$; $Q_{\mathbf{X}}$ is the marginal distribution of Q on $\mathbf{X}$ (defined by $Q_{\mathbf{X}}(A) := Q(A \times \mathbf{Y})$ for all measurable $A \subseteq \mathbf{X}$), and $\mathbf{U}$ is the uniform distribution on $[0, 1]$;
- *N-optimal* if, for any idealised conformity measure B and any significance level ϵ,

$$\mathbb{E}_{x,\tau} \left|\Gamma^{\epsilon}_{A}(x)\right| \leq \mathbb{E}_{x,\tau} \left|\Gamma^{\epsilon}_{B}(x)\right| ;$$

- *OF-optimal* if, for any idealised conformity measure B,

$$\mathbb{E}_{(x,y),\tau} \sum_{y' \neq y} p_A(x, y') \leq \mathbb{E}_{(x,y),\tau} \sum_{y' \neq y} p_B(x, y') ,$$

  where the lower index (x, y) in $\mathbb{E}_{(x,y),\tau}$ refers to averaging over $(x, y) \sim Q$ (with (x, y) and $\tau \sim \mathbf{U}$ independent);
- *OE-optimal* if, for any idealised conformity measure B and any significance level ϵ,

$$\mathbb{E}_{(x,y),\tau}\left|\Gamma_A^{\epsilon}(x)\setminus\{y\}\right| \le \mathbb{E}_{(x,y),\tau}\left|\Gamma_B^{\epsilon}(x)\setminus\{y\}\right| .$$

We will define later (in Sect. 3.1.6) the idealised versions of the other six criteria listed in Table 3.1.

3.1.5 Conditionally Proper Criteria of Efficiency

In this subsection we will characterise the optimal idealised conformity measures for the four criteria of efficiency that are set in italics in Table 3.1. We will assume in the rest of the section that the set $\mathbf{X}$ is finite (with the discrete σ-algebra; from the practical point of view, being finite is not a restriction). Since we consider the case of classification, $|\mathbf{Y}| < \infty$, this implies that the whole example space $\mathbf{Z}$ is finite. Without loss of generality, we also assume that the data-generating probability distribution Q satisfies $Q_{\mathbf{X}}(x) > 0$ for all $x \in \mathbf{X}$ (we often omit curly braces in expressions such as $Q_{\mathbf{X}}(\{x\})$): we can always omit the xs for which $Q_{\mathbf{X}}(x) = 0$.

The *conditional probability (CP) idealised conformity measure* is

$$A(x, y) = Q(y \mid x) = Q_{\mathbf{Y}|\mathbf{X}}(y \mid x) := \frac{Q(x, y)}{Q_{\mathbf{X}}(x)} . \tag{3.11}$$

(In this section, we will invariably use the shorter notation $Q(y \mid x)$ instead of the more precise $Q_{\mathbf{Y}|\mathbf{X}}(y \mid x)$; we will not need $Q_{\mathbf{X}|\mathbf{Y}}$, which is defined analogously, before Chap. 8.) We say that an idealised conformity measure A is a *refinement* of an idealised conformity measure B if

$$B(z_1) < B(z_2) \Longrightarrow A(z_1) < A(z_2) \tag{3.12}$$

for all $z_1, z_2 \in \mathbf{Z}$. Let $\mathcal{R}(\mathrm{CP})$ be the set of all refinements of the CP idealised conformity measure. If C is a criterion of efficiency (one of the ten criteria in Table 3.1), we let $\mathcal{O}(C)$ stand for the set of all C-optimal idealised conformity measures.

Theorem 3.1 $\mathcal{O}(\mathrm{S}) = \mathcal{O}(\mathrm{OF}) = \mathcal{O}(\mathrm{N}) = \mathcal{O}(\mathrm{OE}) = \mathcal{R}(\mathrm{CP})$.

We say that an efficiency criterion is *conditionally proper* if the CP idealised conformity measure is always optimal for it. We will also use two modifications of this definition: an efficiency criterion is *conditionally strongly proper* if any refinement of the CP idealised conformity measure is optimal for it, and it is *conditionally weakly proper* if some refinement of the CP idealised conformity measure is optimal for it. We will say that it is *binary conditionally weakly proper* if some refinement of the CP idealised conformity measure is optimal for it whenever $|\mathbf{Y}| = 2$. Theorem 3.1 shows that four of our ten criteria are conditionally strongly proper, namely S, N, OF, and OE (they are set in italics in Table 3.1). In the next section we will see that in general the other six criteria are not conditionally proper (they are only binary conditionally weakly proper). The intuition behind conditionally proper criteria will be discussed in Sect. 3.1.7.

Theorems 3.1–3.4 will be proved in Sect. 3.4.1 below.

3.1.6 Criteria of Efficiency that Are not Conditionally Proper

Now we define the idealised analogues of the six criteria that are not set in italics in Table 3.1, slightly modifying the U and F criteria in the direction of the M and E criteria, respectively (this only concerns breaking the ties, which are rare for the U and F criteria). An idealised conformity measure A is:

- *U-optimal* if, for any idealised conformity measure B, we have either

$$\mathbb{E}_{x,\tau} \min_{y} \max_{y' \neq y} p_A(x, y') < \mathbb{E}_{x,\tau} \min_{y} \max_{y' \neq y} p_B(x, y') \tag{3.13}$$

  or both

$$\mathbb{E}_{x,\tau} \min_{y} \max_{y' \neq y} p_A(x, y') = \mathbb{E}_{x,\tau} \min_{y} \max_{y' \neq y} p_B(x, y') \tag{3.14}$$

  and

$$\mathbb{E}_{x,\tau} \max_{y} p_A(x, y) \leq \mathbb{E}_{x,\tau} \max_{y} p_B(x, y) \, ; \tag{3.15}$$

- *M-optimal* if, for any idealised conformity measure B and any significance level ϵ, we have either

$$\mathbb{P}_{x,\tau}(\left|\Gamma_A^{\epsilon}(x)\right| > 1) < \mathbb{P}_{x,\tau}(\left|\Gamma_B^{\epsilon}(x)\right| > 1) \tag{3.16}$$

  or both

$$\mathbb{P}_{x,\tau}(\left|\Gamma_A^{\epsilon}(x)\right| > 1) = \mathbb{P}_{x,\tau}(\left|\Gamma_B^{\epsilon}(x)\right| > 1) \tag{3.17}$$

  and

$$\mathbb{P}_{x,\tau}(\left|\Gamma_A^{\epsilon}(x)\right| = 0) \geq \mathbb{P}_{x,\tau}(\left|\Gamma_B^{\epsilon}(x)\right| = 0) \tag{3.18}$$

  (we are using $\mathbb{P}_{x,\tau}$ and $\mathbb{P}_{(x,y),\tau}$ in the same sense as $\mathbb{E}_{x,\tau}$ and $\mathbb{E}_{(x,y),\tau}$);
- *F-optimal* if, for any idealised conformity measure B, we have either

$$\mathbb{E}_{x,\tau}\Big(\sum_{y} p_A(x, y) - \max_{y} p_A(x, y)\Big) < \mathbb{E}_{x,\tau}\Big(\sum_{y} p_B(x, y) - \max_{y} p_B(x, y)\Big) \tag{3.19}$$

or both

$$\mathbb{E}_{x,\tau}\Big(\sum_y p_A(x,y) - \max_y p_A(x,y)\Big) = \mathbb{E}_{x,\tau}\Big(\sum_y p_B(x,y) - \max_y p_B(x,y)\Big) \tag{3.20}$$

and (3.15);

- *E-optimal* if, for any idealised conformity measure B and any significance level ϵ, we have either

$$\mathbb{E}_{x,\tau}\big(\big(|\Gamma_A^\epsilon(x)| - 1\big)^+\big) < \mathbb{E}_{x,\tau}\big(\big(|\Gamma_B^\epsilon(x)| - 1\big)^+\big) \tag{3.21}$$

or both

$$\mathbb{E}_{x,\tau}\big(\big(|\Gamma_A^\epsilon(x)| - 1\big)^+\big) = \mathbb{E}_{x,\tau}\big(\big(|\Gamma_B^\epsilon(x)| - 1\big)^+\big) \tag{3.22}$$

and (3.18);

- *OU-optimal* if, for any idealised conformity measure B,

$$\mathbb{E}_{(x,y),\tau} \max_{y' \neq y} p_A(x,y') \le \mathbb{E}_{(x,y),\tau} \max_{y' \neq y} p_B(x,y') ; \tag{3.23}$$

- *OM-optimal* if, for any idealised conformity measure B and any significance level ϵ,

$$\mathbb{P}_{(x,y),\tau}(\Gamma_A^\epsilon(x) \setminus \{y\} \neq \emptyset) \le \mathbb{P}_{(x,y),\tau}(\Gamma_B^\epsilon(x) \setminus \{y\} \neq \emptyset) . \tag{3.24}$$

Suitable idealised conformity measures for the U and M criteria use the notion of the *predictability* of $x \in \mathbf{X}$ defined as

$$f(x) := \max_{y \in \mathbf{Y}} Q(y \mid x) . \tag{3.25}$$

A *choice function* $\hat{y} : \mathbf{X} \to \mathbf{Y}$ is defined by the condition

$$\forall x \in \mathbf{X} : f(x) = Q(\hat{y}(x) \mid x) . \tag{3.26}$$

Define the *signed predictability idealised conformity measure* corresponding to a choice function $\hat{y}$ by

$$A(x,y) := \begin{cases} f(x) & \text{if } y = \hat{y}(x) \\ -f(x) & \text{if not ;} \end{cases} \tag{3.27}$$

a *signed predictability (SP) idealised conformity measure* is the signed predictability idealised conformity measure corresponding to some choice function.

For some of the following theorems we will need to modify the operator $\mathcal{R}$ of refinement. Let $\mathcal{R}'(\text{SP})$ be the set of all idealised conformity measures A such that there exists an SP idealised conformity measure B that satisfies both (3.12) and

$$B(x, y_1) = B(x, y_2) \Longrightarrow A(x, y_1) = A(x, y_2) \tag{3.28}$$

for all $x \in \mathbf{X}$ and $y_1, y_2 \in \mathbf{Y}$.

Theorem 3.2 $\mathcal{O}(\text{U}) = \mathcal{O}(\text{M}) = \mathcal{R}'(\text{SP})$.

Define the *ACP (adjusted conditional probability) idealised conformity measure* corresponding to a choice function $\hat{y}$ by

$$A(x, y) := \begin{cases} Q(y \mid x) & \text{if } y = \hat{y}(x) \\ Q(y \mid x) - 1 & \text{if not ;} \end{cases}$$

an *ACP idealised conformity measure* is an idealised conformity measure corresponding to some choice function; $\mathcal{R}(\text{ACP})$ is defined analogously to $\mathcal{R}(\text{CP})$ but using ACP idealised conformity measures rather than the CP idealised conformity measure.

Theorem 3.3 $\mathcal{O}(\text{F}) = \mathcal{O}(\text{E}) = \mathcal{R}(\text{ACP})$.

Of course, Theorems 3.2 and 3.3 are equivalent when $|\mathbf{Y}| = 2$ (in which case the premise of (3.28) is always false).

The *adjusted signed predictability (ASP) idealised conformity measure* is defined by

$$A(x, y) := \begin{cases} f(x) & \text{if } f(x) > 1/2 \text{ and } y = \hat{y}(x) \\ 0 & \text{if } f(x) \le 1/2 \\ -f(x) & \text{if } f(x) > 1/2 \text{ and } y \ne \hat{y}(x) \, , \end{cases}$$

where f is the predictability function (3.25); notice that this definition is unaffected by the choice of the choice function. Let $\mathcal{R}''(\text{ASP})$ be the set of all refinements A of the ASP idealised conformity measure such that, for all $x \in \mathbf{X}$ and all $y_1, y_2 \in \mathbf{Y}$:

$$f(x) \ge 0.5 \ \ \& \ \ Q(y_1 \mid x) < 0.5 \ \ \& \ \ Q(y_2 \mid x) < 0.5 \Longrightarrow A(x, y_1) = A(x, y_2)$$
$$f(x) < 0.5 \Longrightarrow A(x, y_1) = A(x, y_2) \, .$$

Theorem 3.4 $\mathcal{O}(\text{OU}) = \mathcal{O}(\text{OM}) = \mathcal{R}''(\text{ASP})$.

Table 3.2 summarises the results given above. For each of the criteria listed in Table 3.1 it gives an optimal idealised conformity measure and cites the result asserting the optimality of that idealised conformity measure.

Table 3.2 Idealised conformity measures that are optimal for the ten criteria of efficiency given in Table 3.1; the arrangement of the criteria is the same as in Table 3.1

ϵ-free	ϵ-dependent
S: CP (Theorem 3.1)	*N: CP (Theorem 3.1)*
U: SP (Theorem 3.2)	M: SP (Theorem 3.2)
F: ACP (Theorem 3.3)	E: ACP (Theorem 3.3)
OU: ASP (Theorem 3.4)	OM: ASP (Theorem 3.4)
OF: CP (Theorem 3.1)	*OE: CP (Theorem 3.1)*

Theorems 3.2–3.4 show that the six criteria that are not set in italics in Table 3.1 are not conditionally proper (however, we will see in Corollary 3.5 below that they are binary conditionally weakly proper). These are simple explicit examples (inevitably involving label spaces $\mathbf{Y}$ with $|\mathbf{Y}| > 2$) showing that they are not even conditionally weakly proper:

- Let $\mathbf{X} = \{1\}$ (and so $Q_{\mathbf{X}}(1) = 1$), $\mathbf{Y} = \{1, 2, 3\}$, and

$$Q(1 \mid 1) = 0.2 \quad Q(2 \mid 1) = 0.3 \quad Q(3 \mid 1) = 0.5 \,. \tag{3.29}$$

  (Remember that, in this chapter, $Q(y \mid x)$ always means $Q_{\mathbf{Y}|\mathbf{X}}(y \mid x)$.) In this case, all refinements of the CP idealised conformity measure are equivalent. The U criterion is not conditionally proper since the expression

$$\mathbb{E}_{x,\tau} \min_y \max_{y' \neq y} p(x, y')$$

  (cf. (3.13)) is 0.35 for the CP idealised conformity measure and is smaller, 0.25, for the SP idealised conformity measure. The M criterion is not conditionally proper since at significance level $\epsilon = 0.2$ the CP idealised conformity measure gives the predictor $\Gamma^\epsilon(1) = \{2, 3\}$ (a.s.), and so

$$\mathbb{P}_{x,\tau}(|\Gamma^\epsilon_{\mathrm{CP}}(x)| > 1) = 1 > 0.6 = \mathbb{P}_{x,\tau}(|\Gamma^\epsilon_{\mathrm{SP}}(x)| > 1)$$

  (cf. (3.16)).
- Let $\mathbf{X} = \{1, 2\}$, $\mathbf{Y} = \{1, 2, 3\}$, $Q_{\mathbf{X}}(1) = Q_{\mathbf{X}}(2) = 0.5$, and, for a small $\delta > 0$,

$$\begin{aligned} &Q(1 \mid 1) = 1/3 - \delta \qquad Q(2 \mid 1) = 1/3 \qquad\quad\; Q(3 \mid 1) = 1/3 + \delta \\ &Q(1 \mid 2) = 1/3 - 5\delta \quad\; Q(2 \mid 2) = 1/3 + 2\delta \quad Q(3 \mid 2) = 1/3 + 3\delta \,. \end{aligned}$$

  The CP idealised conformity measure again has only equivalent refinements. The F criterion is not conditionally proper since the expression

$$\mathbb{E}_{x,\tau}\Bigl(\sum_y p(x, y) - \max_y p(x, y)\Bigr)$$

(cf. (3.19)) is $3/4 + O(\delta)$ for the CP idealised conformity measure and is smaller (provided δ is sufficiently small), $2/3 + O(\delta)$, for the ACP idealised conformity measure (which is unique). The E criterion is not conditionally proper since at significance level $\epsilon = 2/3$ the CP idealised conformity measure has a larger expected excess (for small δ) than the ACP idealised conformity measure (whose expected excess is zero):

$$\mathbb{E}_{x,\tau}\left(\left(|\Gamma^{\epsilon}_{\text{CP}}(x)| - 1\right)^{+}\right) = 0.5 + O(\delta) > 0 = \mathbb{E}_{x,\tau}\left(\left(|\Gamma^{\epsilon}_{\text{ACP}}(x)| - 1\right)^{+}\right)$$

(cf. (3.21)).

- Let us again set $\mathbf{X} = \{1\}$ and $\mathbf{Y} = \{1, 2, 3\}$, and define Q by (3.29). The OU criterion is not conditionally proper since the expression

$$\mathbb{E}_{(x,y),\tau} \max_{y' \neq y} p(x, y')$$

(cf. (3.23)) is 0.55 for the CP idealised conformity measure and is smaller, 0.5, for the ASP idealised conformity measure. The OM criterion is not conditionally proper since at significance level $\epsilon = 0.2$ the CP idealised conformity measure gives the predictor $\Gamma^{\epsilon}(1) = \{2, 3\}$ (a.s.), and so

$$\mathbb{P}_{(x,y),\tau}(\Gamma^{0.2}_{\text{CP}}(x) \setminus \{y\} \neq \emptyset) = 1 > 0.8 = \mathbb{P}_{(x,y),\tau}(\Gamma^{0.2}_{\text{ASP}}(x) \setminus \{y\} \neq \emptyset)$$

(cf. (3.24)).

Corollary 3.5 *All ten criteria of efficiency in Table 3.1 are binary conditionally weakly proper.*

Proof Criteria S, N, OF, and OE are binary conditionally weakly proper by Theorem 3.1. Criteria OU and OM are identical to OF and OE, respectively, in the binary case, and so are also binary conditionally weakly proper. Criteria F and E are identical to U and M, respectively, in the binary case, and so our task reduces to proving that U and M are binary conditionally weakly proper. By Theorem 3.2, it suffices to check $\mathcal{R}(\text{CP}) \cap \mathcal{R}'(\text{SP}) \neq \emptyset$, which is obvious: SP is in both $\mathcal{R}(\text{CP})$ and $\mathcal{R}'(\text{SP})$ when $|\mathbf{Y}| = 2$. □

3.1.7 Discussion

Criteria of efficiency that are not conditionally proper are analogous to improper loss functions in probability forecasting (see Sects. 6.4.3 and 10.5.3 below or, e.g., [72] and [133]; discussions in literature are often couched in terms of "scoring rules" rather than loss functions, but we do not use this terminology in this book, in part because "scoring rules" can be confused with "scoring functions", which we use in a different sense in Chaps. 4 and 6). The optimal idealised conformity measures for the criteria of efficiency given in this section that are not conditionally proper have clear disadvantages, such as:

- They usually depend on the arbitrary choice of a choice function. In many cases there is a unique choice function, but the possibility of non-uniqueness is still awkward.
- They encourage "strategic behaviour" (such as ignoring the differences, which may be very substantial, between potential labels other than $\hat{y}(x)$ for a test object x when using the M criterion in the case $|\mathbf{Y}| > 2$).

However, we do not use the expression "proper criterion of efficiency" since it is conceivable that some criteria of efficiency that are not conditionally proper may turn out to be useful in situations where we do not insist on conditioning on the objects.

Protocol 3.1 essentially coincides with Protocol 2.1 in Chap. 2, but this time we include not only the variables Err_n^ϵ (the total number of errors made up to and including trial n at significance level ϵ) but also the analogous variables OE_n^ϵ and OF_n associated with the OE and OF criteria of efficiency. These are our preferred criteria: first, they are conditionally proper, and second, they produce small values for clean datasets (unlike the S and N criteria, which are also conditionally proper: for the former, the p-values for the true labels are not small, and for the latter, the sizes of even good prediction sets can be expected to be at least one).

We will sometimes represent the OE criterion in the form of the *average false p-value* OF_n/n. In the offline setting, the average false p-value is the modification of (3.7) given by

$$\frac{1}{k(|\mathbf{Y}|-1)} \sum_{i=l+1}^{l+k} \sum_{y \neq y_i} p_i^y . \tag{3.30}$$

(A possible generalization, discussed in Remark 3.15 below, is to replace p_i^y by $\phi(p_i^y)$ in (3.30) for a strictly increasing smooth function $\phi : [0,1] \to \mathbb{R}$.) Similarly,

Protocol 3.1 Online confidence prediction with new notation

1: $\mathrm{Err}_0^\epsilon := 0$, $\mathrm{OE}_0^\epsilon := 0$, for all $\epsilon \in (0,1)$, and $\mathrm{OF}_0 := 0$
2: **for** $n = 1, 2, \ldots$
3: Reality outputs $x_n \in \mathbf{X}$
4: Predictor outputs $\Gamma_n^\epsilon \subseteq \mathbf{Y}$ for all $\epsilon \in (0,1)$
5: Transducer outputs p_n^y for all $y \in \mathbf{Y}$
6: Reality outputs $y_n \in \mathbf{Y}$
7: $\mathrm{Err}_n^\epsilon := \mathrm{Err}_{n-1}^\epsilon + 1_{y_n \notin \Gamma_n^\epsilon}$ for all $\epsilon \in (0,1)$
8: $\mathrm{OE}_n^\epsilon := \mathrm{OE}_{n-1}^\epsilon + \left|\Gamma_n^\epsilon \setminus \{y_n\}\right|$ for all $\epsilon \in (0,1)$
9: $\mathrm{OF}_n := \mathrm{OF}_{n-1} + \sum_{y \in \mathbf{Y} \setminus \{y_n\}} p_n^y$.

we will sometimes represent the OE criterion in the form of the *average observed excess* OE_n/n.

3.2 More Ways of Computing Nonconformity Scores

First of all we notice that the general scheme (2.31) discussed in Sect. 2.2.9 is applicable generally, including the case of classification. For classification, it is especially important to allow $\hat{\mathbf{Y}} \neq \mathbf{Y}$.

Another general remark is that any procedure of computing nonconformity scores for regression can be used for computing nonconformity scores in binary classification (and there are standard ways to reduce general classification problems to binary ones, as we will see in Sect. 3.2.3). Indeed, if $\mathbf{Y}$ consists of just two elements, we can encode them by two different real numbers and run the regression procedure for computing nonconformity scores. In particular, we can use the nonconformity scores produced by ridge regression and by nearest neighbours regression, as discussed in the previous chapter, in classification problems. In this section we will discuss other ways, which are often more efficient.

Remember that, when defining specific nonconformity measures A, we often do it in terms of their deleted versions (2.35). As always, z_i often stands for (x_i, y_i).

3.2.1 Nonconformity Scores from Nearest Neighbours

Assuming the objects are vectors in a Euclidean space, the nonconformity scores can be defined, in the spirit of the 1-nearest neighbour algorithm, as

$$A^{\mathrm{del}}\left(\lbag (x_1, y_1), \dots, (x_l, y_l) \rbag, (x, y)\right) := \frac{\min_{i=1,\dots,l: y_i = y} d(x, x_i)}{\min_{i=1,\dots,l: y_i \neq y} d(x, x_i)}, \tag{3.31}$$

where d is the Euclidean distance (i.e., an object is considered nonconforming if it is close to an object labelled in a different way and far from any object labelled in the same way). It is possible for (3.31) to be equal to ∞ (if the denominator in (3.31) is zero).

We can simplify the definition (3.31) and consider only the numerator of (3.31) as nonconformity measure. Alternatively, we can consider the denominator of (3.31) as conformity measure.

Figures 3.1 and 3.2 show the online performance of the *1-nearest neighbour conformal predictor* (determined by (3.31) as nonconformity measure) on the USPS dataset (the original 9298 hand-written digits, as described in Appendix B, but randomly permuted) for the significance levels 1–5%. The left panel of Fig. 3.1

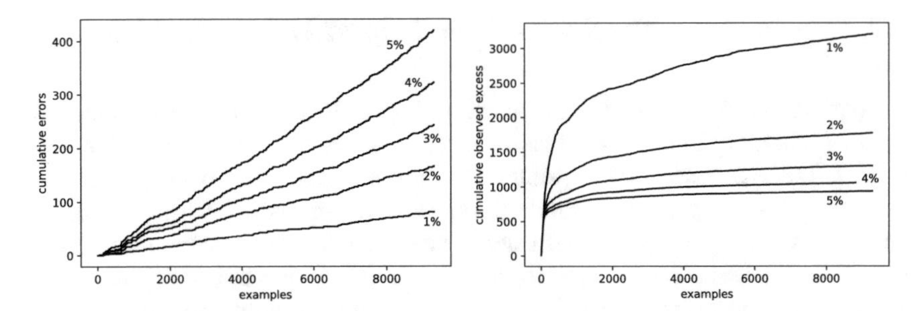

Fig. 3.1 Results for the 1-nearest neighbour conformal predictor on the USPS dataset (9298 handwritten digits, randomly permuted) for the significance levels from $\epsilon = 1\%$ to $\epsilon = 5\%$. Left panel: the cumulative number of errors Err_n^ϵ suffered over the first n examples plotted against n. Right panel: the cumulative number OE_n^ϵ of the false predicted labels

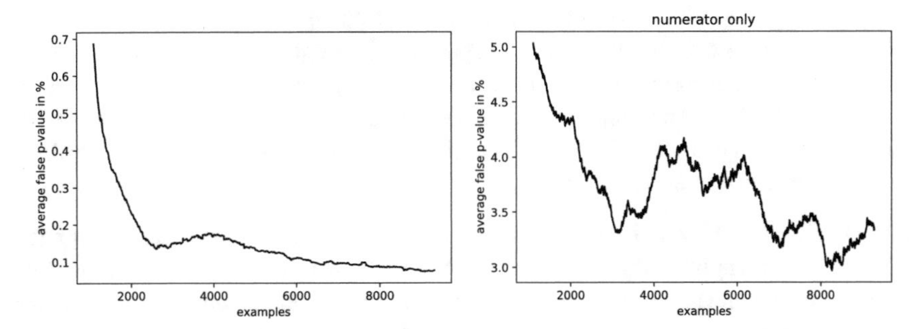

Fig. 3.2 The average false p-value OF_n/n starting from example 1100 and averaged over the last 1000 examples, output by the 1-nearest neighbour conformal predictor on the USPS dataset. Left panel: with the standard nonconformity measure (3.31). Right panel: using the numerator only

again confirms empirically the validity of conformal predictors; the cumulative numbers of errors at $\epsilon = 1\%$ and $\epsilon = 5\%$ were already given in Chap. 1 (Fig. 1.6). The right panel of Fig. 3.1 shows that for a majority of examples the prediction set contains precisely one label at the considered significance levels. At the significance levels 2% and, especially, 3% the graphs for OE_n^ϵ become almost horizontal towards the end of the dataset. (This, however, should not be directly compared with the error rate of 2.5% for humans reported in Vapnik [366, Table 12.1] with a reference to Jane Bromley and Eduard Sackinger, since our experiment has been carried out on a randomly permuted dataset, whereas the test part of the USPS dataset is known to be especially hard.) The left panel of Fig. 3.2 shows that a typical average false p-value is close to 0.1% towards the end of the USPS dataset. The right panel shows that the results become much worse if we ignore the denominator in (3.31) (and the results becomes even worse if we ignore the numerator in (3.31)).

3.2.2 Nonconformity Scores from Support Vector Machines

Support vector machines were proposed by Vapnik [366, Part II]. We concentrate on the problem of binary classification, assuming that the set $\mathbf{Y}$ of possible labels is $\{-1, 1\}$. The basic idea of support vector machine (SVM) is, given a bag $\Lbag(x_1, y_1), \ldots, (x_n, y_n)\Rbag$, to separate the objects x_i labelled as 1 from the objects labelled as -1 by a hyperplane, usually after mapping the objects to a "feature space" (as in Sect. 2.3.4). The separation should be performed in an optimal manner, and the task is formalized as an optimization problem with constraints corresponding to examples; roughly, the ith constraint says that x_i should lie on the correct side of the hyperplane (with a cushion if possible, and with a slack allowed if needed).

As usual, the Lagrange method is applied, and the resulting Lagrange multipliers α_i reflect the importance of the constraints, and so α_i reflects the importance of the example z_i in defining the optimal separating hyperplane. The examples z_i with $\alpha_i = 0$ can be interpreted as typical (they are not *support vectors*). The examples z_i with $\alpha_i > 0$ are unusual, and the size of α_i can serve as a measure of z_i's lack of conformity as an element of the bag. Therefore, we can use the Lagrange multipliers as nonconformity scores $\alpha_i = A(\Lbag z_1, \ldots, z_n\Rbag, z_i)$.

Remark 3.6 Actually, it is quite possible that the Lagrange multipliers computed by a given computer implementation of SVM will not provide a bona fide nonconformity measure, with α_i depending on the order in which the examples $z_1, \ldots, z_n$ are presented. The order of the examples may be especially important in so-called "chunking", a standard feature of SVM implementations. To ensure the required invariance (and so the validity of the resulting conformal predictor), the examples $z_1, \ldots, z_n$ can be sorted in some way (e.g., in the lexicographic order of their ASCII representations) obtaining $z_{\pi(1)}, \ldots, z_{\pi(n)}$, where π is a permutation of the set $\{1, \ldots, n\}$. The Lagrange multipliers computed from $z_{\pi(1)}, \ldots, z_{\pi(n)}$ should then be permuted using π^{-1} to obtain $\alpha_1, \ldots, \alpha_n$.

3.2.3 Reducing Classification Problems to the Binary Case

The original SVM method can only deal with binary classification problems, but we will now see that there are ways to use it (or any other method designed for binary classification) for solving *multiclass* classification problems (i.e., those with $|\mathbf{Y}| > 2$).

There are two standard ways to reduce multiclass classification problems to the binary case: the "one-against-the-rest" procedure and the "one-against-one" procedure. Suppose we have a reasonable nonconformity measure A for binary classification but are confronted with a multiclass classification problem. For concreteness we will assume that the label space in the binary classification problem is $\{0, 1\}$; if it is $\{a, b\}$ (e.g., $a = -1$ and $b = 1$ in the case of SVM), the reduction will be achieved by a further scaling, $y \mapsto (y - a)/(b - a)$.

The *one-against-the-rest procedure* gives the nonconformity measure

$$\begin{aligned} A_1^{\text{del}} &\left(\wr (x_1, y_1), \ldots, (x_l, y_l) \int, (x, y) \right) \\ &:= \lambda A^{\text{del}} \left(\wr (x_1, 1_{y_1 = y}), \ldots, (x_l, 1_{y_l = y}) \int, (x, 1) \right) \\ &+ \frac{1 - \lambda}{|\mathbf{Y}| - 1} \sum_{y' \neq y} A^{\text{del}} \left(\wr (x_1, 1_{y_1 = y'}), \ldots, (x_l, 1_{y_l = y'}) \int, (x, 0) \right) , \end{aligned} \tag{3.32}$$

where $\lambda \in [0, 1]$ is a constant (parameter of the procedure). Intuitively, we consider $|\mathbf{Y}|$ auxiliary binary classification problems and compute the nonconformity score of an example (x, y) as the weighted average of the scores this example receives in the auxiliary problems.

The *one-against-one procedure* gives the nonconformity measure

$$A_2^{\text{del}} \left(\wr (x_1, y_1), \ldots, (x_l, y_l) \int, (x, y) \right) := \frac{1}{|\mathbf{Y}| - 1} \sum_{y' \neq y} A^{\text{del}} \left(B_{y,y'}, (x, 1) \right) ,$$

where $B_{y,y'}$ is the bag obtained from $\wr (x_1, y_1), \ldots, (x_l, y_l) \int$ as follows: remove all (x_i, y_i) with $y_i \notin \{y, y'\}$; replace each (x_i, y) by $(x_i, 1)$; replace each (x_i, y') by $(x_i, 0)$. We now have $|\mathbf{Y}| - 1$ auxiliary binary classification problems.

The numbers $|\mathbf{Y}|$ and $|\mathbf{Y}| - 1$ of auxiliary binary classification problems given above refer to computing only one nonconformity score. When using nonconformity measures for conformal prediction, we have to compute all n nonconformity scores (2.23) for all $y \in \mathbf{Y}$. With the one-against-the-rest procedure, we have to consider $2 |\mathbf{Y}|$ auxiliary binary classification problems altogether, whereas with the one-against-one procedure $|\mathbf{Y}| (|\mathbf{Y}| - 1)$ auxiliary binary classification problems are required.[1] When $|\mathbf{Y}| = 3$, the numbers of auxiliary problems coincide, $2 |\mathbf{Y}| = |\mathbf{Y}| (|\mathbf{Y}| - 1)$, but for $|\mathbf{Y}| > 3$ the one-against-the-rest procedure requires fewer auxiliary problems, $2 |\mathbf{Y}| < |\mathbf{Y}| (|\mathbf{Y}| - 1)$.

3.3 Weak Teachers

In the pure online setting, we get an immediate feedback (the true label) for every example that we predict. This makes practical applications of this scenario questionable. Imagine, for example, a mail sorting centre using an online prediction

[1] To see, e.g., that the number of auxiliary binary classification problems is $|\mathbf{Y}| (|\mathbf{Y}| - 1)$ for the one-against-one procedure, notice that to each of the $\frac{1}{2} |\mathbf{Y}| (|\mathbf{Y}| - 1)$ unordered pairs of different labels correspond two auxiliary binary classification problems.

algorithm for zip code recognition; suppose the feedback about the "true" label comes from a human expert. If the feedback is given for every object x_i, there is no point in having the prediction algorithm: we can just as well use the label provided by the expert. It would help if the prediction algorithm could still work well, in particular be valid, if only every, say, tenth object were classified by a human expert. Alternatively, even if the prediction algorithm requires the knowledge of all labels, it might still be useful if the labels were allowed to be given not immediately but with a delay (in our mail sorting example, such a delay might make sure that we hear from local post offices about any mistakes made before giving a feedback to the algorithm). In this section we will see that asymptotic validity still holds in many cases where missing labels and delays are allowed.

In the pure online protocol we had validity in the strongest possible sense: at each significance level ϵ each smoothed conformal predictor made errors independently with probability ϵ. Now we will not have validity in this strongest sense, and so we will consider three natural asymptotic definitions, requiring only that $\mathrm{Err}_n^\epsilon/n \to \epsilon$ in a certain sense: weak validity, strong validity, and validity in the sense of the law of the iterated logarithm. Finally, we will prove a simple result about asymptotic efficiency.

3.3.1 Imperfectly Taught Predictors

We are interested in the protocol where the predictor receives the true labels y_n only for a subset of trials n, and even for this subset, y_n may be given with a delay. This is formalized by a function $\mathcal{L} : N \to \mathbb{N}$ defined on an infinite set $N \subseteq \mathbb{N}$ and required to satisfy

$$\mathcal{L}(n) \leq n$$

for all $n \in N$ and

$$m \neq n \Longrightarrow \mathcal{L}(m) \neq \mathcal{L}(n)$$

for all $m \in N$ and $n \in N$; a function satisfying these properties will be called a *teaching schedule*. A teaching schedule $\mathcal{L}$ describes the way the data is disclosed to the predictor: at the end of trial $n \in N$ it is given the label $y_{\mathcal{L}(n)}$ for the object $x_{\mathcal{L}(n)}$. The elements of $\mathcal{L}$'s domain N in the ascending order will be denoted n_i: $N = \{n_1, n_2, \dots\}$ and $n_1 < n_2 < \dots$. We denote the total number of labels disclosed by the beginning of trial n to a predictor taught according to the teaching schedule $\mathcal{L}$ by $s(n) := |\{i : i \in N, i < n\}|$.

Let Γ be a confidence predictor and $\mathcal{L}$ be a teaching schedule. The *$\mathcal{L}$-taught version* $\Gamma^{\mathcal{L}}$ of Γ is

$$\Gamma^{\mathcal{L},\epsilon}(z_1, \ldots, z_{n-1}, x_n) := \Gamma^{\epsilon}(z_{\mathcal{L}(n_1)}, \ldots, z_{\mathcal{L}(n_{s(n)})}, x_n) ,$$

where $z_i = (x_i, y_i)$, as usual. Intuitively, at the end of trial n the predictor $\Gamma^{\mathcal{L}}$ learns the label $y_{\mathcal{L}(n)}$ if $n \in N$ and learns nothing otherwise. An *$\mathcal{L}$-taught (smoothed) conformal predictor* is a confidence predictor that can be represented as $\Gamma^{\mathcal{L}}$ for some (smoothed) conformal predictor Γ.

Let us now consider several examples of teaching schedules.

Ideal teacher. If $N = \mathbb{N}$ and $\mathcal{L}(n) = n$ for each $n \in N$, then $\Gamma^{\mathcal{L}} = \Gamma$.

Slow teacher with a fixed lag. If $N = \{l + 1, l + 2, \ldots\}$ for some $l \in \mathbb{N}$ and $\mathcal{L}(n) = n - l$ for all $n \in N$, then $\Gamma^{\mathcal{L}}$ is a predictor which learns true labels with a delay of l.

Slow teacher. The previous example can be generalized as follows. Let $l(n) = n + \text{lag}(n)$ where $\text{lag} : \mathbb{N} \to \mathbb{N}$ is an increasing function. Define $N := l(\mathbb{N})$ and $\mathcal{L}(n) := l^{-1}(n)$, $n \in N$. Then $\Gamma^{\mathcal{L}}$ is a predictor which learns the true label for each object x_n with a delay of $\text{lag}(n)$.

Lazy teacher. If $N \neq \mathbb{N}$ and $\mathcal{L}(n) = n$, $n \in N$, then $\Gamma^{\mathcal{L}}$ is given the true labels immediately but not for every object.

All results of this section (as will be clear from the proofs, which are given in Sect. 3.4.2) use only the following properties of smoothed conformal predictors Γ:

- At each significance level ϵ, the errors $\text{err}_n^{\epsilon}(\Gamma)$, $n = 1, 2, \ldots$, are independent Bernoulli random variables with parameter ϵ.
- The predictions do not depend on the order of the examples learnt so far:

$$\Gamma^{\epsilon}(z_1, \ldots, z_{n-1}, x_n) = \Gamma^{\epsilon}(z_{\pi(1)}, \ldots, z_{\pi(n-1)}, x_n)$$

  for any permutation π of $\{1, \ldots, n - 1\}$. (Remember that we call confidence predictors satisfying this property *invariant*.)

3.3.2 Weak Validity

In this subsection we state a necessary and sufficient condition for a teaching schedule to preserve a weak asymptotic property of validity of conformal predictors. The condition turns out also to be rather weak: feedback should be given at more than a logarithmic fraction of trials. The proofs for this section are postponed to Sect. 3.4.2.

We start from the definitions, assuming, for simplicity, randomness rather than exchangeability. A randomized confidence predictor Γ is *asymptotically exact in probability* if, for all significance levels ϵ and all probability distributions Q on $\mathbf{Z}$,

$$\frac{1}{n}\sum_{i=1}^{n}\mathrm{Err}_n^\epsilon(\Gamma)-\epsilon\to 0$$

in probability (assuming the examples generated from Q independently and with independent coin tosses in Γ). Similarly, a confidence predictor Γ is *asymptotically conservative in probability* if, for all significance levels ϵ and all probability distributions Q on $\mathbf{Z}$,

$$\left(\frac{1}{n}\sum_{i=1}^{n}\mathrm{Err}_n^\epsilon(\Gamma)-\epsilon\right)^+\to 0$$

in probability.

Theorem 3.7 *Let $\mathcal{L}$ be a teaching schedule with domain $N=\{n_1,n_2,\dots\}$, where $n_1,n_2,\dots$ is a strictly increasing infinite sequence of positive integers.*

- *If $\lim_{k\to\infty}(n_k/n_{k-1})=1$, any $\mathcal{L}$-taught smoothed conformal predictor is asymptotically exact in probability.*
- *If $\lim_{k\to\infty}(n_k/n_{k-1})=1$ does not hold, there exists an $\mathcal{L}$-taught smoothed conformal predictor which is not asymptotically exact in probability.*

In words, this theorem asserts that an $\mathcal{L}$-taught smoothed conformal predictor is guaranteed to be asymptotically exact in probability if and only if the growth rate of n_k is sub-exponential.

Corollary 3.8 *If $\lim_{k\to\infty}(n_k/n_{k-1})=1$, any $\mathcal{L}$-taught conformal predictor is asymptotically conservative in probability.*

3.3.3 Strong Validity

In Sect. 2.1.3 (see (2.13) and (2.14)), we defined asymptotically exact and asymptotically conservative confidence predictors, in which the convergence is a.s. rather than in probability.

Theorem 3.9 *Suppose the example space $\mathbf{Z}$ is standard Borel. Let Γ be a smoothed conformal predictor and $\mathcal{L}$ be a teaching schedule whose domain is $N=\{n_1,n_2,\dots\}$, where $n_1<n_2<\dots$. If*

$$\sum_k\left(\frac{n_k}{n_{k-1}}-1\right)^2<\infty \tag{3.33}$$

then $\Gamma^{\mathcal{L}}$ is asymptotically exact.

This theorem shows that $\Gamma^{\mathcal{L}}$ is asymptotically exact when n_k grows as $\exp(\sqrt{k}/\ln k)$; on the other hand, it does not guarantee that it is asymptotically exact if n_k grows as $\exp(\sqrt{k})$.

Corollary 3.10 *Let Γ be a conformal predictor and $\mathcal{L}$ be a teaching schedule with domain $N = \{n_1, n_2, \dots\}$, $n_1 < n_2 < \dots$. Under condition* (3.33)*, $\Gamma^{\mathcal{L}}$ is an asymptotically conservative confidence predictor.*

3.3.4 Iterated Logarithm Validity

The following result asserts, in particular, that when n_k are equally spaced a stronger version of asymptotic validity, in the spirit of the law of the iterated logarithm, holds.

Theorem 3.11 *Suppose the domain $\{n_1, n_2, \dots\}$, $n_1 < n_2 < \dots$, of a teaching schedule satisfies $n_k = O(k)$. Each $\mathcal{L}$-taught smoothed conformal predictor $\Gamma^{\mathcal{L}}$ satisfies*

$$\left|\frac{\mathrm{Err}_n^{\epsilon}(\Gamma^{\mathcal{L}})}{n} - \epsilon\right| = O\left(\sqrt{\frac{\ln\ln n}{n}}\right) \quad \text{a.s.}$$

and each $\mathcal{L}$-taught conformal predictor $\Gamma^{\mathcal{L}}$ satisfies

$$\left(\frac{\mathrm{Err}_n^{\epsilon}(\Gamma^{\mathcal{L}})}{n} - \epsilon\right)^{+} = O\left(\sqrt{\frac{\ln\ln n}{n}}\right) \quad \text{a.s.}\ ,$$

at each significance level ϵ and for each $Q \in \mathbf{P}(\mathbf{Z})$ generating individual examples.

3.3.5 Efficiency

Now we consider the case of classification and take the average observed excess as our criterion of efficiency, as discussed in Sect. 3.1.7. If Γ is a confidence predictor and ϵ a significance level, we set

$$I^{\epsilon}(\Gamma) = \left[\liminf_{n\to\infty}\frac{\mathrm{OE}_n^{\epsilon}}{n}, \limsup_{n\to\infty}\frac{\mathrm{OE}_n^{\epsilon}}{n}\right],$$

with OE_n^{ϵ} defined as in Protocol 3.1. The intervals $I^{\epsilon}(\Gamma)$ characterize the asymptotic efficiency of Γ; of course, these are random intervals, since they depend on the actual examples output by Reality. It turns out, however, that in important cases (including conformal predictors and many $\mathcal{L}$-taught conformal predictors, smoothed and deterministic) these intervals are close to being deterministic under the assumption of randomness.

Lemma 3.12 *For each invariant confidence predictor Γ (randomized or deterministic), significance level ϵ, and probability distribution Q on $\mathbf{Z}$ there exists an interval $[a, b] \subseteq [0, |\mathbf{Y}| - 1]$ such that $I^\epsilon(\Gamma) = [a, b]$ a.s., provided the examples and random numbers τ (if applicable) are generated from Q^∞ and $\mathbf{U}^\infty$ independently.*

Proof The statement of this lemma is an immediate consequence of the Hewitt–Savage zero-one law (see, e.g., [330, Theorem 4.1.3]). □

We will use the notation $I^\epsilon(\Gamma, Q)$ for the interval whose existence is asserted in the lemma; it characterizes the asymptotic efficiency of Γ at significance level ϵ with examples distributed according to Q.

Theorem 3.13 *Let Γ be a (smoothed) conformal predictor and $\mathcal{L}$ be a teaching schedule defined on $N = \{n_1, n_2, \dots\}$, where $n_1 < n_2 < \dots$ is an increasing sequence. If, for some $c \in \mathbb{N}$, $n_{k+1} - n_k = c$ from some k on, then $I^\epsilon(\Gamma^\mathcal{L}, Q) = I^\epsilon(\Gamma, Q)$ for all significance levels $\epsilon \in (0, 1)$ and all probability distributions Q on $\mathbf{Z}$.*

Theorems 3.9 and 3.13 can be illustrated with the following simple example. Suppose only every mth label is revealed to a conformal predictor, and even this is done with a delay of l, where m and l are positive integer constants. Then (smoothed) conformal predictors will remain asymptotically valid, and asymptotically their average observed excess will not deteriorate.

3.4 Proofs

3.4.1 Proofs for Sect. 3.1

Proof of Theorem 3.1

We start from proving $\mathcal{R}(\text{CP}) = \mathcal{O}(\text{N})$. Let A be any idealised conformity measure. Fix for a moment a significance level ϵ. For each example $(x, y) \in \mathbf{Z}$, let $P(x, y)$ be the probability that the idealised conformal predictor determined by A makes an error on the example (x, y) at the significance level ϵ, i.e., the probability (over τ) of $y \notin \Gamma_A^\epsilon(x)$. It is clear from (3.8) and (3.9) that P takes at most three possible values (0, 1, and an intermediate value) and that

$$\sum_{x,y} Q(x, y) P(x, y) = \epsilon \tag{3.34}$$

(which just reflects the fact that the probability of error is ϵ). Vice versa, any P satisfying these properties will also satisfy

$$\forall(x, y) : P(x, y) = \mathbb{P}_\tau \left(y \notin \Gamma_A^\epsilon(x, \tau)\right)$$

for some A, $\mathbb{P}_\tau$ standing for the probability when $\tau \sim \mathbf{U}$. Let us see when we will have $A \in \mathcal{O}(\mathrm{N})$ (A is an N-optimal idealised conformity measure). Define Q' to be the probability measure on $\mathbf{Z}$ such that $Q'_\mathbf{X} = Q_\mathbf{X}$ and $Q'(y \mid x) = 1/|\mathbf{Y}|$ does not depend on y. The N criterion at significance level ϵ for A can be evaluated as

$$\mathbb{E}_{x,\tau} \left|\Gamma_A^\epsilon(x)\right| = |\mathbf{Y}| \left(1 - \sum_{(x,y)\in\mathbf{Z}} Q'(x, y)P(x, y)\right) ;$$

this expression should be minimised, i.e., $\sum_{(x,y)} Q'(x, y)P(x, y)$ should be maximised, under the restriction (3.34). Let us apply the Neyman–Pearson fundamental lemma [222, Theorem 3.2.1], using Q as the null and Q' as the alternative hypotheses. We can see that $\mathbb{E}_{x,\tau} \left|\Gamma_A^\epsilon(x)\right|$ takes its minimal value if and only if there exist thresholds $k_1 = k_1(\epsilon)$, $k_2 = k_2(\epsilon)$, and $k_3 = k_3(\epsilon)$ such that:

- $Q\{(x, y) : Q(y \mid x) < k_1\} < \epsilon \le Q\{(x, y) : Q(y \mid x) \le k_1\}$,
- $k_2 < k_3$,
- $A(x, y) < k_2$ if $Q(y \mid x) < k_1$,
- $k_2 < A(x, y) < k_3$ if $Q(y \mid x) = k_1$,
- $A(x, y) > k_3$ if $Q(y \mid x) > k_1$.

This will be true for all ϵ if and only if $Q(y \mid x)$ is a function of $A(x, y)$ (meaning that there exists a function F such that, for all (x, y), $Q(y \mid x) = F(A(x, y))$). This completes the proof of $\mathcal{R}(\mathrm{CP}) = \mathcal{O}(\mathrm{N})$.

Next we show that $\mathcal{O}(\mathrm{N}) = \mathcal{O}(\mathrm{S})$. The chain of equalities

$$\sum_{y\in\mathbf{Y}} p(x, y) = \sum_{y\in\mathbf{Y}} \int_0^1 1_{p(x,y)>\epsilon} \mathrm{d}\epsilon = \int_0^1 \sum_{y\in\mathbf{Y}} 1_{p(x,y)>\epsilon} \mathrm{d}\epsilon = \int_0^1 \left|\Gamma^\epsilon(x)\right| \mathrm{d}\epsilon \tag{3.35}$$

(which will be used as the model in several other proofs in the rest of this subsection) implies, by Fubini's theorem,

$$\mathbb{E}_{x,\tau} \sum_{y\in\mathbf{Y}} p(x, y) = \int_0^1 \mathbb{E}_{x,\tau} \left|\Gamma^\epsilon(x)\right| \mathrm{d}\epsilon . \tag{3.36}$$

We can see that $A \in \mathcal{O}(\mathrm{S})$ whenever $A \in \mathcal{O}(\mathrm{N})$: indeed, any N-optimal idealised conformity measure minimises the expectation $\mathbb{E}_{x,\tau} |\Gamma^\epsilon(x)|$ on the right-hand side of (3.36) for all ϵ simultaneously, and so minimises the whole right-hand-side, and so minimises the left-hand-side. On the other hand, $A \notin \mathcal{O}(\mathrm{S})$ whenever $A \notin \mathcal{O}(\mathrm{N})$: indeed, if an idealised conformity measure fails to minimise the expectation $\mathbb{E}_{x,\tau} |\Gamma^\epsilon(x)|$ on the right-hand side of (3.36) for some ϵ, it fails to do so for all ϵ in a nonempty open interval (because of the right-continuity of $\mathbb{E}_{x,\tau} |\Gamma^\epsilon(x)|$ in ϵ, which is proved in Lemma 3.14(b) below), and therefore, it does not minimise the right-hand side of (3.36) (any N-optimal idealised conformity measure, such as the CP idealised conformity measure, will give a smaller value), and therefore, it does not minimise the left-hand side of (3.36).

The equality $\mathcal{O}(\mathrm{S}) = \mathcal{O}(\mathrm{OF})$ follows from

$$\mathbb{E}_{x,\tau} \sum_{y} p(x, y) = \mathbb{E}_{(x,y),\tau} \sum_{y'\neq y} p(x, y') + \frac{1}{2} ,$$

where we have used the fact that $p(x, y)$ is distributed uniformly on $[0, 1]$ when $((x, y), \tau) \sim Q \times \mathbf{U}$ (by Proposition 2.11).

Finally, we notice that $\mathcal{O}(\mathrm{N}) = \mathcal{O}(\mathrm{OE})$. Indeed, for any significance level ϵ,

$$\mathbb{E}_{x,\tau}|\Gamma^{\epsilon}(x)| = \mathbb{E}_{(x,y),\tau}|\Gamma^{\epsilon}(x) \setminus \{y\}| + (1-\epsilon) ,$$

again using the fact that $p(x, y)$ is distributed uniformly on [0, 1] and so $\mathbb{P}_{(x,y),\tau}(y \in \Gamma^{\epsilon}(x)) = 1 - \epsilon$.

To complete the proof of Theorem 3.1 we need to establish the following lemma.

Lemma 3.14

(a) The function $\Gamma^{\epsilon}(x) = \Gamma^{\epsilon}(x, \tau)$ of ϵ is right-continuous for fixed x and τ.
(b) The function $\mathbb{E}_{x,\tau} |\Gamma^{\epsilon}(x, \tau)|$ is right-continuous in ϵ.

Proof Let us first check (a). We have (i) $p(x, y, \tau) > \epsilon$ for all $y \in \Gamma^{\epsilon}(x, \tau)$, and (ii) $p(x, y, \tau) \le \epsilon$ for all $y \notin \Gamma^{\epsilon}(x, \tau)$. If we increase ϵ, (ii) will be still satisfied, and if the increase is sufficiently small, (i) will be also satisfied and, therefore, $\Gamma^{\epsilon}(x, \tau)$ will not change. As for (b), the right-continuity of $\Gamma^{\epsilon}(x, \tau)$ in ϵ implies the right-continuity of $|\Gamma^{\epsilon}(x, \tau)|$ in ϵ, which implies the right-continuity of $\mathbb{E}_{x,\tau} |\Gamma^{\epsilon}(x, \tau)|$ in ϵ by the Lebesgue dominated convergence theorem. □

Remark 3.15 The statement $\mathcal{O}(\mathrm{S}) = \mathcal{R}(\mathrm{CP})$ of Theorem 3.1 can be generalised to the criterion S_{ϕ} preferring small values of

$$\frac{1}{k} \sum_{i=l+1}^{l+k} \sum_{y} \phi(p_i^y) \text{ or } \mathbb{E}_{x,\tau} \sum_{y} \phi(p(x, y)) \tag{3.37}$$

(instead of (3.1) or either side of (3.10)), where $\phi : [0, 1] \to \mathbb{R}$ is a fixed continuously differentiable strictly increasing function, not necessarily the identity function. Namely, we still have $\mathcal{O}(\mathrm{S}_{\phi}) = \mathcal{R}(\mathrm{CP})$. Indeed, we can assume, without loss of generality, that $\phi(0) = 0$ and $\phi(1) = 1$ and replace (3.35) by

$$\begin{aligned}\sum_{y\in\mathbf{Y}} \phi(p(x, y)) = \sum_{y\in\mathbf{Y}} \int_0^1 1_{\phi(p(x,y))>\epsilon} \mathrm{d}\epsilon &= \int_0^1 \sum_{y\in\mathbf{Y}} 1_{p(x,y)>\phi^{-1}(\epsilon)} \mathrm{d}\epsilon \\ &= \int_0^1 \left|\Gamma^{\phi^{-1}(\epsilon)}(x)\right| \mathrm{d}\epsilon = \int_0^1 \left|\Gamma^{\epsilon'}(x)\right| \phi'(\epsilon')\mathrm{d}\epsilon' ,\end{aligned}$$

where ϕ' is the (continuous) derivative of ϕ, and then use the same argument as before. Notice that this argument does not cover averaging Greenland's [140] S-values $-\log_2(p_i^y)$ instead of averaging the p-values in (3.1): the function $\phi(p) := -\log_2 p$ may take value ∞.

Remark 3.16 It is interesting that the two expressions in (3.37) can be considered as natural generalizations of the N criterion. Consider, e.g., the first of these expressions,

$$\frac{1}{k} \sum_{i=l+1}^{l+k} \sum_{y} \phi(p_i^y) , \tag{3.38}$$

only assuming that $\phi : [0, 1] \to \mathbb{R}$ is an increasing function. It contains (3.2) as a special case corresponding to $\phi(p) := 1_{p>\epsilon}$. However, as discussed earlier (see, e.g., Fig. 1.5), it is usually unwise to consider only one significance level, and we might want to replace (3.2) by the average

$$\frac{1}{k}\sum_{i=l+1}^{l+k}\frac{|\Gamma_i^{\epsilon_1}|+|\Gamma_i^{\epsilon_2}|+|\Gamma_i^{\epsilon_3}|}{3},$$

which coincides with (3.38) for $\phi(p) := (1_{p>\epsilon_1} + 1_{p>\epsilon_2} + 1_{p>\epsilon_3})/3$. Taking sufficiently many significance levels ϵ (and optionally, replacing arithmetic average by weighted average), we can approximate any increasing ϕ. Therefore, properly generalized, the S and N criteria become almost indistinguishable. The OF and OE criteria also become almost indistinguishable after similar generalizations.

Proof of Theorem 3.2

The proofs of Theorems 3.2–3.4 will be slightly less formal than the proof of Theorem 3.1; in particular, all references to the Neyman–Pearson lemma will be implicit.

For Theorem 3.2, we start from checking that $\mathcal{O}(\mathrm{M}) = \mathcal{R}'(\mathrm{SP})$. We will analyze the requirements imposed by being M-optimal on the prediction set Γ^ϵ starting from small values of $\epsilon \in (0, 1)$. (Now we only consider ϵ in the interval (0, 1), even if this restriction is not mentioned explicitly.)

Let $f_1 > f_2 > \cdots > f_n > 0$ be the list of the predictabilities (see (3.25)) of all objects $x \in \mathbf{X}$, with all duplicates removed and the remaining predictabilities sorted in the descending order. It is clear that an M-optimal idealised conformity measure will assign the lowest conformity to the group of examples (x, y) with $f(x) = f_1$ and $y \neq \hat{y}(x)$ for some choice function $\hat{y}$ (see (3.26)). The conformity of such examples can be different unless they contain the same object (in which case it must be the same); the conformity of any example in any other group must be higher than the conformity of the examples in this first group. If these conditions are satisfied for some idealised conformity measure A, A will satisfy (3.16) or (3.17) for any idealised conformity measure B and any

$$\epsilon \in \left(0, Q\left\{(x, y) : f(x) = f_1 \;\;\&\;\; y \neq \hat{y}(x)\right\}\right].$$

The second least conforming group of examples consists of (x, y) with $f(x) = f_2$ and $y \neq \hat{y}(x)$ for some choice function $\hat{y}$. The conformity of examples in the second group can again be different unless they contain the same object. These and previous conditions ensure that A will satisfy (3.16) or (3.17) for any

$$\epsilon \in \left(0, Q\left\{(x, y) : f(x) \geq f_2 \;\;\&\;\; y \neq \hat{y}(x)\right\}\right].$$

Continuing in such a way, we will obtain a choice function $\hat{y}$ and the conformity ordering for the examples whose label is not chosen by that choice function $\hat{y}$. All these examples are divided into n groups, and each element of the ith group is coming before each element of the jth group when $i < j$; in the end we will get $2n$ groups satisfying this property. The first n groups take care of

$$\epsilon \in \left(0, Q\left\{(x, y) : y \neq \hat{y}(x)\right\}\right].$$

The next, $(n + 1)$th, group of examples are $(x, \hat{y}(x)) \in \mathbf{Z}$ with $f(x) = f_n$; they can be ordered in any way between themselves. If the conditions listed so far are satisfied for an idealised conformity measure A, A will satisfy (3.16)–(3.18) for any idealised conformity measure B and any

$$\epsilon \in \left(0, Q\left\{(x, y) : y \neq \hat{y}(x) \text{ or } \left(y = \hat{y}(x) \;\;\&\;\; f(x) = f_n\right)\right\}\right].$$

The following, $(n+2)$th, group consists of $(x, \hat{y}(x)) \in \mathbf{Z}$ with $f(x) = f_{n-1}$. Continuing in the same way until all examples are exhausted, we will obtain a refinement of the SP idealised conformity measure that belongs to $\mathcal{R}'(\text{SP})$.

This proof of $\mathcal{O}(\text{M}) = \mathcal{R}'(\text{SP})$ demonstrates the following property of M-optimal idealised conformity measures.

Corollary 3.17 *If $A \in \mathcal{O}(\text{M})$,*

$$\mathbb{P}_{x,\tau}(|\Gamma_A^\epsilon(x)| > 1)\mathbb{P}_{x,\tau}(|\Gamma_A^\epsilon(x)| = 0) = 0$$

at each significance level ϵ.

Let us now check that $\mathcal{O}(\text{U}) = \mathcal{O}(\text{M})$. Analogously to (3.35) and (3.36), we have, for a given idealised conformity measure A (omitted from our notation),

$$\begin{aligned} \mathbb{E}_{x,\tau} \min_y \max_{y' \neq y} p(x, y', \tau) &= \mathbb{E}_{x,\tau} \int_0^1 1_{\min_y \max_{y' \neq y} p(x,y',\tau) > \epsilon} \mathrm{d}\epsilon \\ &= \mathbb{E}_{x,\tau} \int_0^1 1_{|\Gamma^\epsilon(x)|>1} \mathrm{d}\epsilon = \int_0^1 \mathbb{P}_{x,\tau} \left(|\Gamma^\epsilon(x)| > 1\right) \mathrm{d}\epsilon\,. \end{aligned} \tag{3.39}$$

Similarly, we have

$$\begin{aligned} \mathbb{E}_{x,\tau} \max_y p(x, y, \tau) &= \mathbb{E}_{x,\tau} \int_0^1 1_{\max_y p(x,y,\tau) > \epsilon} \mathrm{d}\epsilon \\ &= \mathbb{E}_{x,\tau} \int_0^1 1_{|\Gamma^\epsilon(x)|>0} \mathrm{d}\epsilon = \int_0^1 \mathbb{P}_{x,\tau} \left(|\Gamma^\epsilon(x)| > 0\right) \mathrm{d}\epsilon \\ &= 1 - \int_0^1 \mathbb{P}_{x,\tau} \left(|\Gamma^\epsilon(x)| = 0\right) \mathrm{d}\epsilon\,. \end{aligned} \tag{3.40}$$

Our argument will also use the following continuity property for idealised conformal predictors. (For now, we only need parts (a) and (b).)

Corollary 3.18 *The functions*

(a) $\mathbb{P}_{x,\tau}\left(|\Gamma^\epsilon(x)| > 1\right)$
(b) $\mathbb{P}_{x,\tau}\left(|\Gamma^\epsilon(x)| = 0\right)$
(c) $\mathbb{E}_{x,\tau}\left((|\Gamma^\epsilon(x)| - 1)^+\right)$
(d) $\mathbb{P}_{(x,y),\tau}\left(\Gamma^\epsilon(x) \setminus \{y\} \neq \emptyset\right)$

are right-continuous in ϵ.

Proof All these statements can be deduced from part (a) of Lemma 3.14 in the same way as in the proof of part (b) of that lemma. The right-continuity of the function $\Gamma^\epsilon(x, \tau)$ implies the right-continuity of $1_{|\Gamma^\epsilon(x)|>1}$ (remember that $|\Gamma^\epsilon(x)|$ takes only integer values). Therefore, the right-continuity of $\mathbb{P}_{x,\tau}\left(|\Gamma^\epsilon(x)| > 1\right)$ follows by the Lebesgue dominated convergence theorem. This proves (a), and proofs of (b)–(d) are analogous. □

First suppose that A is M-optimal. Let B be any idealised conformity measure. From (3.39), it is clear that (3.13) holds with $<$ replaced by $\leq$. If, furthermore, we have (3.14): by Corollary 3.18 we also have (3.17) for all ϵ; therefore, we also have (3.18) for all ϵ; in combination with (3.40), we obtain (3.15). Therefore, A is U-optimal.

Now suppose that A is U-optimal. Let B be the SP idealised conformity measure, which we know to be not only M-optimal but also U-optimal (as shown in the previous paragraph). By the definition ((3.13)–(3.15)) of U-optimality, we have (3.14) and (3.15) with $=$ in place of $\leq$. This implies that (3.17) holds for all ϵ (had the equality been violated for some $\epsilon \in (0, 1)$, it would have been violated for a range of ϵ by Corollary 3.18, which would have contradicted (3.14)). In the same way, it implies that (3.18) holds (even with $=$ in place of $\geq$) for all ϵ. Therefore, A is M-optimal.

Proof of Theorem 3.3

Our argument for $\mathcal{O}$(E) $=$ $\mathcal{R}$(ACP) will be similar to the argument for $\mathcal{O}$(M) $=$ $\mathcal{R}'$(SP) given earlier; we will again analyze the requirements imposed by being E-optimal starting from small values of $\epsilon \in (0, 1)$. Let $g_1 < g_2 < \cdots < g_n$ be the list of the conditional probabilities $Q(y \mid x)$ of all examples $(x, y) \in \mathbf{Z}$, with all duplicates removed and the remaining conditional probabilities sorted in the ascending order. All examples will be split into $2n$ groups, with the examples in the ith and $(n+i)$th groups satisfying $Q(y|x) = g_i, i = 1, \dots, n$. Initially the ith group, $i = 1, \dots, n$, contains all examples satisfying $Q(y \mid x) = g_i$, and the other groups are empty. (Later some of the examples will be moved into the groups numbered $n + 1, n + 2, \dots$, and as a result some of the first n groups may become empty.) It will be true that each element of the ith group will be coming (in the order of conformity) before each element of the jth group when $1 \leq i < j \leq 2n$.

Any F-optimal idealised conformity measure will assign the lowest conformity to the first group of examples, perhaps except for examples (x, y) for which $Q(y \mid x) = \max_{y'} Q(y' \mid x)$. If for some $x \in \mathbf{X}$, the first group contains (x, y) with $Q(y \mid x) = \max_{y'} Q(y' \mid x)$, we choose one such (x, y) for each such x and move it to the $(n + 1)$th group. The rest of the examples in the group can be ordered in their conformity in any way (with ties allowed). The examples in the $(n + 1)$th group can also be ordered arbitrarily. Process the 2nd, 3rd,..., nth groups in the same way. It is clear that in the end we will obtain a refinement of an ACP idealised conformity measure.

Next we prove $\mathcal{O}$(E) $=$ $\mathcal{O}$(F). Defining a *p-choice function* $\tilde{y} : \mathbf{X} \to \mathbf{Y}$ (for a given idealised conformity measure) by the requirement

$$p(x, \tilde{y}(x)) = \max_y p(x, y) ,$$

we have the following analogue of (3.35):

$$\begin{aligned}\sum_{y \in \mathbf{Y}} p(x, y) - \max_{y \in \mathbf{Y}} p(x, y) &= \sum_{y \in \mathbf{Y} \setminus \{\tilde{y}(x)\}} p(x, y) = \sum_{y \in \mathbf{Y} \setminus \{\tilde{y}(x)\}} \int_0^1 1_{p(x,y) > \epsilon} d\epsilon \\ &= \int_0^1 \sum_{y \in \mathbf{Y} \setminus \{\tilde{y}(x)\}} 1_{p(x,y) > \epsilon} d\epsilon = \int_0^1 \left(\left|\Gamma^\epsilon(x)\right| - 1\right)^+ d\epsilon .\end{aligned}$$

This implies, similarly to (3.36),

$$\mathbb{E}_{x,\tau} \left(\sum_{y \in \mathbf{Y}} p(x, y) - \max_{y \in \mathbf{Y}} p(x, y) \right) = \int_0^1 \mathbb{E}_{x,\tau} \left(\left(\left|\Gamma^\epsilon(x)\right| - 1\right)^+ \right) d\epsilon . \quad (3.41)$$

Suppose that A is E-optimal, and let B be any idealised conformity measure. From (3.41), it is clear that (3.19) holds with $<$ replaced by $\leq$. If, furthermore, we have (3.20): by Corollary 3.18(c) we also have (3.22) for all ϵ; therefore, we also have (3.18) for all ϵ; in combination with (3.40), we obtain (3.15). Therefore, A is F-optimal.

Now suppose that A is F-optimal. Let B be any ACP idealised conformity measure, which we know to be both E-optimal and F-optimal. By the definition of F-optimality, we have (3.20) and (3.15) with $=$ in place of $\leq$. As in the proof of Theorem 3.2, this implies that (3.22) holds for all ϵ, and also implies that (3.18) holds (even with $=$ in place of $\geq$) for all ϵ. Therefore, A is E-optimal.

Proof of Theorem 3.4

The proof is similar to the proofs of Theorems 3.2 and 3.3. First we check that $\mathcal{O}(\mathrm{OM}) = \mathcal{R}''(\mathrm{ASP})$, analysing the requirement of OM-optimality starting from small values of $\epsilon \in (0, 1)$. Let $f_1 > f_2 > \cdots > f_n > 0.5$ be the list of the predictabilities of all objects $x \in \mathbf{X}$ whose predictability exceeds 0.5, with all duplicates removed and the remaining predictabilities sorted in the descending order. All examples are split into $2n + 1$ groups (perhaps some of them empty) in such a way that each element of the ith group is coming (in the conformity order) before each element of the jth group when $1 \leq i < j \leq 2n + 1$. The ith group, $i = 1, \ldots, n$, contains all examples (x, y) with predictability f_i and $Q(y \mid x) < 1/2$, the $(n + 1)$th group contains all examples with predictability 0.5 or less, and the $(n + 1 + i)$th group, $i = 1, \ldots, n$, contains all examples (x, y) with $Q(y \mid x) = f_i$ (there is, however, at most one such example for each x); it is possible that $n = 0$.

Any OM-optimal idealised conformity measure will assign the lowest conformity to the first group of examples (assuming $n \geq 1$), and those examples can be ordered arbitrarily in their conformity, except that any examples sharing their objects should have the same conformity. This group takes care of the values

$$\epsilon \in (0, Q\{(x, y) : f(x) = f_1 \;\; \& \;\; Q(y \mid x) \neq f_1\}] \; .$$

Proceed in the same way through groups $2, \ldots, n$. The $(n+1)$th group is most complicated (when nonempty). It contains the following kinds of examples:

- Examples whose predictability is less than 0.5. All such examples should have the same conformity if they share the same object.
- Examples (x, y) whose predictability is exactly 0.5 and which satisfy $Q(y \mid x) < 0.5$. All such examples should have the same conformity if they share the same object.
- Examples (x, y) whose predictability is exactly 0.5 and which satisfy $Q(y \mid x) = 0.5$.

Otherwise, the examples in the $(n + 1)$th group can be ordered arbitrarily in their conformity. Groups $n + 2, \ldots, 2n + 1$ do not cause any problems; the conformity order within those groups is arbitrary. Therefore, an idealised conformity measure is OM-optimal if and only if it is in $\mathcal{R}''(\mathrm{ASP})$.

Next we check that $\mathcal{O}(\mathrm{OU}) = \mathcal{O}(\mathrm{OM})$. Similarly to (3.39), we have, for a given idealised conformity measure,

$$\begin{aligned} \mathbb{E}_{(x,y),\tau} \max_{y' \neq y} p(x, y', \tau) &= \mathbb{E}_{(x,y),\tau} \int_0^1 1_{\max_{y' \neq y} p(x,y',\tau) > \epsilon} \mathrm{d}\epsilon \\ &= \mathbb{E}_{(x,y),\tau} \int_0^1 1_{\Gamma^\epsilon(x) \setminus \{y\} \neq \emptyset} \mathrm{d}\epsilon = \int_0^1 \mathbb{P}_{x,\tau} \left(\Gamma^\epsilon(x) \setminus \{y\} \neq \emptyset \right) \mathrm{d}\epsilon \; . \end{aligned} \tag{3.42}$$

By (3.42), OM-optimality immediately implies OU-optimality.

Now suppose that A is OU-optimal. Let B be the ASP idealised conformity measure, which is both OM-optimal and OU-optimal. If (3.24) is violated for some ϵ, it is violated for a range of ϵ (by Corollary 3.18(d)), which, by (3.42), contradicts the OU-optimality of A. Therefore, A is OM-optimal.

3.4.2 Proofs for Sect. 3.3

Proof of Theorem 3.7, Part I

First we prove that the condition $n_k/n_{k-1} \to 1$ is sufficient, starting from a simple general lemma about martingale differences.

Lemma 3.19 *If $\xi_1, \xi_2, \ldots$ is a martingale difference w.r. to a filtration of σ-algebras $\mathcal{F}_1, \mathcal{F}_2, \ldots$, $w_1, w_2, \ldots$ is a sequence of positive numbers such that, for all $i = 1, 2, \ldots$,*

$$\mathbb{E}(\xi_i^2 \mid \mathcal{F}_{i-1}) \le w_i^2 ,$$

and $n \in \mathbb{N}$, then

$$\mathbb{E}\left(\left(\frac{\xi_1 + \cdots + \xi_n}{w_1 + \cdots + w_n}\right)^2\right) \le \frac{w_1^2 + \cdots + w_n^2}{(w_1 + \cdots + w_n)^2} .$$

Proof Since elements of a martingale difference sequence are uncorrelated, we have

$$\mathbb{E}\left((\xi_1 + \cdots + \xi_n)^2\right) = \sum_{1 \le i \le n} \mathbb{E}(\xi_i^2) + 2 \sum_{1 \le i < j \le n} \mathbb{E}(\xi_i \xi_j) \le \sum_{1 \le i \le n} w_i^2 .$$

□

Fix a significance level ϵ and a power probability distribution Q^∞ on $\mathbf{Z}^\infty$ generating the examples $z_i = (x_i, y_i)$; the $\mathcal{L}$-taught smooth conformal predictor $\Gamma^{\mathcal{L}}$ is fed with the examples z_i and random numbers $\tau_i \in [0, 1]$. The error sequence and *predictable error sequence* of $\Gamma^{\mathcal{L}}$ will be denoted

$$e_n := \mathrm{err}_n^\epsilon(\Gamma^{\mathcal{L}}) = \begin{cases} 1 & \text{if } y_n \notin \Gamma^{\mathcal{L},\epsilon}(x_1, \tau_1, y_1, \ldots, x_{n-1}, \tau_{n-1}, y_{n-1}, x_n, \tau_n) \\ 0 & \text{otherwise} \end{cases}$$

and

$$d_n := \overline{\mathrm{err}}_n^\epsilon(\Gamma^{\mathcal{L}}) := (Q \times \mathbf{U})\Big\{(x, y, \tau) \in \mathbf{Z} \times [0, 1] :$$
$$y \notin \Gamma^{\mathcal{L},\epsilon}(x_1, \tau_1, y_1, \ldots, x_{n-1}, \tau_{n-1}, y_{n-1}, x, \tau)\Big\} .$$

Along with the original predictor $\Gamma^{\mathcal{L}}$ we also consider the *ghost predictor*, which is Γ fed with the examples

$$z_1' = (x_1', y_1') := z_{\mathcal{L}(n_1)}, z_2' = (x_2', y_2') := z_{\mathcal{L}(n_2)}, \ldots$$

and random numbers $\tau_1', \tau_2', \ldots$ (independent from each other and from the sequences z_i and τ_i). The ghost predictor is given all labels, and each label is given without delay. Notice that its input sequence $z_{\mathcal{L}(n_1)}, z_{\mathcal{L}(n_2)}, \ldots$ is also distributed according to Q^∞. The error and predictable error sequences of the ghost predictor are

$$e_n' := \mathrm{err}_n^\epsilon(\Gamma, (z_1', z_2', \ldots)) = \begin{cases} 1 & \text{if } y_n' \notin \Gamma^\epsilon(x_1', \tau_1', y_1', \ldots, x_{n-1}', \tau_{n-1}', y_{n-1}', x_n', \tau_n') \\ 0 & \text{otherwise} \end{cases}$$

and

$$d'_n := \overline{\text{err}}^{\epsilon}_n(\Gamma, (z'_1, z'_2, \dots)) = (Q \times \mathbf{U})\Big\{(x, y, \tau) \in \mathbf{Z} \times [0, 1] :$$
$$y \notin \Gamma^{\epsilon}(x'_1, \tau'_1, y'_1, \dots, x'_{n-1}, \tau'_{n-1}, y'_{n-1}, x, \tau)\Big\} .$$

It is clear that, for each k, d_n is the same for all $n = n_{k-1} + 1, \dots, n_k$, their common value being

$$d_{n_k} = d'_k . \tag{3.43}$$

Corollary 3.20 *For each k,*

$$\mathbb{E}\left(\left(\frac{(e'_1 - \epsilon)n_1 + (e'_2 - \epsilon)(n_2 - n_1) + \cdots + (e'_k - \epsilon)(n_k - n_{k-1})}{n_k}\right)^2\right)$$
$$\leq \frac{n_1^2 + (n_2 - n_1)^2 + \cdots + (n_k - n_{k-1})^2}{n_k^2} .$$

Proof It is sufficient to apply Lemma 3.19 to $w_i := n_i - n_{i-1}$ (n_0 is understood to be 0 in this subsection), the independent zero-mean (by Proposition 2.4) random variables $\xi_i := (e'_i - \epsilon)w_i$, and the σ-algebras $\mathcal{F}_i$ generated by $\xi_1, \dots, \xi_i$. □

Corollary 3.21 *For each k,*

$$\mathbb{E}\left(\left(\frac{(e'_1 - d'_1)n_1 + (e'_2 - d'_2)(n_2 - n_1) + \cdots + (e'_k - d'_k)(n_k - n_{k-1})}{n_k}\right)^2\right)$$
$$\leq \frac{n_1^2 + (n_2 - n_1)^2 + \cdots + (n_k - n_{k-1})^2}{n_k^2} .$$

Proof Use Lemma 3.19 for $w_i := n_i - n_{i-1}$, $\xi_i := (e'_i - d'_i)w_i$, and the σ-algebras $\mathcal{F}_i$ generated by $z'_1, \dots, z'_i$ and $\tau'_1, \dots, \tau'_i$. □

Corollary 3.22 *For each k,*

$$\mathbb{E}\left(\frac{(e_1 - d_1) + (e_2 - d_2) + \cdots + (e_{n_k} - d_{n_k})}{n_k}\right)^2 \leq \frac{1}{n_k} .$$

Proof Apply Lemma 3.19 to $w_i := 1$, $\xi_i := e_i - d_i$, and the σ-algebras $\mathcal{F}_i$ generated by $z_1, \dots, z_i$ and $\tau_1, \dots, \tau_i$. □

Lemma 3.23 *If $\lim_{k\to\infty}(n_k/n_{k-1}) = 1$ for some strictly increasing sequence of positive integers $n_1, n_2, \dots$, then*

$$\lim_{k\to\infty} \frac{n_1^2 + (n_2 - n_1)^2 + \cdots + (n_k - n_{k-1})^2}{n_k^2} = 0 .$$

Proof For any $\delta > 0$, there exists a K such that $\frac{n_k - n_{k-1}}{n_{k-1}} < \delta$ for any $k > K$. Therefore,

$$\frac{n_1^2 + (n_2 - n_1)^2 + \cdots + (n_k - n_{k-1})^2}{n_k^2} \leq \frac{n_K^2}{n_k^2} + \frac{(n_{K+1} - n_K)^2 + \cdots + (n_k - n_{k-1})^2}{n_k^2}$$

$$\leq \frac{n_K^2}{n_k^2} + \frac{n_{K+1} - n_K}{n_K} \frac{n_{K+1} - n_K}{n_k} + \frac{n_{K+2} - n_{K+1}}{n_{K+1}} \frac{n_{K+2} - n_{K+1}}{n_k} + \ldots$$

$$+ \frac{n_k - n_{k-1}}{n_{k-1}} \frac{n_k - n_{k-1}}{n_k} \leq \frac{n_K^2}{n_k^2} + \delta \frac{(n_{K+1} - n_K) + \cdots + (n_k - n_{k-1})}{n_k}$$

$$= \frac{n_K^2}{n_k^2} + \delta \frac{n_k - n_K}{n_k} \leq 2\delta$$

from some k on. □

Now it is easy to finish the proof of the first part of the theorem. In combination with Chebyshev's inequality and Lemma 3.23, Corollary 3.20 implies that

$$\frac{(e_1' - \epsilon)n_1 + (e_2' - \epsilon)(n_2 - n_1) + \cdots + (e_k' - \epsilon)(n_k - n_{k-1})}{n_k} \to 0$$

in probability; using the notation $k(i) := \min\{k : n_k \geq i\} = s(i) + 1$, we can rewrite this as

$$\frac{1}{n_k} \sum_{i=1}^{n_k} \left(e'_{k(i)} - \epsilon \right) \to 0 . \tag{3.44}$$

Similarly, (3.43) and Corollary 3.21 imply

$$\frac{1}{n_k} \sum_{i=1}^{n_k} \left(e'_{k(i)} - d'_{k(i)} \right) = \frac{1}{n_k} \sum_{i=1}^{n_k} \left(e'_{k(i)} - d_i \right) \to 0 , \tag{3.45}$$

and Corollary 3.22 implies

$$\frac{1}{n_k} \sum_{i=1}^{n_k} (e_i - d_i) \to 0 \tag{3.46}$$

(all convergences are in probability). Combining (3.44)–(3.46), we obtain

$$\frac{1}{n_k} \sum_{i=1}^{n_k} (e_i - \epsilon) \to 0 ; \tag{3.47}$$

the condition $n_k / n_{k-1} \to 1$ allows us to replace n_k with n in (3.47).

Proof of Theorem 3.7, Part II

Now let us check that the condition $n_k / n_{k-1} \to 1$ is necessary.

Fix $\epsilon := 5\%$. As a first step, we construct the example space $\mathbf{Z}$, the probability distribution Q on $\mathbf{Z}$, and a smoothed conformal predictor for which d'_k deviate consistently from ϵ. Let $\mathbf{X} = \{0\}$, $\mathbf{Y} = \{0, 1\}$, so that the examples z_i are, essentially, 0 or 1; we will identify $\mathbf{Z}$ with $\mathbf{Y}$, i.e., $\{0, 1\}$. The probability distribution Q is uniform on $\mathbf{Z}$: $Q\{0\} = Q\{1\} = 1/2$. The nonconformity measure is

$$\alpha_i = A\left(\langle \zeta_1, \dots, \zeta_k \rangle, \zeta_i\right) := \begin{cases} \zeta_i & \text{if } \zeta_1 + \dots + \zeta_k \text{ is even} \\ 1 - \zeta_i & \text{if } \zeta_1 + \dots + \zeta_k \text{ is odd} . \end{cases}$$

It follows from the central limit theorem that

$$\frac{\left|\{i = 1, \dots, k : z'_i = 1\}\right|}{k} \in (0.4, 0.6) \tag{3.48}$$

with probability at least 99% for k large enough. We will show that d'_k deviates significantly from ϵ with probability at least 99% for sufficiently large k. Let $\alpha_i := A(\langle z'_1, \dots, z'_k \rangle, z'_i)$. There are two possibilities:

- If $z'_1 + \dots + z'_{k-1}$ is odd, then

$$z'_k = 1 \Longrightarrow z'_1 + \dots + z'_{k-1} + z'_k \text{ is even} \Longrightarrow \alpha_k = z'_k = 1$$
$$z'_k = 0 \Longrightarrow z'_1 + \dots + z'_{k-1} + z'_k \text{ is odd} \Longrightarrow \alpha_k = 1 - z'_k = 1 .$$

In both cases we have $\alpha_k = 1$ and, therefore, outside an event of probability at most 1%,

$$\begin{aligned} d'_k &= (Q \times \mathbf{U})\{(y, \tau) : \tau\,|\{i = 1, \dots, k : \alpha_i = 1\}| \le k\epsilon\} \\ &= \int_{\mathbf{Y}} 1 \wedge \frac{k\epsilon}{|\{i = 1, \dots, k : \alpha_i = 1\}|} Q(\mathrm{d}y) \ge \frac{k\epsilon}{0.7k} = \frac{10}{7}\epsilon . \end{aligned}$$

- If $z'_1 + \dots + z'_{k-1}$ is even, then

$$z'_k = 1 \Longrightarrow z'_1 + \dots + z'_{k-1} + z'_k \text{ is odd} \Longrightarrow \alpha_k = 1 - z'_k = 0$$
$$z'_k = 0 \Longrightarrow z'_1 + \dots + z'_{k-1} + z'_k \text{ is even} \Longrightarrow \alpha_k = z'_k = 0 .$$

In both cases $\alpha_k = 0$ and, therefore, outside an event of probability at most 1%,

$$\begin{aligned} d'_k = (Q \times \mathbf{U})\big\{(y, \tau) : &|\{i = 1, \dots, k : \alpha_i = 1\}| \\ &+ \tau\,|\{i = 1, \dots, k : \alpha_i = 0\}| \le k\epsilon\big\} \le (Q \times \mathbf{U})\{(y, \tau) : 0.3k \le k\epsilon\} = 0 . \end{aligned}$$

To summarize, for large enough k,

$$\left|d'_k - \epsilon\right| = \left|d_{n_k} - \epsilon\right| > \epsilon/3 \tag{3.49}$$

with probability at least 99% (cf. (3.43); we write 99% rather than 98% since the two exceptional events of probability 1% coincide: both are the complement of (3.48)).

Suppose that

$$\frac{1}{n}\sum_{i=1}^{n} e_i \to \epsilon \tag{3.50}$$

in probability; we will deduce that $n_k/n_{k-1} \to 1$. By (3.46) (remember that Corollary 3.22 and, therefore, (3.46) do not depend on the condition $n_k/n_{k-1} \to 1$) and (3.50) we have

$$\frac{1}{n_k}\sum_{i=1}^{n_k} d_i \to \epsilon \;;$$

we can rewrite this in the form

$$\sum_{i=1}^{n_k} d_i = n_k(\epsilon + o(1))$$

(all $o(1)$ are in probability). This equality implies

$$\sum_{k=0}^{K} d_{n_k}(n_k - n_{k-1}) = n_K(\epsilon + o(1))$$

and

$$\sum_{k=0}^{K-1} d_{n_k}(n_k - n_{k-1}) = n_{K-1}(\epsilon + o(1)) \;;$$

subtracting the last equality from the penultimate one we obtain

$$d_{n_K}(n_K - n_{K-1}) = (n_K - n_{K-1})\epsilon + o(n_K) \,,$$

i.e.,

$$\left(d_{n_K} - \epsilon\right)(n_K - n_{K-1}) = o(n_K) \,.$$

In combination with (3.49), this implies $n_K - n_{K-1} = o(n_K)$, i.e., $n_K/n_{K-1} \to 1$ as $K \to \infty$.

Proof of Theorem 3.9

This proof is similar to the proof of Theorem 3.7. (The definition of $\Gamma^{\mathcal{L}}$ being asymptotically exact involves the assumption of exchangeability rather than randomness; however, since we assumed that $\mathbf{Z}$ is a standard Borel space, de Finetti's theorem, stated in Sect. A.5.1, shows that these assumptions are equivalent in our current context.) Instead of Corollaries 3.20, 3.21, and 3.22 we now have:

Corollary 3.24 *As $k \to \infty$,*

$$\frac{(e'_1 - \epsilon)n_1 + (e'_2 - \epsilon)(n_2 - n_1) + \cdots + (e'_k - \epsilon)(n_k - n_{k-1})}{n_k} \to 0 \quad a.s.$$

Proof It is sufficient to apply Kolmogorov's strong law of large numbers (stated in Sect. A.6.2) to the independent zero-mean random variables $\xi_i = (e'_i - \epsilon)(n_i - n_{i-1})$. Condition (A.6) follows from

$$\sum_{i=1}^{\infty} \frac{(n_i - n_{i-1})^2}{n_i^2} < \infty \,,$$

which is equivalent to (3.33). □

Corollary 3.25 *As $k \to \infty$,*

$$\frac{(e'_1 - d'_1)n_1 + (e'_2 - d'_2)(n_2 - n_1) + \cdots + (e'_k - d'_k)(n_k - n_{k-1})}{n_k} \to 0 \quad a.s.$$

Proof Apply the martingale strong law of large numbers (Sect. A.6.2) to the martingale difference $\xi_i = (e'_i - d'_i)(n_i - n_{i-1})$ w.r. to the σ-algebras $\mathcal{F}_i$ generated by $z'_1, \ldots, z'_i$ and $\tau'_1, \ldots, \tau'_i$. □

Corollary 3.26 *As $k \to \infty$,*

$$\frac{(e_1 - d_1) + (e_2 - d_2) + \cdots + (e_{n_k} - d_{n_k})}{n_k} \to 0 \quad a.s.$$

Proof Apply the martingale strong law of large numbers to the martingale difference $\xi_i = e_i - d_i$ w.r. to the σ-algebras $\mathcal{F}_i$ generated by $z_1, \ldots, z_i$ and $\tau_1, \ldots, \tau_i$. □

Corollary 3.24 can be rewritten as (3.44), Corollary 3.25 as (3.45), and Corollary 3.26 as (3.46); all convergences are now a.s. Combining (3.44)–(3.46), we obtain (3.47). It remains to replace n_k with n, as before.

We do not spell out the proof of Theorem 3.11, but it is very similar and uses the martingale law of the iterated logarithm stated in Sect. A.6.2.

Proof of Theorem 3.13

We will only consider the case where Γ is a smoothed conformal predictor (the proof for deterministic Γ is almost identical: just ignore all random numbers τ). As before, we use the device of a "ghost predictor" introduced near the beginning of this subsection.

Fix a significance level ϵ and define

$$\overline{\mathrm{oe}}_n(\Gamma^{\mathcal{L}}) := \mathbb{E}_{(x,y,\tau)\sim Q\times\mathbf{U}} \left|\Gamma^{\mathcal{L},\epsilon}(x_1, \tau_1, y_1, \ldots, x_{n-1}, \tau_{n-1}, y_{n-1}, x, \tau) \setminus \{y\}\right|,$$

$$\overline{\mathrm{oe}}_k(\Gamma) := \mathbb{E}_{(x,y,\tau)\sim Q\times\mathbf{U}} \left|\Gamma^{\epsilon}(x'_1, \tau'_1, y'_1, \ldots, x'_{k-1}, \tau'_{k-1}, y'_{k-1}, x, \tau) \setminus \{y\}\right|,$$

$$\overline{\mathrm{OE}}_n(\Gamma^{\mathcal{L}}) := \sum_{i=1}^{n} \overline{\mathrm{oe}}_i(\Gamma^{\mathcal{L}}), \quad \overline{\mathrm{OE}}_k(\Gamma) := \sum_{i=1}^{k} \overline{\mathrm{oe}}_i(\Gamma), \quad \mathrm{oe}^{\epsilon}_n := \mathrm{OE}^{\epsilon}_n - \mathrm{OE}^{\epsilon}_{n-1}$$

(where oe^{ϵ}_n and OE^{ϵ}_n can be applied to both $\Gamma^{\mathcal{L}}$ and the ghost predictor Γ fed with $z'_1, z'_2, \ldots$ and $\tau'_1, \tau'_2, \ldots$). Since $\mathrm{OE}^{\epsilon}_n(\Gamma^{\mathcal{L}}) - \overline{\mathrm{OE}}_n(\Gamma^{\mathcal{L}})$ is a martingale and

$$\left|\mathrm{oe}^{\epsilon}_n(\Gamma^{\mathcal{L}}) - \overline{\mathrm{oe}}_n(\Gamma^{\mathcal{L}})\right| \le |\mathbf{Y}| - 1\,,$$

the martingale strong law of large numbers (see Sect. A.6.2) implies that

$$\lim_{n\to\infty} \frac{\mathrm{OE}^{\epsilon}_n(\Gamma^{\mathcal{L}}) - \overline{\mathrm{OE}}_n(\Gamma^{\mathcal{L}})}{n} = 0 \quad \text{a.s.} \tag{3.51}$$

Analogously,

$$\lim_{k\to\infty} \frac{\mathrm{OE}^{\epsilon}_k(\Gamma, (z'_1, z'_2, \ldots)) - \overline{\mathrm{OE}}_k(\Gamma)}{k} = 0 \quad \text{a.s.} \tag{3.52}$$

By (3.51) and (3.52), we can replace OE^{ϵ} with $\overline{\mathrm{OE}}$ in the definitions of $I^{\epsilon}(\Gamma^{\mathcal{L}}, Q)$ and $I^{\epsilon}(\Gamma, Q)$.

It is clear that $\overline{\mathrm{oe}}_n(\Gamma^{\mathcal{L}}) = \overline{\mathrm{oe}}_{k(n)}(\Gamma)$ for all n. Combining this with $n_{k+1} - n_k = c$ (from some k on), we obtain

$$\sum_{i=1}^{n} \overline{\mathrm{oe}}_i(\Gamma^{\mathcal{L}}) = c \sum_{i=1}^{\lfloor n/c \rfloor} \overline{\mathrm{oe}}_i(\Gamma) + O(1) ,$$

and so $\overline{\mathrm{OE}}_n(\Gamma^{\mathcal{L}}) = c\overline{\mathrm{OE}}_{\lfloor n/c \rfloor}(\Gamma) + o(n)$. The statement of the theorem immediately follows.

3.5 Context

3.5.1 Criteria of Efficiency

In Sect. 3.1, we follow [401] (however, that paper also discusses criteria of efficiency for label-conditional conformal predictors, introduced in Sect. 4.6 below).

When using the M criterion of efficiency, it is very natural to summarize the range of possible prediction sets Γ^ϵ, $\epsilon \in (0, 1)$, by reporting the *confidence*

$$\sup\{1 - \epsilon : |\Gamma^\epsilon| \le 1\} \tag{3.53}$$

(this is 1 minus the unconfidence as defined in Sect. 3.1.2), the *credibility*

$$\inf\{\epsilon : |\Gamma^\epsilon| = 0\} , \tag{3.54}$$

and the *prediction* Γ^ϵ, where $1-\epsilon$ is the confidence (in this case Γ^ϵ is never multiple for conformal predictors and usually contains exactly one label). Reporting the prediction (in the form of an element of $\mathbf{Y}$), confidence, and credibility was suggested in [400] and [303]. The U criterion as defined in Sect. 3.1 was introduced in [102], but it is equivalent to using the average confidence.

The S criterion was introduced in [102], and the N criterion was introduced independently in [179] and [102], although the analogue of the N criterion for regression (where the size of a prediction set is defined to be its Lebesgue measure) had been used earlier in [225].

A criterion that is very similar to the M and E criteria is used by Lei in [223, Sect. 2.2]; that paper considers the binary case, in which the difference between the M and E criteria disappears. The difference of the criterion used in [223] is that it prohibits empty predictions (an intermediate approach would be to prefer smaller values for the number (3.6) of empty predictions). Lei's criterion is extended to the multi-class case in [300], which proposes a modification of the E criterion with a different treatment of empty predictions.

The OU and OM criteria were introduced in [401] (which raised the question of improper criteria of efficiency). That paper used "probabilistic" instead of our "conditionally proper".

The CP idealised conformity measure was introduced by an anonymous referee of the conference version of [102], but its non-idealised analogue in the case of regression had been used in [225] (following [227] and literature on minimum volume prediction).

3.5.2 Examples of Nonconformity Measures

The idea of reducing binary classification to regression is an old one. In the case of simple prediction the procedure is as follows: encode the labels as real numbers, one negative and the other positive, apply a regression algorithm, and define $\hat{y}_n$ to be the "positive" label if the value predicted for y_n by the regression algorithm is positive, to be the "negative" label if the value predicted for y_n by the regression algorithm is negative, and define $\hat{y}_n$ arbitrarily if the value predicted for y_n by the regression algorithm is zero. Probably the earliest suggestion of this kind was Fisher's [111, Sect. 49.2] *discriminant analysis*: if there are, say, l_1 males and l_2 females in the training set and $l_1 + l_2 = l$, encode males as l_2/l and encode females as $-l_1/l$ (so that the mean of the encodings over the training set is 0), and use the least squares algorithm as the regression algorithm.

The precursor of conformal predictor suggested in [127] used the SVM method as the underlying algorithm. Later it was noticed (see [302]) that the Lagrange method applied to ridge regression in analogy with SVM leads to α_i equivalent to the residuals, and this in turn led to the realization that almost any machine learning algorithm can be adapted, often in more than one way, to obtain a nonconformity measure. However, the first genuine conformal predictor (then called "transductive confidence machine") introduced in [400] and [303] still used the Lagrange multipliers α_i as the nonconformity measure. The original conformal predictors for multiclass classification problems using binary SVMs were based on (3.32) with $\lambda = 1$, but it was quickly noticed that taking $\lambda < 1$ improves results dramatically.

We mentioned two methods of reducing multiclass classification problems to binary ones: "one-against-the-rest" and "one-against-one". Another popular method, based on error-correcting coding, was proposed by Dietterich and Bakiri [85] in 1995.

Instead of reducing a multiclass classification problem to the binary case and then applying the SVM method, it is possible to use directly known multiclass generalizations of SVM. First such generalization was proposed by Blanz and Vapnik [366, Sect. 10.10], and later but independently it was found by Watkins and Weston [424]. See [360, Sect. 1] for a review of several existing approaches to multiclass SVM.

In Sect. 2.9.5 we mentioned Romano et al.'s [293] conformity measure based on quantile regression. Similar ideas were applied to multiclass and multilabel classification by Cauchois et al. [44]. A different way of approximating the predictive quantile function in classification problems was proposed by Romano et al. [294]. These methods lead to adaptive prediction sets.

3.5.3 *Universal Predictors*

In this edition we have removed the material about the universal predictor constructed in Chap. 3 (Sects. 3.2–3.5) of the first edition of this book [402]. The reason is that our definition of universality used a criterion of efficiency that was not conditionally proper, and so at this time it appears to us less interesting than an analogous notion based on a conditionally proper criterion of efficiency. There is no doubt that it is possible to construct a universal predictor based on such a notion of efficiency, but we do not do it in this chapter since we will have other examples of universal conformal predictors: namely, in Chap. 6 in the context of probabilistic classification (Sect. 6.3) and in Chap. 7 in the context of probabilistic regression (Sect. 7.6).

3.5.4 *Weak Teachers*

The characterization of lazy teachers for which conformal prediction is valid in probability was obtained by Ilia Nouretdinov. The general notion of a teaching schedule and the device of a "ghost predictor" is due to Daniil Ryabko [299]. Nouretdinov's result (our Theorem 3.7), generalized in light of [299], appeared in [262]. Theorem 3.9 and a version of Theorem 3.13 are given in [299].

Chapter 4
Modifications of Conformal Predictors

Abstract So far we have emphasized desirable properties of conformal predictors (also known as full conformal predictors): validity, asymptotic efficiency, and flexibility (ability to incorporate a wide range of machine-learning methods); we have also mentioned that the hedged predictions output by good conformal predictors are "conditional", in the sense that they take full account of the object to be predicted. In this chapter we will discuss some limitations of full conformal prediction, starting from their relative computational inefficiency, and ways to overcome or alleviate these limitations.

Keywords Inductive conformal prediction · Inductive conformity measure · Cross-conformal prediction · Transductive conformal prediction · Conditionality · Mondrian prediction

4.1 The Topics of This Chapter

The first problem, dealt with in Sect. 4.2, is the relative computational inefficiency of full conformal predictors. In that section we construct "inductive conformal predictors" (ICPs), whose computational efficiency is often much better; the price is some loss in predictive efficiency (which was called simply "efficiency" in the previous chapters). In a natural online setting, inductive conformal predictors are exact.

In Sect. 4.3 we introduce several new nonconformity measures, which are especially natural when used with ICPs (and with cross-conformal predictors discussed in Sect. 4.4).

Both full and inductive conformal predictors are automatically valid, and what is at stake is their efficiency, predictive and computational. While inductive conformal predictors typically vastly improve computational efficiency, their predictive efficiency may suffer. In Sect. 4.4 we try to achieve the best of both worlds, but the price to pay is the much weaker guaranteed property of validity.

V. Vovk et al., *Algorithmic Learning in a Random World*,
https://doi.org/10.1007/978-3-031-06649-8_4

A short Sect. 4.5 is devoted to transductive conformal prediction, where our task is to predict several future examples at once. Remember that in the standard offline prediction the examples in the test set are processed separately, one by one.

The issue of conditionality is taken up in Sects. 4.6 and 4.7. The potentially serious problem with conformal predictors is that they are not automatically *conditionally valid*: e.g., in the USPS dataset some digits (such as "5") are more difficult to recognize correctly than other digits (such as "0"), and it is natural to expect that at the confidence level 95% the error rate will be significantly greater than 5% for the difficult digits; our usual, unconditional, notion of validity only ensures that the average error rate over all digits will be close to 5%. The notions of Venn-conditional and Mondrian conformal predictors are introduced to address concerns of this kind.

4.2 Inductive Conformal Predictors

We start by looking more closely at the reasons for the relative computational inefficiency of full conformal predictors for large datasets. Consider a conformal predictor determined by the nonconformity measure (2.31). Continuing the discussion that we started in Sect. 1.3.4, one can usually assign a simple predictor to one of two types: "inductive" or "transductive". For inductive predictors, a prediction rule $D_{\lbag z_1,\dots,z_n \rbag}$ can be computed, in some sense, from the training data $z_1, \dots, z_n$: e.g., $D_{\lbag z_1,\dots,z_n \rbag}$ may be described by a polynomial, and computing $D_{\lbag z_1,\dots,z_n \rbag}$ may mean computing the coefficients of the polynomial; as soon as $D_{\lbag z_1,\dots,z_n \rbag}$ is computed, computing $D_{\lbag z_1,\dots,z_n \rbag}(x)$ for a new object x takes very little time. For transductive predictors (such as 1-nearest neighbour), relatively little can be done before seeing the new object x; even allowing considerable time for pre-processing $\lbag z_1, \dots, z_n \rbag$, computing $D_{\lbag z_1,\dots,z_n \rbag}(x)$ will be a difficult task.

Notice that, even when D is an inductive algorithm, the conformal predictor determined by the generic nonconformity measure (2.31) (and, even more so, by its deleted version) will still be computationally inefficient: for every new object x_n and for every postulated label for it, computing $\Gamma^\epsilon(z_1, \dots, z_{n-1}, x_n)$ will require constructing new prediction rules. Inductive conformal predictors will be defined in such a way that they can make significant computational savings when the underlying simple predictor D is inductive.

4.2.1 *Inductive Conformal Predictors in the Online Mode*

We start from defining an ICP in the online mode given a nonconformity measure A (which we will use in the deleted form (2.35)). First fix a finite or infinite sequence of positive integer parameters $m_1, m_2, \dots$ (called *update trials*); it is required that $m_1 < m_2 < \dots$. If the sequence $m_1, m_2, \dots$ is finite, $(m_1, m_2, \dots) =$

$(m_1, \ldots, m_r)$, we set $m_i := \infty$ for $i > r$. The *ICP* determined by A and the sequence $m_1, m_2, \ldots$ of update trials is defined to be the confidence predictor Γ such that the prediction sets $\Gamma^\epsilon(z_1, \ldots, z_{n-1}, x_n)$ are computed as follows:

- if $n \le m_1$, $\Gamma^\epsilon(z_1, \ldots, z_{n-1}, x_n)$ is found using a fixed full conformal predictor;
- otherwise, find the k such that $m_k < n \le m_{k+1}$ and set

$$\Gamma^\epsilon(z_1, \ldots, z_{n-1}, x_n) := \left\{ y \in \mathbf{Y} : \frac{\left|\{j = m_k + 1, \ldots, n : \alpha_j \ge \alpha_n\}\right|}{n - m_k} > \epsilon \right\}, \qquad (4.1)$$

where the nonconformity scores α_j are defined by

$$\begin{aligned} \alpha_j &:= A^{\text{del}}\left(\lbag z_1, \ldots, z_{m_k} \rbag, (x_j, y_j)\right) \text{ for } j = m_k + 1, \ldots, n - 1, \\ \alpha_n &:= A^{\text{del}}\left(\lbag z_1, \ldots, z_{m_k} \rbag, (x_n, y)\right) . \end{aligned} \qquad (4.2)$$

Smoothed ICPs can be defined analogously to smoothed conformal predictors: instead of (4.1) we have

$$\Gamma^\epsilon(x_1, \tau_1, y_1, \ldots, x_{n-1}, \tau_{n-1}, y_{n-1}, x_n, \tau_n) := \left\{ y \in \mathbf{Y} : \frac{\left|\{j : \alpha_j > \alpha_n\}\right| + \tau_n \left|\{j : \alpha_j = \alpha_n\}\right|}{n - m_k} > \epsilon \right\}, \qquad (4.3)$$

where $j = m_k + 1, \ldots, n$ and $\tau_n \in [0, 1]$ are the random numbers. The following result (which is also a special case of Theorem 11.1) shows that Propositions 2.3 and 2.4 continue to hold in the case of ICPs and smoothed ICPs, respectively.

Proposition 4.1 *All ICPs are conservatively valid. All smoothed ICPs are exactly valid.*

4.2.2 Inductive Conformal Predictors in the Offline and Semi-Online Modes

Similarly to offline conformal predictors, we can define *offline ICPs*. Now we can slightly generalize the notion of a nonconformity measure: an *inductive nonconformity measure* is a function $A : \mathbf{Z}^* \times \mathbf{Z} \to \overline{\mathbb{R}}$ (we can identify a nonconformity measure with an inductive nonconformity measure that is invariant w.r. to the permutations of its first argument). Let us fix an inductive nonconformity measure A.

The training set of size l is first split into two parts: the *proper training set* $z_1, \dots, z_m$ of size $m < l$ and the *calibration set* $z_{m+1}, \dots, z_l$ of size $l - m$. For every test object x_i, $i = l + 1, \dots, l + k$, compute the prediction sets

$$\Gamma^\epsilon(z_1, \dots, z_l, x_i) := \left\{ y \in \mathbf{Y} : \frac{\left|\{j = m + 1, \dots, l : \alpha_j \geq \alpha_i\}\right| + 1}{l - m + 1} > \epsilon \right\} , \tag{4.4}$$

where the nonconformity scores are defined by

$$\begin{aligned} \alpha_j &:= A\left((z_1, \dots, z_m), z_j\right), \quad j = m + 1, \dots, l , \\ \alpha_i &:= A\left((z_1, \dots, z_m), (x_i, y)\right) , \end{aligned} \tag{4.5}$$

as in (4.1)–(4.2) (so that we are using the inductive nonconformity measure in the deleted mode).

Remark 4.2 A more explicit representation of (4.4) is

$$\Gamma^\epsilon(z_1, \dots, z_l, x_i) := \left\{ y \in \mathbf{Y} : \alpha_i \leq \alpha_{(\lfloor \epsilon (l-m+1) \rfloor)} \right\} , \tag{4.6}$$

where $\alpha_{(\cdot)}$ refers to the elements of the "calibration" bag

$$\wr \alpha_{m+1}, \dots, \alpha_l \int = \wr \alpha_{(1)}, \dots, \alpha_{(l-m)} \int$$

sorted in the descending order: $\alpha_{(1)} \geq \dots \geq \alpha_{(l-m)}$.

For both full conformal predictors and ICPs, it is true that

$$Q^\infty \left\{ (x_1, y_1, x_2, y_2, \dots) : y_i \notin \Gamma^\epsilon(x_1, y_1, \dots, x_l, y_l, x_i) \right\} \leq \epsilon \tag{4.7}$$

for every $i = l + 1, \dots, l + k$, provided all examples are drawn independently from the distribution Q, but the events in (4.7) are not independent, and the percentage of errors

$$\frac{|\{i = l + 1, \dots, l + k : y_i \notin \Gamma^\epsilon(x_1, y_1, \dots, x_l, y_l, x_i)\}|}{k} \tag{4.8}$$

can be significantly above (or below) ϵ with a high probability even when k is very large.

To ensure a stronger notion of validity of the offline ICP, we can modify the application of the ICP constructed from the training set to the test set: after processing each test example z_i, $i = l + 1, \dots, l + k$, the corresponding nonconformity score α_i should be added to the pool of nonconformity scores used in generating the prediction sets for the following test examples. Formally, redefine

$$\Gamma^{\epsilon}(z_1,\ldots,z_{i-1},x_i) := \left\{ y \in \mathbf{Y} : \frac{|\{j = m+1,\ldots,i : \alpha_j \geq \alpha_i\}|}{i-m} > \epsilon \right\}, \quad (4.9)$$

where the nonconformity scores are defined by

$$\alpha_j := A\left((z_1,\ldots,z_m), z_j\right), \quad j = m+1,\ldots,i-1,$$
$$\alpha_i := A\left((z_1,\ldots,z_m),(x_i,y)\right). \quad (4.10)$$

Proposition 4.1 says that this modification is conservatively valid, and so (4.8) will not exceed ϵ, up to statistical fluctuations.

Notice that in the case $k \ll (l-m)$ the *semi-online ICP* (4.9) differs so little from the offline ICP that the latter can be expected to be "nearly conservative".

Let us give a formal definition. A confidence predictor Γ is *(δ_n)-conservative*, where $\delta_1, \delta_2, \ldots$ is a sequence of nonnegative numbers, if for any exchangeable probability distribution P on $\mathbf{Z}^\infty$ there exists a family (2.12) of $\{0,1\}$-valued random variables on $(\mathbf{Z}^\infty \times [0,1], P \times \mathbf{U})$ such that:

- for all n and ϵ, $\xi_n^{(\epsilon)}$ depends only on the first n examples;
- for a fixed ϵ, $\xi_1^{(\epsilon)}, \xi_2^{(\epsilon)}, \ldots$ is a sequence of independent Bernoulli random variables with parameter ϵ;
- for all n and ϵ, $\mathrm{err}_n^{(\epsilon-\delta_n)}(\Gamma) \leq \xi_n^{(\epsilon)}$.

The definition of conservative validity is a special case corresponding to $\delta_n = 0$, $n = 1, 2, \ldots$; we are now interested in the case where δ_n are small (at least for a range of n) positive numbers.

Proposition 4.3 *The confidence predictor*

$$\tilde{\Gamma}^{\epsilon}(z_1,\ldots,z_{i-1},x_i) := \begin{cases} \Gamma^{\epsilon}(z_1,\ldots,z_l,x_i) & \text{if } i > l \\ \mathbf{Y} & \text{otherwise}, \end{cases}$$

where $\Gamma^{\epsilon}(z_1,\ldots,z_l,x_i)$ is defined by (4.4), *is (δ_i)-conservative, where*

$$\delta_i := \begin{cases} \frac{i-l}{l-m} & \text{if } i > l \\ 0 & \text{otherwise}. \end{cases}$$

Proof Let $\Gamma^{\dagger}$ be the smoothed semi-online ICP corresponding to Γ (with the random numbers τ_n obtained by, say, reversing the merging function (2.25)), and let $i > l$. Define ξ_i by the requirement that $\xi_i^{(\epsilon)} = 1$ if and only if $\Gamma^{\dagger}$ makes a mistake at the significance level ϵ when fed with $z_1,\ldots,z_i$. To ensure $\mathrm{err}_i^{(\epsilon-\delta_i)} \leq \xi_i^{(\epsilon)}$, i.e., $\mathrm{err}_i^{(\epsilon-\delta_i)} = 1 \Rightarrow \xi_i^{(\epsilon)} = 1$, it is sufficient that the inequality in (4.9) always imply the inequality in (4.4) with $\epsilon - \delta_i$ in place of ϵ. Therefore, it is sufficient to have

$$\forall N : \frac{N+1}{l-m+1} \leq \epsilon - \delta_i \Longrightarrow \frac{N+i-l}{i-m} \leq \epsilon ,$$

i.e.,

$$\delta_i \geq (1-\epsilon)\frac{i-l-1}{l-m+1} .$$

□

This proof shows that the fraction of errors made by the offline ICP at a significance level ϵ on the test set does not exceed $\epsilon + k/(l-m)$, up to statistical fluctuations.

4.2.3 The General Scheme for Defining Nonconformity

For use with inductive conformal predictors, we will rewrite the definition (2.31) as

$$A\left(\{z_1, \ldots, z_m\}, (x, y)\right) := \Delta\left(y, D_{\{z_1,\ldots,z_m\}}(x)\right) , \tag{4.11}$$

where (x, y) is typically not an element of the comparison bag. (In the case of offline ICP the dependence on the order of the elements in the bag is allowed, but we ignore this possibility.)

Let us first discuss the online prediction protocol. We will consider predictors that are given D as an oracle and assume that the label space $\mathbf{Y}$ is finite and fixed. The ICP requires recomputing the prediction rule being used not at every trial but only at the update trials $m_1, m_2, \ldots$; the rate of growth of m_i determines the chosen balance between predictive and computational efficiency. Let a and b be positive numbers such that either $a \geq 1$ and $b \geq 1$ or $a > 1$, and suppose that the prediction rule $D_{\{z_1,\ldots,z_n\}}$ is computable in time $\Theta(n^a \log^b n)$ and the discrepancy measure Δ is computable in constant time. Then the full conformal predictor determined by (4.11) spends time $\Theta(n^{a+1} \log^b n)$ on the computations needed for the first n trials of the online protocol. On the other hand, if the sequence m_i is infinite and grows exponentially fast, the ICP based on D, Δ, and (m_i) spends the same, to within a constant factor, time $\Theta(n^a \log^b n)$. In the case where the sequence m_i is finite, the ICP's computation time becomes

$$\Theta(n \log n) \tag{4.12}$$

(e.g., use red-black trees for storing the nonconformity scores, but augment them with information needed to find the rank of an element in time $O(\log n)$—see [55, Sect. 14.1]).

ICPs applied in both offline and semi-online modes are also computationally efficient. Let us see what the computation time will be if standard algorithms for standard computation tasks are used. In the case of simple predictions, the application of the inductive algorithm D found from the training set of size l to the test set of size k requires time

$$\Theta\left(T_{\text{train}} + kT_{\text{appl}}\right) ,$$

where T_{train} is the time required for computing the prediction rule $D_{\lbag z_1,\dots,z_l \rbag}$ and T_{appl} is the time needed to apply this prediction rule to a new object. The offline ICP (see (4.4) and (4.5)) requires time

$$\Theta\left(T^{\dagger}_{\text{train}} + (l-m+k)T^{\dagger}_{\text{appl}} + (l-m)\log(l-m) + k\log(l-m)\right) , \tag{4.13}$$

where $T^{\dagger}_{\text{train}}$ is the time required for computing the prediction rule $D_{\lbag z_1,\dots,z_m \rbag}$ and $T^{\dagger}_{\text{appl}}$ is the time needed to apply this prediction rule to a new object (we assume that computing Δ is fast); we allow time $(l-m)\log(l-m)$ for sorting the nonconformity scores obtained from the calibration set [55, Part II] and time $\log(l-m)$ for finding the rank of a test nonconformity score in the set of the calibration nonconformity scores [55, Part III].

In the case of semi-online ICP ((4.9), (4.10)), the required time increases only slightly (for moderately large k) to

$$\Theta\left(T^{\dagger}_{\text{train}} + (l-m+k)T^{\dagger}_{\text{appl}} + (l-m)\log(l-m) + k\log(l-m+k)\right) .$$

As $k \to \infty$, we have the same asymptotic computation time, $\Theta(k\log k)$, as in (4.12).

4.2.4 Normalization and Hyper-Parameter Selection

In practical machine learning, the performance of standard prediction algorithms very much depends on preprocessing the data (such as normalizing the attributes, i.e., bringing them to the same scale) and selecting the hyper-parameters (such as the number K of neighbours to use in nearest neighbours methods). In full conformal prediction, these two standard ways of improving their performance become problematic, since they have be done for each postulated label for each test object. An important advantage of inductive conformal prediction is that this becomes feasible.

Training of prediction algorithms often consists in selecting suitable values of parameters, and this can sometimes be done efficiently even for full conformal prediction. Another advantage of inductive conformal prediction is that this becomes feasible for any underlying prediction algorithm.

4.3 Further Ways of Computing Nonconformity Scores

All nonconformity measures described in the previous chapters can be used in inductive conformal prediction, and all nonconformity measures that will be introduced in this section can be used in full conformal prediction. The nonconformity measures of this section are often computationally feasible for realistic datasets only in the case of ICPs and their modifications (such as those in Sect. 4.4).

Most of the nonconformity measures described in this section have been implemented in standard machine-learning packages, such as `scikit-learn`, `TensorFlow`, and `PyTorch`. In this book we will use the expression *standard machine-learning packages* as the code for this kind of packages. They usually contain plenty of other prediction algorithms that can be used for defining nonconformity measures suitable for inductive conformal prediction and other computationally efficient varieties of conformal prediction that we will introduce later.

Suppose we are given a finite sequence

$$(z_1, \dots, z_m) \in \mathbf{Z}^* \tag{4.14}$$

and an example $z \in \mathbf{Z}$, both fixed for the rest of this section. The problem is to define the inductive nonconformity score

$$A\left((z_1, \dots, z_m), z\right) , \tag{4.15}$$

which we will usually abbreviate to $A(z)$. Sometimes it will be more convenient to define the inductive conformity score $B(z)$ instead. As usual, we write (x_i, y_i) for z_i and (x, y) for z when we need separate notations for the objects and labels.

The terminology that we use in this section often assumes that we are in the situation where the sequence (4.14) is the proper training set in offline conformal prediction, but all of the methods that we discuss can also be applied to defining nonconformity scores for use in full conformal prediction (which, however, will be computationally inefficient). Namely, the inductive conformity scores (4.15) will not depend on the order of the sequence $z_1, \dots, z_m$ and so they can be used as deleted conformity scores $A^{\mathrm{del}}(\lbag z_1, \dots, z_m \rbag, z)$. (Some dependence on the order might appear in practical implementations of prediction algorithms that are supposed to be invariant w.r. to the order, but we can always use the approach of Remark 3.6.)

In the context of online inductive conformal prediction, we are interested in $m \in \{m_1, m_2, \dots\}$.

In this section we will consider both regression and classification. In the case of regression we will only give "two-sided" definitions, in the spirit of (2.30) or (2.32), but they can be easily modified to make them one-sided, as in (2.33) or (2.34).

4.3.1 Nonconformity Measures Considered Earlier

We start from adapting two nonconformity measures introduced earlier and based on least squares to ICP. The two-sided version $\alpha_i := \left|y_i - \hat{y}_{(i)}\right|$ of the nonconformity measure (2.46) used to define the deleted CLS can be rewritten in our present context as

$$A\left((z_1, \ldots, z_m), z\right) = \left|y - \hat{y}\right| , \tag{4.16}$$

where $\hat{y}$ is the least squares prediction for y as computed from the training set $z_1, \ldots, z_m$ and x. The two-sided version of the nonconformity measure (2.47) used to define the studentized CLS can be rewritten as

$$A\left((z_1, \ldots, z_m), z\right) = \frac{\left|y - \hat{y}\right|}{\sqrt{1 + x'(X'X)^{-1}x}} , \tag{4.17}$$

where, in addition, X is the $m \times p$ data matrix $(x_1, \ldots, x_m)'$. This can be checked using the Sherman–Morrison formula (2.66).

The definition (3.31) of nonconformity measures based on the nearest neighbours classification is already given in the form (4.15) convenient for use with ICP. In the case of regression (Sect. 2.4), $A(x, y)$ is defined as $\left|y - \hat{y}\right|$, where $\hat{y}$ is the k-NNR prediction for x computed from (4.14) as training set. Because of the transductive nature of the nearest neighbours method, the computational savings resulting from using ICP are not as substantial as for inductive underlying prediction algorithms.

The online performance of the 1-nearest neighbour ICP on the USPS dataset with update trial 4649 (the middle of the dataset) is shown in Fig. 4.1. It can be seen from this figure (and is obvious anyway) that the ICP's performance (measured by the cumulative observed excess) deteriorates sharply after update trials m_i. (There is a hike of approximately $(|\mathbf{Y}| - 1)/\epsilon = 180$ in the cumulative observed excess, where $\epsilon = 0.05$ is the significance level used.) Perhaps in practice there should be short spells of "learning" after each update trial, when the ICP is provided with fresh "training examples" and its predictions are not used or evaluated.

It is not clear how the way of computing nonconformity scores from SVM, as given in Sect. 3.2.2, could be used by ICP in a computationally efficient way. An easy solution, implemented in standard machine-learning packages, is to compute the SVM prediction rule based on (4.14) as training set and define $A(x, y)$ to be the distance (perhaps in a feature space) between x and the optimal separating hyperplane (taken with the minus sign if the SVM prediction for x is y and with the plus sign otherwise).

A disadvantage of the nonconformity measure (4.16) (to a large degree shared by (4.17)) is that the inductive conformal predictor determined by it outputs prediction intervals of the same length (similar length in the case of (4.17)). To obtain a more adaptive conformal predictor, we can use a nonconformity measure of the type

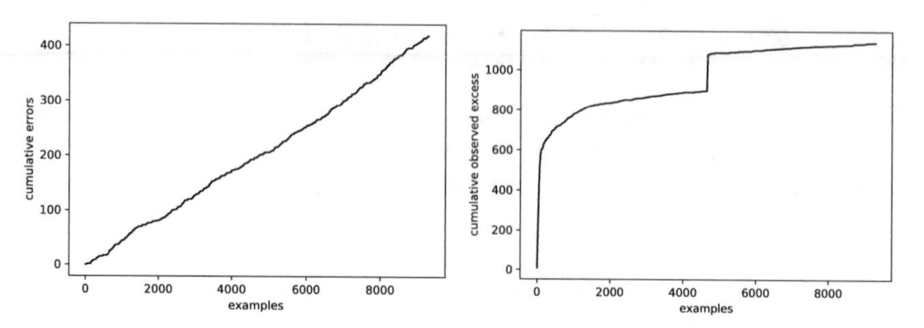

Fig. 4.1 Online performance of the 1-nearest neighbour ICP with the update trial 4649 (= 9298/2) on the USPS dataset for the significance level 5%. In accordance with Proposition 4.1, starting from scratch at trial 4670 does not affect (does not affect much in the deterministic case, as in this figure) the error rate (left panel); however, it affects the observed excess greatly (right panel)

$$A\left((z_1, \ldots, z_m), z\right) = \frac{\left|y - \hat{y}\right|}{\hat{\sigma}(x)}$$

(cf. (2.32)), where $\hat{\sigma} > 0$ is a function computed from (4.14) as training set and estimating the difficulty of the new object x.

4.3.2 De-Bayesing

In the rest of this section we will describe new ways of detecting nonconformity which are especially natural in the case of inductive conformal prediction. We start from a general scheme of using Bayesian prediction algorithms in conformal prediction.

Suppose we have a Bayesian model (compatible with the randomness assumption) for the process of generating the label y given the object x. If we fully trust the model, we can use it for computing, e.g., predictive densities and prediction sets in the form of highest probability density regions (see, e.g., [30, Sect. 5.1]). We are, however, interested in the case where the Bayesian model is plausible, but we do not really believe it. If it happens to be true, we would like our confidence predictor to be efficient. But we also want it to be always valid, even if the Bayesian model is wrong.

A natural definition of inductive conformity measure is as follows: find the posterior (after seeing the old examples $z_1, \ldots, z_m$ and the new object x) conditional distribution p for the label y of x, and define the conformity score for (x, y) as

$$B\left((z_1, \ldots, z_m), (x, y)\right) := p(\{y\}) \tag{4.18}$$

in the case of classification (**Y** is finite) and, e.g.,

$$B\left((z_1, \ldots, z_m), (x, y)\right) := \min(p((-\infty, y]), p([y, \infty)))$$

in the case of regression ($\mathbf{Y} = \mathbb{R}$). In both cases, $B(x, y)$ is small when the label is strange under the Bayesian model, so the corresponding ICP is likely to be predictively efficient. The conditional probability distribution for the next example z given $z_1, \ldots, z_m$ can be computed before seeing x, which may lead to a computationally efficient ICP. And of course, the ICP will be valid automatically.

The ICP determined by one of these conformity measures may be said to be the result of "de-Bayesing" of the original Bayesian algorithm. More generally, we can also say that the CRR algorithm of Sect. 2.3 is a de-Bayesed version of ridge regression (although we did not follow rigidly the scheme described in this subsection).

4.3.3 *Neural Networks and Other Multiclass Scoring Classifiers*

Let $|\mathbf{Y}| < \infty$ (the case of classification). When fed with an object $x \in \mathbf{X}$, a neural network (see, e.g., [151, Chap. 11]) trained on (4.14) as training set outputs a set of numbers o_y, $y \in \mathbf{Y}$, such that o_y reflects the likelihood that y is x's label. We can then use o_y directly as the conformity score $B(x, y)$.

For neural networks the outputs o_y can often be interpreted as probabilities; in particular, they are nonnegative numbers summing to 1. This is achieved by applying the softmax function [137, Sect. 6.2.2.3] to the outputs β_y of the previous layers:

$$o_y := \exp(\beta_y) / \sum_{y' \in \mathbf{Y}} \exp(\beta_{y'}) \, .$$

Conceivably, this step may lead to a loss of information, and alternatively we can use β_y as the conformity score $B(x, y)$.

Neural networks, with or without applying the softmax function at the last step, are examples of *scoring classifiers*, which output as their prediction a score, interpreted as their estimated likelihood, for each possible label $y \in \mathbf{Y}$ of the test object. To obtain a simple prediction (as in Sect. 2.1.2), we can take y with the largest score. Many of the classification algorithms implemented in standard machine-learning packages have an option of outputting a score for each potential label rather than a simple prediction. These scores can be used directly as conformity scores in inductive conformal prediction.

An important special case of scoring classifiers is provided by Bayesian classification algorithms discussed in the previous subsection: see (4.18). And a popular special case of the latter is provided by the naive Bayes algorithms, which often work surprisingly well without elaborate tuning.

4.3.4 Decision Trees and Random Forests

Further examples of scoring classifiers are provided by decision trees and their modifications. A decision tree (for a detailed description see, e.g., [151, Sect. 9.2]) is a way of splitting the objects into a finite number of regions (rectangles in the case of two attributes). The splitting is performed by testing the values of different attributes, but the details will not be important for us.

There are many methods of constructing a decision tree from a training set of examples, but again we do not need the precise details. We will assume that each region contains at least one object from the training set: if this is not the case, the decision tree can always be "pruned" to make sure this property holds.

After a decision tree is constructed from the training set, we can define a conformity score $B(x, y)$ of the new example (x, y) as the percentage of examples labelled as y among the training examples whose objects are in the same region as x.

The predictive performance of decision trees can be greatly improved using Breiman's [35] method of random forests. In this method we create a large number of decision trees (using ideas of bootstrap discussed in Sect. 4.3.7 below) based on the training set ((4.14) in this context). Now we can compute the conformity score $B(x, y)$ for each of those decision trees as described earlier, and then obtain the overall conformity score by averaging them.

4.3.5 Binary Scoring Classifiers

In the case of binary classification, where we will set $\mathbf{Y} := \{-1, 1\}$, it is common for binary classifiers implemented in standard machine-learning packages to have only one score $s(x)$ for a test object x. We will refer to $s(x)$ as the *prediction score*, and we can interpret $s(x)$ as reflecting the likelihood of 1 as the label of x. It is often the case that $-s(x)$ then reflects the likelihood of -1 as the label of x, and the simple prediction is obtained as $\hat{y} := \operatorname{sign}(s(x))$ (with an arbitrary $\hat{y}$ when $s(x) = 0$). We will call such classifiers *binary scoring classifiers*, and we will call s the *scoring function*. The conformity score is then defined by

$$B(x, y) := ys(x) . \tag{4.19}$$

An example of a binary scoring classifier given in Sect. 4.3.1 was the signed distance to the optimal separating hyperplane in a support vector machine. Another example is provided by the method of boosting.

Boosting is, as its name suggests, a method for improving the performance of a given prediction algorithm, usually called the *weak learner*.[1] Typical weak learners are decision trees; let us consider this case for concreteness.

The standard boosting procedure of AdaBoost [306, Sect. 1.2] produces a sequence of decision trees, and each tree is assigned a weight α_t. The result of the boosting procedure is the scoring function $s : \mathbf{X} \to \mathbb{R}$ defined by

$$s(x) := \frac{\sum_{t=1}^{T} \alpha_t h_t(x)}{\sum_{t=1}^{T} \alpha_t} ,$$

where T is the number of decision trees and $h_t(x) \in \{-1, 1\}$ is the simple prediction of the tth tree. As usual, the prediction for a new object x is computed as $\hat{y} := \operatorname{sign} s(x)$. We can use the prediction score $s(x)$ to define the conformity score (4.19) for an example (x, y). The product $ys(x)$ is known as the *margin*, and it plays the paramount role in the theoretical analysis of boosting [306, Sect. 1.4].

Another natural way to define the conformity score of an example (x, y) is

$$B(x, y) := \sum_{t=1}^{T} \alpha_t B_t(x, y) ,$$

where $B_t(x, y)$ is the conformity score of (x, y) computed from h_t, as described in Sect. 4.3.4.

Further examples of binary scoring classifiers are provided by linear and quadratic discriminant analysis.

4.3.6 Logistic Regression

Logistic regression is another example of a binary scoring classifier, but we discuss it separately since it will be used in Sect. 4.3.8 below. The model at the basis of logistic regression assumes that $\mathbf{X} = \mathbb{R}^p$, for some p, and, according to this model, the conditional probability that $y = 1$ given x for an example (x, y) is given by

$$\frac{e^{w \cdot x}}{1 + e^{w \cdot x}}$$

for some weight vector $w \in \mathbb{R}^p$. If $\hat{w}$ is, e.g., the maximum likelihood estimate found from (4.14) as training set, it is natural to use the conformity measure

[1] There is no connection between weak learners and our "weak teachers" considered in Sect. 3.3.

$$B(x, y) := \begin{cases} \hat{w} \cdot x & \text{if } y = 1 \\ -\hat{w} \cdot x & \text{otherwise} \end{cases} \tag{4.20}$$

(i.e., $B(x, y)$ is the estimated log odds ratio for the observed y given the observed x for the current example). This corresponds via (4.19) to the scoring function $s(x) := \hat{w} \cdot x$ (assuming $\mathbf{Y} = \{-1, 1\}$).

4.3.7 Regression and Bootstrap

In this subsection we return to regression. We discussed it at length earlier, and it is clear that any regression algorithm can be used for computing $\hat{y}$ in nonconformity measures such as (2.30) As the next step, we can find a prediction rule $\hat{\sigma}^2 > 0$ for predicting $(y - \hat{y})^2$ and then use (2.32) as nonconformity measure. For an easily interpretable model, we might use the lasso [152].

A very different approach is to define nonconformity scores using the method of bootstrap, which has some common goals with conformal prediction. Bootstrap has valuable properties of validity, but they are approximate or asymptotic. Applying conformal prediction on top of bootstrap adds finite-sample validity.

The basic idea of bootstrap is to use resampling (sampling from the sample, i.e., from the proper training set in this context, obtaining what is called *bootstrap samples*) to get an idea of the variability of the value of interest (for details, see [98]). For a given test example (x, y), the residual $r := y - \hat{y}$ or its modification (such as the studentized residual) is just one number. The method of bootstrap allows us to obtain a large number $r_1, \ldots, r_T$ of simulated residuals whose empirical distribution aims to approximate the true distribution of r. Now we can define a conformity measure B as

$$B(x, y) := \min\left(\left|\{t \in \{1, \ldots, T\} : r \leq r_t\}\right|, \left|\{t \in \{1, \ldots, T\} : r \geq r_t\}\right|\right),$$

where $r := y - \hat{y}$ is the residual for (x, y) with $\hat{y}$ being the prediction for x computed from (4.14) as training set.

4.3.8 Training Inductive Conformal Predictors

In applications of inductive conformal predictors, a standard prediction algorithm, e.g., one of the simple predictors described earlier in this section, is trained on the proper training set, i.e., its parameters are tuned. Let us discuss, for concreteness, logistic regression (Sect. 4.3.6). The label space is $\mathbf{Y} := \{-1, 1\}$ (most of Sect. 4.3.6 is written in such a way that the description in it is applicable to both $\mathbf{Y} = \{-1, 1\}$

and $\mathbf{Y} := \{0, 1\}$, the latter encoding being more common when discussing logistic regression).

The simplest way of tuning the parameter w would be to optimize the predictive performance of the simple predictor on the proper training set, as suggested in Sect. 4.3.6. However, this is an indirect approach, since it does not use the fact that the trained prediction rule will be used as component for an inductive conformal predictor. A more natural approach would be to train an inductive conformal predictor directly. In this subsection we will describe a possible scheme for training inductive conformal predictors.

The conformity measure (4.20) belongs to the family

$$B_w(x, y) := \begin{cases} w \cdot x & \text{if } y = 1 \\ -w \cdot x & \text{if } y = -1 \end{cases}$$

parametrized by $w \in \mathbb{R}^p$. A convenient feature of the conformity measures in this family is that the conformity score does not depend on the comparison bag. We will find an optimal w using the proper training set only (so that we can still use the calibration set for producing valid prediction sets). We will use each element $z_i = (x_i, y_i)$ of the proper training set in turn for evaluating the quality of the p-value computed from the rest of the proper training set using w as parameter value. The p-value for the false label $-y_i$ of x_i is, essentially,

$$p_{w,i} := \frac{1}{m} \sum_{j \in \{1,\dots,m\} \setminus \{i\}} \theta\left(B_w(x_i, -y_i) - B_w(z_j)\right)$$

(cf. (4.4)), where the step function θ is defined by

$$\theta(u) := \begin{cases} 1 & \text{if } u \geq 0 \\ 0 & \text{if } u < 0 \end{cases}, \qquad u \in \mathbb{R}\,.$$

One of the conditionally proper ways of evaluating the quality of conformal predictors that we recommended in Sect. 3.1.7 is to minimize the observed fuzziness

$$\mathrm{OF}_w := \sum_{i=1}^{m} p_{w,i} = \frac{1}{m} \sum_{i,j:i \neq j} \theta\left(B_w(x_i, -y_i) - B_w(z_j)\right) \tag{4.21}$$

in w.

The overall procedure for direct training of the inductive conformal predictor is to find the value $\hat{w}$ of w minimizing (4.21) on the proper training set, and the resulting inductive conformal predictor will use the nonconformity measure (4.20) (with $\hat{w}$ defined in this way) applying it to the calibration set and the test objects.

A serious computational disadvantage of this procedure is the requirement to minimize the expression (4.21) involving the awkward discontinuous function θ. In practice, we could use the familiar strategy of replacing θ by a smooth function, such as a sigmoid, that is only approximately equal to the step function. This makes OF_w a differentiable function of w and facilitates methods such as Gradient Descent.

4.4 Cross-Conformal Prediction

While inductive conformal predictors are computationally efficient and automatically valid under the randomness assumption, they may suffer a drop in predictive efficiency. The two potential sources of predictive inefficiency are using only part of the training set (namely, the proper training set) for forming the prediction rule and using another part of the training set (namely, the calibration set) for computing the p-values given the prediction rule. Full conformal predictors use the full training set for both purposes.

Cross-conformal predictors, the topic of this section, are one of the attempts (see Sect. 4.8.4 for other ones) to obtain both predictive and computational efficiency. Their disadvantage is that their provable property of validity is much weaker than that for full and inductive conformal predictors. However, computational experiments suggest that they are still valid in practice, unless we try hard to ruin the validity (e.g., by using excessive randomization). We will discuss cross-conformal prediction only in the offline mode of prediction.

4.4.1 *Definition of Cross-Conformal Predictors*

The step from inductive to cross-conformal predictors is analogous to the step from using a test set for estimating the quality of a prediction rule (Sect. 1.2.1) to the method of cross-validation [346]. The idea is to split the training set into several parts and use each in turn as calibration set.

Formally, *cross-conformal predictors* (CCPs) are defined as follows. Our training set is $z_1, \dots, z_l$, and we are also given a test object x (an element of the test set). For each set $S \subseteq \{1, \dots, l\}$ of the indices of the training set, let $z_S := \Lbag z_i : i \in S \Rbag$ be the bag consisting of the examples with those indices. The training set is randomly split into K nonempty subsets (*folds*, formally subbags) z_{S_k}, $k = 1, \dots, K$, where $K \in \{2, 3, \dots\}$ is a parameter of the algorithm and $(S_1, \dots, S_K)$ is a partition of $\{1, \dots, l\}$. For each $k \in \{1, \dots, K\}$ and each potential label $y \in \mathbf{Y}$ of the test object x find the nonconformity scores of the examples in z_{S_k} and of (x, y) by

$$\alpha_{i,k} := A(z_{S_{-k}}, z_i), \quad i \in S_k, \qquad \alpha_k^y := A(z_{S_{-k}}, (x, y)) , \tag{4.22}$$

where $S_{-k} := \cup_{j \neq k} S_j = \{1, \dots, l\} \setminus S_k$ and A is a given nonconformity measure (which we are using in the deleted mode). The corresponding p-values are defined by

$$p^y := \frac{\sum_{k=1}^K \left|\left\{i \in S_k : \alpha_{i,k} \geq \alpha_k^y\right\}\right| + 1}{l+1}. \tag{4.23}$$

The prediction set Γ^ϵ is defined as for conformal predictors, by

$$\Gamma^\epsilon(z_1, \ldots, z_l, x) := \{y : p^y > \epsilon\} \tag{4.24}$$

(cf. (2.22)), where the significance level $\epsilon > 0$ is another parameter.

The definition of CCPs parallels that of ICPs, except that now we use the whole training set for calibration. The nonconformity scores (4.22) are computed as for inductive conformal predictors but using the current fold as the calibration set and the union of all the folds except for the current one as the proper training set. Calibration (4.23) is done by combining the ranks of the test example (x, y) with a postulated label in all the folds. As in cross-validation, the folds should be chosen equal or approximately equal.

4.4.2 Computational Efficiency

Cross-conformal predictors inherit the computation efficiency of inductive conformal predictors, assuming the number K of folds is a small constant. The same nonconformity measures of the form (4.11) that we discussed in Sect. 4.3 in the context of inductive conformal prediction can be used in cross-conformal prediction. For a constant K, we can apply the estimate (4.13) to each fold to obtain the overall computation time

$$\Theta\left(T_{\text{train}}^\dagger + (l+k)T_{\text{appl}}^\dagger + (l+k)\log l\right)$$

for the cross-conformal predictor, where $T_{\text{train}}^\dagger$ is the time it takes to train a prediction rule on a training set of size $(1 - 1/K)l$ (assuming all folds have the same size).

4.4.3 Validity and Lack Thereof for Cross-Conformal Predictors

One way to think of cross-conformal predictors and their validity is as follows. Suppose that, in addition to our training set $z_1, \ldots, z_l$, we have a large supply of fresh data Z. Let us modify the definition of a cross-conformal predictor replacing all $z_{S_{-k}}$ by Z. Then the cross-conformal predictor becomes an inductive conformal predictor, with Z as the proper training set and $\{\!\{z_1, \ldots, z_l\}\!\}$ as the calibration set. Therefore, it becomes provably valid.

The change from using Z for training to using $z_{S_{-k}}$ does not appear substantial, and indeed numerous computational experiments have demonstrated the approximate empirical validity of cross-conformal prediction for standard datasets and underlying algorithms.

It is easy, however, to give examples where training using $z_{S_{-k}}$ is very different from training using Z, especially if the word "training" is used broadly. If we define the separate p-value

$$p_k^y := \frac{\left|\left\{i \in S_k : \alpha_{i,k} \geq \alpha_k^y\right\}\right| + 1}{|S_k| + 1}$$

for each fold, we can see that p^y is essentially an average of p_k^y. In particular, if each fold has the same size, $|S_1| = \cdots = |S_K|$, a simple calculation gives

$$p^y = \bar{p}^y + \frac{K-1}{l+1}\left(\bar{p}^y - 1\right) \approx \bar{p}^y \,, \tag{4.25}$$

where $\bar{p}^y := \frac{1}{K}\sum_{k=1}^K p_k^y$ is the arithmetic mean of p_k^y and the $\approx$ assumes $K \ll l$. If we use IID nonconformity scores with continuous distribution (so that there is no nontrivial training), we obtain independent p-values for different folds with the uniform distribution $\mathbf{U}$ on $[0, 1]$ (exactly in the smoothed case and approximately for our official definition (4.23)). In this case the average p-value will not be uniformly distributed, and for $K \gg 1$, the average p-value will be concentrated around 0.5 (by the law of large numbers).

The statement of validity for cross-conformal predictors that we can make is that the probability of error at significance level ϵ does not exceed approximately 2ϵ. This follows from the known result ([397, Table 1], the result going back to Rüschendorf [296]) that, to turn the arithmetic average of K p-values into a valid p-value, it has to be multiplied by 2, and the factor of 2 cannot be improved. Setting y in (4.25) to the true label of the test object, we obtain

$$p = \bar{p} + \frac{K-1}{l+1}(\bar{p} - 1) \,,$$

in the obvious notation. Expressing $\bar{p}$ via p and using $2\bar{p}$ being a p-value, we can obtain

$$\mathbb{P}\left(p \leq \epsilon\right) \leq 2\epsilon + 2\frac{K-1}{l+K}(1-\epsilon) \tag{4.26}$$

for any $\epsilon > 0$. The last equation, (4.26), gives the precise version of our statement of validity: the probability of error at significance level ϵ does not exceed the right-hand side of (4.26).

4.5 Transductive Conformal Predictors

As we discussed in Sect. 1.3.4, the word "transduction" as used in machine learning usually implies that predictions are to be made for several new objects, not one. So far we have considered only single-object transduction, with the test objects processed one by one. In this section we discuss predicting all test labels at once.

4.5.1 Definition

Transductive conformal predictors are determined by their transductive nonconformity measures, which generalize deleted nonconformity measures. A *transductive nonconformity measure* is a measurable function $A : \mathbf{Z}^{(*)} \times \mathbf{Z}^{*} \to \mathbb{R}$. (An important case is where $A(\zeta_1, \zeta_2)$ does not depend on the ordering of ζ_2, so that ζ_2 may also be taken to be a bag.) The intuition is that $A(\zeta_1, \zeta_2)$ (the *transductive nonconformity score*) measures the lack of conformity of the sequence ζ_2 to the "comparison bag" ζ_1.

Let $z_1 = (x_1, y_1), \dots, z_l = (x_l, y_l)$ be the training set and $x_{l+1}, \dots, x_{l+k}$ be the test objects. To compute the prediction set for the test objects $x_{l+1}, \dots, x_{l+k}$ at a significance level $\epsilon \in (0, 1)$, the *transductive conformal predictor* (*TCP*) determined by A proceeds according to these steps:

- For each possible sequence of labels $(\upsilon_1, \dots, \upsilon_k) \in \mathbf{Y}^k$:
  - set $y_j := \upsilon_{j-l}$ and $z_j := (x_j, y_j)$ for $j = l+1, \dots, l+k$;
  - compute the transductive nonconformity scores

    $$\alpha_S := A(z_{\{1,\dots,l+k\}\setminus S}, z_S) ,$$

    where S ranges over all $(l+k)!/l!$ ordered subsets $(s_1, \dots, s_k)$ of size k of the set $\{1, \dots, l+k\}$, z_S stands for the sequence $(z_{s_1}, \dots, z_{s_k})$ (when $S = (s_1, \dots, s_k)$), and $z_{\{1,\dots,l+k\}\setminus S}$ stands for z_B, B being any ordering of $\{1, \dots, l+k\} \setminus S'$ and S' being the set of all elements of S (it does not matter which ordering is chosen, by the definition of a transductive nonconformity measure);
  - compute the p-value

    $$p(\upsilon_1, \dots, \upsilon_k) := \frac{\left|\{S : \alpha_S \ge \alpha_{(l+1,\dots,l+k)}\}\right|}{(l+k)!/l!} , \tag{4.27}$$

    where S ranges, as before, over all $(l+k)!/l!$ ordered subsets of $\{1, \dots, l+k\}$ of size k.

- Output the prediction set

$$\Gamma^{\epsilon}(z_1, \ldots, z_l, x_{l+1}, \ldots, x_{l+k}) := \left\{(v_1, \ldots, v_k) \in \mathbf{Y}^k : p(v_1, \ldots, v_k) > \epsilon\right\}.$$

Smoothed TCPs are defined in the same way except that (4.27) is replaced by

$$p(v_1, \ldots, v_k) := \frac{\left|\{S : \alpha_S > \alpha_{(l+1,\ldots,l+k)}\}\right| + \tau\left|\{S : \alpha_S = \alpha_{(l+1,\ldots,l+k)}\}\right|}{(l+k)!/l!}, \tag{4.28}$$

where τ are random variables distributed uniformly on [0, 1] (no independence between different sequences of postulated labels $v_1, \ldots, v_k$ is required, but when we consider the online prediction protocol in the next subsection we will assume that τ are independent between different trials, as usual).

4.5.2 *Validity*

To state the property of validity of transductive conformal predictors in a strong form, we again consider the online mode. Suppose we are given a sequence of positive integer numbers $k_1, k_2, \ldots$ (the sizes of batches in which we will process the examples), and the incoming sequence of examples is $z_1 = (x_1, y_1)$, $z_2 = (x_2, y_2)$,.... Set $l_n := \sum_{i=1}^{n} k_i$ (including $l_0 := 0$). At trial $n = 1, 2, \ldots$ of the online prediction protocol, our task is to predict the k_n labels $y_{l_{n-1}+1}, \ldots, y_{l_n}$ given the l_{n-1} examples $z_1, \ldots, z_{l_{n-1}}$ and k_n objects $x_{l_{n-1}+1}, \ldots, x_{l_n}$. The prediction is a subset Γ_n of $\mathbf{Y}^{k_n}$. The predictor *makes an error* if $(y_{l_{n-1}+1}, \ldots, y_{l_n}) \notin \Gamma_n$. Suppose the examples $z_1, z_2, \ldots$ are produced from an exchangeable probability distribution on $\mathbf{Z}^{\infty}$ and the random numbers $\tau = \tau_n$ (see (4.28)) used at different trials n are independent of each other and the examples.

The following result stating the validity of transductive conformal predictors will be proved in Sect. 11.5.2.

Proposition 4.4 *In the online mode, a smoothed TCP makes errors with probability ϵ (the significance level) independently at different trials.*

For deterministic conformal predictors we replace the τ in (4.28) by 1. Proposition 4.4 implies that they satisfy a property of conservative validity analogous to the one introduced in Sect. 2.1.3.

4.6 Conditional Conformal Predictors

In this section we discuss several kinds of conditional conformal predictors, i.e., modifications of conformal predictors that satisfy conditional properties of validity.

We start from a one-off version and then move on to a version that satisfies stronger conditional properties of validity in the online setting.

As we already discussed in Sect. 1.4.4, conformal predictors do not guarantee validity within categories (such as classes): the fraction of errors can be much larger than the nominal significance level for some categories, if this is compensated by a smaller fraction of errors for other categories. The stronger kind of validity, validity within categories, is the main property of conditional conformal predictors, constructed in this section. As usual, we will demonstrate validity, in this stronger sense, under the exchangeability assumption; this assumption, however, will be relaxed in Chaps. 11 and 12.

4.6.1 One-Off Conditional Conformal Predictors

A *taxonomy* (or, more fully, *Venn taxonomy*) is a measurable function $K : \mathbf{Z}^{(*)} \times \mathbf{Z} \to \mathbf{K}$, where $\mathbf{K}$ is a countable set with the discrete σ-algebra. The set $\mathbf{K}$ is often even finite, and we refer to its elements as *categories*. Usually the *category* $\kappa_i := K(\lbag z_1, \ldots, z_n \rbag, z_i)$ of an example z_i is a kind of classification of z_i, which may depend on the other examples in the data sequence $z_1, \ldots, z_n$.

Fix the size l of the training set. The *Venn-conditional conformal predictor* determined by a nonconformity measure A and taxonomy K is defined by (4.24), where for each $(x, y) \in \mathbf{Z}$ the corresponding *Venn-conditional p-value* p^y is

$$p^y := \frac{\left|\left\{i = 1, \ldots, l+1 : \kappa_i^y = \kappa_{l+1}^y \ \& \ \alpha_i^y \geq \alpha_{l+1}^y\right\}\right|}{\left|\left\{i = 1, \ldots, l+1 : \kappa_i^y = \kappa_{l+1}^y\right\}\right|}, \tag{4.29}$$

the categories κ and nonconformity scores α being defined by

$$\kappa_i^y := K(\lbag z_1, \ldots, z_{l+1} \rbag, z_i), \qquad \alpha_i^y := A(\lbag z_1, \ldots, z_{l+1} \rbag, z_i),$$

and z_{l+1} standing for (x, y) (test example). To obtain the smoothed version, as usual, we treat the ties in (4.29) more carefully and replace it by

$$p^y := \frac{\left|\left\{i : \kappa_i^y = \kappa_{l+1}^y \ \& \ \alpha_i^y > \alpha_{l+1}^y\right\}\right|}{\left|\left\{i : \kappa_i^y = \kappa_{l+1}^y\right\}\right|} + \tau \frac{\left|\left\{i : \kappa_i^y = \kappa_{l+1}^y \ \& \ \alpha_i^y = \alpha_{l+1}^y\right\}\right|}{\left|\left\{i : \kappa_i^y = \kappa_{l+1}^y\right\}\right|},$$

where i ranges over the set $\{1, \ldots, l+1\}$ and τ is uniformly distributed on $[0, 1]$, independently of the examples.

Venn-conditional conformal predictors automatically satisfy the following property of conditional validity, as explained, in a more general context, in Chap. 12 (see Theorem 12.2).

Proposition 4.5 *If examples $z_1, \ldots, z_{l+1}$ are generated from an exchangeable probability distribution on $\mathbf{Z}^{l+1}$, a smoothed Venn-conditional conformal predictor* *outputs a p-value that is uniformly distributed on* $[0, 1]$*, even conditionally on the category of the test example. Therefore, for both smoothed and deterministic versions of the Venn-conditional conformal predictor, the conditional probability of error,* $y_{l+1} \notin \Gamma^{\epsilon}(z_1, \ldots, z_l, x_{l+1})$*, given that the category of* z_{l+1} *is* κ *will not exceed* ϵ *for any* ϵ*, any Venn-conditional conformal predictor* Γ*, and any category* $\kappa \in \mathbf{K}$.

The problem of extending Proposition 4.5 to the online setting might not have clean solutions similar to Proposition 2.4 in the case of conformal prediction. However, for important classes of Venn taxonomies such clean extensions do exist. One such class is where, in the problem of multiclass classification, the category of an example is its label:

$$K(\{z_1, \ldots, z_n\}, (x, y)) = y .$$

Such taxonomies give rise to *label-conditional conformal predictors*. Another class is where $K(\{z_1, \ldots, z_n\}, (x, y))$ is a function of the object x; these give rise to *object-conditional conformal predictors*. Even for the class of all taxonomies for which $K(\{z_1, \ldots, z_n\}, z)$ depends only on z, but does not depend on the comparison bag, a strong online property of validity is straightforward. In the next subsection we will extend this class further to what we will call Mondrian conformal predictors.

In conclusion of this subsection we will give an example of a Venn-conditional predictor that will not be covered by the notion of a Mondrian conformal predictor that we are about to introduce. Let the category $\kappa := K(\{z_1, \ldots, z_n\}, z)$ of z in a comparison bag $\{z_1, \ldots, z_n\}$ depend only on their objects x and $\{x_1, \ldots, x_n\}$, respectively. The taxonomy K clusters $x_1, \ldots, x_n$ using a standard clustering algorithm, and the category κ is the cluster assigned to x. This is a very natural definition, which gives rise to a Venn-conditional conformal predictor but not to a Mondrian conformal predictor.

4.6.2 Mondrian Conformal Predictors and Transducers

Mondrian taxonomies will go beyond Venn taxonomies in that the category of an example will be allowed to depend on, or even be determined by, the ordinal number of the example. We will apply Mondrian taxonomies to generalize the notion of a conformal transducer, which in turn will be applied to generalize conformal predictors. We start from transducers since validity within categories (or *conditional validity*, as we will say) is especially relevant in situations, such as *asymmetric classification*, where errors for different categories of examples have different consequences; in this case we cannot allow low error rates for some categories to compensate excessive error rates for other categories. Because of our interest in such situations, we will mainly use the language of conformal transducers in our exposition. The standard translation into the language of conformal predictors is straightforward (cf. Sect. 2.7), but in the case of asymmetric classification one might prefer to add flexibility to this translation: instead of comparing all p-values with the same threshold ϵ we might take different ϵs for different categories.

We are given a division of the Cartesian product $\mathbb{N} \times \mathbf{Z}$ into *categories*: a measurable function

$$\kappa : \mathbb{N} \times \mathbf{Z} \to \mathbf{K}$$

maps each pair (n, z) (z is an example and n will be, in our applications, the ordinal number of this example in the data sequence $z_1, z_2, \ldots$) to its category; $\mathbf{K}$ is the measurable space (at most countable with the discrete σ-algebra) of all categories. It is required that the elements $\kappa^{-1}(k)$ of each category $k \in \mathbf{K}$ form a rectangle $A \times B$, for some $A \subseteq \mathbb{N}$ and $B \subseteq \mathbf{Z}$. Such a function κ will be called a *Mondrian taxonomy*. Given a Mondrian taxonomy κ, we first define Mondrian nonconformity measures and then Mondrian conformal transducers (MCTs).

A *Mondrian nonconformity measure* based on κ is a measurable function A of the type

$$A : \mathbf{K}^* \times \left(\mathbf{Z}^{(*)}\right)^{\mathbf{K}} \times \mathbf{Z} \to \overline{\mathbb{R}} . \tag{4.30}$$

The *smoothed Mondrian conformal transducer (smoothed MCT)* determined by the Mondrian nonconformity measure A is the randomized confidence transducer producing the p-values

$$\begin{aligned} p_n &= f(x_1, \tau_1, y_1, \ldots, x_n, \tau_n, y_n) \\ &:= \frac{|\{i : \kappa_i = \kappa_n \;\&\; \alpha_i > \alpha_n\}| + \tau_n \, |\{i : \kappa_i = \kappa_n \;\&\; \alpha_i = \alpha_n\}|}{|\{i : \kappa_i = \kappa_n\}|} , \end{aligned} \tag{4.31}$$

where i ranges over $\{1, \ldots, n\}$, $\kappa_i := \kappa(i, z_i)$, $z_i := (x_i, y_i)$, and

$$\alpha_i := A\big(\kappa_1, \ldots, \kappa_n, \big(k \in \mathbf{K} \mapsto \Lbag z_j : j \in \{1, \ldots, n\} \;\&\; \kappa_j = k \Rbag\big), z_i\big) \tag{4.32}$$

for $i = 1, \ldots, n$ such that $\kappa_i = \kappa_n$. As usual, the definition of a *Mondrian conformal transducer (MCT)* is obtained by replacing (4.31) with

$$p_n = f(x_1, y_1, \ldots, x_n, y_n) := \frac{|\{i : \kappa_i = \kappa_n \;\&\; \alpha_i \geq \alpha_n\}|}{|\{i : \kappa_i = \kappa_n\}|} .$$

In general, a *(smoothed) MCT* based on a Mondrian taxonomy κ is the (smoothed) MCT determined by some Mondrian nonconformity measure based on κ.

Of course, instead of a Mondrian nonconformity measure we could have used a regular nonconformity measure A, replacing (4.32) by

$$\alpha_i := A\big(\Lbag z_j : j \in \{1, \ldots, n\} \Rbag, z_i\big) .$$

However, using Mondrian nonconformity measures adds flexibility.

We say that a randomized confidence transducer f is *category-wise exact w.r. to a Mondrian taxonomy* κ if, for all n, the conditional probability distribution of p_n given $\kappa(1, z_1), p_1, \ldots, \kappa(n-1, z_{n-1}), p_{n-1}, \kappa(n, z_n)$ is uniform on $[0, 1]$, where $z_1, z_2, \ldots$ are examples generated from an exchangeable distribution on $\mathbf{Z}^\infty$ and $p_1, p_2, \ldots$ are the p-values output by f (as usual, the random numbers τ are assumed to be independent).

Proposition 4.6 *Any smoothed MCT based on a Mondrian taxonomy* κ *is category-wise exact w.r. to* κ.

This proposition generalizes Proposition 4.1 but is a special case of Theorem 11.2 (the finitary version of Theorem 11.1). It implies the category-wise property of conservative validity for MCT, whose p-values are always bounded above by the p-values from the corresponding smoothed MCT.

4.6.3 Using Mondrian Conformal Transducers for Prediction

An example of asymmetric classification is distinguishing between useful messages and spam in the problem of e-mail filtering: classifying a useful message as spam is a more serious error than vice versa. In this case we might want to have different significance levels ϵ_k for different categories k.

Let f be a (smoothed) MCT. Given a finite set of significance levels ϵ_k, $k \in \mathbf{K}$ (assuming a finite number of categories, $|\mathbf{K}| < \infty$), we can define the prediction set for the label y_n of a new object x_n given old examples $z_1, \ldots, z_{n-1}$ as

$$\Gamma^{(\epsilon_k : k \in \mathbf{K})}(z_1, \ldots, z_{n-1}, x_n) := \left\{ y \in \mathbf{Y} : f(z_1, \ldots, z_{n-1}, (x_n, y)) > \epsilon_{\kappa(n,(x_n,y))} \right\} .$$

Proposition 4.6 now implies that the long-run frequency of errors made by this predictor (*Mondrian conformal predictor*, or *MCP*) on examples of category k does not exceed (approaches, in the case of smoothed transducer) ϵ_k, for each k.

As in the case of conformal prediction, in applications it is usually not wise to fix thresholds ϵ_k, $k \in \mathbf{K}$, in advance. One possibility would be to suitably choose three sets of significance levels (ϵ_k^1), (ϵ_k^2), and (ϵ_k^3) such that $\epsilon_k^1 \le \epsilon_k^2 \le \epsilon_k^3$ for all $k \in \mathbf{K}$, and say that $\Gamma^{\epsilon^1}(z_1, \ldots, z_{n-1}, x_n)$ is a highly confident prediction, $\Gamma^{\epsilon^2}(z_1, \ldots, z_{n-1}, x_n)$ is a confident prediction, and $\Gamma^{\epsilon^3}(z_1, \ldots, z_{n-1}, x_n)$ is a casual prediction.

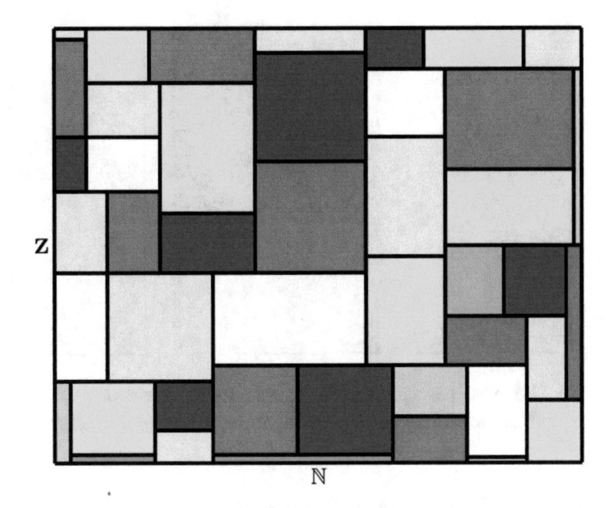

Fig. 4.2 A random Mondrian taxonomy (after Piet Mondrian, 1918)

4.6.4 Generality of Mondrian Taxonomies

We will next consider several classes of MCTs, involving different taxonomies. In this subsection we consider a natural partial order on the taxonomies, which will clarify the relation between different special cases. (We will use the expression "more general than" for this partial order; it might seem strange here but will be explained in Sect. 11.3.) There are many ways to split the rectangle $\mathbb{N} \times \mathbf{Z}$ into smaller rectangles (cf. Fig. 4.2), and it is clearly desirable to impose some order.

We say that a Mondrian taxonomy κ_1 is *more general* than another Mondrian taxonomy κ_2 if, for all pairs (n', z') and (n'', z''),

$$\kappa_1(n', z') = \kappa_1(n'', z'') \Longrightarrow \kappa_2(n', z') = \kappa_2(n'', z'') .$$

We will say that κ_1 and κ_2 are *equivalent* if each of them is more general than the other, and we will sometimes identify equivalent Mondrian taxonomies. Identifying equivalent Mondrian taxonomies means that we are only interested in the equivalence relation a given Mondrian taxonomy κ induces ((n', z') and (n'', z'') are *κ-equivalent* if $\kappa(n', z') = \kappa(n'', z'')$) and not in the chosen labels $\kappa(n, z)$ for the equivalence classes.

Since we are only interested in taxonomies with a countable number of categories, the following proposition immediately follows from the standard properties of conditional expectations (see property 2 in Sect. A.3).

Proposition 4.7 *Let a taxonomy κ_1 be more general than a taxonomy κ_2. If a randomized confidence transducer is category-wise exact w.r.t. to κ_1, it is category-wise exact w.r.t. to κ_2.*

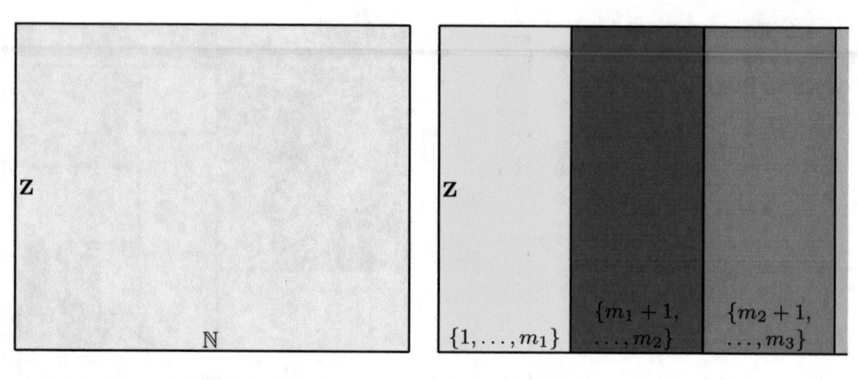

Fig. 4.3 Left panel: Mondrian taxonomy corresponding to conformal transducers. Right panel: Mondrian taxonomy corresponding to inductive conformal transducers

4.6.5 Conformal Prediction

Conformal transducers are MCTs based on the least general (i.e., constant, see the left panel of Fig. 4.3) Mondrian taxonomy. Proposition 2.4, asserting that smoothed conformal predictors are exact, is a special case of Proposition 4.6.

In the rest of this section we will describe several experimental results for the USPS dataset (randomly permuted), using the 1-nearest neighbour ratio (3.31) as the nonconformity measure. We start from results demonstrating the lack of conditional validity for conformal predictors. The USPS dataset is reasonably balanced in the proportion of examples labelled by different digits; for less well-balanced datasets the lack conditional validity of non-Mondrian conformal predictors is often even more pronounced.

Figure 4.4 plots the number of errors Err_n, the expected number of errors ϵn, and the cumulative observed excess OE_n against n for the confidence level 95%; the graphs are almost indistinguishable from the analogous graphs for the deterministic conformal predictor, as shown in Fig. 3.1. It demonstrates that the smoothed conformal predictor is valid "on average" on the USPS dataset (although for our default seed for the random number generator the deviation of Err_n from ϵn is larger than usual).

Figure 4.5 gives similar graphs, but only taking into account the predictions made for the examples labelled "5". It shows that the smoothed conformal predictor is not valid at the 95% confidence level on those examples, giving 9.2% of errors. Since the error rate of 5% is achieved on average, the error rate for some digits is better than 5%; for example, it is below 1% for the examples labelled "0".

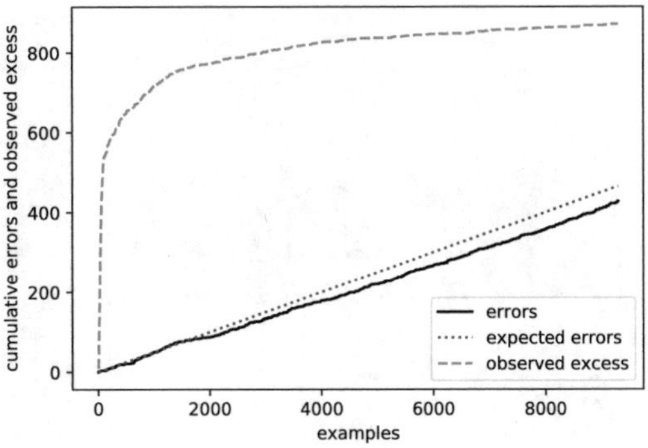

Fig. 4.4 The performance of the smoothed conformal predictor on the USPS dataset at the 95% confidence level

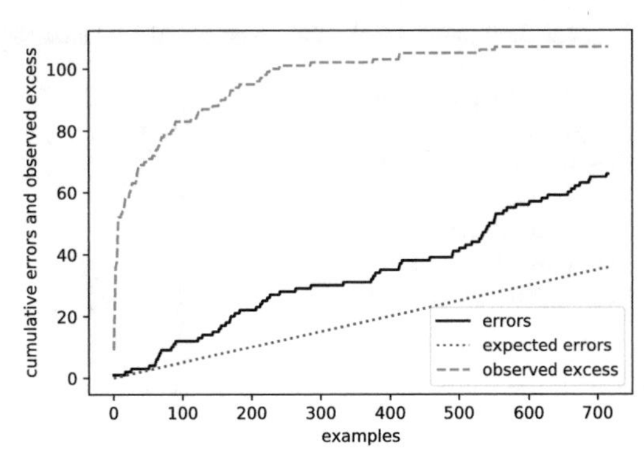

Fig. 4.5 The performance of the smoothed conformal predictor on the USPS dataset for the examples labelled "5" at the 95% confidence level

4.6.6 Inductive Conformal Prediction

Inductive conformal transducers, which output the p-values

$$\frac{\left|\{j : \alpha_j \geq \alpha_n\}\right|}{n - m_k}$$

(deterministic case) or

$$\frac{\left|\{j : \alpha_j > \alpha_n\}\right| + \tau_n \left|\{j : \alpha_j = \alpha_n\}\right|}{n - m_k}$$

(smoothed case), where

$$\alpha_j := A^{\text{del}}\left(\wr(x_1, y_1), \ldots, (x_{m_k}, y_{m_k})\int, (x_j, y_j)\right), \quad j = m_k + 1, \ldots, n$$

(cf. (4.2) and (4.3)), are also a special case of MCTs. The corresponding taxonomy is shown in the right panel of Fig. 4.3. The result of Sect. 4.2.1 that ICPs are valid is a special case of Proposition 4.6. Similarly to conformal predictors, ICPs sometimes violate the property of class-wise validity.

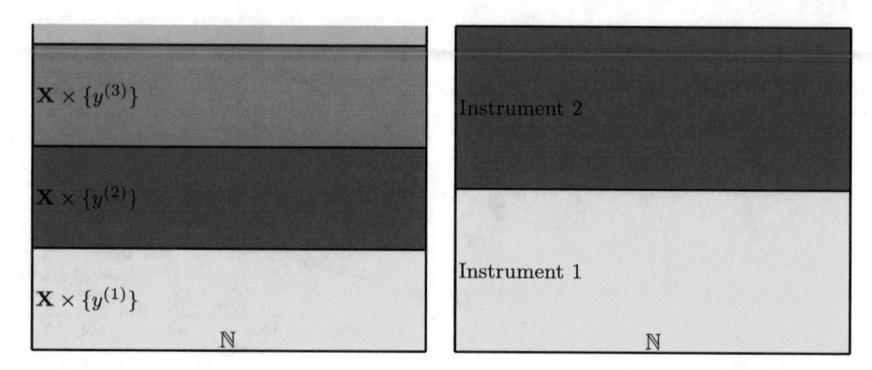

Fig. 4.6 Left panel: label-conditional Mondrian taxonomy. Right panel: Cox's example

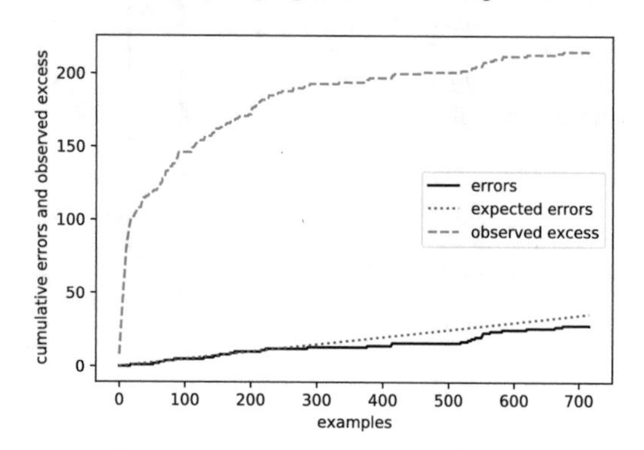

Fig. 4.7 The performance of the label-conditional MCP (based on the taxonomy $\kappa(n, (x, y)) = y$) on the USPS dataset for the examples labelled as "5" at the 95% confidence level

4.6.7 Label-Conditional Conformal Prediction

An important special case is where the category of an example is its label; we called this the label-conditional conformal predictor in Sect. 4.6.1. The corresponding taxonomy is shown in the left panel of Fig. 4.6, where it is assumed that $\mathbf{Y} = \{y^{(1)}, \ldots, y^{(L)}\}$.

Figure 4.7 demonstrates empirically the category-wise (in this case class-wise) validity of MCPs. In contrast to Fig. 4.5, the label-conditional MCP gives approximately 5% of errors when the significance level is set to 5% for the label "5". Figures 4.5 and 4.7 show that the correction in the number of errors results in an increased cumulative observed excess.

4.6.8 Object-Conditional Conformal Prediction

Another situation in which we might want to use a Mondrian, or Venn-conditional, conformal predictor is where the quality of prediction depends very much on a known attribute of the object whose label we are predicting. For instance, if our conformal predictor is valid at the significance level 5% but makes an error with probability 10% for men and 0% for women, both men and women can be unhappy with calling 5% the probability of error.

The statistical conditionality principle [64, Sect. 2.3] is often illustrated using an example (summarized in Sect. 1.4.4) due to Cox [62] in which we have two instruments for measuring an unknown quantity, with one instrument being much more precise. In this case, when predicting the quantity given the measurement, we might want to use the instrument that was used (say, "Instrument 1" or "Instrument 2") as the category of the object. This taxonomy is shown in the right panel of Fig. 4.6. When used with such taxonomies, Mondrian and Venn-conditional conformal predictors become what we call *object-conditional conformal predictors*.

4.7 Training-Conditional Validity

In this section we concentrate on the offline mode of prediction and its associated aspect of conditionality. The validity guarantees in this case (see Proposition 4.8 in Sect. 4.7.2) will be of the inductive type, as discussed in Sect. 1.3.4 (see Fig. 1.4), and so will involve two parameters, ϵ and δ. Inductive validity guarantees are particularly apposite in the context of inductive conformal predictors, and this is indeed the context in which such guarantees will be derived in Sect. 4.7.2, but of course they could also be developed for certain categories of full conformal predictors (and even transductive conformal predictors of Sect. 4.5).

4.7.1 Conditional Validity

First we place various discussions of Sect. 4.6 in a wider context of conditional inference. The idea of conditional inference is that we want our conclusions to be conditional as much as possible on the available information. Full conditionality is not attainable, even if we know the true probability distribution generating the data (e.g., are willing to use a subjective or postulated probability distribution, as in Bayesian theory). The requirement of conditional validity is weaker: for "important" events E we would like to know or control the conditional probability of error given E, e.g., we do not want it to be very different from the given significance level ϵ. Numerous examples were given in Sect. 4.6. It is clear that whenever the size of the training set is sufficient for making conditional claims, this is what we should aim for.

Of course, we cannot achieve conditional validity for all events E, and so we need to decide which events are "important". If we are interested in very few events E whose probabilities are not small, we can achieve conservative conditional validity by decreasing the significance level: the conditional probability of error given E will not exceed $\epsilon/\mathbb{P}(E)$, where $\mathbb{P}(E)$ is the probability of E. (But of course, the true

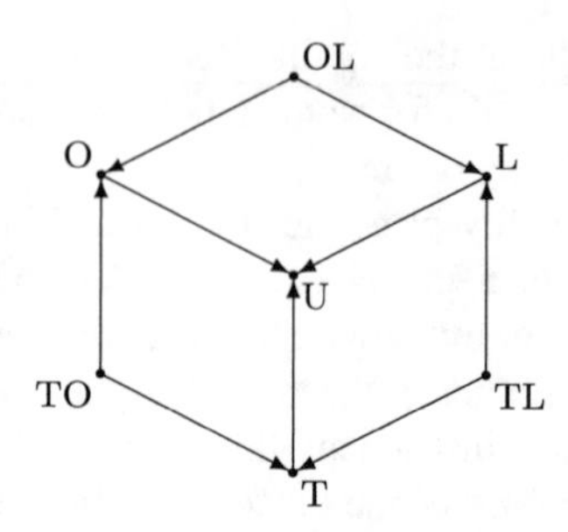

Fig. 4.8 Eight notions of conditional validity. The visible vertices of the cube are U (unconditional), T (training-conditional), O (object-conditional), L (label-conditional), OL (example-conditional), TL (training- and label-conditional), TO (training- and object-conditional). The invisible vertex is TOL (and corresponds to conditioning on everything)

conditional probability of error can be much smaller than this upper bound; besides, we might not know the probability of E, even approximately.) The requirement of conditional validity goes more and more beyond unconditional validity as we become interested in more events E and as the probability of some of them becomes smaller.

It will be convenient to give a crude classification of the conditioning events E that we might be interested in. Figure 4.8 is one possibility. At its centre we have the unconditional notion of validity, which is achieved by conformal predictors automatically. The vertex O corresponds to *object-conditional validity*, where E is a property of the test object; it was discussed in Sect. 4.6.8. The vertex L corresponds to *label-conditional validity*, where E is a property of the label of the test object; we discussed it in Sect. 4.6.7. The vertex T corresponds to *training-conditional validity*, where E is a property of the training set; this will be the topic of this section. We also have various combinations: e.g., OL (*example-conditional validity*) allows E that are properties of both the test object and its label (i.e., of the test example).

Each vertex of the cube in Fig. 4.8 can stand for different notions. For example, object-conditional validity can mean:

- Precise object-conditional validity, where we allow E to be any property of the test example. In Sect. 4.8.6 we will see that precise object-conditional validity is impossible to achieve in a useful way unless the test object is an atom of the data-generating distribution.
- Partial object-conditional validity, where E can be any property of the test example in a given set of properties deemed important. We usually omit "partial". Partial example-conditional validity (including object-conditional validity) was considered in detail in the previous section.
- Two-parameter (as in Fig. 1.4) object-conditional validity: the conditional probability of error given the test object does not exceed a prespecified level with a high probability. We do not discuss this type of object-conditional validity in this book (but we do discuss two-parameter training-conditional validity in Sect. 4.7.2).

- Asymptotic object-conditional validity, where E can be any property of the test example but the conditional probability of error is only required to converge to the significance level as the sample size goes to infinity. Asymptotic object-conditional validity can be achieved by universal conformal predictors, as discussed later, in Sect. 7.6.

Precise label-conditional validity is achievable in the case of classification, as discussed in the previous section. Our topic in the rest of this section is training-conditional validity; and for simplicity we will discuss it in the context of inductive conformal prediction.

4.7.2 Training-Conditional Validity of Inductive Conformal Predictors

Let $z_1, \dots, z_l$ be a random training set and $Z = (X, Y)$ be a random test example. Consider a set predictor $\Gamma : \mathbf{Z}^* \times \mathbf{X} \to 2^{\mathbf{Y}}$. It will be convenient to couch our discussion in terms of the function

$$\Gamma(z_1, \dots, z_l) := \{(x, y) \in \mathbf{Z} : y \in \Gamma(z_1, \dots, z_l, x)\} .$$

It is clear that our usual one-parameter definitions of validity (involving only the significance level ϵ) will not work for precise training-conditional validity: the coverage probability (i.e., the probability of the event $Z \in \Gamma(z_1, \dots, z_l)$) conditional on the training set $z_1, \dots, z_l$ of a set predictor Γ under the randomness assumption is $Q(\Gamma(z_1, \dots, z_l))$, where Q is the data generating distribution on $\mathbf{Z}$, and the conditional probability of error $Q(\mathbf{Z} \setminus \Gamma(z_1, \dots, z_l))$ does not have a useful upper bound:

$$\sup_Q Q(\mathbf{Z} \setminus \Gamma(z_1, \dots, z_l)) = 1$$

unless $\Gamma(z_1, \dots, z_l)$ covers all of $\mathbf{Z}$. However, some of the data-generating distributions Q are very unlikely once we know the training set, and the following two-parameter definition captures this intuition. A set predictor Γ is (ϵ, δ)-*valid*, for given $\epsilon, \delta \in (0, 1)$, if, for any probability distribution Q on $\mathbf{Z}$,

$$Q^l \{(z_1, \dots, z_l) : Q(\Gamma(z_1, \dots, z_l)) \ge 1 - \epsilon\} \ge 1 - \delta . \tag{4.33}$$

In words, (ϵ, δ)-validity means that with probability at least $1 - \delta$ the coverage probability of the prediction set will be at least $1 - \epsilon$. The following proposition shows that offline inductive conformal predictors (with the sizes of training and proper training sets l and m, respectively) satisfy this property for some ϵ and δ (see Corollaries 4.9 and 4.10 below for weaker but more interpretable statements).

Proposition 4.8 *Let $\epsilon, \delta, E \in (0, 1)$. If Γ is an inductive conformal predictor, the set predictor Γ^ϵ is (E, δ)-valid provided*

$$\mathrm{bin}_{h,E}(\lfloor \epsilon(h+1) \rfloor - 1) \leq \delta \,, \tag{4.34}$$

where $h := l - m$ is the size of the calibration set and $\mathrm{bin}_{h,E}$ *is the cumulative binomial distribution function with h trials and probability of success E (except that* $\mathrm{bin}_{h,E}(-1) := 0$*).*

Proof *(Sketch)* Suppose, without loss of generality, that $k := \epsilon(h+1)$ is an integer. Fix a proper training set $z_1, \ldots, z_m$. In this proof sketch we will only check that (4.33) holds for a given data-generating distribution Q if Γ is determined by an inductive nonconformity measure A such that $A((z_1, \ldots, z_m), Z)$ has a continuous distribution when Z is distributed as Q. Moreover, we will see that in this case the condition (4.34) is necessary and sufficient.

Define α^* as the largest value such that $A((z_1, \ldots, z_m), Z) \geq \alpha^*$ with Q-probability E. The probability of error $Q(\mathbf{Z} \setminus \Gamma(z_1, \ldots, z_l))$ exceeds E if and only if at most $k-1$ of the α_i, $i = m+1, \ldots, l$, are in the interval $[\alpha^*, \infty)$. The probability of the last event is $\mathrm{bin}_{h,E}(k-1)$. □

For a full self-contained proof of Proposition 4.8, see [385, proof of Theorem 1]; and for further references, see Sect. 13.3.3. Proposition 4.8 can also be deduced from classical results about tolerance regions: see [15, end of Sect. 2.5].

In combination with Hoeffding's inequality (see Sect. A.6.3), Proposition 4.8 gives:

Corollary 4.9 *Let $\epsilon, \delta \in (0, 1)$. If Γ is an inductive conformal predictor, the set predictor Γ^ϵ is (E, δ)-valid, where*

$$E := \epsilon + \sqrt{\frac{\ln \frac{1}{\delta}}{2h}} \tag{4.35}$$

and $h := l - m$ is the size of the calibration set.

Proof Combining (4.34) with Hoeffding's inequality, we can see that the probability of error $Q(\mathbf{Z} \setminus \Gamma^\epsilon(z_1, \ldots, z_l))$ for an ICP will exceed E with probability at most

$$\mathbb{P}(B \leq \lfloor \epsilon(h+1) - 1 \rfloor) \leq \mathbb{P}(B \leq \epsilon h) \leq \mathrm{e}^{-2(E-\epsilon)^2 h} \,,$$

where $B \sim \mathrm{bin}_{h,E}$. Solving $\mathrm{e}^{-2(E-\epsilon)^2 h} = \delta$ in E we obtain that Γ^ϵ is (E, δ)-valid for (4.35). □

The corollary allows us to construct (ϵ, δ)-valid set predictors when the training set is sufficiently large: for any inductive conformal predictor Γ with the size h of the calibration set, the set predictor

$$\Gamma^{\epsilon-\sqrt{\frac{\ln\frac{1}{\delta}}{2h}}} \tag{4.36}$$

will be (ϵ, δ)-valid. The expression (4.36) assumes that the difference in (4.36) is nonnegative, i.e.,

$$h \geq \frac{1}{2\epsilon^2} \ln \frac{1}{\delta} .$$

For $\epsilon = \delta = 0.05$, we need $h \geq 600$, and for $\epsilon = \delta = 0.01$, we need $h \geq 23{,}026$. The dependence of this lower bound on ϵ is heavy.

Applying a stronger inequality, 2. in [211, p. 278], instead of Hoeffding's inequality to Proposition 4.8 we obtain:

Corollary 4.10 *Let* $\epsilon, \delta \in (0, 1)$. *If* Γ *is an inductive conformal predictor, the set predictor* Γ^ϵ *is* (E, δ)*-valid, where*

$$E := \epsilon + \sqrt{\frac{2\epsilon \ln\frac{1}{\delta}}{h} + \frac{2\ln\frac{1}{\delta}}{h}} .$$

Figure 4.9 gives an idea of the relative accuracy of the bounds in Proposition 4.8 and Corollaries 4.9 and 4.10; remember that (as we saw in the proof) the bound in Proposition 4.8 is optimal.

Proposition 4.8 and Corollaries 4.9 and 4.10 are particularly relevant in the offline mode of prediction (although they are also applicable in the one-off mode): if a set

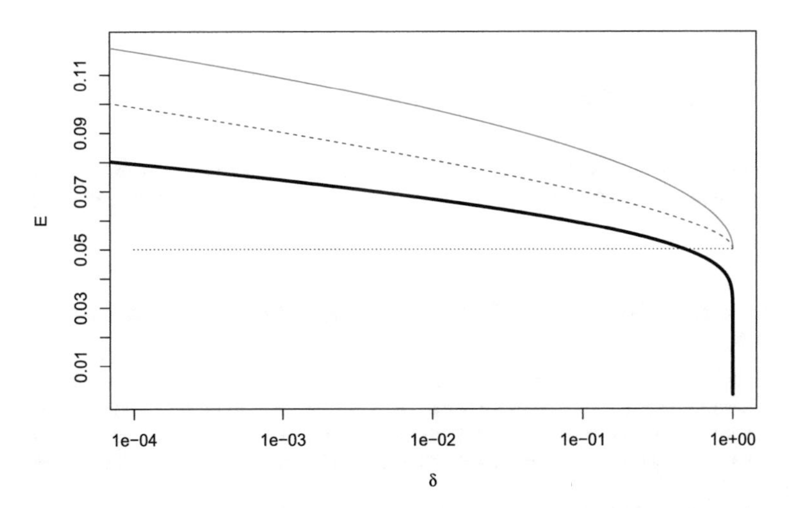

Fig. 4.9 The probability of error E vs δ from Proposition 4.8 (the thick solid black line), Corollary 4.9 (the thin solid red line), and Corollary 4.10 (the thin dashed blue line), where $\epsilon = 0.05$ and $h = 999$

predictor is (ϵ, δ)-valid, the percentage of errors on the test set will be bounded above by ϵ up to statistical fluctuations (whose typical size is the square root of the number of test examples) unless we are unlucky with the training set (which can happen with probability at most δ).

4.8 Context

4.8.1 *Computationally Efficient Hedged Prediction*

To cope with the relative computational inefficiency of conformal predictors, inductive conformal predictors were introduced in [269] and [270] in the offline setting and in [383] in the online setting. They were rediscovered by Lei and Wasserman and called "split conformal predictors". The representation (4.6) is used in, e.g., [226, Algorithm 1 and Lemma 1].

4.8.2 *Specific Learning Algorithms and Nonconformity Measures*

The de-Bayesing procedure has also been called "frequentizing" [423, Sect. 3], "conformal Bayes" [115], and "Bayes-optimal prediction with frequentist coverage control" [170]. In this book we use *conformalizing* as the generic term.

In this edition we shortened and streamlined the descriptions of many ways of producing inductive nonconformity scores, since they are so standard. The number of known machine learning algorithms is huge, and potentially any of them can be used as a source of nonconformity measures.

4.8.3 *Training Conformal Predictors*

Our discussion of training conformal predictors in Sect. 4.3.8 is a simplified version of the approach suggested in [53]. Other papers on this topic include Bellotti [25, 26] and Stutz et al. [350].

4.8.4 *Cross-Conformal Predictors and Alternative Approaches*

The simple idea of cross-conformal prediction was first described in [386]. That paper pointed out that cross-conformal predictors may violate the usual property of validity for conformal prediction and gave an example of such violation [386, Appendix]. The much simpler and more intuitive argument used in Sect. 4.4.3 (based on excessive randomization) was first given by Linusson et al. [233].

A more general procedure than cross-conformal prediction was proposed in [42] under the name of "aggregated conformal predictor". Another important advance in this direction was the introduction of the jackknife+ procedure by Barber et al. [19]. It is interesting that Barber et al.

also obtain the same factor of 2 (see their Theorems 1 and 2 and our (4.26)) needed to ensure the validity of their procedure.

For experiments demonstrating the approximate empirical validity of cross-conformal predictors see [386]. Linusson et al. [233] give both examples of approximate empirical validity and (in the presence of excessive randomization) examples where empirical validity is violated.

4.8.5 *Transductive Conformal Predictors*

In this book we use "transduction" in a wide sense including single-object transduction, according to Vapnik's philosophical description of this term and Johnson's description of eduction. As we mentioned in Chap. 3, before the first edition of this book conformal predictors had often been called "transductive confidence machines". This expression may have been misleading in that "transductive" in it refers to single-object transduction, whereas the most standard mode of using transduction in machine learning is in situations of a sizeable test set, as in Sect. 13.3.5. Transductive conformal predictors are more general; they were introduced in [388].

A key practical problem is to define efficient transductive nonconformity measures. Transductive nonconformity measures are more general than nonconformity measures: an equivalent definition of a deleted nonconformity measure is as the restriction of a transductive nonconformity measure to the domain $\mathbf{Z}^* \times \mathbf{Z}$ (we identify a 1-element sequence with its only element). On the other hand, there are many ways of turning (aggregating) nonconformity measures into transductive nonconformity measures: see [388].

4.8.6 *Conditional Conformal Predictors*

In this book we use Venn taxonomies both for Venn predictors in Chap. 6 and for one-off conditional conformal predictors in Sect. 4.6.1 of this chapter. The more general notion introduced in [394, Sect. 2] and later used in [393, Definitions 8 and 9] satisfies the same properties of validity as our Venn taxonomies, but it is slightly more complicated.

In Sect. 4.6.8 we discussed simple ways of achieving object-conditional validity for rough taxonomies of objects. Lei and Wasserman proved [225, Lemma 1] an interesting negative result which says that the requirement of precise object-conditional validity cannot be satisfied in a nontrivial way for rich object spaces (such as $\mathbb{R}$). As defined earlier, if Q is a probability distribution on $\mathbf{Z}$, $Q_{\mathbf{X}}$ is its marginal distribution on $\mathbf{X}$: $Q_{\mathbf{X}}(A) := Q(A \times \mathbf{Y})$. To cover randomized predictors, we have, as usual, a random number $\tau \sim \mathbf{U}$ modelling a random number generator.

Let us say that a set predictor Γ *has* $1-\epsilon$ *object-conditional validity*, where $\epsilon \in (0,1)$, if, for all probability distributions Q on $\mathbf{Z}$ and $Q_{\mathbf{X}}$-almost all $x \in \mathbf{X}$,

$$\mathbb{P}(Y \in \Gamma(Z_1, \dots, Z_l, X, \tau) \mid X = x) \geq 1 - \epsilon ,$$

where $Z_1, \dots, Z_l, (X, Y) \sim Q$ and $\tau \sim \mathbf{U}$ are all independent. If P is a probability distribution on $\mathbf{X}$, we say that a property F of elements of $\mathbf{X}$ holds for P-*almost all* elements of a measurable set $E \subseteq \mathbf{X}$ if $P(E \setminus F) = 0$; a P-*non-atom* is an element $x \in \mathbf{X}$ such that $P\{x\} = 0$. The Lebesgue measure on $\mathbb{R}$ will be denoted Λ, and the convex hull of $E \subseteq \mathbb{R}$ will be denoted co$E$.

We state Lei and Wasserman's result [225, Lemma 1] in the form given in [385, Theorem 2] (where a proof can be found).

Theorem 4.11 *Suppose* **X** *is a separable metric space equipped with the Borel* σ*-algebra. Let* $\epsilon \in (0, 1)$. *Suppose that a set predictor* Γ *has* $1 - \epsilon$ *object-conditional validity. In the case of regression, we have, for all probability distributions* Q *on* **Z** *and for* $Q_{\mathbf{X}}$*-almost all* $Q_{\mathbf{X}}$*-non-atoms* $x \in \mathbf{X}$,

$$\mathbb{P}(\Lambda(\Gamma(Z_1, \ldots, Z_l, x, \tau)) = \infty) \geq 1 - \epsilon \tag{4.37}$$

and

$$\mathbb{P}(\mathrm{co}\Gamma(Z_1, \ldots, Z_l, x, \tau) = \mathbb{R}) \geq 1 - 2\epsilon , \tag{4.38}$$

where $Z_1, \ldots, Z_l, (X, Y) \sim Q$ *and* $\tau \sim \mathbf{U}$ *are all independent. In the case of classification, we have, instead of* (4.37) *and* (4.38),

$$\mathbb{P}(y \in \Gamma(Z_1, \ldots, Z_l, x, \tau)) \geq 1 - \epsilon \tag{4.39}$$

for all Q, *all* $y \in \mathbf{Y}$, *and* $Q_{\mathbf{X}}$*-almost all* $Q_{\mathbf{X}}$*-non-atoms* x. *The constant* ϵ *in each of* (4.37), (4.38), *and* (4.39) *is optimal, in the sense that it cannot be replaced by a smaller constant.*

We are mainly interested in the case of a small ϵ (corresponding to high confidence), and in this case (4.37) implies that, in the case of regression, the prediction interval (i.e., the convex hull of the prediction set) can be expected to be infinitely long unless the test object is an atom. Even an infinitely long prediction interval can be somewhat informative providing a one-sided bound on the label of the test example; (4.38) says that, with probability at least $1 - 2\epsilon$, the prediction interval is completely uninformative unless the test object is an atom. In the case of classification, (4.39) says that each particular $y \in \mathbf{Y}$ is likely to be included in the prediction set, and so the prediction set is likely to be large. In particular, (4.39) implies that the expected size of the prediction set is a least $(1 - \epsilon)\,|\mathbf{Y}|$.

Of course, the condition that the test object x be a non-atom is essential: if $Q_{\mathbf{X}}\{x\} > 0$, an inductive conformal predictor that ignores all examples with objects different from the current test object can have $1 - \epsilon$ object-conditional validity and still produce a small prediction set for a test object x if the training set is big enough to contain many examples with x as their object.

Part II
Probabilistic Prediction

Chapter 5
Impossibility Results

Abstract In this part of the book (this and following two chapters) we will discuss probabilistic prediction under unconstrained randomness. This chapter concentrates on impossibility results, but in the following two chapters we will discuss positive results for classification and regression.

Keywords Diverse datasets · Probabilistic prediction

5.1 Introduction

As we noticed in Chap. 1, there are several possible goals associated with probabilistic prediction; some of these are attainable and some are not. This chapter concentrates on unattainable goals.

Two important factors that determine the feasibility of probabilistic prediction are:

- Are we interested in asymptotic results or are we interested in the finite world of our experience?
- Do we want to estimate the true probabilities or do we just want some numbers that can pass for probabilities?

In some sections of this book (e.g., when discussing efficiency of conformal predictors in Chap. 2 and universal Venn predictors in the next chapter, Chap. 6) we are interested in asymptotic results, but in this chapter we will emphasize the limited interest of such results for practical learning problems. Most of this chapter is about estimating the true probabilities, but in the last section (Sect. 5.5.4) we also state and prove a result showing that already in the simplest problem of probabilistic prediction it is not possible to produce numbers that can pass for probabilities, if "can pass" is understood in the sense of the algorithmic theory of randomness.

In Sects. 5.2 and 5.3 we state the main negative result of this chapter: probabilistic prediction in the sense of estimating true probabilities from a given finite training set is impossible under unconstrained randomness unless one can use precise repetition of objects in the training set. As we explained in Chap. 1, we are primarily interested

V. Vovk et al., *Algorithmic Learning in a Random World*,
https://doi.org/10.1007/978-3-031-06649-8_5

in learning methods that work in high-dimensional environment, where precise repetitions are hardly possible. In Sect. 5.2 we briefly discuss the nature of the assumption of no repetitions, and in Sect. 5.3 state the mathematical result.

To state our results in their strongest form, in this chapter we use the assumption of randomness (rather than exchangeability): the examples $z_1, z_2, \dots$ are generated from a power distribution Q^∞ on $\mathbf{Z}^\infty$. In most of the chapter we even fix an n and concentrate on the first n examples $z_1, \dots, z_n$; i.e., we consider what we call the one-off setting (see Sect. 1.3.3).

5.2 Diverse Datasets

This section continues the discussion of learning under unconstrained randomness started in Chap. 1 (Sect. 1.1.2).

Important features of even moderately interesting machine-learning problems (such as hand-written digit recognition) are:

1. It might be reasonable to assume that different objects are drawn from the same probability distribution independently of each other, but we cannot make any further assumptions beyond randomness.
2. The objects we are presented with will typically have a fairly complicated structure (in the case of the USPS dataset, every object is a 16×16 grey-scale matrix).
3. In general, we do not expect the objects to be repeated *precisely*.

Features 2 and 3 are closely connected but still different: one can imagine even complicated patterns repeated precisely (such as two twins' genetic code), and in many cases one can expect that even simple unstructured objects, such as real numbers, are never repeated (if they, e.g., are generated by a continuous probability distribution).

Of course, no learning method can work unless the probability distribution generating the data is benign in some respects; for example, different instances of the same digit in the USPS dataset look reasonably similar. But we may say that a type of learning problems is infeasible under unconstrained randomness if those problems can be solved *only* in the case where the dataset has repeated objects.

In some cases estimation of probabilities is possible: for example, if objects are absent ($|\mathbf{X}| = 1$) and $\mathbf{Y} = \{0, 1\}$, estimation of the probability that $y_n = 1$ given $y_1, \dots, y_{n-1}$ is easy for a large n (this was one of the first problems solved by the mathematical theory of probability; see Sect. 13.2.1). But we will see that in general the problem of estimation of probabilities should be classified as infeasible in learning under unconstrained randomness.

5.3 Impossibility of Estimation of Probabilities

The prediction problem considered in this chapter is more challenging than that of the preceding chapters in that the prediction algorithm is required not just to predict the next label y_n but to estimate the conditional probabilities $Q_{\mathbf{Y}|\mathbf{X}}(y \mid x_n)$ for $y \in \mathbf{Y}$. The label space $\mathbf{Y}$ will be assumed to be finite, and we start from the simplest binary case, $\mathbf{Y} = \{0, 1\}$. We will also assume that the object space $\mathbf{X}$ is finite. This is not a restrictive assumption from the practical point of view (all datasets we are aware of allocate a fixed number of bits for each object) but will allow us to avoid complications coming from the foundations of probability.

5.3.1 *Binary Case*

Suppose $\mathbf{Y} = \{0, 1\}$; in this case the estimation of probabilities $Q_{\mathbf{Y}|\mathbf{X}}(y \mid x_n)$, $y \in \mathbf{Y}$, boils down to estimating $Q_{\mathbf{Y}|\mathbf{X}}(1 \mid x_n)$. The notions introduced in this subsection will be redefined (essentially, generalized) in the next one. Fix a positive integer n (we are interested in the case where n is large).

A (one-off) *probability estimator* is a measurable function

$$\Gamma^{\epsilon} : (x_1, y_1, \ldots, x_{n-1}, y_{n-1}, x_n) \mapsto A \,, \tag{5.1}$$

where the significance level ϵ ranges over $(0, 1)$ and A ranges over subsets of the interval $[0, 1]$ (typically A will be an interval, open, closed, or mixed), which satisfies, for all incomplete data sequences and all significance levels $\epsilon_1 > \epsilon_2$,

$$\Gamma^{\epsilon_1}(x_1, y_1, \ldots, x_{n-1}, y_{n-1}, x_n) \subseteq \Gamma^{\epsilon_2}(x_1, y_1, \ldots, x_{n-1}, y_{n-1}, x_n) \,. \tag{5.2}$$

The measurability of (5.1) means that the set

$$\left\{(\epsilon, x_1, y_1, \ldots, x_{n-1}, y_{n-1}, x_n, p) : p \in \Gamma^{\epsilon}(x_1, y_1, \ldots, x_{n-1}, y_{n-1}, x_n)\right\} \tag{5.3}$$

is a measurable subset of $(0, 1) \times \mathbf{Z}^{n-1} \times \mathbf{X} \times [0, 1]$.

We say that such a probability estimator is *weakly valid* if, for any probability distribution Q on $\mathbf{Z}$ and any ϵ,

$$Q^n \left\{(x_1, y_1, \ldots, x_n, y_n) : Q_{\mathbf{Y}|\mathbf{X}}(1 \mid x_n) \in \Gamma^{\epsilon}(x_1, y_1, \ldots, x_{n-1}, y_{n-1}, x_n)\right\} \geq 1 - \epsilon \,. \tag{5.4}$$

In other words, a probability estimator should cover the true conditional probability of 1 with probability at least $1 - \epsilon$, under any power probability distribution generating the data. The word "weakly" is to emphasize our one-off setting (another respect in which the requirement (5.4) is weak is that the outer probability in (5.4) is not conditional on, say, the object x_n [18]). The following result is our formalization of the impossibility of estimation of probabilities in the binary case.

Proposition 5.1 *For any weakly valid probability estimator Γ there is another weakly valid probability estimator $\tilde{\Gamma}$ such that, for any incomplete data sequence $x_1, y_1, \ldots, x_n$ that does not contain repeated objects and for any $\epsilon \in (0, 1)$,*

$$\Gamma^{\epsilon}(x_1, y_1, \ldots, x_{n-1}, y_{n-1}, x_n) \subseteq (0, 1)$$
$$\Longrightarrow \tilde{\Gamma}^{\epsilon}(x_1, y_1, \ldots, x_{n-1}, y_{n-1}, x_n) = \emptyset \,. \qquad (5.5)$$

Let us see why the conclusion (5.5) can be interpreted as saying that nontrivial probability estimation is impossible. Let us assume for simplicity that the set output by Γ is an interval $[a, b]$. The case where $[a, b]$ contains 0 or 1 corresponds to classification: for example, if $a = 0$, the prediction $[a, b] = [0, b]$ essentially says that we expect the new label to be 0, with b quantifying our confidence in this prediction. Genuine probability estimation corresponds to the case $[a, b] \subseteq (0, 1)$, and the theorem says that if we can output such an estimate, we can also output an empty (which is better, because more precise) estimate. An empty estimate is analogous to a contradiction in logic: the foundation for the inference is not sound (the examples are not typical under the randomness assumption) and everything can be deduced. Slightly abusing the standard statistical terminology, we may say that an estimate $[a, b] \subseteq (0, 1)$ is not admissible: it can be improved to being false.

5.3.2 *Multiclass Case*

Let now $\mathbf{Y}$ be an arbitrary finite set (with the discrete σ-algebra). The notions defined in the binary case in the previous subsection can be carried over to the general case as follows. A *probability estimator* is a function (5.1), where $\epsilon \in (0, 1)$ and A ranges over subsets of the family $\mathbf{P}(\mathbf{Y})$ of all probability distributions on $\mathbf{Y}$, which is required to satisfy the conditions of consistency (5.2) (whenever $\epsilon_1 > \epsilon_2$) and measurability (i.e., (5.3) being a measurable subset of $(0, 1) \times \mathbf{Z}^{n-1} \times \mathbf{X} \times \mathbf{P}(\mathbf{Y})$). Such a probability estimator is *weakly valid* if, for any probability distribution Q on $\mathbf{Z}$, any incomplete data sequence $x_1, y_1, \ldots, x_n$, and any ϵ,

$$Q^n \left\{(x_1, y_1, \ldots, x_n, y_n) : Q(\cdot \mid x_n) \in \Gamma^{\epsilon}(x_1, y_1, \ldots, x_{n-1}, y_{n-1}, x_n)\right\} \geq 1 - \epsilon \,,$$

where $Q(\cdot|x_n)$ is the probability distribution on $\mathbf{Y}$ assigning probability $Q_{\mathbf{Y}|\mathbf{X}}(y|x_n)$ to each $y \in \mathbf{Y}$.

Let $\mathbf{P}^{\circ}(\mathbf{Y})$ be the subset of $\mathbf{P}(\mathbf{Y})$ consisting of the *non-degenerate* probability distributions p on $\mathbf{Y}$, i.e., those distributions p that satisfy $\max_{y \in \mathbf{Y}} p(y) < 1$. The following theorem is a generalization of Proposition 5.1.

Theorem 5.2 *For any weakly valid probability estimator Γ there is another weakly valid probability estimator $\tilde{\Gamma}$ such that, for any incomplete data sequence $x_1, y_1, \ldots, x_n$ that does not contain repeated objects and any $\epsilon \in (0, 1)$,*

$$\Gamma^\epsilon(x_1, y_1, \ldots, x_{n-1}, y_{n-1}, x_n) \subseteq \mathbf{P}^\circ(\mathbf{Y}) \\ \Longrightarrow \tilde{\Gamma}^\epsilon(x_1, y_1, \ldots, x_{n-1}, y_{n-1}, x_n) = \emptyset \,. \quad (5.6)$$

5.4 Proof of Theorem 5.2

We first sketch the idea behind the proof. Let Q be the true probability distribution on $\mathbf{Z}$ generating the individual examples. We can imagine that the incomplete data sequence $(x_1, y_1, \ldots, x_n)$, supposed not to contain repeated objects, is generated in two steps. First for each $x \in \mathbf{X}$ we choose randomly $g(x) \in \mathbf{Y}$ setting $g(x) := y$ with probability $Q_{\mathbf{Y}|\mathbf{X}}(y \mid x)$; this is done for each x independently of the other xs. After such a function $g : \mathbf{X} \to \mathbf{Y}$ is generated, we generate $x_i \in \mathbf{X}$, $i = 1, \ldots, n$, independently from $Q_\mathbf{X}$ and finally set $y_i := g(x_i)$.

There is no way to tell from the data sequence $x_1, y_1, \ldots, x_n$ whether it was generated from Q or from $Q_\mathbf{X}$ and g and so, even if the true distribution $Q_{\mathbf{Y}|\mathbf{X}}(y \mid x_n)$, $y \in \mathbf{Y}$, is not degenerate, we will not be able to exclude the corners of the simplex $\mathbf{P}(\mathbf{Y})$ from our prediction $\Gamma^\epsilon(x_1, y_1, \ldots, x_n)$ (unless the sequence $x_1, y_1, \ldots, x_n$ itself is untypical of Q).

5.4.1 *Probability Estimators and Statistical Tests*

Our first step will be to reduce Theorem 5.2 to a statement about "incomplete statistical tests". Complete statistical tests, which are not really needed in this book but clarify the notion of incomplete statistical tests, will be discussed in the next subsection. In this chapter statistical testing serves merely as a technical tool; it will be discussed more systematically later (mainly in Part III).

An *incomplete statistical test* is a measurable function $t : \mathbf{Z}^{n-1} \times \mathbf{X} \times \mathbf{P}(\mathbf{Z}) \to [0, 1]$ such that, for any $Q \in \mathbf{P}(\mathbf{Z})$ and $\epsilon \in (0, 1)$,

$$Q^n\{(x_1, y_1, \ldots, x_n, y_n) : t_Q(x_1, y_1, \ldots, x_{n-1}, y_{n-1}, x_n) \leq \epsilon\} \leq \epsilon$$

(we write Q as a lower index: $t_Q(x_1, y_1 \ldots, x_n)$ instead of $t(x_1, y_1, \ldots, x_n, Q)$). If $t_Q(x_1, y_1 \ldots, x_n) \leq \epsilon$, we say that the test t *rejects* Q at level ϵ (the incomplete data sequence $x_1, y_1 \ldots, x_n$ will be clear from the context).

There is a close connection (although by no means equivalence) between weakly valid probability estimators and incomplete statistical tests:

- If Γ is a weakly valid probability estimator,

  $$t_Q(x_1, y_1, \ldots, x_n) := \sup\left\{\epsilon : Q_{\mathbf{Y}|\mathbf{X}}(\cdot \mid x_n) \in \Gamma^\epsilon(x_1, y_1, \ldots, x_n)\right\} \quad (5.7)$$

  (with $\sup \emptyset := 0$) is an incomplete statistical test.
- If t is an incomplete statistical test,

$$\Gamma^{\epsilon}(x_1, y_1, \ldots, x_n) := \{p \in \mathbf{P}(\mathbf{Y}) : \\ \exists Q \in \mathbf{P}(\mathbf{Z}) : t_Q(x_1, y_1, \ldots, x_n) > \epsilon \;\&\; p = Q_{\mathbf{Y}|\mathbf{X}}(\cdot \mid x_n)\} \tag{5.8}$$

is a weakly valid probability estimator.

The second statement is obvious, and the first follows from

$$Q^n \{(x_1, y_1, \ldots, x_n, y_n) : t_Q(x_1, y_1, \ldots, x_n) \le \epsilon_0\} \\ = Q^n \{(x_1, y_1, \ldots, x_n, y_n) : \forall \epsilon > \epsilon_0 : Q_{\mathbf{Y}|\mathbf{X}}(\cdot \mid x_n) \notin \Gamma^{\epsilon}(x_1, y_1, \ldots, x_n)\} \\ \le \inf_{\epsilon > \epsilon_0} \epsilon = \epsilon_0 .$$

5.4.2 Complete Statistical Tests

To help the reader's intuition, we briefly discuss a more natural notion of statistical test. A function $t : \mathbf{Z}^n \times \mathbf{P}(\mathbf{Z}) \to [0, 1]$ is called a *complete statistical test* if, for any $Q \in \mathbf{P}(\mathbf{Z})$ and $\epsilon \in (0, 1)$:

$$Q^n\{(z_1, \ldots, z_n) : t_Q(z_1, \ldots, z_n) \le \epsilon\} \le \epsilon .$$

Complete statistical tests provide a means of testing whether a data sequence $z_1, \ldots, z_n$ could have been generated from a power distribution Q^n.

To any incomplete statistical test $t_Q(x_1, y_1, \ldots, x_n)$ corresponds the complete statistical test

$$\tilde{t}_Q(x_1, y_1, \ldots, x_n, y_n) := t_Q(x_1, y_1, \ldots, x_n)$$

and to any complete statistical test $t_Q(x_1, y_1, \ldots, x_n, y_n)$ corresponds the incomplete statistical test

$$\tilde{t}_Q(x_1, y_1, \ldots, x_n) := \max_{y \in \mathbf{Y}} t_Q(x_1, y_1, \ldots, x_n, y) .$$

5.4.3 Restatement of the Theorem in Terms of Statistical Tests

Now we can start implementing the idea of the proof sketched above. For any probability distribution $Q \in \mathbf{P}(\mathbf{Z})$ and any $g : \mathbf{X} \to \mathbf{Y}$ define the weight

$$\pi_Q(g) := \prod_{x \in \mathbf{X}} Q_{\mathbf{Y}|\mathbf{X}}(g(x) \mid x)$$

(this is the probability of generating g according to the procedure in the proof sketch) and define the probability distribution Q_g on $\mathbf{Z}$ by the conditions that $(Q_g)_{\mathbf{X}} = Q_{\mathbf{X}}$ and that $(Q_g)_{\mathbf{Y}|\mathbf{X}}(\cdot \mid x)$ is concentrated at the point $g(x)$ for all $x \in \mathbf{X}$ (Q_g is the probability distribution governing the second stage of the procedure of generating the incomplete data sequence in the proof sketch). Notice that

$$\sum_{g : \mathbf{X} \to \mathbf{Y}} \pi_Q(g) = 1 .$$

Theorem 5.2 will be deduced from the following statement.

Proposition 5.3 *For any incomplete statistical test t there exists an incomplete statistical test T such that the following holds. If $(x_1, y_1, \ldots, x_n)$ is an incomplete data sequence without repeated objects and $Q \in \mathbf{P}(\mathbf{Z})$, then there exists $g : \mathbf{X} \to \mathbf{Y}$ such that*

$$T_Q(x_1, y_1, \ldots, x_n) \le t_{Q_g}(x_1, y_1, \ldots, x_n) . \tag{5.9}$$

To prove Proposition 5.3, we will need the following lemma, in which $[x_1, y_1, \ldots, x_n]$ stands for the set of all continuations in $\mathbf{Z}^n$ of $x_1, y_1, \ldots, x_n$.

Lemma 5.4 *For any incomplete data sequence $x_1, y_1, \ldots, x_n$ without repeated objects and any $Q \in \mathbf{P}(\mathbf{Z})$,*

$$Q^n[x_1, y_1, \ldots, x_n] = \sum_{g:\mathbf{X}\to\mathbf{Y}} Q_g^n[x_1, y_1, \ldots, x_n]\boldsymbol{\pi}_Q(g) . \tag{5.10}$$

Proof Let $\tilde{Q}$ be the mixture $\sum_{g:\mathbf{X}\to\mathbf{Y}} Q_g^n \boldsymbol{\pi}_Q(g)$. In this proof we will use the notation X_i for the ith random object and Y_i for the ith random label chosen by Reality; x_i and y_i will be the values taken by X_i and Y_i, respectively. (In the rest of the book we mostly use x_i and y_i to serve both goals.) We have, for any $y_n \in \mathbf{Y}$:

$$\begin{aligned}\tilde{Q}(x_1, y_1, \ldots, x_n, y_n) &= \prod_{i=1}^n Q_{\mathbf{X}}(x_i) \prod_{i=1}^n \tilde{Q}(Y_i = y_i \\ &\qquad \mid X_1 = x_1, \ldots, X_n = x_n, Y_1 = y_1, \ldots, Y_{i-1} = y_{i-1}) \\ &= \prod_{i=1}^n Q_{\mathbf{X}}(x_i) \prod_{i=1}^n \tilde{Q}(Y_i = y_i \mid X_i = x_i) \\ &= \prod_{i=1}^n Q_{\mathbf{X}}(x_i) \prod_{i=1}^n Q_{\mathbf{Y}|\mathbf{X}}(y_i \mid x_i) = Q(x_1, y_1, \ldots, x_n, y_n)\end{aligned}$$

(of course, the second equality is true only because the objects $x_1, \ldots, x_n$ are all different). It remains to sum over $y_n \in \mathbf{Y}$. □

The proof shows that, if there are repeated objects in ω, we still have an inequality between the two sides of (5.10): "$\le$" if the same object is always labelled in the same way in ω, and "$\ge$" otherwise (the latter is obvious, since in this case the right-hand side of (5.10) is zero).

It is easy to derive the statement of Proposition 5.3 from Lemma 5.4.

Proof of Proposition 5.3 Define

$$T_Q(x_1, y_1, \ldots, x_n) := \begin{cases} \max_{g:\mathbf{X}\to\mathbf{Y}} t_{Q_g}(x_1, y_1, \ldots, x_n) & \text{if } x_1, \ldots, x_n \text{ are all different} \\ 1 & \text{otherwise} . \end{cases}$$

This is an incomplete statistical test since, by Lemma 5.4,

$$Q^n \left\{(x_1, y_1, \ldots, x_n, y_n) : T_Q(x_1, y_1, \ldots, x_n) \le \epsilon\right\}$$

$$= \sum_{g:\mathbf{X}\to\mathbf{Y}} Q_g^n \left\{(x_1, y_1, \ldots, x_n, y_n) : T_Q(x_1, y_1, \ldots, x_n) \le \epsilon\right\} \pi_Q(g)$$

$$\le \sum_{g:\mathbf{X}\to\mathbf{Y}} Q_g^n \left\{(x_1, y_1, \ldots, x_n, y_n) : t_{Q_g}(x_1, y_1, \ldots, x_n) \le \epsilon\right\} \pi_Q(g) \le \epsilon .$$

It is obvious that T will satisfy the requirement of Proposition 5.3. □

5.4.4 The Proof of the Theorem

It remains to derive Theorem 5.2 from Proposition 5.3. Let Γ be a weakly valid probability estimator. Define an incomplete statistical test t by (5.7). Let T be an incomplete statistical test whose existence is guaranteed by Proposition 5.3. Define a probability estimator $\tilde{\Gamma}$ as Γ in (5.8) with t replaced by T.

Let $x_1, y_1, \ldots, x_n$ be an incomplete data sequence. If the antecedent of (5.6) holds, t will reject at level ϵ all Q for which $Q_{\mathbf{Y}|\mathbf{X}}(\cdot \mid x_n)$ is degenerate. By (5.9), T will reject all Q at level ϵ. By the definition of $\tilde{\Gamma}$, $\tilde{\Gamma}^\epsilon(x_1, y_1, \ldots, y_n)$ will be empty.

5.5 Context

Theorem 5.2 (in the binary case, i.e., Proposition 5.1) was first proved in [263]. The original statement of this theorem was given in terms of algorithmic randomness, but later it was strengthened by restating it in more traditional terms (as mentioned in Sect. 2.9.2, this is a typical development). The algorithmic result and the idea of its proof were first mentioned in [400].

Our proof of Theorem 5.2 was based on the notion of a statistical test; this is, of course, a standard notion of statistics, but in the form used here it was first introduced, perhaps, by Per Martin-Löf [238] in his version of Kolmogorov's theory of algorithmic randomness. (An alternative recent term is "p-variable" [397, 398]; see also Sect. 9.5.1.)

5.5.1 More Advanced Results

The intuition behind Proposition 5.1 (and indirectly Theorem 5.2) is expressed by Barber [18, Theorem 1] differently: assuming that the marginal distribution $Q_{\mathbf{X}}$ is atomless, she proves that any weakly valid probability estimator provides valid prediction sets for the label y_n at the same significance level ϵ. This result suggests that the expected size of the confidence set Γ^ϵ is substantial unless $Q_{\mathbf{Y}|\mathbf{X}}(1 \mid \cdot)$ tends to be concentrated close to the end-points of the interval $[0, 1]$. Asymptotically matching upper and lower bounds for this expected size are given in [18, Theorems 2 and 3].

In this chapter, we have discussed two cases in which the problem of the possibility of learning the conditional probabilities for y_n given x_n is simple: if a significant number of $x_1, \ldots, x_{n-1}$ coincide with x_n, we can estimate the conditional probabilities from the available statistics; and if the objects $x_1, \ldots, x_n$ are all different, the task is infeasible, unless the probabilities are degenerate. In terms of $Q_{\mathbf{X}}$, learning the conditional probabilities for y_n is easy if $Q_{\mathbf{X}}$ is concentrated on a small finite set and hard if $Q_{\mathbf{X}}$ is atomless. Interesting intermediate cases are studied by Lee and Barber in [220].

5.5.2 *Density Estimation, Regression Estimation, and Regression with Deterministic Objects*

Vapnik [365, 366] lists pattern recognition (called classification in this book), regression estimation, and density estimation as the three main learning problems. In the version of the problem of density estimation relevant to the topic of this book ("conditional density estimation"), we start from a measure μ on the label space $\mathbf{Y}$ and our goal is, given a new object x_n, to estimate the density (which is assumed to exist) of the probability distribution of its label y_n w.r.t. to the measure μ. The standard case is where $\mathbf{Y}$ is a Euclidean space, but Proposition 5.1 shows that this problem is infeasible already in the simplest case $\mathbf{Y} = \{0, 1\}$. Most of the existing literature deals with the case where the objects are absent. For a good exposition of the theory see [81]; Vapnik [366] constructs a version of SVM for density estimation.

There are several possible understandings of the term "regression". In Sect. 2.3 we constructed an efficient confidence predictor for regression under unconstrained randomness. There are, however, two popular understandings (one of them setting a different goal and the other making a different assumption about Reality) that are not feasible for diverse datasets under unconstrained randomness.

One understanding is "regression estimation": we assume that the examples are generated independently from some probability distribution Q on $\mathbf{Z}$ and our goal is, given (x_i, y_i), $i = 1, \dots, n-1$, and x_n, to estimate the conditional expectation of y_n given x_n. This problem is infeasible already in the simple case $\mathbf{Y} = \{0, 1\}$, since it then coincides with that of probability estimation. The situation changes, however, when we replace conditional expectation by conditional median [245], median being a robust counterpart of mean.

Another understanding is "regression with deterministic objects". The classic textbook [65] clearly describes two different assumptions that can be made in regression:

- in one approach [65, Chap. 23] it is assumed that the examples (x_i, y_i) are generated by some power distribution;
- in the other approach [65, Chap. 37] it is assumed that the objects x_i are generated by an unknown mechanism (for example, are chosen arbitrarily by the experimenter) and only the labels y_i are generated stochastically; namely, it is assumed that for every $x \in \mathbf{X}$ there is a probability distribution $Q(x)$ on $\mathbf{Y}$ such that y_i is generated from $Q(x_i)$ (formally, Q is required to be a Markov kernel from $\mathbf{X}$ to $\mathbf{Y}$).

The first model (essentially the model used in Chap. 2) is the combination of the second model (with no assumptions about x_i) and the assumption that x_i are generated by a power distribution. It is easy to see that regression in the second sense is infeasible for diverse datasets. Indeed, if we have a data sequence $(x_1, y_1), \dots, (x_n, y_n)$ such that all x_i are different, it will be perfectly typical (e.g., algorithmically random) w.r.t. to any Markov kernel Q such that $Q(x_i)$ is concentrated at y_i, $i = 1, \dots, n$; therefore, nontrivial prediction of y_n given $x_1, y_1, \dots, x_n$ is not possible. Of course, the situation changes if additional assumptions are imposed on the Markov kernel Q: cf. the discussion of the Gauss linear model in Sect. 11.4.2.

5.5.3 Universal Probabilistic Predictors

In Sect. 1.5.1 we mentioned Stone's [347] result about the existence of a universally consistent probabilistic predictor. Intuitively, the fact that conditional probabilities can be estimated to any accuracy in the limit, under unconstrained randomness and without assuming precise repetitions, appears to be in some conflict with this chapter's results. From the purely mathematical point of view it suffices to say, as we did in Sect. 1.5.2, that convergence in Stone's theorem and its extensions is not uniform. For details, see [82, Chap. 7]; we will only illustrate this with a simple example.

Suppose x_n are generated from the uniform distribution on $[0, 1]$ and $y_n = f(x_n)$, where $f : [0, 1] \to \{0, 1\}$ is a Borel function; we want our inferences to hold without any further assumption about f. Asymptotically, predicting with the majority of the K_n nearest neighbours of x_n (with $K_n \to \infty$ and $K_n = o(n)$) we will ensure that the frequency of errors ($Q_{\mathbf{Y}|\mathbf{X}}(1 \mid x_n) \neq f(x_n)$, in our usual notation) tends to zero in probability (see [347] or [82, Theorem 6.4]). Before infinity, however, we cannot say anything at all about the closeness of our predictions to the true conditional probabilities. For any given value of n, no matter how large, there exists an f agreeing with the available data (i.e., such that $f(x_i) = y_i, i = 1, \ldots, n-1$) for which $f(x_n) = 0$ and there exists an f agreeing with the available data for which $f(x_n) = 1$.

At a more philosophical and controversial level, it might even be argued that the asymptotic results about universal probabilistic prediction (and so, by implication, some of our results in this part, Chaps. 6 and 7) are devoid of empirical meaning. Let $\mathbf{Y} = \{0, 1\}$. If the object space $\mathbf{X}$ is fixed and finite, the objects in an infinite data sequence $z_1, z_2, \ldots$ will eventually start to repeat, no matter how big $\mathbf{X}$ is, and we will then be able to estimate $Q(1 \mid x_n)$ for a new object x_n. Kolmogorov's axioms of probability include the axiom of continuity (A.1) (equivalent, in the presence of the other axioms, to σ-additivity). According to Kolmogorov [199, Sect. II.1], it is almost impossible to elucidate the empirical meaning of this axiom, and its acceptance is an arbitrary, although expedient, choice. (For further details, see [318].) It appears that the main effect of the acceptance of this axiom was to make infinite probability spaces (the subject of Chap. II of [199]) similar to finite probability spaces (the subject of Kolmogorov's Chap. I). Fixing $\mathbf{X}$ (maybe infinite but with the probability distribution generating examples satisfying the axiom of continuity (A.1)) and letting $n \to \infty$ might be just an embellished version of fixing a finite $\mathbf{X}$ and letting $n \to \infty$. More relevant, from the empirical point of view, asymptotic results would consider a variable object space $\mathbf{X}$; e.g., we could consider a triangular array $x_1, y_1, \ldots, x_n, y_n$ where $x_i \in \mathbf{X}_n$, $y_i \in \mathbf{Y}$, and $\mathbf{X}_n$ is an increasing sequence of measurable spaces.

5.5.4 Algorithmic Randomness Perspective

Theorem 5.2 shows that probabilistic prediction is infeasible in the sense that the conditional probability for the label cannot be estimated, under the stated conditions. A related question is: can we find a conditional probability which is as good as the true conditional probability? The former can be far from the latter, but still be as good in explaining the data. The following result by Ilia Nouretdinov formalizes this question using the algorithmic notion of randomness [238] and shows that the answer is "no", even if the labels are binary and the objects are absent. Remember that the notion of algorithmic randomness has nothing to do with statistical randomness and, in particular, the assumption of randomness (see Remark 2.22).

Theorem 5.5 *For any computable probability distribution P on $\{0, 1\}^\infty$ there exist a computable Bernoulli distribution $\mathbf{B}_\theta$ on $\{0, 1\}$ and an infinite sequence in $\{0, 1\}^\infty$ which is algorithmically random w.r. to $\mathbf{B}_\theta^\infty$ but not algorithmically random w.r. to P.*

The probability distribution P in this theorem is the suggested way of finding a good conditional probability (since a conditional probability can be found for any data sequence, these conditional probabilities can be put together to form a probability distribution). The theorem asserts that there exists a power distribution (even a computable one) and an infinite sequence which can be produced by this power distribution but for which P is not a good explanation.

Proof If $\omega = (y_1, y_2, \dots) \in \{0, 1\}^\infty$ is an infinite binary sequence, we set

$$\theta_N(\omega) := \frac{y_1 + \cdots + y_N}{N} .$$

For each $n = 1, 2, \dots$, $N(n)$ is defined constructively as any number N such that, for any $\theta \in [0, 1]$,

$$\mathbf{B}_\theta^\infty \left\{\omega : |\theta_N(\omega) - \theta| > 2 \times 10^{-n-1}\right\} \le 2^{-n-1} .$$

Set

$$[a_0, b_0] := [0, 1] \quad \text{and} \quad U_0 := \{0, 1\}^\infty .$$

Define inductively, for $n = 1, 2, \dots$,

$$[a_n, b_n] := [a_{n-1} + 2 \times 10^{-n}, a_{n-1} + 3 \times 10^{-n}] \tag{5.11}$$

or

$$[a_n, b_n] := [a_{n-1} + 7 \times 10^{-n}, a_{n-1} + 8 \times 10^{-n}] \tag{5.12}$$

and then

$$U_n := \{\omega \in U_{n-1} : \theta_{N(n)}(\omega) \in [a_n, b_n]\} ,$$

where the choice between (5.11) and (5.12) is done effectively and so that $P(U_n)$ decreases significantly (say, $P(U_n) \le \frac{2}{3} P(U_{n-1})$). Finally, set

$$\theta := \lim_{n\to\infty} a_n = \lim_{n\to\infty} b_n \quad \text{and} \quad U := \cap_n U_n .$$

The statement of the theorem follows from the following properties of this construction:

1. U is a constructively null set ($P(U) = 0$ since $P(U_n) \le \frac{2}{3} P(U_{n-1})$ for all n) and θ is computable.
2. U contains some algorithmically random sequences w.r. to $\mathbf{B}_\theta^\infty$.

Only the last property requires a separate proof.

Using $\mathbf{B}_\theta^\infty(U_0) = 1$ and the fact that, for any n,

$$a_n + 2 \times 10^{-n-1} \le a_{n+1} \le \theta \le b_{n+1} \le b_n - 2 \times 10^{-n-1} ,$$

we obtain

$$\begin{aligned}\mathbf{B}_\theta^\infty(U_{n-1} \setminus U_n) &\le \mathbf{B}_\theta^\infty\{\omega : \theta_{N(n)}(\omega) \notin [a_n, b_n]\} \\ &\le \mathbf{B}_\theta^\infty \left\{\omega : \left|\theta_{N(n)}(\omega) - \theta\right| > 2 \times 10^{-n-1}\right\} \le 2^{-n-1} ;\end{aligned}$$

therefore, $\mathbf{B}_\theta^\infty(U) \ge 1/2$.

Chapter 6
Probabilistic Classification: Venn Predictors

Abstract In this chapter we discuss positive results about probabilistic prediction in the case of classification. For that, we introduce a new kind of predictors, Venn predictors, which produce predictions satisfying a natural property of validity at the price of lowering the bar as compared with the previous chapter. To gain some wiggle room, Venn predictors are allowed to announce several probability distributions as their prediction, and the overall prediction is considered valid whenever any of its components is valid. Such a multiprobability prediction is useful if its components are close to each other, in which case we have an approximate probabilistic prediction; if not, we may be in a situation of Knightean uncertainty, and the multiprobability prediction may be considered as a useful warning.

Keywords Classification · Venn prediction · Venn–Abers predictors

6.1 Introduction

We saw in Chap. 1 that there are different types of algorithms for learning under randomness, where examples (objects with labels) are drawn one by one from an unknown probability distribution. Appropriate criteria for validity and efficiency may vary from type to type. In Chaps. 2 and 3, we developed criteria for the validity and efficiency of confidence predictors, and we developed a class of confidence predictors, conformal predictors, that satisfy the criterion for validity while varying in their efficiency. In this chapter, we study algorithms of a new type, multiprobability predictors. We develop a criterion of validity for multiprobability predictors, and we introduce and study a class of multiprobability predictors, Venn predictors, that satisfy the criterion for validity when the label space is finite. Venn predictors also vary in their efficiency, but we will see that there are arguments for the efficiency of some simple Venn predictors; therefore, the impossibility of estimation of non-extreme probabilities in a diverse random environment (Theorem 5.2) does not prevent us from producing probabilities, perhaps quite different from the "true" ones, which perform well in important respects.

V. Vovk et al., *Algorithmic Learning in a Random World*,
https://doi.org/10.1007/978-3-031-06649-8_6

Venn predictors use a familiar idea: divide the old objects into categories, somehow classify the new object into one of the categories, and then use the frequencies of labels in the chosen category as probabilities for the new object's label. We innovate only in a couple of details:

- We divide examples rather than objects into categories. When we compute the frequencies of labels in the category containing the new example, we include the new example along with the old examples already in that category. Since at the time of prediction we do not yet know the new object's label, we compute these frequencies several times, once for each label the new object might have. (This is analogous to the way we treat the new object when we use a nonconformity measure to define a conformal predictor.)
- We interpret each set of frequencies as a probability distribution for the new object's unknown label. Thus we announce several probability distributions for the new label rather than a single one. Fortunately, once the number of old examples in each category is large, these different probability distributions will often be practically identical.

In Sect. 6.2, we define precisely the class of Venn predictors and show that these predictors do provide multiprobability predictions that are valid according to a natural criterion in the one-off mode.

In Sect. 6.3, we give an asymptotic result: there exists a Venn predictor that asymptotically approaches the true conditional probabilities. Unfortunately, this result is impractical in the same sense as the other strong asymptotic results we discuss in this book (see Sect. 5.5.3).

The problem of choosing a good partition ("taxonomy") into categories is a difficult one. In Sect. 6.4 we discuss a procedure, based on the powerful method of isotonic regression [11], for turning standard machine learning algorithms into taxonomies. Similarly to full conformal prediction, Venn predictors, including those based on isotonic regression, are computationally inefficient in general, and in the same section we introduce their computationally efficient versions analogous to inductive conformal and cross-conformal predictors.

In Chap. 12, we will extend some of the concepts introduced in this chapter to online compression models.

6.2 Venn Predictors

In this section we formally define Venn predictors, which were described informally in Sect. 6.1, and we show that they are valid multiprobability predictors. As in Chap. 3, we now deal with the problem of classification, assuming that the label space $\mathbf{Y}$ is finite. (We carry this assumption throughout this chapter.)

The concept of a taxonomy, or Venn taxonomy, was introduced in Sect. 4.6.1. For every taxonomy $K : \mathbf{Z}^{(*)} \times \mathbf{Z} \to \mathbf{K}$ we will define a Venn predictor.

Consider a potential label $y \in \mathbf{Y}$ for a test object x_{l+1}, and let $x_1, y_1, \dots, x_l, y_l$ be the training set. Write (as usual) z_i for (x_i, y_i), $i = 1, \dots, l$, and (just for the moment) z_{l+1} for (x_{l+1}, y) and y_{l+1} for y. Let P^y be the empirical probability distribution of the labels in the category containing the test example:

$$P^y\{y'\} := P^y(\{y'\}) := \frac{\left|\{i = 1, \dots, l+1 : \kappa_i^y = \kappa_{l+1}^y \ \& \ y_i = y'\}\right|}{\left|\{i = 1, \dots, l+1 : \kappa_i^y = \kappa_{l+1}^y\}\right|}, \qquad (6.1)$$

where we drop the parentheses around $\{\dots\}$ and set

$$\kappa_i^y := K(⦃z_1, \dots, z_{l+1}⦄, z_i) .$$

This is a probability distribution on $\mathbf{Y}$. The *Venn predictor* determined by the taxonomy is the multiprobability predictor $P := (P^y : y \in \mathbf{Y})$. The family P consists of between one and $|\mathbf{Y}|$ distinct probability distributions on $\mathbf{Y}$. We will sometimes refer to the family P as the *multiprobability prediction*.

Of course, the multiprobability prediction P is really useful only when all P^y are close to each other. However, having several probabilistic predictions in P may serve as a valuable diagnostic tool. If the elements of a multiprobability prediction are very different, this serves as an indication that we may be in a situation of uncertainty rather than risk, to use Knight's terminology (see Sect. 6.6.1 below) and so do not have a probabilistic prediction, even approximately. If the probabilistic predictions in P are close to each other, they may serve as an overall probabilistic prediction, and we are essentially in the situation of risk.

There are many Venn predictors, one for each taxonomy. Some will be more efficient than others on particular datasets. But all share one virtue: they are valid, in various senses. In this book we concentrate on the simplest notion of validity, one applicable to one-off Venn predictors (cf. Sect. 6.6.3).

6.2.1 Validity of One-Off Venn Predictors

Let us say that a random variable P taking values in $\mathbf{P}(\mathbf{Y})$ is *perfectly calibrated* for a random variable Y taking values in $\mathbf{Y}$ if, for each $y \in \mathbf{Y}$,

$$\mathbb{P}(Y = y \mid P) = P\{y\} \qquad \text{a.s.} \qquad (6.2)$$

Intuitively, P is the prediction made by a probabilistic predictor for Y, and perfect calibration means that the probabilistic predictor gets the probabilities right, at least on average, for each value of the prediction. A probabilistic predictor for Y whose prediction P satisfies (6.2) with an approximate equality is said to be well calibrated [72], or unbiased in the small [72, 255]; this terminology will be used only in informal discussions, of course.

We consider the one-off setting of size $n = l+1$. Let Y be the function that maps $(z_1, \ldots, z_{l+1})$ to the true label y_{l+1} of the test example z_{l+1}; this is what we would like to predict. This function is measurable, and so is a random variable; in this section all random variables will be functions on $\mathbf{Z}^{l+1}$. A *selector* is a measurable function $S : \mathbf{Z}^{l+1} \to \mathbf{Y}$, i.e., a $\mathbf{Y}$-valued random variable. The probability measure P^y in (6.1) is a function of $(z_1, \ldots, z_{l+1}) \in \mathbf{Z}^{l+1}$ (which in fact does not depend on z_{l+1}'s label), so it is a $\mathbf{P}(\mathbf{Y})$-valued random variable (for a fixed $y \in \mathbf{Y}$). The composition P^S, with S plugged in place of y, is also a $\mathbf{P}(\mathbf{Y})$-valued random variable.

Proposition 6.1 *There exists a selector S such that, for any Venn predictor* (6.1)*, the composition P^S is perfectly calibrated for Y, under any power distribution.*

Intuitively, at least one of the at most $|\mathbf{Y}|$ probability measures output by the Venn predictor is perfectly calibrated. Therefore, if all the probability measures P^y tend to be close to each other, we expect them (or, say, their average) to be well calibrated.

It may help intuition to give an interpretation of Proposition 6.1 in the spirit of frequentist statistics. Let us sample an infinite sequence of data sequences $s_k := (z_1^{(k)}, \ldots, z_{l+1}^{(k)})$, $k = 1, 2, \ldots$, from a fixed power distribution Q^{l+1}. For each of these sequences s_k compute the probability measure $P^S = P_k^S \in \mathbf{P}(\mathbf{Y})$ and the actual outcome $Y = Y_k \in \mathbf{Y}$. Notice that there will be only finitely many distinct P_k^S, since (6.1) is a proper fraction or 1, and its denominator is at most $l + 1$. If we take all entries P_k^S of any $P \in \mathbf{P}(\mathbf{Y})$ (if any, there will be infinitely many such entries almost surely) in the order of increasing k, the empirical distribution of the first K among the corresponding $Y_k \in \mathbf{Y}$ will tend to P as $K \to \infty$ almost surely.

Proposition 6.1 is a special case of Theorem 12.1, which will be proved in Chap. 12. The selector will be $S := Y$.

6.2.2 Are There Other Ways to Achieve Perfect Calibration?

Our following result, Theorem 6.3, will say that under mild regularity conditions perfect calibration (6.2) holds only for Venn predictors (perhaps weakened by adding irrelevant probabilistic predictions).

To state Theorem 6.3 we need a few further definitions. A *multiprobability predictor* is a function that maps each training set $(z_1, \ldots, z_l) \in \mathbf{Z}^l$ and each test object $x \in \mathbf{X}$ to a subset of $\mathbf{P}(\mathbf{Y})$ (not required to be measurable in any sense). Venn predictors are an example. Let us say that a multiprobability predictor is *invariant* if it is independent of the ordering of the training set $(z_1, \ldots, z_l)$. An *invariant selector* for an invariant multiprobability predictor F is a measurable function $f : \mathbf{Z}^{l+1} \to \mathbf{P}(\mathbf{Y})$ such that $f(z_1, \ldots, z_{l+1})$ does not change when $z_1, \ldots, z_l$ are permuted and such that $f(z_1, \ldots, z_{l+1}) \in F(z_1, \ldots, z_l, x_{l+1})$ for all $(z_1, \ldots, z_{l+1})$.

Remark 6.2 It is natural to consider only invariant statistics under the assumption of randomness because of the principle of sufficiency [64, Chap. 2]. We have two close but different notions of a selector: an invariant selector selects a probabilistic prediction directly, while a selector selects it via selecting a label in the context of Venn prediction; in fact, the selector Y used in the proof of Proposition 6.1 also satisfies the property of invariance.

We say that an invariant multiprobability predictor F is *invariantly perfectly calibrated* if it has an invariant selector f such that, for each $y \in \mathbf{Y}$,

$$\mathbb{P}\big(Y = y \mid f\big) = f\{y\} \quad \text{a.s.} \tag{6.3}$$

under any power distribution Q^{l+1} for the examples $z_1, \ldots, z_{l+1}$. (As before, our probability space is $(\mathbf{Z}^{l+1}, Q^{l+1})$, and so f is a $\mathbf{P}(\mathbf{Y})$-valued random variable, which makes $f\{y\}$ a $[0, 1]$-valued random variable.)

Theorem 6.3 *If an invariant multiprobability predictor F is invariantly perfectly calibrated, then it contains a Venn predictor V in the sense that $V(z_1, \ldots, z_l, x_{l+1}) \subseteq F(z_1, \ldots, z_l, x_{l+1})$ for all $z_1, \ldots, z_l, x_{l+1}$.*

Proof Let f be an invariant selector of F satisfying the condition (6.3) of being invariantly perfectly calibrated, and let $y \in \mathbf{Y}$. By definition,

$$\mathbb{E}\big(1_{Y=y} - f\{y\} \mid f\big) = 0 \quad \text{a.s.}\,,$$

which implies

$$\mathbb{E}\big((1_{Y=y} - f\{y\})1_{f\in I}\big) = 0 \ \text{a.s.} \tag{6.4}$$

for all sets $I \subseteq \mathbf{P}(\mathbf{Y})$ of the form

$$I := \big\{P \in \mathbf{P}(\mathbf{Y}) : P\{y\} \in [a_y, b_y] \text{ for all } y \in \mathbf{Y}\big\}\ ,$$

where all the intervals $[a_y, b_y]$ have rational end-points. The expected value in (6.4) can be obtained in two steps: first we average

$$(1_{y'_{l+1}=y} - f(z'_1, \ldots, z'_{l+1})\{y\})1_{f(z'_1,\ldots,z'_{l+1})\in I}$$

over the orderings $(z'_1, \ldots, z'_{l+1})$ of each bag $\Lbag z_1, \ldots, z_{l+1} \Rbag$, where $z_i = (x_i, y_i)$ and $z'_i = (x'_i, y'_i)$, and then we average over the bags $\Lbag z_1, \ldots, z_{l+1} \Rbag$ generated according to the image of Q^{l+1} under the mapping of forgetting the order $(z_1, \ldots, z_{l+1}) \mapsto \Lbag z_1, \ldots, z_{l+1} \Rbag$. The first operation is discrete: the average over the orderings of $\Lbag z_1, \ldots, z_{l+1} \Rbag$ is the arithmetic mean of $(1_{y_i=y} - p_i\{y\})1_{p_i\in I}$ over $i = 1, \ldots, l + 1$, where $p_i := f(\ldots, z_i)$ and the dots stand for $z_1, \ldots, z_{i-1}$ and $z_{i+1}, \ldots, z_{l+1}$ arranged in any order (since f is invariant, the order does not matter). By the completeness of the mapping of forgetting the order [222, Example 4.3.4], this average is zero for all I and almost all bags. In other words, for almost all bags $\wr z_1, \ldots, z_{l+1} \int$, the arithmetic mean of $1_{y_i=y} - P\{y\}$ over all $i \in \{1, \ldots, l+1\}$ such that $p_i = P$ is zero when defined (i.e., does not involve division by 0).

Now let us get rid of "almost" in "almost all bags". Suppose there exists a bag $\wr z_1, \ldots, z_{l+1} \int$ such that the arithmetic mean of $1_{y_i=y} - P\{y\}$ over those i is (defined and) different from 0. We arrive at a contradiction by setting $Q \in \mathbf{P}(\mathbf{Z})$ to the uniform probability measure on the set $\{z_1, \ldots, z_{l+1}\}$: the arithmetic mean of $1_{y_i=y} - P\{y\}$ over those i will be different from 0 with a positive Q^{l+1}-probability.

Therefore, the arithmetic mean of $1_{y_i=y} - P\{y\}$ over those i is zero when defined, for all bags, all $y \in \mathbf{Y}$, and all P. It remains to take

$$K(\wr z_1, \ldots, z_{l+1} \int, z_i) := p_i$$

as Venn taxonomy, where p_i is defined as above. The corresponding Venn predictor will satisfy the requirement in Theorem 6.3. □

The invariance assumption in Theorem 6.3 is essential [394, Sect. 2]. (However, as the proof shows, it is not essential for the multiprobability predictor F, as long as it admits an invariant selector.)

6.2.3 Venn Prediction with Binary Labels and No Objects

Let us now discuss a particularly simple special case of Venn prediction. Suppose there are no objects and that the label space is binary: $\mathbf{Y} = \{0, 1\}$. In other words, we have the *Bernoulli problem* (cf. Sect. 13.2.1): having observed a sequence of bits $y_1, \ldots, y_l$ we are to predict the probability $p \in [0, 1]$ that $y_{l+1} = 1$. (Formally, Venn predictors output probability distributions on $\mathbf{Y}$, but in the binary case a probability distribution P on $\{0, 1\}$ carries the same information as the number $p := P\{1\} \in [0, 1]$.)

Let k be the number of 1s among $y_1, \ldots, y_l$. The most popular point estimates (see, e.g., [141, Sect. 9.1]) of the probability that $y_{l+1} = 1$ are the maximum likelihood estimate k/l (the frequency of 1s in the training set, used already by Jacob Bernoulli)), the Jeffreys, or Krichevsky–Trofimov, estimate $(k+1/2)/(l+1)$, and the Laplace estimate $(k + 1)/(l + 2)$ (*Laplace's rule of succession*).

Since we do not have objects in the Bernoulli problem, the most natural Venn predictor V corresponds to the taxonomy assigning all examples to the same category. This Venn predictor outputs

$$P := \left\{ \frac{k}{l+1}, \frac{k+1}{l+1} \right\} \tag{6.5}$$

as its multiprobability prediction. The convex hull $[k/(l + 1), (k + 1)/(l + 1)]$ of this set contains the three standard estimates of the probability that $y_{l+1} = 1$. The upper Venn estimate $(k+1)/(l+1)$ and the lower Venn estimate $k/(l+1)$ are dual in the sense that the lower Venn estimate of the probability that $y_{l+1} = 1$ is equal to 1 minus the upper Venn estimate of the probability of the complementary event $y_{l+1} = 0$ (cf., e.g., [317, (1.7)]).

6.3 A Universal Venn Predictor

The following result asserts the existence of a universal Venn predictor. As the proof (given in Sect. 6.5.1) shows, such a predictor can be constructed quite easily, using the histogram approach to probability estimation [82].

Theorem 6.4 *Suppose* $\mathbf{X}$ *is a standard Borel space and* $\mathbf{Y}$ *is finite. There exists a Venn predictor that satisfies the following property of consistency for any probability distribution* Q *on* $\mathbf{Z}$ *with regular conditional probabilities* $Q_{\mathbf{Y}|\mathbf{X}}(\cdot|\cdot)$*. If the examples are generated from* Q^∞*, then*

$$\max_{p \in P_n} \rho(p, Q_{\mathbf{Y}|\mathbf{X}}(\cdot \mid x_n)) \to 0 \quad (n \to \infty)$$

in probability, where P_n *are the multiprobability predictions produced by the Venn predictor and* ρ *is the variation distance,*

$$\rho(p, q) := \sum_{y \in \mathbf{Y}} |p\{y\} - q\{y\}| \ .$$

This theorem can be interpreted by saying that some Venn predictors have asymptotically optimal efficiency.

6.4 Venn–Abers Predictors

Venn–Abers predictors only work in the case of a binary label, at least in the current state of the theory, so in this section we set $\mathbf{Y} := \{0, 1\}$. Remember that a function f is *increasing* if its domain is an ordered set and $t_1 \le t_2 \Rightarrow f(t_1) \le f(t_2)$; such functions are also called *isotonic*.

6.4.1 Full Venn–Abers Predictors

As discussed in Sect. 4.3.5, many machine-learning algorithms for binary classification are in fact binary scoring classifiers (we usually drop "binary" in this section): when trained on a training set of examples and fed with a test object x, they output a prediction score $S(x)$, and we call $S : \mathbf{X} \to \mathbb{R}$ the scoring function for that training set. The actual prediction rule is then obtained by fixing a threshold c (we discussed $c = 0$ in Sect. 4.3.5) and predicting the label of x to be 1 if $S(x) > c$ and 0 if $S(x) < c$. Several examples were given in Sect. 3.2. Alternatively, one could apply an increasing function g to $S(x)$ in an attempt to "calibrate" the scores, so that $g(S(x))$ can be used as the predicted probability that the label of x is 1.

Fix a scoring classifier and let $z_1, \ldots, z_l$ be a training set of examples $z_i = (x_i, y_i)$, $i = 1, \ldots, l$. The most direct application [430] of the method of isotonic regression [11] to the problem of prediction score calibration is as follows. Train the scoring classifier on the training set and compute the prediction score $S(x_i)$ for each training example (x_i, y_i), where S is the scoring function for $z_1, \ldots, z_l$. Let g be the increasing function on the set $\{S(x_1), \ldots, S(x_l)\}$ that maximizes the likelihood

$$\prod_{i=1}^{l} p_i, \quad \text{where } p_i := \begin{cases} g(S(x_i)) & \text{if } y_i = 1 \\ 1 - g(S(x_i)) & \text{if } y_i = 0 \, . \end{cases} \tag{6.6}$$

Such a function g is unique [11, Corollary 2.1] and can be easily found using the "pair-adjacent violators algorithm" (PAVA), which is the topic of Sect. 6.5.2. We will say that g is the *isotonic calibrator for* $(S(x_1), y_1), \ldots, (S(x_l), y_l)$. To predict the label of a test object x, the direct method finds the closest (or one of the closest) $S(x_i)$ to $S(x)$ and outputs $g(S(x_i))$ as its prediction. We will refer to this method as *direct isotonic regression* (DIR).

The DIR method is prone to overfitting as the same examples $z_1, \ldots, z_l$ are used both for training the scoring classifier and for calibration without taking any precautions. Our next step can be interpreted as "Vennizing" the DIR method, and it is analogous to conformalizing, which we did in Sects. 2.3 and 2.4.

The *Venn–Abers predictor* corresponding to the given scoring classifier is the multiprobability predictor that is defined as follows. Try the two possible labels, 0 and 1, for the test object x. Let S^0 be the scoring function for $z_1, \ldots, z_l, (x, 0)$, S^1 be the scoring function for $z_1, \ldots, z_l, (x, 1)$, g^0 be the isotonic calibrator for $(S^0(x_1), y_1), \ldots, (S^0(x_l), y_l), (S^0(x), 0)$, and g^1 be the isotonic calibrator for $(S^1(x_1), y_1), \ldots, (S^1(x_l), y_l), (S^1(x), 1)$. The *multiprobability prediction* output by the Venn–Abers predictor is (p^0, p^1), where $p^0 := g^0(S^0(x))$ and $p^1 := g^1(S^1(x))$. (And we can expect p^0 and p^1 to be close to each other unless DIR overfits grossly.) The Venn–Abers predictor is shown as Algorithm 6.1.

Algorithm 6.1 Venn–Abers predictor

Input: a training set $z_1, \ldots, z_l$ and a test object x.
Output: multiprobability prediction (p^0, p^1).
1: **for** $y \in \{0, 1\}$
2: set S^y to the scoring function for $z_1, \ldots, z_l, (x, y)$
3: set g^y to the isotonic calibrator for $(S^y(x_1), y_1), \ldots, (S^y(x_l), y_l), (S^y(x), y)$
4: set $p^y := g^y(S^y(x))$.

The intuition behind Algorithm 6.1 is that it tries to evaluate the robustness of the DIR prediction. To see how sensitive the scoring function and, eventually, the probabilistic prediction are to the training set we extend the latter by adding the test object labelled in both different ways. For large datasets and inflexible scoring functions, we will have $p^0 \approx p^1$, and both numbers will be close to the DIR prediction. However, even if the dataset is very large but the scoring function is very flexible, p^0 can be far from p^1 (the extreme case is where the scoring function is so flexible that it ignores all examples apart from a few that are most similar to the test object, and in this case it does not matter how big the dataset is). We rarely know in advance how flexible our scoring function is relative to the size of the dataset, and the difference between p^0 and p^1 gives us some indication of this.

The following proposition says that Venn–Abers predictors are Venn predictors and, therefore, inherit all properties of validity of the latter, such as the one given by Proposition 6.1 (see also Sect. 6.6.3 below).

Proposition 6.5 *Venn–Abers predictors are Venn predictors.*

Proof Fix a Venn–Abers predictor. The corresponding Venn taxonomy is

$$K(\lbag z_1, \ldots, z_n \rbag, (x_i, y_i)) := g(S(x_i)) \ ,$$

where S is the scoring function for $z_1, \ldots, z_n$, and g is the isotonic calibrator for $(S(x_1), y_1), \ldots, (S(x_n), y_n)$. Lemma 6.6 below shows that the Venn predictor determined by this taxonomy gives predictions identical to those given by the original Venn–Abers predictor. This proves the proposition. □

The following lemma is an obvious property of the PAVA, and its proof will be given in Sect. 6.5.2, where we introduce the PAVA.

Lemma 6.6 *Let g be the isotonic calibrator for $(t_1, y_1), \ldots, (t_n, y_n)$, where $t_i \in \mathbb{R}$ and $y_i \in \{0, 1\}$, $i = 1, \ldots, n$. Any $p \in \{g(t_1), \ldots, g(t_n)\}$ is equal to the arithmetic mean of the labels y_i of the t_i, $i = 1, \ldots, n$, satisfying $g(t_i) = p$.*

6.4.2 Inductive Venn–Abers Predictors

Inductive Venn–Abers predictors are a computationally efficient modification of Venn–Abers predictors similar to inductive conformal predictors (Sect. 4.2). They can be implemented very efficiently, but as a function, the *inductive Venn–Abers predictor* (IVAP) can be defined in terms of a scoring classifier (see the beginning of Sect. 6.4.1) as follows:

- Divide the training set of size l into two subsets, the *proper training set* of size m and the *calibration set* $z_1, \ldots, z_k$ of size k, so that $l = m + k$.
- Train the scoring classifier on the proper training set obtaining a scoring function S.

- Find the scores $s_1, \ldots, s_k$, where $s_i := S(x_i)$ and $z_i = (x_i, y_i)$, of the calibration objects $x_1, \ldots, x_k$.
- When a new test object x arrives, compute its score $s := S(x)$. Fit isotonic regression to $(s_1, y_1), \ldots, (s_k, y_k), (s, 0)$ obtaining an isotonic calibrator f^0. Fit isotonic regression to $(s_1, y_1), \ldots, (s_k, y_k), (s, 1)$ obtaining an isotonic calibrator f^1. The multiprobability prediction for the label y of x is the pair $(p^0, p^1) := (f^0(s), f^1(s))$ (intuitively, the prediction is that the probability that $y = 1$ is either $f^0(s)$ or $f^1(s)$).

When the PAVA is used (see Sect. 6.5.2), the multiprobability prediction (p^0, p^1) output by an IVAP always satisfies $p^0 < p^1$, and so p^0 and p^1 can be interpreted as the lower and upper probabilities, respectively; in practice, they are close to each other for large training sets.

If we fix the proper training set, an IVAP becomes a (full) Venn–Abers predictor based on the scoring classifier which outputs the scoring function S regardless of the training set. Therefore, the IVAPs inherit the property of validity of Venn–Abers predictors, which in turn inherit the property of validity of Venn predictors stated in Proposition 6.1: for some selector, their probabilistic prediction is perfectly calibrated. This holds conditionally on the proper training set; in other words, we do not make any assumptions about the probability distribution generating the proper training set, and only make the assumption of randomness about the calibration and test examples.

The next proposition, to be proved in Sect. 6.5.3, shows that IVAPs are computationally efficient.

Proposition 6.7 *Given the prediction scores $s_1, \ldots, s_k$ of the calibration objects, the prediction rule for computing the IVAP's predictions can be computed in time $O(k \log k)$ and space $O(k)$. Its application to each test object takes time $O(\log k)$. Given the sorted prediction scores of the calibration objects, the prediction rule can be computed in time and space $O(k)$.*

6.4.3 Probabilistic Predictors Derived from Venn Predictors

Venn–Abers predictors output multiprobability predictions, but in some cases (e.g., when combining probabilities with utilities to arrive at optimal decisions, as discussed in Sects. 7.7 and 7.10.6) it is useful to have a probabilistic prediction, i.e., one probability measure as prediction. In this subsection we will discuss how to turn a pair of probabilities output by a Venn–Abers predictor into a single probability. So our goal here is to merge p^0 and p^1 into one probability p, and we would like our procedure to be optimal in some sense. It should be kept in mind, however, that p does not make much sense if p^0 and p^1 are very different.

We will use two popular loss functions, log loss and square loss, to formalize the optimality, in a minimax sense, of our merging procedures. The *log loss* suffered when predicting $p \in [0, 1]$ whereas the true label is y is

$$\lambda_{\log}(y, p) := \begin{cases} -\log(1-p) & \text{if } y = 0 \\ -\log p & \text{if } y = 1 \,. \end{cases} \tag{6.7}$$

This is the most fundamental loss function, since the cumulative loss $\sum_{i=1}^{n} \lambda_{\log}(y_i, p_i)$ over a test set of size n is equal to the minus log of the probability that the predictor assigns to the sequence of labels (this assumes either the offline mode of prediction with independent test examples or the online mode of prediction); therefore, a smaller cumulative log loss corresponds to a larger probability. The *square loss* suffered when predicting $p \in [0, 1]$ for the true label y is

$$\lambda_{\mathrm{sq}}(y, p) := (y - p)^2 \,. \tag{6.8}$$

The main advantage of this loss function is that it is *proper* (see, e.g., [72]): the function $\mathbb{E}_{y \sim \mathbf{B}_p} \lambda_{\mathrm{sq}}(y, q)$ of $q \in [0, 1]$, where $\mathbf{B}_p$ is the Bernoulli distribution with parameter p, attains its minimum at $q = p$. (Of course, the log loss function is proper as well.) See also Sect. 10.5.3.

First suppose that our loss function is $\lambda_{\log}$ and we are given a multiprobability prediction (p^0, p^1); let us find the corresponding minimax probabilistic prediction p. If the true outcome is $y = 0$, our regret for using p instead of the appropriate p^0 is $-\log(1-p)+\log(1-p^0)$. If $y = 1$, our regret for using p instead of the appropriate p^1 is $-\log p + \log p^1$. The first regret as a function of $p \in [0, 1]$ strictly increases from a nonpositive value to ∞ as p changes from 0 to 1. The second regret as a function of p strictly decreases from ∞ to a nonpositive value as p changes from 0 to 1. Therefore, the minimax regret is the solution to

$$-\log(1-p) + \log(1-p^0) = -\log p + \log p^1 \,,$$

which is

$$p = \frac{p^1}{1 - p^0 + p^1} \,. \tag{6.9}$$

The intuition behind this minimax value of p is that we can interpret the multiprobability prediction (p^0, p^1) as the unnormalized probability distribution P on $\{0, 1\}$ such that $P\{0\} = 1 - p^0$ and $P\{1\} = p^1$; we then normalize P to get a genuine probability distribution $P' := P/P(\{0, 1\})$, and the p in (6.9) is equal to $P'\{1\}$. (If $p^0 < p^1$, there is an excess of mass in P: $P\{0\} + P\{1\} > 1$.) Of course, it is always true that $p \in \mathrm{co}(p^0, p^1)$ (co standing for the convex hull).

In the case of the square loss function, the regret is

$$\begin{cases} p^2 - (p^0)^2 & \text{if } y = 0 \\ (1-p)^2 - (1-p^1)^2 & \text{if } y = 1 \end{cases}$$

and the two regrets are equal when

$$p := p^1 + (p^0)^2/2 - (p^1)^2/2 \,. \tag{6.10}$$

To see how natural this expression is notice that (6.10) is equivalent to

$$p = \bar{p} + (p^1 - p^0)\left(\frac{1}{2} - \bar{p}\right),$$

where $\bar{p} := (p^0 + p^1)/2$. Therefore, p is a regularized version of $\bar{p}$: we move $\bar{p}$ towards the neutral value $1/2$ in the typical (for the Venn–Abers method) case where $p^0 < p^1$. In any case, we always have $p \in \operatorname{co}(p^0, p^1)$.

The following lemma shows that log loss is never infinite for probabilistic predictors derived from Venn predictors.

Lemma 6.8 *Neither* (6.9) *nor* (6.10) *ever produces* $p \in \{0, 1\}$ *when applied to Venn–Abers predictors.*

Proof Lemma 6.6 implies that $p^0 < 1$ and that $p^1 > 0$. It remains to notice that both (6.9) and (6.10) produce p in the interior of $\operatorname{co}(p^0, p^1)$ if $p^0 \neq p^1$ and produce $p = p^0 = p^1$ if $p^0 = p^1$ (and this is true for any sensible averaging method). □

6.4.4 Cross Venn–Abers Predictors

A *cross Venn–Abers predictor (CVAP)* is just a combination of $K > 1$ IVAPs, where K is the parameter of the algorithm. It is described as Algorithm 6.2, where IVAP(A, B, x) stands for the output of IVAP applied to A as proper training set, B as calibration set, and x as test object, and GM stands for geometric mean (so that GM(p^1) is the geometric mean of $p^{1,1}, \dots, p^{1,K}$ and GM$(1 - p^0)$ is the geometric mean of $1 - p^{0,1}, \dots, 1 - p^{0,K}$). The folds should be of approximately equal size, and usually the training set is split into folds at random. One way to obtain a random assignment of the training examples to folds (see line 1) is to start from a regular array in which the first l_1 examples are assigned to fold 1, the following l_2 examples are assigned to fold 2, up to the last l_K examples which are assigned to fold K, where $|l_k - l/K| < 1$ for all k, and then to apply a random permutation. Remember that the procedure RANDOMIZE-IN-PLACE [55, Sect. 5.3] can do the last step in time $O(l)$.

Algorithm 6.2 CVAP(T, x) ▷ cross-Venn–Abers predictor

Input: training set T and test object x.
1: Split the training set T randomly into K folds $T_1, \dots, T_K$
2: **for** $k \in \{1, \dots, K\}$
3: $(p^{0,k}, p^{1,k}) := \text{IVAP}(T \setminus T_k, T_k, x)$
4: **return** $\text{GM}(p^1)/(\text{GM}(1 - p^0) + \text{GM}(p^1))$.

Algorithm 6.2 outputs probabilistic predictions rather than multiprobability ones. The expression $\text{GM}(p^1)/(\text{GM}(1 - p^0) + \text{GM}(p^1))$ used for merging the IVAPs' outputs is justified by the arguments of the previous subsection; see the next subsection for details.

6.4.5 Merging Multiprobability Predictions into a Probabilistic Prediction

In CVAP (Algorithm 6.2) we merge the K multiprobability predictions output by K IVAPs. In this section we design a minimax way (for the log and square loss functions) for merging them. For the log loss function the result is especially simple, $\text{GM}(p^1)/(\text{GM}(1-p^0)+\text{GM}(p^1))$; this is what we used in line 4 of Algorithm 6.2.

Let us check that $\text{GM}(p^1)/(\text{GM}(1 - p^0) + \text{GM}(p^1))$ is indeed the minimax expression under log loss. Suppose the pairs of lower and upper probabilities to be merged are $(p^{0,1}, p^{1,1}), \dots, (p^{0,K}, p^{1,K})$ and the merged probability is p. The extra cumulative loss suffered by p over the correct members $p^{1,1}, \dots, p^{1,K}$ of the pairs when the true label is 1 is

$$\log \frac{p^{1,1}}{p} + \cdots + \log \frac{p^{1,K}}{p}, \tag{6.11}$$

and the extra cumulative loss of p over the correct members of the pairs when the true label is 0 is

$$\log \frac{1 - p^{0,1}}{1 - p} + \cdots + \log \frac{1 - p^{0,K}}{1 - p}. \tag{6.12}$$

Equalizing the two expressions we obtain

$$\frac{p^{1,1} \dots p^{1,K}}{p^K} = \frac{(1 - p^{0,1}) \dots (1 - p^{0,K})}{(1 - p)^K},$$

which gives the required minimax expression for the merged probability (since (6.11) is decreasing and (6.12) is increasing in p).

In the case of the square loss function, we solve the linear equation

$$(1-p)^2-(1-p^{1,1})^2+\cdots+(1-p)^2-(1-p^{1,K})^2 = p^2-(p^{0,1})^2+\cdots+p^2-(p^{0,K})^2$$

in p; the result is

$$p = \frac{1}{K}\sum_{k=1}^{K}\left(p^{1,k} + \frac{1}{2}(p^{0,k})^2 - \frac{1}{2}(p^{1,k})^2\right).$$

This expression is more natural than it looks: see Sect. 6.4.3; notice that it reduces to arithmetic mean when $p^0 = p^1$.

6.5 Proofs

6.5.1 *Proof of Theorem 6.4*

Since $\mathbf{X}$ is a standard Borel space, we assume, without loss of generality, that $\mathbf{X} = [0, 1]$. For each $n = 1, 2, \ldots$ consider the partition of the interval $[0, 1]$ into the bins

$$B_{n,k} := \left[\frac{k-1}{K_n}, \frac{k}{K_n}\right), \quad k = 1, \ldots, K_n - 1, \quad B_{n,K_n} := \left[\frac{K_n - 1}{K_n}, 1\right]$$

of equal width $1/K_n$; the number of bins K_n is allowed to depend on n. We will be interested in the case where $K_n \to \infty$ but $K_n/n \to 0$ as $n \to \infty$.

Let (x_i, y_i), $i = 1, 2, \ldots$, be the examples output by Reality. Define, for every $(x, y) \in \mathbf{Z}$,

$$Q_n(y \mid x) := \frac{N_n(x, y)}{N_n(x)}, \quad Q_n^*(y \mid x) := \frac{N_n(x, y)}{n Q_{\mathbf{X}}(B_n(x))},$$

where $B_n(x)$ is the bin in the nth partition (consisting of the bins $B_{n,k}$, $k = 1, \ldots, K_n$) containing x, $N_n(x)$ is the number of $i = 1, \ldots, n$ such that $x_i \in B_n(x)$, $N_n(x, y)$ is the number of $i = 1, \ldots, n$ such that $x_i \in B_n(x)$ and $y_i = y$, and the uncertainty $0/0$ is resolved to, say, $1/|\mathbf{Y}|$. We will write $Q(y \mid x)$ for $Q_{\mathbf{Y}|\mathbf{X}}(y \mid x)$.

Lemma 6.9 *Suppose $K_n \to \infty$, $K_n = o(n)$, and $\mathbf{Y} = \{0, 1\}$. For any $\delta > 0$ and large enough n,*

$$\mathbb{P}\left\{\int \left|Q(1 \mid x) - Q_n^*(1 \mid x)\right| Q_{\mathbf{X}}(\mathrm{d}x) > \delta\right\} \le \mathrm{e}^{-n\delta^2/8}$$

where the outermost probability distribution $\mathbb{P} = Q^\infty$ generates the examples (x_i, y_i), which determine the empirical distributions Q_n and "semi-empirical distributions" Q_n^*.

Proof See [82], Theorem 9.4 and the displayed equation preceding (9.1). □

Lemma 6.10 *Suppose $K_n \to \infty$ and $K_n = o(n)$. For any $\delta > 0$ there exists a $\delta^* > 0$ such that, for large enough n,*

$$\mathbb{P}\left\{Q_{\mathbf{X}}\left\{x : \max_{y\in\mathbf{Y}} \left|Q_n^*(y \mid x) - Q(y \mid x)\right| > \delta\right\} > \delta\right\} \leq \mathrm{e}^{-\delta^* n} . \tag{6.13}$$

Proof We apply Lemma 6.9 to the binary classification problem obtained from our classification problem by replacing label $y \in \mathbf{Y}$ with 1 and replacing all other labels with 0:

$$\mathbb{P}\left\{\int \left|Q(y \mid x) - Q_n^*(y \mid x)\right| Q_{\mathbf{X}}(\mathrm{d}x) > \delta\right\} \leq \mathrm{e}^{-n\delta^2/8} .$$

By Markov's inequality this implies

$$\mathbb{P}\left\{Q_{\mathbf{X}}\{\left|Q(y \mid x) - Q_n^*(y \mid x)\right| > \sqrt{\delta}\} > \sqrt{\delta}\right\} \leq \mathrm{e}^{-n\delta^2/8} ,$$

which, in turn, implies

$$\mathbb{P}\left\{Q_{\mathbf{X}}\left\{\max_{y\in\mathbf{Y}} |Q(y \mid x) - Q_n(y \mid x)| > \sqrt{\delta}\right\} > |\mathbf{Y}|\sqrt{\delta}\right\} \leq \mathrm{e}^{-n\delta^2/8} .$$

This completes the proof, since we can take the δ in the last equation arbitrarily small as compared with the δ in the statement of the lemma. □

□

Lemma 6.10 implies the analogous statement (Corollary 6.12) for the empirical distributions Q_n, but we need an intermediate step.

Lemma 6.11 *Suppose $K_n \to \infty$ and $K_n = o(n)$. For any $\delta > 0$ there exists a $\delta^* > 0$ such that, for large enough n,*

$$\mathbb{P}\left\{Q_{\mathbf{X}}\left\{x : \left|\frac{N_n(x)/n}{Q_{\mathbf{X}}(B_n(x))} - 1\right| > \delta\right\} > \delta\right\} \leq \mathrm{e}^{-\delta^* n} .$$

Proof Replacing $\max_{y\in\mathbf{Y}}$ by $\sum_{y\in\mathbf{Y}}$ in (6.13), we obtain

$$\mathbb{P}\left\{Q_{\mathbf{X}}\left\{x : \sum_{y\in\mathbf{Y}} \left|Q_n^*(y \mid x) - Q(y \mid x)\right| > |\mathbf{Y}|\,\delta\right\} > \delta\right\} \leq \mathrm{e}^{-\delta^* n} ;$$

it remains to notice that

$$\sum_{y\in\mathbf{Y}} \left|Q_n^*(y \mid x) - Q(y \mid x)\right| \geq \left|\sum_{y\in\mathbf{Y}} Q_n^*(y \mid x) - \sum_{y\in\mathbf{Y}} Q(y \mid x)\right| = \left|\frac{N_n(x)/n}{Q_{\mathbf{X}}(B_n(x))} - 1\right| .$$

□

The two preceding lemmas immediately give

Corollary 6.12 *Suppose $K_n \to \infty$ and $K_n = o(n)$. For any $\delta > 0$ there exists a $\delta^* > 0$ such that, for large enough n,*

$$\mathbb{P}\left\{Q_{\mathbf{X}}\left\{x : \max_{y\in\mathbf{Y}} |Q_n(y \mid x) - Q(y \mid x)| > \delta\right\} > \delta\right\} \leq \mathrm{e}^{-\delta^* n} .$$

The following result is proved in [82] (Theorem 6.2 and its proof):

Lemma 6.13 *Suppose $K_n \to \infty$ and $K_n = o(n)$. For any constant C,*

$$Q_{\mathbf{X}}\{x : N_n(x) > C\} \to 1$$

in probability as $n \to \infty$.

Now it is easy to prove Theorem 6.4. Consider the Venn predictor determined by the taxonomy

$$K(D, (x, y)) := B_n(x)$$

(so that $K(D, (x, y))$ does not depend on the bag $D \in \mathbf{Z}^{(*)}$ or the label $y \in \mathbf{Y}$). It suffices to show that

$$\operatorname{diam}(P_n) \to 0 , \tag{6.14}$$

where P_n is the Venn prediction, and

$$\rho\,(Q_n(\cdot \mid x_n), Q(\cdot \mid x_n)) \to 0 \tag{6.15}$$

in probability as $n \to \infty$ (remember that, by the definition of Venn predictor, $Q_n(\cdot \mid x_n) \in P_n$). But this is simple: (6.14) follows from Lemma 6.13 and (6.15) follows from Corollary 6.12.

6.5.2 *PAVA and the Proof of Lemma 6.6*

In this subsection we will describe the PAVA, and Lemma 6.6 will be seen as an obvious property of the algorithm. We will use the notation of the lemma but will not assume that t_i are prediction scores. In general, we will call the function that maximizes the likelihood (6.6) with $t_1, \dots, t_n$ in place of $S(x_1), \dots, S(x_l)$ the *isotonic regressor*. It is unique by Ayer et al. [11, Corollary 2.1], which is a statement about isotonic regressors, not just isotonic calibrators; Lemma 6.6 also holds for any isotonic regressor. Now we will see how the isotonic regressor g can be found.

Arrange the numbers t_i in the strictly ascending order $t'_1 < \cdots < t'_{n'}$, where $n' \le n$ is the number of distinct elements among t_i; therefore, t'_i, $i = 1, \dots, n'$, is the ith smallest element of the set $\{t_1, \dots, t_n\}$ (with all duplicates removed, of course). We would like to find the increasing function g on the set $\{t'_1, \dots, t'_{n'}\} = \{t_1, \dots, t_n\}$ maximizing the likelihood. The procedure is recursive. At each step the set $\{t'_1, \dots, t'_{n'}\}$ is partitioned into a number of disjoint cells consisting of adjacent elements of the set; to each cell is assigned a ratio a/W (formally, a pair of integers, with $a \ge 0$ and $W > 0$); the function g defined at this step (perhaps to be redefined at the following steps) is constant on each cell. For $j = 1, \dots, n'$, let a_j be the number of i such that $y_i = 1$ and $t_i = t'_j$, and let W_j be the number of i such that $t_i = t'_j$. Start from the partition of $\{t'_1, \dots, t'_{n'}\}$ into one-element cells, assign the ratio a_j/W_j to $\{t'_j\}$, and set

$$g(t'_j) := \frac{a_j}{W_j} \tag{6.16}$$

(in the notation used in this description, a/W is a pair of integers, whereas $\frac{a}{W}$ is a rational number, the result of the division). If the function g is increasing, we are done. If not, there is a pair C_1, C_2 of adjacent cells ("violators") such that C_1 is to the left of C_2 and $g(C_1) > g(C_2)$ (where $g(C)$ stands for the common value of $g(t'_j)$ for $t'_j \in C$); in this case redefine the partition by merging C_1 and C_2 into one cell C, assigning the ratio $(a_1 + a_2)/(W_1 + W_2)$ to C, where a_1/W_1 and a_2/W_2 are the ratios assigned to C_1 and C_2, respectively, and setting

$$g(t'_j) := \frac{W_1}{W_1 + W_2} g(C_1) + \frac{W_2}{W_1 + W_2} g(C_2) = \frac{a_1 + a_2}{W_1 + W_2} \tag{6.17}$$

for all $t'_j \in C$. Repeat the process until g becomes increasing (the number of cells decreases by 1 at each iteration, so the process will terminate in at most n' steps). The final function g is the one that maximizes the likelihood. The statement of the lemma follows from this recursive definition: it is true by definition for the initial function (6.16) and remains true when g is redefined by (6.17).

6.5.3 Proof of Proposition 6.7

The proof relies on an elegant geometric representation of the PAVA (described in the previous subsection). Let us see how to fit the isotonic regressor to $(s_1, y_1), \ldots, (s_k, y_k)$ in geometric terms (now we make our notation closer to the one used for IVAPs, but still do not necessarily assume that s_i are the calibration scores and y_i are the calibration labels). As before, we start from sorting all scores $s_1, \ldots, s_k$ in the ascending order and removing the duplicates. (This is the most computationally expensive step in our calibration procedure, taking time $O(k \log k)$ in the worst case.) Let $k' \leq k$ be the number of distinct elements among $s_1, \ldots, s_k$, i.e., the cardinality of the set $\{s_1, \ldots, s_k\}$. Define s'_j, $j = 1, \ldots, k'$, to be the jth smallest element of $\{s_1, \ldots, s_k\}$, so that $s'_1 < s'_2 < \cdots < s'_{k'}$. Define $w_j := \left|\left\{i = 1, \ldots, k : s_i = s'_j\right\}\right|$ to be the number of times s'_j occurs among $s_1, \ldots, s_k$. Finally, define

$$y'_j := \frac{1}{w_j} \sum_{i=1,\ldots,k: s_i = s'_j} y_i$$

to be the average label corresponding to $s_i = s'_j$.

The *cumulative sum diagram* (CSD) of $(s_1, y_1), \ldots, (s_k, y_k)$ is the set of points

$$P_i := \left(\sum_{j=1}^{i} w_j, \sum_{j=1}^{i} y'_j w_j \right), \quad i = 0, 1, \ldots, k' ; \tag{6.18}$$

in particular, $P_0 = (0, 0)$. The *greatest convex minorant* (GCM) is the greatest convex minorant of the CSD, i.e., the largest convex function of the type $[0, k] \to [0, \infty)$ such that every point P_i in the CSD lies on or above its graph (its domain is $[0, k]$ since $\sum_j w_j = k$). The value at s'_i, $i = 1, \ldots, k'$, of the isotonic regressor fitted to $(s_1, y_1), \ldots, (s_k, y_k)$ is defined to be the slope of the GCM between $\sum_{j=1}^{i-1} w_j$ and $\sum_{j=1}^{i} w_j$; the values at other s are somewhat arbitrary (namely, the value at $s \in (s'_i, s'_{i+1})$ can be set to anything between the left and right slopes of the GCM at $\sum_{j=1}^{i} w_j$), but we never need them (unlike in the standard use of isotonic regression in machine learning, [430]): e.g., $f^1(s)$ is the value of the isotonic regression fitted to a sequence that already contains $(s, 1)$.

The idea behind computing the pair $(f^0(s), f^1(s))$ efficiently is to pre-compute two vectors F^0 and F^1 storing $f^0(s)$ and $f^1(s)$, respectively, for all possible values of s. Let k' and s'_i be as defined above in the case where $s_1, \ldots, s_k$ are the calibration scores and $y_1, \ldots, y_k$ are the corresponding labels. The vectors F^0 and F^1 are of length k', and for all $i = 1, \ldots, k'$ and both $\epsilon \in \{0, 1\}$, F^ϵ_i is the value of $f^\epsilon(s)$ when $s = s'_i$. Therefore, for all $i = 1, \ldots, k'$:

- F_i^1 is also the value of $f^1(s)$ when s is just to the left of s'_i;
- F_i^0 is also the value of $f^0(s)$ when s is just to the right of s'_i.

Since f^0 and f^1 can change their values only at the points s'_i, the vectors F^0 and F^1 uniquely determine the functions f^0 and f^1, respectively.

Let k', s'_i, and w_i be as defined above in the case where $s_1, \ldots, s_k$ and $y_1, \ldots, y_k$ are the calibration scores and labels. The *corners* of a GCM are the points on the GCM where the slope of the GCM changes. It is clear that the corners belong to the CSD, and we also add the extreme points (P_0 and $P_{k'}$ in the case of (6.18)) of the CSD to the list of corners.

We will only explain in detail how to compute F^1; the computation of F^0 is analogous and will be explained only briefly. First we explain how to compute F_1^1.

Extend the CSD as defined above (in the case where $s_1, \ldots, s_k$ and $y_1, \ldots, y_k$ are the calibration scores and labels) by adding the point $P_{-1} := (-1, -1)$. The corresponding GCM will be referred to as the *initial GCM*; it has at most $k' + 2$ corners. Algorithm 6.3, which operates with a stack S (initially empty), computes the corners; it is a trivial modification of Graham's scan ([139] and [55, Sect. 33.3]). We are using the notation of [55], which should, however, be self-explanatory. The corners are returned on the stack S, and they are ordered from left to right (P_{-1} being at the bottom of S and $P_{k'}$ at the top). The operator "and" in line 4 is, as usual, short circuiting. The expression "the angle formed by points a, b, and c makes a nonleft (resp. nonright) turn" may be taken to mean that $(b - a) \times (c - b) \leq 0$ (resp. ≥ 0), where $\times$ stands for cross product of planar vectors; this avoids computing angles and divisions (see, e.g., [55, Sect. 33.1]).

Algorithm 6.3 allows us to compute F_1^1 as the slope of the line between the two bottom corners in S, but this will be done by the next algorithm.

The rest of the procedure for computing the vector F^1 is shown as Algorithm 6.4. The main data structure in Algorithm 6.4 is a stack S', which is initialized (in lines 1–2) by putting in it all corners of the initial GCM in reverse order as compared with S (so that $P_{-1} = (-1, -1)$ is initially at the top of S').

At each point in the execution of Algorithm 6.4 we will have a length-1 *active interval* and the *active corner*, which will nearly always be at the top of the stack S'. The initial CSD can be visualized by connecting each pair of adjacent points: P_{-1} and P_0, P_0 and P_1, etc. It stretches over the interval $[-1, k]$ of the horizontal axis; the subinterval $[-1, 0]$ corresponds to the test score s (assumed to be to the left of all s'_i) and each subinterval $\left[\sum_{j=1}^{i-1} w_j, \sum_{j=1}^{i} w_j\right]$ corresponds to the calibration score s'_i, $i = 1, \ldots, k'$. The active corner is initially at $P_{-1} = (-1, -1)$; the corners to the left of the active corner are irrelevant and ignored (not remembered in S'). The active interval is always between the first coordinate of $\text{TOP}(S')$ and the first coordinate of $\text{NEXT-TO-TOP}(S')$. At each iteration $i = 1, \ldots, k'$ of the main loop 3–11 we are computing F_i^1, i.e., $f^1(s)$ for the situation where s is between s'_{i-1} and s'_i (meaning to the left of s'_1 if $i = 1$), and after that we swap the active interval (corresponding to s) and the interval corresponding to s'_i; of course, after swapping pieces of CSD are adjusted vertically in order to make the CSD as a whole continuous.

Algorithm 6.3 Initializing the corners for computing F^1

1: $\text{PUSH}(P_{-1}, S)$
2: $\text{PUSH}(P_0, S)$
3: **for** $i \in \{1, 2, \ldots, k'\}$
4:     **while** S.size > 1 and the angle formed by points $\text{NEXT-TO-TOP}(S)$, $\text{TOP}(S)$, and P_i makes a nonleft turn
5:         $\text{POP}(S)$
6:     $\text{PUSH}(P_i, S)$
7: **return** S.

Algorithm 6.4 Computing F^1

```
1: while ¬STACK-EMPTY(S)
2:     PUSH(POP(S), S')
3: for i ∈ {1, 2, ..., k'}
4:     set F_i^1 to the slope of →(TOP(S'), NEXT-TO-TOP(S'))
5:     P_{i-1} = P_{i-2} + P_i − P_{i-1}
6:     if P_{i-1} is at or above →(TOP(S'), NEXT-TO-TOP(S'))
7:         continue
8:     POP(S')
9:     while S'.size > 1 and the angle formed by points P_{i-1}, TOP(S'), and
           NEXT-TO-TOP(S') makes a nonleft turn
10:        POP(S')
11:    PUSH(P_{i-1}, S')
12: return F^1.
```

At the beginning of each iteration i of the loop 3–11 we have the CSD

$$P_{-1}, P_0, P_1, \ldots, P_{k'} \tag{6.19}$$

corresponding to

$$\text{the points } s'_1, \ldots, s'_{i-1}, s, s'_i, s'_{i+1}, \ldots, s'_{k'}$$
$$\text{with the weights } w_1, \ldots, w_{i-1}, 1, w_i, w_{i+1}, \ldots, w_{k'}$$

(respectively); the active interval is the projection of $\overrightarrow{P_{i-2}, P_{i-1}}$ (onto the horizontal axis, here and later). At the end of that iteration we have the CSD which looks identical to (6.19) but in fact contains a different point P_{i-1} (cf. line 5 of the algorithm) and corresponds to

$$\text{the points } s'_1, \ldots, s'_{i-1}, s'_i, s, s'_{i+1}, \ldots, s'_{k'}$$
$$\text{with the weights } w_1, \ldots, w_{i-1}, w_i, 1, w_{i+1}, \ldots, w_{k'}$$

(respectively); the active interval becomes the projection of $\overrightarrow{P_{i-1}, P_i}$. To achieve this, in line 5 we redefine P_{i-1} to be the reflection of the old P_{i-1} across the mid-point $(P_{i-2} + P_i)/2$. The stack S' always consists of corners of the GCM of the current CSD, and it contains all the corners to the right of the active interval (plus one more corner, which is the active corner).

At each iteration i of the loop 3–11:

- We report the slope of the GCM over the active interval as F_i^1 (line 4).
- We then swap the fragments of the CSD corresponding to the active interval and to s'_i leaving the rest of the CSD intact. This way the active interval moves to the right (from the projection of $\overrightarrow{P_{i-2}, P_{i-1}}$ to the projection of $\overrightarrow{P_{i-1}, P_i}$).
- If the point P_{i-1} above the left end-point of the active interval is above (or at) the GCM, move to the next iteration of the loop. (The active corner does not change.) The rest of this description assumes that P_{i-1} is strictly below.
- Make P_{i-1} the active corner. Redefine the GCM to the right of the active corner by connecting the active corner to the right-most corner C such that the slope of the line connecting the active corner and that corner is minimal; all the corners between the active corner and that right-most corner C are then forgotten.

Lemma 6.14 *The worst-case computation time of Algorithms 6.3 and 6.4 is $O(k')$.*

Proof In the case of Algorithm 6.3, see [55, Sect. 33.3]. In the case of Algorithm 6.4, it suffices to notice that the total number of iterations for the **while** loop does not exceed the total number of elements pushed onto S' (since at each iteration we pop an element off S'); and the total number of elements pushed onto S' is at most k' (in the first **for** loop) plus k' (in the second **for** loop). □

For convenience of the reader wishing to program IVAPs and CVAPs, we also give the counterparts of Algorithms 6.3 and 6.4 for computing F^0: see Algorithms 6.5 and 6.6. In those algorithms, we do not need the point P_{-1} any more; however, we need a new point $P_{k'+1} := P_{k'} + (1, 0)$. The stacks S and S' that they use are initially empty.

After computing F^0 and F^1 we can arrange the calibration scores $s'_1, \dots, s'_{k'}$ into a binary search tree: see Algorithm 6.7, where F^0_0 is defined to be 0 and $F^1_{k'+1}$ is defined to be 1; we will refer to s'_i as the *keys* of the corresponding nodes (only internal nodes will have keys). Algorithm 6.7 is in fact more general than what we need: it computes the binary search tree for the scores $s'_a, s'_{a+1}, \dots, s'_b$ for $a \le b$; therefore, we need to run BST$(1, k')$. The size of the binary search tree is $2k' + 1$; k' of its nodes are internal nodes corresponding to different values of s'_i, $i = 1, \dots, k'$, and the other $k' + 1$ of its nodes are leaves corresponding to the $k' + 1$ intervals formed by the points $s'_1, \dots, s'_{k'}$.

Once we have the binary search tree it is easy to compute the prediction for a test object x in time logarithmic in k': see Algorithm 6.8, which passes x through the tree and uses N to denote the current node. Formally, we give the test object x, the proper training set T', and the calibration set T'' as the inputs of Algorithm 6.8; however, the algorithm uses for prediction the binary search tree built from T' and T'', and the bulk of work is done in Algorithms 6.3–6.7.

Algorithm 6.5 Initializing the corners for computing F^0

1: PUSH$(P_{k'+1}, S)$
2: PUSH$(P_{k'}, S)$
3: **for** $i \in \{k' - 1, k' - 2, \dots, 0\}$
4:     **while** S.size > 1 and the angle formed by points NEXT-TO-TOP(S), TOP(S), and P_i makes a nonright turn
5:         POP(S)
6:     PUSH(P_i, S)
7: **return** S.

Algorithm 6.6 Computing F^0

1: **while** ¬STACK-EMPTY(S)
2:     PUSH(POP(S), S')
3: **for** $i \in \{k', k' - 1, \dots, 1\}$
4:     set F^0_i to the slope of $\overrightarrow{\text{TOP}(S'), \text{NEXT-TO-TOP}(S')}$
5:     $P_i = P_{i-1} + P_{i+1} - P_i$
6:     **if** P_i is at or above $\overrightarrow{\text{TOP}(S'), \text{NEXT-TO-TOP}(S')}$
7:         **continue**
8:     POP(S')
9:     **while** S'.size > 1 and the angle formed by points P_i, TOP(S'), and NEXT-TO-TOP(S') makes a nonright turn
10:         POP(S')
11:     PUSH(P_i, S')
12: **return** F^0.

Algorithm 6.7 BST(a, b) ▷ binary search tree

Input: to create the binary search tree, run BST$(1, k')$.

1: **if** $b = a$
2: construct the binary tree whose root has key s'_a and payload $\{F^0_a, F^1_a\}$,
left child is a leaf with payload $\{F^0_{a-1}, F^1_a\}$,
and right child is a leaf with payload $\{F^0_a, F^1_{a+1}\}$
3: **return** its root
4: **elif** $b = a + 1$
5: construct the binary tree whose root has key s'_a and payload $\{F^0_a, F^1_a\}$,
left child is a leaf with payload $\{F^0_{a-1}, F^1_a\}$,
and right child is BST(b, b)
6: **return** its root
7: **elif**
8: $c = \lfloor (a + b)/2 \rfloor$
9: construct the binary tree whose root has key s'_c and payload $\{F^0_c, F^1_c\}$,
left child is BST$(a, c - 1)$,
and right child is BST$(c + 1, b)$
10: **return** its root.

Algorithm 6.8 IVAP(T', T'', x) ▷ inductive Venn–Abers predictor

1: Set N to the root of the binary search tree and compute the prediction score s of x.
2: **while** N is not a leaf
3: **if** $s <$ key(N)
4: set N to N's left child
5: **elif** $s >$ key(N)
6: set N to N's right child
7: **else** ▷ if $s =$ key(N)
8: **return** payload(N)
9: **return** payload(N).

The worst-case computational complexity of the overall procedure involves the following components:

- Training the algorithm on the proper training set, computing the scores of the calibration objects, and computing the scores of the test objects; at this stage the computation time is determined by the underlying algorithm.
- Sorting the scores of the calibration objects takes time $O(k \log k)$.
- Running our procedure for pre-computing f^0 and f^1 takes time $O(k)$ (by Lemma 6.14).
- Processing each test object takes an additional time of $O(\log k)$ (using binary search).

In principle, using binary search does not require an explicit construction of a binary search tree (cf. [55], Exercise 2.3-5), but once we have a binary search tree we can easily transform it into a red-black tree, which allows us to add new examples to (and remove old examples from) the calibration set in time $O(\log k)$ [55, Chap. 13]).

6.6 Context

Venn predictors were introduced in [404], Venn–Abers predictors in [394], and IVAPs in [408]; see those papers for numerous empirical results demonstrating their predictive efficiency. In discussing one-off Venn predictors we follow [394] generalizing from the binary case considered in [394] to multiclass classification.

6.6.1 Risk and Uncertainty

The distinction between risk and uncertainty, as used in Sect. 6.2, was made by the Chicago economist Frank Knight in his 1921 book [194]. Risk is quantifiable while Knightean uncertainty is not. This distinction quickly became very standard in economics. See, e.g., [195] for further information.

6.6.2 John Venn, Frequentist Probability, and the Problem of the Reference Class

According to the frequentist conception of probability, the probability of an event is the frequency of its occurrence in some population. But how do we choose this population? This is known as the "problem of the reference class". John Venn wrote what is often regarded as the first systematic account [372] of the frequentist theory of probability, and he was the first to formulate and analyze this problem with due depth [146, 191]. The term "problem of the reference class" is due to Reichenbach [288, Sect. 72].

In the chapter devoted to induction [372, Chap. XII], Venn asks for the probability that John Smith, aged fifty, would live to sixty-one. Should we count how many men of the age of John Smith, respectively do and do not live for eleven years? Or, if we know that John Smith is a consumptive man and a native of a northern climate, should we find the frequency in one of these narrower categories?

On the one hand, we want the categories into which we divide the examples to be large, in order to have a reasonable sample size for estimating the probabilities. But we also want them to

John Venn (1834–1923).

The portrait by Maull & Fox. Used with permission of The Royal Society. ©The Royal Society

be small and homogeneous. According to Reichenbach, the reference class should be chosen as "the narrowest class for which reliable statistics can be compiled" [288, Sect. 72].

The primitive solution to the problem of the reference class used in Sect. 6.2.3 was to include *all* examples into the reference class. All three popular point estimates that we discussed, maximum likelihood, Jeffreys, and Laplace, are special cases of the Bayes estimate $(k+\alpha)/(l+\alpha+\beta)$, where $\alpha \geq 0$ and $\beta \geq 0$ (based on the Beta(α, β) prior, with $\alpha = 0$ and $\beta = 0$ added as limiting cases); see, e.g., [141, Sect. 9.1] for a derivation. It can be checked that the Bayes estimate is in the convex hull of (6.5) for all l and k if and only if $(\alpha, \beta) \in [0, 1]^2$. The maximum likelihood and Laplace estimates have parameters on the boundary of this square, whereas for the Jeffreys estimate they are at the centre.

An advantage of the Venn–Abers predictors is that they provide a natural automatic solution to the problem of the reference class.

6.6.3 Online Venn Predictors Are Calibrated

Online Venn predictors were a major topic in the first edition of this book, where we proved a result about their validity in an online protocol with finite horizon. The result shows that online Venn predictors are automatically calibrated in a quite satisfactory sense, but its statement is complicated and given in a language that would be unfamiliar to many of our potential readers (namely, it uses the language and intuition of game-theoretic probability [320]). Therefore, we have decided to content ourselves with a simple statement of validity for one-off predictors (Proposition 6.1). The reader interested in online Venn predictors may consult [402, Sects. 6.1–6.3].

6.6.4 Isotonic Regression

The method of isotonic regression was popularized in statistics by Hugh D. Brunk, and the PAVA algorithm was proposed in [11]; "Abers" in "Venn–Abers prediction" refers to the initial letters of the surnames of the authors of that paper (Ayer, Brunk, Ewing, Reid, and Silverman). Our description of the PAVA in Sect. 6.5.2 follows the summary of [11] and [21, Sect. 1.2]. Less efficient but more general algorithms were proposed in Brunk's paper [39] published in the same issue of the *Annals of Mathematical Statistics* as [11]. In the early days of isotonic regression, the procedure was referred to as "Brunkizing" [305].

The geometric representation of the PAVA that served for us as the basis of the efficient implementation of the IVAP in Sect. 6.5.3 is described in [21, pp. 9–13] (especially Theorem 1.1).

Zadrozny and Elkan [430] realized that the method of isotonic regression (namely, the PAVA) can be used on top of scoring classifiers with scores in the interval [0, 1] in order to improve their performance as probabilistic predictors. Applied in this fashion, the method has become very popular in machine learning.

Chapter 7
Probabilistic Regression: Conformal Predictive Systems

Abstract In this chapter we discuss probabilistic regression. Namely, we apply conformal prediction to derive predictive distributions that are valid, in an important sense, under unconstrained randomness. The advantage of these conformal predictive distributions over the usual conformal prediction intervals is that the former contain more information; in particular, a conformal predictive distribution can produce a plethora of conformal prediction intervals. A key application will be to decision making.

Keywords Conformal predictive system · Least squares prediction machine · Kernel ridge regression prediction machine · Conformal predictive decision making

7.1 Introduction

We start our formal exposition in Sect. 7.2 by defining conformal predictive distributions (CPDs), which are just p-values arranged into a distribution function. An unusual feature of CPDs is that they are randomized, although they are typically affected by randomness very little. Not all conformity measures give rise to CPDs, but we give simple sufficient conditions. (In this chapter it will be convenient to base our exposition on conformity rather than nonconformity measures.)

In the rest of the chapter we discuss in detail several conformal predictive systems (CPSs), i.e., ways of producing CPDs. In Sect. 7.3 we apply our method to the classical least squares procedure obtaining what we call the least squares prediction machine (LSPM). The LSPM is defined in terms of regression residuals; accordingly, it has three main versions: ordinary, deleted, and studentized. The most useful version appears to be studentized, which does not require any assumptions on how influential any of the individual examples is. We state the studentized version (and, more briefly, the ordinary version) as an explicit algorithm. Next we discuss the validity and efficiency of the LSPM. Whereas the LSPM, as any CPS, is valid under unconstrained randomness, for investigating its efficiency we assume a parametric model, namely the standard Gaussian linear model. The question that

V. Vovk et al., *Algorithmic Learning in a Random World*,
https://doi.org/10.1007/978-3-031-06649-8_7

we try to answer is how much we should pay (in terms of efficiency) for the validity under unconstrained randomness enjoyed by the LSPM. We compare the LSPM with three kinds of oracles under the parametric model; the oracles are adapted to the parametric model and are only required to be valid under it. The weakest oracle (Oracle I) only knows the parametric model, and the strongest one (Oracle III) also knows the parameters of the model.

When the number of attributes is 1 and the only attribute is identical 1, the LSPM turns into the classical procedure that we refer to as the Dempster–Hill procedure and discuss in detail in Sect. 13.3.4. But already in this chapter the Dempster–Hill procedure serves as a useful benchmark.

The soft model underlying the LSPM includes a linearity assumption and, therefore, lacks flexibility. In Sect. 7.4 we introduce the kernel ridge regression prediction machine (KRRPM) combining a slight generalization of the LSPM with the kernel trick, so that the CPDs can take any shape, especially when used with universal kernels (discussed in Sect. 2.9.6). Despite an arbitrary shape of the predictive distributions, KRRPMs remain insufficiently flexible, since their predictive distributions do not depend much on the test object (they only adapt to the training set). In Sect. 7.5 we introduce a simple but very flexible CPS based on the nearest neighbours idea, and in Sect. 7.6 we state a theoretical result about the existence of a universal, i.e., fully adaptive, CPS.

A significant advantage of conformal predictive distributions over traditional conformal prediction is that the former can be combined with a utility function to arrive at optimal decisions. We make first steps in this direction in Sect. 7.7.

Similarly to full conformal prediction, full CPSs are computationally efficient only for a narrow class of underlying prediction algorithms. Therefore, in Sect. 7.8 we introduce computationally efficient modifications of CPSs, along the lines of ICPs and CCPs.

All proofs are postponed to Sect. 7.9, and for some proofs of asymptotic results we refer to research papers in Sect. 7.10.

7.2 Conformal Predictive Systems

We consider the one-off, or online, prediction task: given a training set of $n-1$ examples $z_i = (z_i, y_i)$, $i = 1, \dots, n-1$, and a test object $x_n \in \mathbf{X}$, to predict the label $y_n \in \mathbb{R}$ of x_n. Our notions of predictive systems will not contain any requirements of validity (similarly to confidence predictors in Chap. 2), and those will be introduced separately.

7.2.1 *Basic Definitions*

For now let us fix $n \in \mathbb{N}$. A measurable function $\Pi : \mathbf{Z}^n \to [0, 1]$ is called a *deterministic predictive system* (DPS) if it satisfies the following two requirements:

r1 For each training set $(z_1, \ldots, z_{n-1}) \in \mathbf{Z}^{n-1}$ and each test object $x_n \in \mathbf{X}$, the function $\Pi(z_1, \ldots, z_{n-1}, (x_n, y))$ is monotonically increasing in $y \in \mathbb{R}$.
r2 For each training set $(z_1, \ldots, z_{n-1}) \in \mathbf{Z}^{n-1}$ and each test object $x_n \in \mathbf{X}$,

$$\lim_{y \to -\infty} \Pi(z_1, \ldots, z_{n-1}, (x_n, y)) = 0$$

$$\lim_{y \to \infty} \Pi(z_1, \ldots, z_{n-1}, (x_n, y)) = 1 .$$

The deterministic predictive system Π outputs the predictive distribution function

$$\Pi_n : y \in \mathbb{R} \mapsto \Pi(z_1, \ldots, z_{n-1}, (x_n, y))$$

on any training set $z_1, \ldots, z_{n-1}$ and any test object x_n. Items r1 and r2 just say that Π's output should be a genuine distribution function except that we do not formally require the right-continuity of Π_n (but when we define the integrals w.r. to such functions, our definition will be equivalent to the integration w.r. to their right-continuous modifications).

To be able to attain interesting properties of validity, let us relax this definition slightly. A function $\Pi : \mathbf{Z}^n \times [0, 1] \to [0, 1]$ is a *randomized predictive system* (RPS) if:

R1 For all $(z_1, \ldots, z_{n-1}) \in \mathbf{Z}^{n-1}$ and $x_n \in \mathbf{X}$, $\Pi(z_1, \ldots, z_{n-1}, (x_n, y), \tau)$ is monotonically increasing both in $y \in \mathbb{R}$ and in $\tau \in [0, 1]$. In other words, for each $\tau \in [0, 1]$, the function

$$y \in \mathbb{R} \mapsto \Pi(z_1, \ldots, z_{n-1}, (x_n, y), \tau)$$

is monotonically increasing, and for each $y \in \mathbb{R}$, the function

$$\tau \in [0, 1] \mapsto \Pi(z_1, \ldots, z_{n-1}, (x_n, y), \tau)$$

is monotonically increasing.
R2 For all $(z_1, \ldots, z_{n-1}) \in \mathbf{Z}^{n-1}$ and $x_n \in \mathbf{X}$,

$$\lim_{y \to -\infty} \Pi(z_1, \ldots, z_{n-1}, (x_n, y), 0) = 0 \tag{7.1}$$

and

$$\lim_{y \to \infty} \Pi(z_1, \ldots, z_{n-1}, (x_n, y), 1) = 1 . \tag{7.2}$$

Apart from relaxing the definition of a distribution function we allow randomization.

The output of the randomized predictive system Π on a training set $z_1, \ldots, z_{n-1}$ and a test object x_n is the function

$$\Pi_n : (y, \tau) \in \mathbb{R} \times [0, 1] \mapsto \Pi(z_1, \ldots, z_{n-1}, (x_n, y), \tau) , \tag{7.3}$$

which will be called the *randomized predictive distribution (function)* (RPD) output by Π. The *thickness* of an RPD Π_n is the infimum of the numbers $\epsilon \geq 0$ such that the diameter

$$\Pi_n(y, 1) - \Pi_n(y, 0) \tag{7.4}$$

of the set

$$\{\Pi_n(y, \tau) : \tau \in [0, 1]\} \tag{7.5}$$

is at most ϵ for all $y \in \mathbb{R}$ except for finitely many values. The *exception size* of Π_n is the cardinality of the set of y for which the diameter (7.4) exceeds the thickness of Π_n. Notice that *a priori* the exception size can be infinite.

In this chapter we will be interested in RPDs of thickness $\frac{1}{n}$ with exception size at most $n - 1$, for typical training sets of size $n - 1$ (cf. (7.17) below). In all our examples, $\Pi(z_1, \ldots, z_n, \tau)$ will be a continuous function of τ. Therefore, the set (7.5) will be a closed interval in $[0, 1]$. However, we do not include these requirements in our official definition.

An example of a randomized predictive distribution is shown in Fig. 7.2 below as the shaded area. The size of the training set for that plot is $n - 1 = 10$; see later for details. Therefore, we are discussing an instance of Π_{11}, of thickness $1/11$ with exception size 10. The shaded area is $\{(y, \Pi_{11}(y, \tau)) : y \in \mathbb{R}, \tau \in [0, 1]\}$. We can regard (y, τ) as a coordinate system for the shaded area. The cut of the shaded area by the vertical line passing through a point y of the horizontal axis is the closed interval $[\Pi_{11}(y, 0), \Pi_{11}(y, 1)]$.

According to our general definition, the conformal transducer determined by a conformity measure A is defined as

$$\begin{aligned}\Pi(z_1, \ldots, z_{n-1}, (x_n, y), \tau) := \frac{1}{n} &\left|\left\{i = 1, \ldots, n : \alpha_i^y < \alpha^y\right\}\right| \\ &+ \frac{\tau}{n} \left|\left\{i = 1, \ldots, n : \alpha_i^y = \alpha_n^y\right\}\right| ,\end{aligned} \tag{7.6}$$

where $(z_1, \ldots, z_{n-1}) \in \mathbf{Z}^{n-1}$ is a training set, $x_n \in \mathbf{X}$ is a test object, and for each $y \in \mathbb{R}$ the corresponding conformity scores α_i^y are defined by

$$\begin{aligned}\alpha_i^y &:= A(\lbag z_1, \ldots, z_{n-1}, (x_n, y) \rbag, z_i), \qquad i = 1, \ldots, n-1, \\ \alpha_n^y &:= A(\lbag z_1, \ldots, z_{n-1}, (x_n, y) \rbag, (x_n, y)) .\end{aligned} \tag{7.7}$$

A *conformal predictive system* (CPS) is a function which is both a conformal transducer (determined by some conformity measure) and a randomized predictive system. A *conformal predictive distribution* (CPD) is a function Π_n defined by (7.3) for a conformal predictive system Π. The shaded area in Fig. 7.2 still serves as an example.

7.2.2 Properties of Validity

Any conformal predictive system Π and Borel set $A \subseteq [0, 1]$ define the generalized conformal predictor

$$\Gamma^A(z_1, \dots, z_{n-1}, x_n, \tau) := \{y \in \mathbb{R} : \Pi(z_1, \dots, z_{n-1}, (x_n, y), \tau) \in A\} . \qquad (7.8)$$

The standard property of validity for conformal prediction implies that the p-values $\Pi(z_1, \dots, z_n, \tau)$ are distributed uniformly on $[0, 1]$ when $z_1, \dots, z_n$ are IID and τ is generated independently of $z_1, \dots, z_n$ from the uniform probability distribution $\mathbf{U}$ on $[0, 1]$ (see Propositions 2.4 and 2.11). This property of validity is usually referred to as *calibration in probability* in the literature on probability forecasting. It implies that the coverage probability, i.e., the probability of $y_n \in \Gamma^A(z_1, \dots, z_{n-1}, x_n)$, for the generalized conformal predictor (7.8) is $\mathbf{U}(A)$.

A function $\Pi : (\mathbf{Z}^* \setminus \{\Box\}) \times [0, 1] \to [0, 1]$ is a *randomized predictive system* (resp. *conformal predictive system*) if its restriction to any $\mathbf{Z}^n \times [0, 1] \to [0, 1]$, $n \in \mathbb{N}$, is an RPS (resp. CPS). We know that, when applied in the online mode, a CPS satisfies a stronger property of validity: if the examples $z_1, z_2, \dots \sim Q$ and random numbers $\tau_1, \tau_2, \dots \sim \mathbf{U}$ are all independent, the values $\Pi(z_1, \dots, z_n, \tau_n) \sim \mathbf{U}$, $n = 1, 2, \dots$, are also independent; here Q is any probability measure on $\mathbf{Z}$. This may be called *strong calibration in probability*.

In Sect. 2.2.9 we discussed two notions of conformity: conformity to a bag and conformity to a property. In this chapter we will only be interested in conformity to a property. A standard example, already discussed on several occasions, is (2.33). In this example α_i scores how well y_i conforms to the property of being large. A label can only be strange (nonconforming) if it is too small; large is never strange (unless we are in an anomalous situation where an increase in y_i leads to an even larger increase in $\hat{y}_i$).

Next we discuss why the definition of a randomized predictive system, combined with the requirement of calibration in probability, is natural. The key elements of this combination are that (1) the distribution function Π is monotonically increasing, and (2) its value is uniformly distributed. The following lemma shows that these are defining properties of distribution functions of probability measures on the real line. For simplicity, we only consider the case of a continuous distribution function.

Lemma 7.1 *Suppose F is a continuous distribution function on $\mathbb{R}$, and Y is a random variable distributed as F. If $\Pi : \mathbb{R} \to \mathbb{R}$ is a monotonically increasing function such that the distribution of $\Pi(Y)$ is uniform on $[0, 1]$, then $\Pi = F$.*

This lemma suggests that requirement R1 in the definition of RPSs and the requirement of calibration in probability are the important ones. However, requirement R2 is formally independent in our case of unrestricted randomness (rather than a single probability measure on $\mathbb{R}$): consider, e.g., a conformity measure A that depends only on the objects x_i but does not depend on their labels y_i; in this case the left-hand side of (7.1) will be close to 1 for large n and highly conforming x_n.

7.2.3 Simplest Example: Monotonic Conformity Measures

We start from a simple but very restrictive condition on a conformity measure making the corresponding conformal transducer satisfy requirement R1. Consider $n \geq 2$ (the statements in the rest of this section can be interpreted in two different ways: either for a fixed n or with a quantifier over n). A conformity measure A is *monotonic* if $A^{\text{del}}(\lbag z_1, \ldots, z_{n-1} \rbag, z_n)$ is:

- monotonically increasing in y_n,

$$y_n \leq y'_n \Longrightarrow A^{\text{del}}(\lbag z_1, \ldots, z_{n-1} \rbag, (x_n, y_n)) \leq A^{\text{del}}(\lbag z_1, \ldots, z_{n-1} \rbag, (x_n, y'_n)) ;$$

- monotonically decreasing in y_1,

$$y_1 \leq y'_1 \Longrightarrow A^{\text{del}}(\lbag (x_1, y_1), z_2, \ldots, z_{n-1} \rbag, z_n) \\ \geq A^{\text{del}}(\lbag (x_1, y'_1), z_2, \ldots, z_{n-1} \rbag, z_n) .$$

  (By the definition of a bag, being decreasing in y_1 is equivalent to being decreasing in y_i for any $i = 2, \ldots, n-1$.)

This condition implies that the corresponding conformal transducer (7.6) satisfies requirement R1 by Proposition 7.2 below.

An example of a monotonic conformity measure is (2.33), where the predictions $\hat{y}_i$ are produced by the K-nearest neighbours regression algorithm:

$$A^{\text{del}}(\lbag (x_1, y_1), \ldots, (x_{n-1}, y_{n-1}) \rbag, (x_n, y_n)) := y_n - \hat{y}_n := y_n - \frac{1}{K} \sum_{k=1}^{K} y_{(k)} ,$$

where $y_{(1)}, \ldots, y_{(n-1)}$ is the sequence $y_1, \ldots, y_{n-1}$ sorted in the order of increasing distances from x_n (we assume $n > K$ and ignore the possibility of ties), and so $\hat{y}_n$

is the average label of the K nearest neighbours of x_n. This conformity measure satisfies, additionally,

$$\lim_{y \to \pm\infty} A^{\mathrm{del}}(\lbag z_1, \ldots, z_{n-1} \rbag, (x_n, y)) = \pm\infty$$

and, therefore, the corresponding conformal transducer also satisfies R2 and so is an RPS and a CPS.

7.2.4 Criterion of Being a CPS

Unfortunately, many important conformity measures are not monotonic, and the next lemma introduces a weaker sufficient condition for a conformal transducer to be an RPS.

Proposition 7.2 *The conformal transducer determined by a conformity measure A satisfies requirement R1 if, for each $i \in \{1, \ldots, n-1\}$, each comparison bag $\lbag z_1, \ldots, z_{n-1} \rbag \in \mathbf{Z}^{n-1}$, and each object $x_n \in \mathbf{X}$, $\alpha_n^y - \alpha_i^y$ is a monotonically increasing function of $y \in \mathbb{R}$ (in the notation of (7.7)).*

Of course, we can fix i to, say, $i := 1$ in Proposition 7.2. We can strengthen the conclusion of the proposition to the conformal transducer determined by A being an RPS (and, therefore, a CPS) if, e.g.,

$$\lim_{y \to \pm\infty} \left(\alpha_n^y - \alpha_1^y\right) = \pm\infty .$$

More generally, we have the following corollary. Let A_n be the restriction of A to $\mathbf{Z}^{(n)} \times \mathbf{Z}$ (so that $A_n = A$ if n is fixed).

Corollary 7.3 *Suppose a conformity measure A satisfies the condition of Proposition 7.2 (e.g., is monotonic) and the following three conditions:*

- *for all comparison bags $\lbag z_1, \ldots, z_{n-1} \rbag$, and all objects x_n,*

$$\inf_y A^{\mathrm{del}}(\lbag z_1, \ldots, z_{n-1} \rbag, (x_n, y)) = \inf A_n , \tag{7.9}$$

$$\sup_y A^{\mathrm{del}}(\lbag z_1, \ldots, z_{n-1} \rbag, (x_n, y)) = \sup A_n ; \tag{7.10}$$

- *the $\inf_y$ in (7.9) is either attained for all $\lbag z_1, \ldots, z_{n-1} \rbag$ and x_n or not attained for all $\lbag z_1, \ldots, z_{n-1} \rbag$ and x_n;*
- *the $\sup_y$ in (7.10) is either attained for all $\lbag z_1, \ldots, z_{n-1} \rbag$ and x_n or not attained for all $\lbag z_1, \ldots, z_{n-1} \rbag$ and x_n.*

Then the conformal transducer determined by A is a randomized predictive system.

7.3 Least Squares Prediction Machine

In this and following sections we develop specific conformal predictive systems starting from strong soft models. In particular, in this section we use the method of least squares as our underlying algorithm. We will introduce three versions of what we call the least squares prediction machine (LSPM). They are analogous to conformalized ridge regression of Sect. 2.3, but produce (at least usually) distribution functions rather than prediction intervals. We consider the regression problem with p attributes. Correspondingly, the object space is $\mathbf{X} := \mathbb{R}^p$ and the example space is $\mathbf{Z} := \mathbb{R}^{p+1} = \mathbb{R}^p \times \mathbb{R}$.

7.3.1 Three Kinds of LSPM

The *ordinary LSPM* is defined to be the conformal transducer determined by the conformity measure

$$A(\lbag z_1, \ldots, z_n \rbag, z_n) := y_n - \hat{y}_n \tag{7.11}$$

(cf. (2.33)), where y_n is the label in z_n and $\hat{y}_n$ is the prediction for y_n computed using least squares from x_n (the object in z_n) and $z_1, \ldots, z_n$ (including z_n) as training set. The right-hand side of (7.11) is the ordinary residual, but in Chap. 2 we also discussed two other kinds of residuals; see (2.46) and (2.47). The *deleted LSPM* is determined by the conformity measure

$$A(\lbag z_1, \ldots, z_n \rbag, z_n) := y_n - \hat{y}_{(n)}\ , \tag{7.12}$$

whose difference from (7.11) is that $\hat{y}_n$ is replaced by the prediction $\hat{y}_{(n)}$ for y_n computed using least squares from x_n and $z_1, \ldots, z_{n-1}$ as training set (so that the training set does not include the last example z_n). The version that will be most useful in this book will be the half-way studentized LSPM.

Unfortunately, the ordinary and deleted LSPM are not RPS, because their output Π_n (see (7.3)) is not necessarily monotonically increasing in y (remember that, for conformal transducers, $\Pi_n(y, \tau)$ is monotonically increasing in τ automatically). However, we will see that this can happen only in the presence of high-leverage points.

The best choice, from the point of view of predictive distributions, seems to be the *studentized LSPM* determined by the conformity measure

$$A(\lbag z_1, \ldots, z_n \rbag, z_n) := \frac{y_n - \hat{y}_n}{\sqrt{1 - \bar{h}_n}} \tag{7.13}$$

(cf. (2.47)). An important advantage of studentized LSPM is that to get predictive distributions we do not need any assumptions of low leverage.

Let $z_1, \ldots, z_{n-1}$ be a training set, $z_i = (x_i, y_i)$, and x_n be a test object with an unknown label y_n. In this chapter it will be convenient to use $\bar{X}$ as the notation for the data matrix (2.38) that was denoted X_n in Chap. 2 (we will later use the shorter notation X for the training data matrix X_{n-1}). Therefore, $\bar{X}$ is the $n \times p$ matrix whose ith row is the transpose x_i' to the ith object (training object for $i = 1, \ldots, n-1$ and test object for $i = n$). The hat matrix for the n examples $z_1, \ldots, z_n$ is

$$\bar{H} = \bar{X}(\bar{X}'\bar{X})^{-1}\bar{X}' . \tag{7.14}$$

Our notation for the elements of this matrix will be $\bar{h}_{i,j}$, i standing for the row and j for the column. For the diagonal elements $\bar{h}_{i,i}$ we will use the shorthand $\bar{h}_i$, as before (except that in Sect. 2.3 we wrote h_i in place of $\bar{h}_i$).

Proposition 7.4 *The function Π_n output by the ordinary LSPM (see* (7.3)*) is monotonically increasing in y provided $\bar{h}_n < 0.5$.*

The condition needed for Π_n to be monotonically increasing, $\bar{h}_n < 0.5$, means that the test object x_n is not a very influential point. An overview of high-leverage points is given in [46, Sect. 4.2.3.1], which starts from Huber's 1981 [172] proposal to regard points x_i with $\bar{h}_i > 0.2$ as influential.

The assumption $\bar{h}_n < 0.5$ in Proposition 7.4 is essential:

Proposition 7.5 *Proposition 7.4 ceases to be true if the constant* 0.5 *in it is replaced by a larger constant.*

The next two propositions show that for the deleted LSPM, determined by (7.12), the situation is even worse than for the ordinary LSPM: we have to require $\bar{h}_i < 0.5$ for all $i = 1, \ldots, n-1$.

Proposition 7.6 *The function Π_n output by the deleted LSPM according to* (7.3) *is monotonically increasing in y provided* $\max_{i=1,\ldots,n-1} \bar{h}_i < 0.5$.

We have the following analogue of Proposition 7.5 for the deleted LSPM.

Proposition 7.7 *Proposition 7.6 ceases to be true if the constant* 0.5 *in it is replaced by a larger constant.*

And finally, a statement about the studentized LSPM.

Proposition 7.8 *The function Π_n output by the studentized LSPM according to* (7.3) *is monotonically increasing in y provided* $\max_{i=1,\ldots,n} \bar{h}_i < 1$.

Our interpretation of Proposition 7.8 is that, for practical purposes, the studentized LSPM is an RPS and, therefore, a CPS. We will discuss this further at the end of Sect. 7.3.2. Without the condition $\max_{i=1,\ldots,n} \bar{h}_i < 1$ some of the conformity scores $A(\lbag z_1, \ldots, z_n \rbag, z_i)$ in the definition of the studentized LSPM are undefined. (Propositions 7.4 and 7.6 implicitly assume that all conformity scores are defined.)

7.3.2 The Studentized LSPM in an Explicit Form

In this and next subsections we will give two explicit forms for the studentized LSPM (Algorithms 7.1 and 7.2); the versions for the ordinary and deleted LSPM are similar (we will give an explicit form for the former, which is particularly intuitive, in Sect. 7.3.4). Predictive distributions (7.3) will be represented in the form

$$\Pi_n(y) := [\Pi_n(y, 0), \Pi_n(y, 1)] ;$$

this function Π_n maps each potential label $y \in \mathbb{R}$ to a closed interval of $\mathbb{R}$. It is clear that in the case of conformal transducers this interval-valued version of Π_n carries the same information as the original one: each original value $\Pi_n(y, \tau)$ can be restored as a convex mixture of the end-points of $\Pi_n(y)$; namely, $\Pi_n(y, \tau) = (1 - \tau)a + \tau b$ if $\Pi_n(y) = [a, b]$.

Remember that the vector $(\hat{y}_1, \dots, \hat{y}_n)'$ of ordinary least squares predictions is the product of the hat matrix $\bar{H}$ and the vector $(y_1, \dots, y_n)'$ of labels. For the studentized residuals (7.13), we can easily obtain

$$\alpha_n^y - \alpha_i^y = B_i y - A_i, \quad i = 1, \dots, n-1 ,$$

in the notation of (7.7), where y is the label of the nth object x_n and

$$B_i := \sqrt{1 - \bar{h}_n} + \frac{\bar{h}_{i,n}}{\sqrt{1 - \bar{h}_i}}, \tag{7.15}$$

$$A_i := \frac{\sum_{j=1}^{n-1} \bar{h}_{j,n} y_j}{\sqrt{1 - \bar{h}_n}} + \frac{y_i - \sum_{j=1}^{n-1} \bar{h}_{i,j} y_j}{\sqrt{1 - \bar{h}_i}} \tag{7.16}$$

(for details of computations, see (7.66) below). We will assume that all A_i are defined and all B_i are defined and positive; these assumptions, which are satisfied almost automatically, will be discussed further at the end of this subsection.

Set $C_i := A_i / B_i$ for all $i = 1, \dots, n-1$. These are the solutions to the equations $\alpha_n^y = \alpha_i^y$; as y increases, the values (7.6) of the predictive distribution function change as y crosses C_i. Sort all C_i in the ascending order and let the resulting sequence be $C_{(1)} \leq \dots \leq C_{(n-1)}$. Set $C_{(0)} := -\infty$ and $C_{(n)} := \infty$. The predictive distribution (7.6) can now be written as

$$\Pi_n(y) := \begin{cases} [\frac{i}{n}, \frac{i+1}{n}] & \text{if } y \in (C_{(i)}, C_{(i+1)}) \text{ for } i \in \{0, 1, \dots, n-1\} \\ [\frac{i'-1}{n}, \frac{i''+1}{n}] & \text{if } y = C_{(i)} \text{ for } i \in \{1, \dots, n-1\} , \end{cases} \tag{7.17}$$

where $i' := \min\{j : C_{(j)} = C_{(i)}\}$ and $i'' := \max\{j : C_{(j)} = C_{(i)}\}$. We can see that the thickness of this CPD is $1/n$ with the exception size equal to the number of distinct C_i, at most $n - 1$.

Algorithm 7.1 Least squares prediction machine

Input: a training set $(x_i, y_i) \in \mathbb{R}^p \times \mathbb{R}$, $i = 1, \dots, n-1$, and a test object $x_n \in \mathbb{R}^p$.
1: Set $\bar{X}$ to the data matrix for the given n objects
2: define the hat matrix $\bar{H}$ by (7.14)
3: **for** $i \in \{1, 2, \dots, n-1\}$
4: define A_i and B_i by (7.16) and (7.15), respectively
5: set $C_i := A_i / B_i$
6: sort $C_1, \dots, C_{n-1}$ in the ascending order obtaining $C_{(1)} \le \dots \le C_{(n-1)}$
7: return the predictive distribution (7.17) for y_n.

The overall algorithm is summarized as Algorithm 7.1. Remember that the data matrix $\bar{X}$ has x'_i, $i = 1, \dots, n$, as its ith row; its size is $n \times p$.

Finally, let us discuss the conditions that all A_i are defined and all B_i are defined and positive. All these quantities will be defined (i.e., all the denominators will be non-zero) when $\bar{h}_i < 1$ for all $i \in \{1, \dots, n\}$; and this is exactly the condition for all conformity scores in the augmented training set $z_1, \dots, z_n$ to be defined. Another equivalent condition [251, Lemma 2.1(iii)] is that the rank of the extended data matrix $\bar{X}$ is p, and it remains p after removal of any one of its n rows.

As for the condition that all B_i are positive, it is sufficient to assume that the rank of the extended data matrix $\bar{X}$ remains p after removal of row n and any of the remaining $n-1$ rows. This follows from [46, Property 2.13(c)].

7.3.3 The Offline Version of the Studentized LSPM

There is a much more efficient implementation of the LSPM in situations where we have a large test set of objects $x_n, \dots, x_{n+m-1}$ instead of just one test object x_n. In this case we can precompute the hat matrix for the training objects $x_1, \dots, x_{n-1}$, and then, when processing each test object x_{n+j}, use the standard updating formulas based on the Sherman–Morrison formula: see Remark 2.8 and [46, (2.18)–(2.18c)]. We spell out the formulas for the use in Algorithm 7.2, which implements such an offline procedure. Let X be the $(n-1) \times p$ data matrix for the first $n-1$ examples: its ith row is x'_i, $i = 1, \dots, n-1$. Set

$$g_i := x'_i (X'X)^{-1} x_n, \quad i = 1, \dots, n . \tag{7.18}$$

Finally, let H be the $(n-1) \times (n-1)$ hat matrix

$$H := X(X'X)^{-1}X' \tag{7.19}$$

for the first $n-1$ objects; its entries will be denoted $h_{i,j}$, with $h_{i,i}$ abbreviated to h_i. The full hat matrix $\bar{H}$ is larger than H, with the extra entries

$$\bar{h}_{i,n} = \bar{h}_{n,i} = \frac{g_i}{1 + g_n}, \quad i = 1, \dots, n . \tag{7.20}$$

Algorithm 7.2 Least squares prediction machine (offline version)

Input: a training set $(x_i, y_i) \in \mathbb{R}^p \times \mathbb{R}$, $i = 1, \dots, n-1$, and a test set $x_{n+j} \in \mathbb{R}^p$, $j = 0, \dots, m-1$.

1: Set X to the data matrix for the $n-1$ training objects
2: set $H = (h_{i,j})$ to the hat matrix (7.19)
3: **for** $j \in \{0, 1, \dots, m-1\}$
4: set $x_n := x_{n+j}$
5: define an $n \times n$ matrix $\bar{H} = (\bar{h}_{i,j})$ by (7.18), (7.20), and (7.21)
6: **for** $i \in \{1, 2, \dots, n-1\}$
7: define A_i and B_i by (7.16) and (7.15), respectively
8: set $C_i := A_i / B_i$
9: sort $C_1, \dots, C_{n-1}$ in the ascending order obtaining $C_{(1)} \le \dots \le C_{(n-1)}$
10: return the predictive distribution (7.17) for the label of x_{n+j}.

The other entries of $\bar{H}$ are

$$\bar{h}_{i,j} = h_{i,j} - \frac{g_i g_j}{1 + g_n}, \quad i, j = 1, \dots, n-1 . \tag{7.21}$$

The overall algorithm is summarized as Algorithm 7.2. The two steps before the outer **for** loop are preprocessing; they do not depend on the test set.

7.3.4 The Ordinary LSPM

A routine calculation (for details, see the end of Sect. 7.9.2) shows that the ordinary LSPM has a particularly useful and intuitive representation:

$$C_i = \frac{A_i}{B_i} = \hat{y}_n + (y_i - \hat{y}_i) \frac{1 + g_n}{1 + g_i} , \tag{7.22}$$

where $\hat{y}_n$ and $\hat{y}_i$ are the least squares predictions for y_n and y_i, respectively, computed from the test objects x_n and x_i, respectively, and the examples $z_1, \dots, z_{n-1}$ as the training set. The predictive distribution is still defined by (7.17). The fraction $\frac{1+g_n}{1+g_i}$ in (7.22) is typically and asymptotically (see, e.g., Remark 7.9 below) close to 1, and can usually be ignored. The two other versions of the LSPM also typically have

$$C_i \approx \hat{y}_n + (y_i - \hat{y}_i) . \tag{7.23}$$

We have defined a procedure producing a "fuzzy" distribution function Π_n given a training set $z_i = (x_i, y_i)$, $i = 1, \dots, n-1$, and a test object x_n. In this and following sections we will use both notation $\Pi_n(y)$ (for an interval) and $\Pi_n(y, \tau)$ (for a point inside that interval, as above).

7.3.5 Asymptotic Efficiency of the LSPM

In this subsection we state some basic results about the LSPM's efficiency. The LSPM has a property of validity under unconstrained randomness, but a natural question is how much we should pay for it in terms of efficiency in situations where narrow parametric or even Bayesian assumptions are also satisfied. This is a special case of the Burnaev–Wasserman programme discussed in Sects. 2.5 and 2.9.7.

Our narrow parametric model is that, given the objects $x_1, x_2, \ldots$, the labels $y_1, y_2, \ldots$ are generated by the rule (2.49), where w is a vector in $\mathbb{R}^p$ and ξ_i are independent random variables distributed as $\mathbf{N}_{0,\sigma^2}$. There are two parameters: vector w and positive number σ. We assume an infinite sequence of examples $(x_1, y_1), (x_2, y_2), \ldots$ but take only the first n of them as our training set (of size $n-1$) and test example and let $n \to \infty$. These are all the assumptions used in our efficiency results:

A1 The sequence $x_1, x_2, \ldots$ is IID and bounded: $\sup_i \|x_i\| < C$ a.s. for a constant C.
A2 The first component of each vector x_i is 1.
A3 The second-moment matrix $\mathbb{E} x_1 x_1'$ is non-singular.
A4 The labels $y_1, y_2, \ldots$ are generated according to (2.49): $y_i = w' x_i + \xi_i$, where ξ_i are independent (among themselves and of the objects x_i) Gaussian noise random variables distributed as $\mathbf{N}_{0,\sigma^2}$.

Remark 7.9 Under conditions A1–A4, as $n \to \infty$, we have $\max_{i=1,\ldots,n} |g_i| = O(n^{-1})$ a.s. and so $\frac{1+g_n}{1+g_i} = 1 + O(n^{-1})$ a.s. in (7.22). Indeed,

$$\max_{i=1,\ldots,n} |g_i| \le \frac{\|x_n\| \max_{i=1,\ldots,n} \|x_i\|}{\lambda_{\min}(X'X)} < \frac{\|x_n\| \max_{i=1,\ldots,n} \|x_i\|}{n\epsilon} = O(n^{-1})$$

a.s., with $\lambda_{\min}$ standing for the smallest eigenvalue and the inequality holding for some $\epsilon > 0$ from some n on.

Alongside the three versions of the LSPM, we will consider three "oracles". Intuitively, all three oracles know that the data is generated from the model (2.49). Oracle I knows neither w nor σ (and has to estimate them from the data or somehow manage without them). Oracle II does not know w but knows σ. Finally, Oracle III knows both w and σ.

Formally, *proper Oracle I* outputs the standard predictive distribution for the label y_n of the test object x_n given the training set of the first $n-1$ examples and x_n, namely it predicts with

$$\hat{y}_n + \sqrt{1+g_n}\,\hat{\sigma}_n t_{n-1-p}\,, \tag{7.24}$$

where g_n is defined in (7.18),

$$\hat{y}_n := x_n'(X'X)^{-1}X'Y,$$

$$\hat{\sigma}_n := \sqrt{\frac{1}{n-1-p}\sum_{i=1}^{n-1}(y_i - \hat{y}_i)^2}, \quad \hat{y}_i := x_i'(X'X)^{-1}X'Y\ , \tag{7.25}$$

X is the data matrix for the training set (the $(n-1) \times p$ matrix whose ith row is x_i', $i = 1, \dots, n-1$), Y is the vector $(y_1, \dots, y_{n-1})'$ of the training labels, and t_{n-1-p} is Student's t-distribution with $n-1-p$ degrees of freedom; see, e.g., [313, Sect. 5.3.1]. (By condition A3, $(X'X)^{-1}$ exists from some n on a.s.) The version that is more popular in the literature on empirical processes for residuals is *simplified Oracle I* outputting

$$\mathbf{N}\left(\hat{y}_n, \hat{\sigma}_n^2\right) . \tag{7.26}$$

The difference between the two versions, however, is asymptotically negligible [278], and the results stated below will be applicable to both versions.

Proper Oracle II outputs the predictive distribution

$$\mathbf{N}\left(\hat{y}_n, (1+g_n)\sigma^2\right) . \tag{7.27}$$

Correspondingly, *simplified Oracle II* outputs the predictive distribution

$$\mathbf{N}\left(\hat{y}_n, \sigma^2\right) ; \tag{7.28}$$

the difference between the two versions of Oracle II is again asymptotically negligible under our assumptions. *Oracle III* outputs the predictive distribution

$$\mathbf{N}\left(w'x_n, \sigma^2\right) . \tag{7.29}$$

Our notation is Π_n for the conformal predictive distribution (7.3) (as before), Π_n^{I} for simplified or proper Oracle I's predictive distribution, (7.26) or (7.24) (Theorem 7.10 will hold for both), Π_n^{II} for simplified or proper Oracle II's predictive distribution, (7.28) or (7.27) (Theorem 7.11 will hold for both), and Π_n^{III} for Oracle III's predictive distribution, (7.29). Theorems 7.10–7.12 are applicable to all three versions of the LSPM. The distribution function and density of the standard normal distribution $\mathbf{N}_{0,1}$ are denoted Φ and ϕ, respectively.

A remarkable property of the limiting processes for Oracles I and II (unlike Oracle III) is that they do not depend on the number p of the attributes.

Theorem 7.10 *The random functions $G_n : \mathbb{R} \to \mathbb{R}$ defined by*

$$G_n(t) := \sqrt{n}\left(\Pi_n(\hat{y}_n + \hat{\sigma}_n t, \tau) - \Pi_n^{\mathrm{I}}(\hat{y}_n + \hat{\sigma}_n t)\right)$$

converge in law to a Gaussian process Z with mean zero and covariance function

$$\operatorname{cov}(Z(s), Z(t)) = \Phi(s)\,(1 - \Phi(t)) - \left(1 + \frac{st}{2}\right)\phi(s)\phi(t), \quad s \le t\,. \tag{7.30}$$

Theorem 7.11 *The random functions $G_n : \mathbb{R} \to \mathbb{R}$ defined by*

$$G_n(t) := \sqrt{n}\left(\Pi_n(\hat{y}_n + \sigma t, \tau) - \Pi_n^{\mathrm{Ii}}(\hat{y}_n + \sigma t)\right)$$

converge in law to a Gaussian process Z with mean zero and covariance function

$$\operatorname{cov}(Z(s), Z(t)) = \Phi(s)\,(1 - \Phi(t)) - \phi(s)\phi(t), \quad s \le t\,.$$

For the simplified oracles, we have $\Pi_n^{\mathrm{I}}(\hat{y}_n + \hat{\sigma}_n t) = \Phi(t)$ in Theorem 7.10 and $\Pi_n^{\mathrm{Ii}}(\hat{y}_n + \sigma t) = \Phi(t)$ in Theorem 7.11.

Theorem 7.12 *The random functions $G_n : \mathbb{R} \to \mathbb{R}$ defined by*

$$\begin{aligned} G_n(t) &:= \sqrt{n}\left(\Pi_n(w'x_n + \sigma t, \tau) - \Pi_n^{\mathrm{Iii}}(w'x_n + \sigma t)\right) \\ &= \sqrt{n}\left(\Pi_n(w'x_n + \sigma t, \tau) - \Phi(t)\right) \end{aligned}$$

converge in law to a Gaussian process Z with mean zero and covariance function

$$\operatorname{cov}(Z(s), Z(t)) = \Phi(s)\,(1 - \Phi(t)) + (p - 1)\phi(s)\phi(t), \quad s \le t\,.$$

We state Theorems 7.10–7.12 without proofs, but in Sect. 7.10.1 give references to published proofs.

In the case $p = 1$ the limiting process Z in Theorem 7.12 is known as *Brownian bridge*. This is the case of the *Dempster–Hill procedure*, which is much simpler than the general LSPM. It will be spelled out and discussed in detail in Sect. 13.3.4.

In Theorems 7.10–7.12 we have $\tau \sim \mathbf{U}$; alternatively, they will remain true if we fix τ to any value in $[0, 1]$. Applying those theorems to a fixed argument t, we obtain (dropping τ altogether) the following corollary.

Corollary 7.13 *For a fixed $t \in \mathbb{R}$,*

$$\sqrt{n}\left(\Pi_n(\hat{y}_n + \hat{\sigma}_n t) - \Pi_n^{\mathrm{I}}(\hat{y}_n + \hat{\sigma}_n t)\right) \xrightarrow{\text{law}} \mathbf{N}\left(0, \Phi(t)(1 - \Phi(t)) - \left(1 + \frac{t^2}{2}\right)\phi(t)^2\right),$$

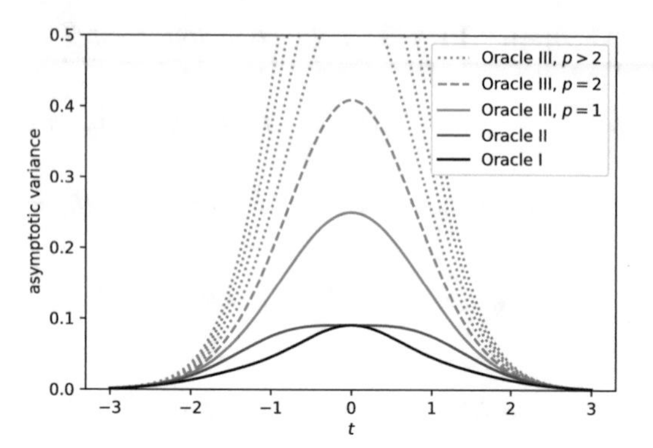

Fig. 7.1 The asymptotic variances for the LSPM procedure as compared with the truth (Oracle III, red) and as compared with the oracular procedures for known σ (Oracle II, blue) and unknown σ (Oracle I, black)

$$\sqrt{n}\left(\Pi_n(\hat{y}_n + \sigma t) - \Pi_n^{\mathrm{II}}(\hat{y}_n + \sigma t)\right) \xrightarrow{\text{law}} \mathbf{N}\left(0, \Phi(t)(1-\Phi(t)) - \phi(t)^2\right),$$

$$\sqrt{n}\left(\Pi_n(\hat{y}_n + \sigma t) - \Pi_n^{\mathrm{III}}(\hat{y}_n + \sigma t)\right) \xrightarrow{\text{law}} \mathbf{N}\left(0, \Phi(t)(1-\Phi(t)) + (p-1)\phi(t)^2\right).$$

Figure 7.1 presents graphs for the asymptotic variances, given in Corollary 7.13, for three oracular predictive distributions: black for Oracle I ($\Phi(t)(1-\Phi(t)) - \phi(t)^2 - \frac{1}{2}t^2\phi(t)^2$ vs t), blue for Oracle II ($\Phi(t)(1-\Phi(t)) - \phi(t)^2$ vs t), and red for Oracle III ($\Phi(t)(1-\Phi(t)) + (p-1)\phi(t)^2$ vs t). The first two asymptotic variances coincide at $t = 0$, where they attain their maximum of between 0.0908 and 0.0909. They are dwarfed by the third one unless p is very small. Interestingly, the LSPM for $p \gg 1$ is as efficient as the Dempster–Hill procedure when comparing with Oracles I and II.

We can see that under the Gaussian model (2.49) complemented by other natural assumptions (including a large n), the LSPM is asymptotically close to the oracular predictive distributions for Oracles I–III, and therefore is approximately conditionally valid and efficient (namely, valid and efficient given $x_1, x_2, \ldots$). On the other hand, the marginal validity of the LSPM is guaranteed under unconstrained randomness, regardless of whether (2.49) holds.

7.3.6 Illustrations

In this subsection we illustrates the efficiency of the studentized LSPM using simple simulations. Figure 7.2 compares the conformal predictive distribution with the true (Oracle III's) distribution for a randomly generated test object and a randomly generated training set of size 10 with 2 attributes. The first attribute is a dummy all-1 attribute; remember that Theorems 7.10–7.12 assume that one of the attributes is an identical 1 (without it, the graphs may become qualitatively different: cf. [47, Corollary 2.4.1]). The second attribute is generated from the standard Gaussian distribution, and the labels are generated as $y_n \sim 2x_{n,2} + \mathbf{N}_{0,1}$, $x_{n,2}$ being the second

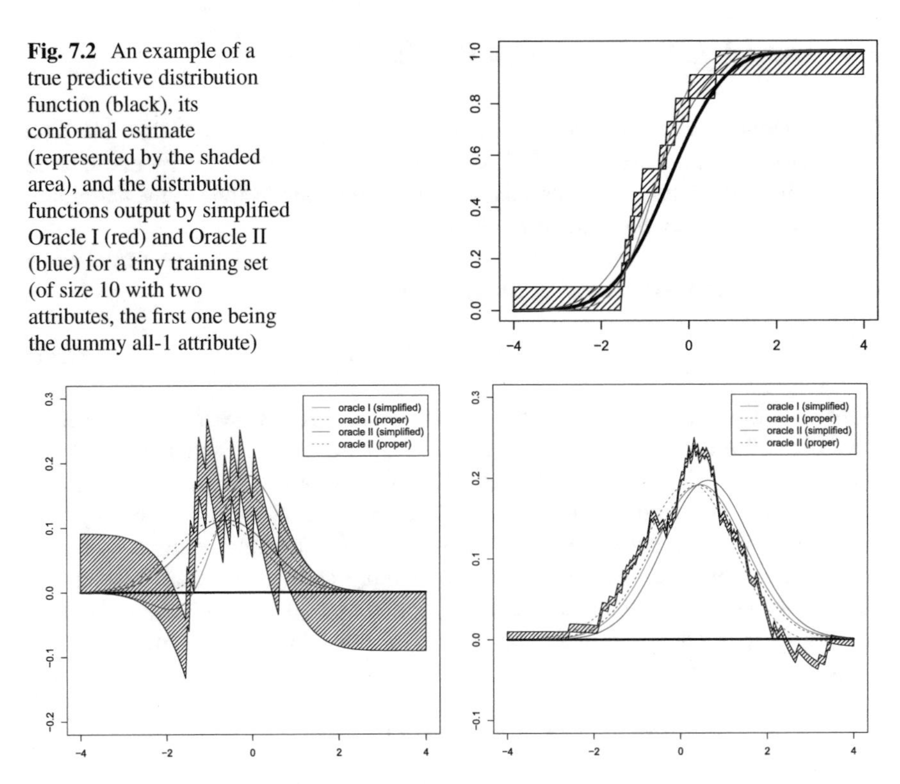

Fig. 7.2 An example of a true predictive distribution function (black), its conformal estimate (represented by the shaded area), and the distribution functions output by simplified Oracle I (red) and Oracle II (blue) for a tiny training set (of size 10 with two attributes, the first one being the dummy all-1 attribute)

Fig. 7.3 Left panel: the plot of Fig. 7.2 normalized by subtracting the true distribution function (the thick black line in Fig. 7.2, which now coincides with the x-axis) and with the outputs of the proper oracles added; the right-hand plot is an analogous plot for a larger training set (of size 100 with 20 attributes, the first one being the dummy all-1 attribute)

attribute. We also show (with thinner lines) the output of Oracle I and Oracle II, but only for the simplified versions, in order not to clutter the plots.

In the left-hand plot of Fig. 7.3 we show the plot of Fig. 7.2 that is normalized by subtracting the true distribution function; this time, we show the output of both simplified and proper Oracles I and II; the difference is not large but noticeable. The right-hand plot of Fig. 7.3 is similar except that the training set is of size 100 and there are 20 attributes generated independently from the standard Gaussian distribution except for the first one, which is the dummy all-1 attribute; the labels are generated as before, $y_n \sim 2x_{n,2} + \mathbf{N}_{0,1}$.

Since Oracle III is more powerful than Oracles I and II (it knows the true data-generating distribution), it is more difficult to compete with; therefore, the black lines are farther from the shaded areas than the blue and red lines for the plots in Figs. 7.2 and 7.3.

7.4 Kernel Ridge Regression Prediction Machine

In this section we introduce the kernel ridge regression prediction machine (KRRPM), which, unlike the LSPM, does not depend on the linearity assumption (even in its soft model). As before, there are three natural versions of the definition, but we concentrate on the studentized version.

Given a training set $z_1, \ldots, z_n \in \mathbf{Z}$ and a test object $x \in \mathbf{X}$, the kernel ridge regression, which we discussed in Sect. 2.3.4, predicts

$$\hat{y} = k_n'(K_n + aI_n)^{-1}Y_n \tag{7.31}$$

in the notation of (2.53) for the label y of x. Let us write $\bar{K} := K_n$ for the kernel matrix for all n examples, $\bar{K}_{i,j} := \mathcal{K}(x_i, x_j)$, $i, j = 1, \ldots, n$, and $I = I_n$ for the $n \times n$ unit matrix. As discussed in Sect. 2.3, the case of least squares corresponds to ridge regression with $a = 0$. Whereas it was feasible to restrict ourselves to the case $a = 0$ in the previous section, it is not feasible now. Increasing a increases the amount of what is known in machine learning as regularization (making the prediction rule less complex and less noisy), and $a > 0$ is indispensable in non-linear cases, particularly for universal kernels (discussed in Sect. 2.9.6).

According to (7.31),

$$\bar{H} := \bar{K}(\bar{K} + aI)^{-1} = (\bar{K} + aI)^{-1}\bar{K} \tag{7.32}$$

is the hat matrix, which "puts hats on the ys" (cf. (2.55)). As before, we refer to the entries of the matrix $\bar{H}$ as $\bar{h}_{i,j}$ and abbreviate $\bar{h}_{i,i}$ to $\bar{h}_i$.

In this section we will only consider the conformal transducer determined by the studentized conformity measure

$$A(\lbag z_1, \ldots, z_n \rbag, z_n) := \frac{y_n - \hat{y}_n}{\sqrt{1 - \bar{h}_n}} \tag{7.33}$$

and will refer to it as the (studentized) *KRRPM*. It follows from the results about the LSPM (namely, Proposition 7.8) that the KRRPM is an RPS; in particular, the expression (7.33) is always defined. This is the main reason why this is the main version considered in this section, with "studentized" usually omitted.

7.4.1 Explicit Forms of the KRRPM

Let the training set be $z_1, \ldots, z_{n-1}$, $z_i = (x_i, y_i)$, and the test object be x_n. According to the definition of p-values, to compute the predictive distributions produced by the KRRPM, we need to solve the equation $\alpha_i^y = \alpha_n^y$ (and the corresponding inequality $\alpha_i^y < \alpha_n^y$) for $i = 1, \ldots, n-1$. This is done exactly

Algorithm 7.3 Kernel ridge regression prediction machine

Input: a training set $(x_i, y_i) \in \mathbf{X} \times \mathbb{R}$, $i = 1, \ldots, n-1$, and a test object $x_n \in \mathbf{X}$.

1: Define the hat matrix $\bar{H}$ by (7.32), $\bar{K}$ being the $n \times n$ kernel matrix
2: **for** $i \in \{1, 2, \ldots, n-1\}$
3: define A_i and B_i by (7.16) and (7.15), respectively
4: set $C_i := A_i / B_i$
5: sort $C_1, \ldots, C_{n-1}$ in the ascending order obtaining $C_{(1)} \le \cdots \le C_{(n-1)}$
6: return the predictive distribution (7.17) for y_n.

as in Sect. 7.3.2 but with a different definition of the hat matrix. The situation is now slightly simpler, since it is always true that all $B_i > 0$. Solving $\alpha_i^y = \alpha_n^y$, which is a linear equation, we obtain $y = C_i := A_i / B_i$, where A_i and B_i are defined by (7.16) and (7.15). This gives Algorithm 7.3 for computing the conformal predictive distribution.

Algorithm 7.3 is not computationally efficient for a large test set, since the hat matrix $\bar{H}$ needs to be computed from scratch for each test object. To obtain a more efficient version, we start from the kernel matrix $K := K_{n-1}$ for the training objects ($K_{i,j} := \mathcal{K}(x_i, x_j)$, $i, j = 1, \ldots, n-1$) and the training hat matrix

$$H := K(K + aI)^{-1} = (K + aI)^{-1} K .$$

Given the test object x_n, we can find the vector $k := k_{n-1}$ (in the notation of (2.53)) for the training objects (the ith component of k is $\mathcal{K}(x_i, x_n)$, $i = 1, \ldots, n-1$) and the scalar $\kappa := \mathcal{K}(x_n, x_n)$, and then use the standard formula (2.56) for inverting partitioned matrices to obtain

$$\begin{aligned}
\bar{H} = (\bar{K} + aI)^{-1} \bar{K} &= \begin{pmatrix} K + aI & k \\ k' & \kappa + a \end{pmatrix}^{-1} \begin{pmatrix} K & k \\ k' & \kappa \end{pmatrix} \\
&= \begin{pmatrix} (K + aI)^{-1} + d(K + aI)^{-1} k k' (K + aI)^{-1} & -d(K + aI)^{-1} k \\ -d k' (K + aI)^{-1} & d \end{pmatrix} \begin{pmatrix} K & k \\ k' & \kappa \end{pmatrix} \\
&= \left(\begin{matrix} H + d(K + aI)^{-1} k k' H - d(K + aI)^{-1} k k' \\ -d k' H + d k' \end{matrix} \right. \qquad (7.34)
\end{aligned}$$

$$\left. \begin{matrix} (K + aI)^{-1} k + d(K + aI)^{-1} k k' (K + aI)^{-1} k - d\kappa (K + aI)^{-1} k \\ -d k' (K + aI)^{-1} k + d\kappa \end{matrix} \right) \qquad (7.35)$$

$$\begin{aligned}
&= \begin{pmatrix} H + d(K + aI)^{-1} k k' (H - I) & d(I - H) k \\ d k' (I - H) & -d k' (K + aI)^{-1} k + d\kappa \end{pmatrix} &(7.36) \\
&= \begin{pmatrix} H - ad(K + aI)^{-1} k k' (K + aI)^{-1} & ad(K + aI)^{-1} k \\ ad k' (K + aI)^{-1} & d\kappa - d k' (K + aI)^{-1} k \end{pmatrix} , &(7.37)
\end{aligned}$$

where

$$d := \frac{1}{\kappa + a - k'(K + aI)^{-1}k} . \tag{7.38}$$

The equality in line (7.36) follows from $\bar{H}$ being symmetric (which allows us to ignore the upper right block of the matrix (7.34)–(7.35)), and the equality in line (7.37) follows from

$$I - H = (K + aI)^{-1}(K + aI) - (K + aI)^{-1}K = a(K + aI)^{-1} .$$

The important components in the expressions for A_i and B_i (cf. (7.16) and (7.15)) are, according to (7.37),

$$\begin{aligned} 1 - \bar{h}_n &= 1 + dk'(K + aI)^{-1}k - d\kappa = 1 + \frac{k'(K + aI)^{-1}k - \kappa}{\kappa + a - k'(K + aI)^{-1}k} \\ &= \frac{a}{\kappa + a - k'(K + aI)^{-1}k} = ad , \qquad (7.39) \\ 1 - \bar{h}_i &= 1 - h_i + ade_i'(K + aI)^{-1}kk'(K + aI)e_i \\ &= 1 - h_i + ad(e_i'(K + aI)^{-1}k)^2 , \qquad (7.40) \end{aligned}$$

where e_i is the ith vector in the standard basis of $\mathbb{R}^{n-1}$ (so that the jth component of e_i is $1_{i=j}$ for $j = 1, \ldots, n-1$). Let $\hat{y}_i := e_i' HY$ be the prediction for y_i computed from the training set $z_1, \ldots, z_{n-1}$ and the test object x_i. Using (7.39) (but not using (7.40) for now), we can transform (7.16) as

$$\begin{aligned} A_i &:= \frac{\sum_{j=1}^{n-1} \bar{h}_{n,j} y_j}{\sqrt{1 - \bar{h}_n}} + \frac{y_i - \sum_{j=1}^{n-1} \bar{h}_{ij} y_j}{\sqrt{1 - \bar{h}_i}} = (ad)^{-1/2} \sum_{j=1}^{n-1} ady_j k'(K + aI)^{-1} e_j \\ &+ \frac{y_i - \sum_{j=1}^{n-1} h_{ij} y_j + \sum_{j=1}^{n-1} ady_j e_i'(K + aI)^{-1}kk'(K + aI)^{-1} e_j}{\sqrt{1 - \bar{h}_i}} \\ &= (ad)^{1/2} k'(K + aI)^{-1} Y + \frac{y_i - \hat{y}_i + ade_i'(K + aI)^{-1}kk'(K + aI)^{-1}Y}{\sqrt{1 - \bar{h}_i}} \\ &= \sqrt{ad}\,\hat{y}_n + \frac{y_i - \hat{y}_i + ad\hat{y}_n e_i'(K + aI)^{-1}k}{\sqrt{1 - \bar{h}_i}} , \qquad (7.41) \end{aligned}$$

where $\hat{y}_n$ is the kernel ridge regression prediction for y_n based on the training set and Y is the vector of training labels, and we can transform (7.15) as

$$B_i := \sqrt{1-\bar{h}_n} + \frac{\bar{h}_{i,n}}{\sqrt{1-\bar{h}_i}} = \sqrt{ad} + \frac{adk'(K+aI)^{-1}e_i}{\sqrt{1-\bar{h}_i}}. \quad (7.42)$$

Therefore, we can implement Algorithm 7.3 as follows. Preprocessing the training set takes time $O(n^3)$ (or faster if using, say, versions of the Coppersmith–Winograd algorithm for matrix inversion; we assume that the kernel $\mathcal{K}$ can be computed in time $O(1)$ for any pair of inputs):

1. The $n \times n$ kernel matrix K can be computed in time $O(n^2)$.
2. The matrix $(K+aI)^{-1}$ can be computed in time $O(n^3)$.
3. The diagonal of the training hat matrix $H := (K+aI)^{-1}K$ can be computed in time $O(n^2)$.
4. All $\hat{y}_i$, $i = 1, \dots, n-1$, can be computed by $\hat{y} := HY = (K+aI)^{-1}(KY)$ in time $O(n^2)$ (even without knowing H).

Processing each test object x_n takes time $O(n^2)$:

1. Vector k and number $\kappa := \mathcal{K}(x_n, x_n)$ can be computed in time $O(n)$ and $O(1)$, respectively.
2. Vector $(K+aI)^{-1}k$ can be computed in time $O(n^2)$.
3. Number $k'(K+aI)^{-1}k$ can now be computed in time $O(n)$.
4. Number d defined by (7.38) can be computed in time $O(1)$.
5. For all $i = 1, \dots, n-1$, compute $1 - \bar{h}_i$ as (7.40), in time $O(n)$ overall (given the vector computed in step 2).
6. Compute the number $\hat{y}_n := k'(K+aI)^{-1}Y$ in time $O(n)$ (given the vector computed in step 2).
7. Finally, compute A_i and B_i for all $i = 1, \dots, n-1$ as per (7.41) and (7.42), set $C_i := A_i/B_i$, and output the predictive distribution (7.17). This takes time $O(n)$ except for sorting the C_i, which takes time $O(n \log n)$.

7.4.2 Limitation of the KRRPM

The KRRPM makes a significant step forward as compared with the LSPM: our soft model is no longer linear in the object. In fact, using a universal kernel (see Sect. 2.9.6) allows the prediction rule produced by the kernel ridge regression to approximate any continuous function (arbitrarily well within any compact set in **X**). However, since we are interested in predictive distributions rather than point predictions, even using a universal kernel still results in the KRRPM being restricted. In this subsection we discuss the nature of the restriction.

As we discussed in Sect. 7.3.4, the predictive distribution, for a large training set, will be approximated by the empirical distribution function (7.23). This approximation is inherited by the KRRPM, at least under some regularity conditions. For the same training set but two different test objects, the location of the predictive distribution functions for their labels will be different (and asymptotically we can hope to get $\hat{y}_n$ right when using a universal kernel). But the shapes of the two predictive distribution functions will both be similar to the shape of the empirical distribution function of the residuals $y_i - \hat{y}_i$ for the training set. Therefore, the shape does not depend (or depends weakly) on the test object x_n. This lack of sensitivity of the predictive distribution to the test object prevents the conformal predictive distributions output by the KRRPM from being universally consistent, as defined formally in Sect. 7.6 below. The shape of the predictive distribution can be arbitrary, but it is fitted to all training residuals. In the next section (Sect. 7.5) we will discuss fitting the predictive distribution to the residuals for objects similar to the test object.

7.5 Nearest Neighbours Prediction Machine

In Sect. 7.4.2 we saw that even the KRRPM has serious limitations and cannot fully adapt the shape of the predictive distribution to the test object. In this section we will discuss a conformal predictive system that does not have obvious limitations of this kind.

Let the object space $\mathbf{X}$ be a metric space. Our underlying algorithm will be nearest neighbours, and in order to avoid irrelevant distractions (caused by the possibility of ties) we will assume that all labels and all distances between pairs of objects are different with probability one, and will consider only such datasets $z_1, \ldots, z_n$. This will be the case, e.g., where the example space $\mathbf{Z}$ is Euclidean and the data-generating probability measure $Q \in \mathbf{P}(\mathbf{Z})$ is absolutely continuous.

Fix $K \in \mathbb{N}$; we are interested in the case where K is large but fixed. A simple deterministic predictive system with an adaptive shape of the predictive distribution functions is

$$\Pi(z_1, \ldots, z_{n-1}, (x_n, y)) := \frac{\left|\{j : y_j \leq y\}\right|}{K}, \quad y \in \mathbb{R},$$

where $j \in \{1, \ldots, n-1\}$ ranges over the indices of the K nearest neighbours of x_n among $x_1, \ldots, x_{n-1}$; we are assuming that $n - 1 \geq K$. To ensure validity under unconstrained randomness, let us conformalize this predictive system. We will see that we can implement the conformalized version efficiently.

As the conformity score $A(\{z_1, \ldots, z_n\}, z_i)$, where $z_j = (x_j, y_j)$ for all j, we take the value at y_i of the empirical distribution function for the labels of the K nearest neighbours of x among $x_1, \ldots, x_{i-1}, x_{i+1}, \ldots, x_n$. (This will be well defined in our context.) Equivalently, after multiplication by n, we can set

$$A(\{z_1, \ldots, z_n\}, z_i) := \left|\{j : y_j \leq y_i\}\right|,$$

where $j \in \{1, \ldots, n\}$ ranges over the indices of the K nearest neighbours of x_i (among $x_1, \ldots, x_{i-1}, x_{i+1}, \ldots, x_n$, here and in what follows). This conformity measure is monotonic and attains 0 as its minimum and K as its maximum when

Algorithm 7.4 Nearest neighbours prediction machine

Input: a training set $z_i = (x_i, y_i) \in \mathbf{X} \times \mathbb{R}$, $i = 1, \ldots, n-1$, and a test object $x_n \in \mathbf{X}$.
1: Set K' to the total number of neighbours and semi-neighbours
2: sort the labels of the K' neighbours and semi-neighbours in the ascending order obtaining $y_{i_1} \leq \cdots \leq y_{i_{K'}}$
3: set $y_{i_0} := -\infty$ and $y_{i_{K'+1}} := \infty$
4: initialize the arrays α and N by (7.44) and (7.45), respectively
5: set $L_0 := 0$ and $U_0 := N_0/n$.
6: **for** $k \in \{1, 2, \ldots, K'\}$
7:     **if** z_{i_k} is a neighbour
8:         $N_{\alpha_n} -= 1$; $\alpha_n += 1$; $N_{\alpha_n} += 1$
9:     **if** z_{i_k} is a full neighbour or semi-neighbour
10:         $N_{\alpha_{i_k}} -= 1$; $\alpha_{i_k} -= 1$; $N_{\alpha_{i_k}} += 1$
11:     set $L_k := \sum_{k'=0}^{\alpha_n - 1} N_{k'}/n$ and $U_k := \sum_{k'=0}^{\alpha_n} N_{k'}/n$
12: return (7.43) as the predictive distribution for y_n.

varying y_i, and so, by Corollary 7.3, produces a conformal predictive system. Algorithm 7.4 gives its explicit representation, and in the rest of this section we will go over it (in particular, define the terminology used in it) and discuss the computational resources that it requires.

We are given a training set $z_1, \ldots, z_{n-1}$ and a test object x_n. The algorithm assumes that $n - 1 \geq K$ and outputs the predictive distribution function

$$\Pi_n(y) := \begin{cases} [L_k, U_k] & \text{if } y \in (y_{i_k}, y_{i_{k+1}}) \text{ for } k \in \{0, 1, \ldots, K'\} \\ [L_{k-1}, U_k] & \text{if } y = y_{i_k} \text{ for } k \in \{1, \ldots, K'\}, \end{cases} \tag{7.43}$$

where K', L_k, U_k, and y_{i_k} will be defined later. It maintains two arrays: α_i, $i = 1, \ldots, n$, and N_k, $k = 0, \ldots, K$.

An element α_i of the array α contains the conformity score of the example z_i in the augmented training set $z_1, \ldots, z_n$, where $z_n := (x_n, y)$; initially the postulated label y is $-\infty$, but then we move through the values of y from $-\infty$ to ∞. Therefore, initially

$$\alpha_i := \left|\left\{j : y_j \leq y_i\right\}\right|, \tag{7.44}$$

where j ranges over the indices of the K nearest neighbours of x_i and y_n is understood to be $-\infty$.

The array N is the histogram of the array α:

$$N_k := |\{i \in \{1, \ldots, n\} : \alpha_i = k\}|. \tag{7.45}$$

In Algorithm 7.4, this array is updated as soon as α is updated. Notice that initially $N_0 \geq 1$ (since $\alpha_n = 0$), and so $U_0 \geq 1/n$ in line 5.

The algorithm uses the following classification of the examples $z_1, \ldots, z_{n-1}$ in the training set:

- z_i is a *full neighbour* if x_i is among the K nearest neighbours of x_n and, vice versa, x_n is among the K nearest neighbours of x_i;
- z_i is a *single neighbour* if x_i is among the K nearest neighbours of x_n but x_n is not among the K nearest neighbours of x_i;
- z_i is a *semi-neighbour* if x_i is not among the K nearest neighbours of x_n but x_n is among the K nearest neighbours of x_i;
- otherwise, z_i is a non-neighbour.

By a *neighbour* we mean either a full neighbour or a single neighbour; therefore, it is synonymous with being one of the K nearest neighbours of x_n.

The algorithm starts, in line 2, from sorting the labels of the neighbours and semi-neighbours, so that y_{i_1} is the smallest of those labels, y_{i_2} is the second smallest, etc. In line 3, we complement these definitions by defining y_{i_k} for $k \in \{0, K'+1\}$; notice that i_k is still undefined for those k. In line 5, we initialize the predictive distribution function to the left of y_{i_1}.

In the loop starting in line 6 we go over all labels y_{i_k} from left to right updating the arrays α and N and the values $[L_k, U_k]$ of the predictive distribution function. In lines 8 and 10, we use the notation $a \mathrel{+}= b$ and $a \mathrel{-}= b$ to mean $a := a + b$ and $a := a - b$, respectively. In those lines we update the conformity scores α_{i_k} and α_n as appropriate, so that during the kth iteration the array α contains the correct value for the postulated label in the range $y \in (y_{i_k}, y_{i_{k+1}})$. The two commands updating the array α (in lines 8 and 10) are surrounded by two commands each that update N according to (7.45). Those arrays allow us to compute $[L_k, U_k]$ in line 11; the expression for L_k is understood to be 0 when $\alpha_n = 0$.

When analysing the computation time, we assume that K is a constant. In the one-off mode, Algorithm 7.4 takes time $O(n^2)$, since we need to compute the n^2 distances between the objects. In the online mode, the computation time is $O(n)$ at the nth step of the online protocol, which can be achieved by maintaining the array of K nearest neighbours and K shortest distances for all examples seen so far. Updating the array of neighbours requires time $O(n)$, it takes time $O(n)$ to compute the initial array α for step n and then the same time $O(n)$ to compute the initial N for step n, and finally it takes time $O(1)$ for each iteration of the loop. In the offline mode, the required time is $O(n^2 + nm)$, where m is the size of the test set: in time $O(n^2)$ we can preprocess the training set computing the array of K nearest neighbours and K shortest distances inside the training set, and then we can process each test example in time $O(n)$ (cf. [51, Table 1]).

7.6 Universal Conformal Predictive Systems

The main result of this section (Theorem 7.14) is that there exists a universally consistent, or universal, conformal predictive system, in the sense that it produces predictive distributions that are consistent under any probability distribution for one example. The notion of consistency is used in an unusual situation here, and our formalization is based on Belyaev's [27, 28, 335] notion of weakly approaching sequences of distributions.

The importance of universal consistency will be illustrated in the following section (Sect. 7.7); namely, applying the expected utility maximization principle to the predictive distributions produced by a universal predictive system leads, under natural conditions, to asymptotically optimal decisions.

7.6.1 Definitions

Let us say that a randomized predictive system Π is *consistent* for a probability measure Q on $\mathbf{Z}$ if, for any bounded continuous function $f : \mathbb{R} \to \mathbb{R}$,

$$\int f \, \mathrm{d}\Pi_n - \mathbb{E}_Q(f \mid x_n) \to 0 \qquad (n \to \infty) \tag{7.46}$$

in probability, where:

- Π_n is the predictive distribution $\Pi_n : y \mapsto \Pi(z_1, \ldots, z_{n-1}, (x_n, y), \tau)$ output by Π as its prediction for the label y_n of the test object x_n based on the training data $(z_1, \ldots, z_{n-1})$, where $z_i = (x_i, y_i)$ for all i;
- $\mathbb{E}_Q(f \mid x_n)$ is the conditional expectation of $f(y)$ given $x = x_n$ under $(x, y) \sim Q$;
- $z_i = (x_i, y_i) \sim Q$, $i = 1, \ldots, n$, and $\tau \sim \mathbf{U}$ are all independent.

It is clear that this notion of consistency does not depend on the choice of the version of the conditional expectation $\mathbb{E}_Q(f \mid \cdot)$ in (7.46). The integral in (7.46) is not quite standard since we did not require Π_n to be exactly a distribution function, so we understand $\int f \, \mathrm{d}\Pi_n$ as $\int f \, \mathrm{d}\bar{\Pi}_n$ with the measure $\bar{\Pi}_n$ on $\mathbb{R}$ defined by $\bar{\Pi}_n((u, v]) := \Pi_n(v+) - \Pi_n(u+)$ for any interval $(u, v]$ of this form (nonempty, open on the left, and closed on the right) in $\mathbb{R}$.

A randomized predictive system Π is *universal*, or *universally consistent*, if it is consistent for any probability measure Q on $\mathbf{Z}$. As already mentioned, this definition is based on Belyaev's (see, e.g., [28]). It implies the convergence of, e.g., the Lévy distance between the predictive distribution and true distribution to 0 (as asserted in [28]; the details are spelled out in [393, Appendix A]).

7.6.2 Universal Conformal Predictive Systems

First we introduce a clearly innocuous extension of conformal predictive systems allowing further randomization. In particular, the extension will not affect their properties of validity (such as strong calibration in probability) while simplifying various constructions.

A *randomized conformity measure* is defined in the same way as a conformity measure but depends on the extended training examples $(z, \tau) \in \mathbf{Z} \times [0, 1]$ rather than examples $z \in \mathbf{Z}$. This is essentially the same definition, except that each example is extended by adding a number (later it will be generated randomly from $\mathbf{U}$) that can be used for tie-breaking. We can still use the same definition, given by the right-hand side of (7.6), of the conformal transducer determined by a randomized conformity measure A, except for replacing each example in (7.7) by an extended example.

Notice that our new definition of conformal transducers is a special case of the old definition, in which the original example space $\mathbf{Z}$ is replaced by the extended example space $\mathbf{Z} \times [0, 1]$. An extended example $(z, \tau) = (x, y, \tau)$ can be interpreted to consist of an extended object (x, τ) and a label y. The main difference from the old framework is that now we are only interested in the probability measures on $\mathbf{Z} \times [0, 1]$ that are the product of a probability measure Q on $\mathbf{Z}$ and the uniform probability measure $\mathbf{U}$ on $[0, 1]$.

The definition of randomized predictive systems generalizes by replacing objects x_j by extended objects (x_j, τ_j). Conformal predictive systems are defined literally as before.

Theorem 7.14 *Suppose the object space* $\mathbf{X}$ *is standard Borel. There exists a universal conformal predictive system.*

For the proof of Theorem 7.14, see [393, proof of Theorem 3]. For further information and references, see Sect. 7.10.4.

7.6.3 Universal Deterministic Predictive Systems

Let us now state a simple corollary of Theorem 7.14 for deterministic predictive systems. A deterministic predictive system Π is *universal*, or *universally consistent*, if, for any probability measure Q on $\mathbf{Z}$ and any bounded continuous function $f : \mathbb{R} \to \mathbb{R}$, (7.46) holds in probability, where $\Pi_n : y \mapsto \Pi(z_1, \ldots, z_{n-1}, (x_n, y))$, assuming $z_n \sim Q$ are independent.

Corollary 7.15 *If the object space* $\mathbf{X}$ *is standard Borel, there exists a universal deterministic predictive system.*

7.7 Applications to Decision Making

In this section we discuss how conformal predictive distributions can be used for the purpose of decision-making. A major limitation of conformal predictive distributions is that, at this time, they are only applicable to regression problems, where the label is a real number; however, this does not prevent them from being used in a general problem of decision making. The main theoretical result of the section is that there exists an asymptotically efficient predictive decision-making system which can be obtained by using our methodology (and therefore, based on calibrated probabilities). See Sect. 7.10.6 for a brief discussion of the standard von Neumann–Morgenstern decision theory.

7.7.1 A Standard Problem of Decision Making

In this section, with each label $y \in \mathbf{Y}$ and each decision $d \in \mathbf{D}$ chosen from a *decision space* $\mathbf{D}$ we associate a von Neumann–Morgenstern utility $U(y, d)$. (More generally, we could have allowed U to depend on the object x as well.) Formally, we are given a *utility function* $U : \mathbf{Y} \times \mathbf{D} \to \mathbb{R}$, which will always be assumed measurable. We will also assume that the decision space is finite, $|\mathbf{D}| < \infty$.

Our task is, given a training set $z_1, \ldots, z_{n-1}$, where $z_i = (x_i, y_i)$, and a test object $x_n \in \mathbf{X}$, to choose a suitable decision $d \in \mathbf{D}$ in view of the utility function U; the problem is that the utility $U(y, d)$ depends on the unknown label y of x_n. A *predictive decision-making system* (PDMS) is a measurable function $F : \mathbf{Z}^* \times \mathbf{X} \to \mathbf{D}$; the intuition is that $F(z_1, \ldots, z_{n-1}, x_n)$ is the decision recommended for the new object x_n based on the training set $z_1, \ldots, z_{n-1}$. It is *randomized* when it depends, additionally, on a random number $\tau \in [0, 1]$ (representing internal coin tossing). The *regret* of a PDMS F (perhaps randomized) on an object x_n and training set $z_1, \ldots, z_{n-1}$ under a probability measure Q on $\mathbf{Z}$ is defined to be

$$
\begin{aligned}
R_F(z_1, \ldots, z_{n-1}, x_n) := \max_{d \in \mathbf{D}} \int U(y, d) Q(\mathrm{d}y \mid x_n) \\
- \int U(y, F(z_1, \ldots, z_{n-1}, x_n)) Q(\mathrm{d}y \mid x_n) , \qquad (7.47)
\end{aligned}
$$

where $Q(\mathrm{d}y \mid x)$ is a regular conditional distribution (assumed to exist) for y given x under Q. The F in the last integral in (7.47) may depend on the internal coin tossing

Algorithm 7.5 Conformal predictive decision making

Input: a training set $(x_i, y_i) \in \mathbf{Z}$, $i = 1, \ldots, n-1$, and a test object $x_n \in \mathbf{X}$.
1: **for** $d \in \mathbf{D}$
2: create a new training set $(x_i, U(y_i, d))$, $i = 1, \ldots, n-1$
3: find the CPD Π_d^* by (7.48) from this training set
4: compute the utility of d as $\int u \Pi_d^*(\mathrm{d}u)$
5: return a $d \in \mathbf{D}$ with the largest utility.

(i.e., on $\tau \sim \mathbf{U}$), in which case we write $R_F(z_1, \ldots, z_{n-1}, x_n, \tau)$ for the left-hand side of (7.47). We are interested in PDMSs with small regret.

Our randomized algorithm is given as Algorithm 7.5. The CPD Π_d^* in line 3 is defined as

$$\Pi_d^*(u) := \Pi\big((x_1, U(y_1, d)), \ldots, (x_{n-1}, U(y_{n-1}, d)), (x_n, u), \tau\big) , \tag{7.48}$$

where Π is defined by (7.6). The integral in line 4 is defined as in Sect. 7.6.1, via the measure $\bar{\Pi}_d^*$. The total mass of this measure is less than 1, but it does not matter as we use it merely for comparison of different decisions d.

7.7.2 Examples

In statistical decision theory, it is customary to use the *loss function* $L(y, d) := -U(y, d)$ (see Sect. 7.10.6). Algorithm 7.5 in terms of losses is obtained by replacing U with L, replacing "utility" with "loss", and replacing "largest" with "smallest".

The case of *basic classification* is where $\mathbf{Y}$ is a finite set, $\mathbf{D} = \mathbf{Y}$, and the loss function is

$$L(y, d) := \begin{cases} 0 & \text{if } y = d \\ 1 & \text{otherwise} . \end{cases} \tag{7.49}$$

In this subsection we will be applying Algorithm 7.5 in terms of losses.

Let us check that, in the case of basic classification, Algorithm 7.5 becomes a version of the standard one-against-the-rest procedure (discussed in Sect. 3.2.3). For a fixed $d \in \mathbf{D}$, we create a new training set replacing label d by 0 and replacing all other labels by 1. The resulting CPD Π_d^* is not necessarily concentrated, or nearly concentrated, at one point (intuitively, such a point would represent the probability that the test label is different from d); however, we can still interpret $\int u \Pi_d^*(\mathrm{d}u)$ as the probability that the test label will be different from d. Now we predict the label d with the highest probability (by choosing the smallest $\int u \Pi_d^*(\mathrm{d}u)$).

Next we discuss *asymmetric classification*. There are many cases where, unlike (7.49), different types of classification errors lead to different losses (for example, classifying an ill person as healthy is usually regarded a graver mistake than classifying a healthy person as ill). This corresponds to replacing 1 in (7.49) (which can now be interpreted as a square matrix of size $|\mathbf{Y}| \times |\mathbf{Y}|$) by different positive numbers.

More generally, we can consider arbitrary finite sets $\mathbf{Y}$ and $\mathbf{D}$; then the utility function U (or loss function L) can be identified with a $|\mathbf{Y}| \times |\mathbf{D}|$ matrix.

7.7.3 Asymptotically Efficient Decision Making

In this subsection we state a simple corollary of Theorem 7.14 showing that asymptotically we can choose the best possible decisions under weak assumptions about the object space $\mathbf{X}$ and the label space $\mathbf{Y}$.

A randomized PDMS is *asymptotically efficient* if, for any probability measure Q on $\mathbf{Z} = \mathbf{X} \times \mathbf{Y}$, its regret $R_F(z_1, \dots, z_{n-1}, x_n, \tau_n)$ (see (7.47)) on x_n and $z_1, \dots, z_{n-1}$ under Q tends to 0 in probability when $z_i = (x_i, y_i) \in \mathbf{Z}$, $i = 1, 2, \dots$, are produced from Q and the random numbers τ_i are produced from $\mathbf{U}$, all independently.

Theorem 7.16 *Suppose that, in addition to the decision space* $\mathbf{D}$ *being finite,* $\mathbf{X}$ *is a standard Borel space and the utility function* $U : \mathbf{Y} \times \mathbf{D} \to \mathbb{R}$ *is bounded. Asymptotically efficient PDMSs exist; in particular, Algorithm 7.5 is an asymptotically efficient randomized PDMS whenever it is based on a universally consistent randomized predictive system (which exists under these assumptions).*

Theorem 7.16 is an easy corollary of Theorem 7.14; the details are spelled out in Sect. 7.9.3. Theorem 7.14 asserts the existence of a universally consistent conformal predictive system, which automatically satisfies properties of validity, such as calibration in probability. Theorem 7.16 does not use such properties of validity, but they are likely to improve the quality of predictions as measured by the utility function U.

7.7.4 Dangers of Overfitting

So far in this section we have discussed only the case of a finite decision space $\mathbf{D}$. One problem for an infinite, or finite but very large, $\mathbf{D}$ is the possibility of "overfitting", where one of the decisions $d \in \mathbf{D}$ can lead to a small loss simply by chance. This is a general problem with optimization over a large set of alternatives in machine learning and statistics, and it sometimes shows in our current setting.

As a simple example, consider a probability measure Q on an example space $\mathbf{Z} = \mathbf{X} \times \mathbf{Y} = \mathbf{X} \times \mathbb{R}$ such that the conditional distribution of the label $y \in \mathbf{Y}$ given the object $x \in \mathbf{X}$ is always continuous (for some version of the conditional distribution). The decision space is infinite, namely the set of all finite subsets of the label space $\mathbf{Y} = \mathbb{R}$. The utility function is

$$U(y, d) := \begin{cases} 1 & \text{if } d = \emptyset \\ 2 & \text{if } y \in d \\ 0 & \text{otherwise}. \end{cases}$$

Given a training set $(x_1, y_1), \dots, (x_{n-1}, y_{n-1})$, Algorithm 7.5 will return (for a reasonable conformity measure) a decision $d \supseteq \{y_1, \dots, y_{n-1}\}$, which is clearly suboptimal; the optimal decision would be $d := \emptyset$. (A more cautious way of evaluating the integral in line 4 of Algorithm 7.5 would have helped: namely, we could take the infimum over all distribution functions inside the conformal "prediction band", such as the shaded area in Fig. 7.2.)

7.8 Computationally Efficient Versions

In this section we will modify the definition of conformal predictive systems making them more computationally efficient and concentrating on the one-off and offline modes.

7.8.1 *Inductive Conformal Predictive Systems*

Let us modify the definitions of Sect. 4.2.2. An *inductive conformity measure* is a measurable function $A : \mathbf{Z}^* \times \mathbf{Z} \to \mathbb{R}$. Mathematically it is the same notion as that of inductive nonconformity measure of Sect. 4.2.2, but the intention is that $A(z_1, \dots, z_m, z_{m+1})$ measures how large the label y_{m+1} in z_{m+1} is, as compared with the labels in $z_1, \dots, z_m$. As before, the training set $z_1, \dots, z_l$ is split into the proper training set $z_1, \dots, z_m$ and the calibration set $z_{m+1}, \dots, z_l$; we are given a test object x (which may be an element of a test set). In the rest of this subsection, l and m are fixed. The output of the *inductive conformal transducer* determined by the inductive conformity measure A is defined as

$$\begin{aligned} \Pi(z_1, \dots, z_l, (x, y), \tau) := &\frac{1}{l - m + 1} \left|\left\{i = m + 1, \dots, l : \alpha_i < \alpha^y\right\}\right| \\ &+ \frac{\tau}{l - m + 1} \left|\left\{i = m + 1, \dots, l : \alpha_i = \alpha^y\right\}\right| + \frac{\tau}{l - m + 1}, \end{aligned} \tag{7.50}$$

where the *conformity scores* α_i, $i = m + 1, \dots, l$, and α^y, $y \in \mathbb{R}$, are defined by

$$\begin{aligned} \alpha_i &:= A(z_1, \dots, z_m, (x_i, y_i)), \qquad i = m + 1, \dots, l, \\ \alpha^y &:= A(z_1, \dots, z_m, (x, y)) . \end{aligned} \tag{7.51}$$

An *inductive conformal predictive system* (ICPS) is a function which is both an inductive conformal transducer (determined by some inductive conformity measure) and a randomized predictive system (i.e., satisfies requirements R1 and R2 with $n := l + 1$ and x in place of x_n).

The standard property of validity (satisfied automatically) for inductive conformal transducers is that the values $\Pi(z_1, \ldots, z_l, z, \tau)$ are distributed uniformly on $[0, 1]$ when $z_1, \ldots, z_l, z$ are IID and τ is generated independently of $z_1, \ldots, z_l, z$ from the uniform probability distribution $\mathbf{U}$ on $[0, 1]$.

It is much easier to get an RPS using inductive conformal transducers than using conformal transducers. An inductive conformity measure A is *isotonic* if, for all $z_1, \ldots, z_m$ and x, $A(z_1, \ldots, z_m, (x, y))$ is isotonic in y, i.e.,

$$y \leq y' \Longrightarrow A(z_1, \ldots, z_m, (x, y)) \leq A(z_1, \ldots, z_m, (x, y')) ; \qquad (7.52)$$

this definition is analogous to, but simpler than, the definition of monotonic conformity measures in Sect. 7.2.3. An isotonic inductive conformity measure A is *balanced* if, for any $z_1, \ldots, z_m$, the set

$$\operatorname{co} A(z_1, \ldots, z_m, (x, \mathbb{R})) := \operatorname{co} \{A(z_1, \ldots, z_m, (x, y)) : y \in \mathbb{R}\} \qquad (7.53)$$

does not depend on x, where co stands for the convex hull in $\mathbb{R}$. The set (7.53) then coincides with $\operatorname{co} A(z_1, \ldots, z_m, \mathbf{Z})$ and has one of four forms: (a, b), $[a, b)$, $(a, b]$, or $[a, b]$, where $a \leq b$ are elements of the extended real line $\overline{\mathbb{R}}$ ($a = b$ is only possible for sets of the form $[a, b]$). We will be mainly interested in the case $\operatorname{co} A(z_1, \ldots, z_m, \mathbf{Z}) = (-\infty, \infty)$. The following two propositions will be proved in Sect. 7.9.4.

Proposition 7.17 *The inductive conformal transducer* (7.50) *determined by a balanced isotonic inductive conformity measure is an RPS.*

The next proposition shows that an inductive conformity measure being isotonic and balanced is not only a sufficient but also a necessary condition for the corresponding inductive conformal transducer to be an RPS.

Proposition 7.18 *If the inductive conformal transducer determined by an inductive conformity measure A is an RPS, A is isotonic and balanced.*

The ICPS can be implemented directly using the definition (7.50) and a grid of values of y. Algorithm 7.6 describes another implementation assuming that the inductive conformity measure A is balanced isotonic. It defines the predictive distribution apart from a finite number of points y; we can set the interval $\Pi(z_1, \ldots, z_l, (x, y), [0, 1])$ at those points y to the union of such intervals at the points in a small deleted neighbourhood of y without a substantial change to the predictive system. Some of the α_i, $i = 1, \ldots, l-m$, in Algorithm 7.6 may coincide, so we can only say that $k \in \{1, \ldots, l-m\}$ rather than $k = l - m$ (notice that the sequence $\alpha_{[j]}$, $j = 1, \ldots, k$, is strictly increasing). The predictive distribution that it outputs is

$$\Pi(z_1, \ldots, z_l, (x, y), \tau) =$$

Algorithm 7.6 Inductive conformal predictive system

Input: balanced isotonic inductive conformity measure A, training set $z_i = (x_i, y_i) \in \mathbf{Z}$, $i = 1, \dots, l$, positive integer $m < l$, test object $x \in \mathbf{X}$, and random number $\tau \in [0, 1]$.

1: **for** $i \in \{1, \dots, l-m\}$
2: $\quad \alpha_i := A(z_1, \dots, z_m, z_{m+i})$
3: sort $\alpha_1, \dots, \alpha_{l-m}$ in the ascending order removing the duplicates and obtaining $\alpha_{[1]} < \dots < \alpha_{[k]}$
4: **for** $j \in \{1, \dots, k\}$
5: $\quad n_j := \left|\{i = 1, \dots, l-m : \alpha_i = \alpha_{[j]}\}\right|$
6: $\quad m_j := \sup\{y : A(z_1, \dots, z_m, (x, y)) < \alpha_{[j]}\}$
7: $\quad M_j := \inf\{y : A(z_1, \dots, z_m, (x, y)) > \alpha_{[j]}\}$
8: return the predictive distribution Π given by (7.54) for the label y of x.

$$\begin{cases} \frac{\tau}{l-m+1} & \text{if } y < m_1 \\ \frac{n_1+\dots+n_{j-1}+\tau n_j+\tau}{l-m+1} & \text{if } y \in (m_j, M_j),\ \ j \in \{1, \dots, k\} \\ \frac{n_1+\dots+n_j+\tau}{l-m+1} & \text{if } y \in (M_j, m_{j+1}),\ \ j \in \{1, \dots, k-1\} \\ \frac{n_1+\dots+n_k+\tau}{l-m+1} = \frac{l-m+\tau}{l-m+1} & \text{if } y > M_k\ . \end{cases} \tag{7.54}$$

Let us say that an inductive conformity measure A is *strictly isotonic* if (7.52) holds with both "$\le$" replaced by "$<$". In this case it is convenient to modify Algorithm 7.6 and define the predictive distribution by

$$\Pi(z_1, \dots, z_l, (x, y), \tau) := \begin{cases} \frac{i+\tau}{l-m+1} & \text{if } y \in (C_i, C_{i+1}) \text{ for } i \in \{0, 1, \dots, l-m\} \\ \frac{i'-1+(i''-i'+2)\tau}{l-m+1} & \text{if } y = C_i \text{ for } i \in \{1, \dots, l-m\}\ , \end{cases} \tag{7.55}$$

where

$$\begin{aligned} C_i &:= \sup\{y : A(z_1, \dots, z_m, (x, y)) < \alpha_{(i)}\} \\ &= \inf\{y : A(z_1, \dots, z_m, (x, y)) > \alpha_{(i)}\}\ , \end{aligned}$$

$\alpha_{(i)}$ is the ith smallest element of the bag $\Lbag \alpha_1, \dots, \alpha_{l-m} \Rbag$ (now the repetitions among $\alpha_{(i)}$ are not removed, so there are $l-m$ of $\alpha_{(i)}$, $\alpha_{(1)} \le \dots \le \alpha_{(l-m)}$), $i' := \min\{j : C_j = C_i\}$, $i'' := \max\{j : C_j = C_i\}$, and C_0 and C_{l-m+1} are understood to be $-\infty$ and ∞, respectively. We can see that the thickness of the predictive distribution (7.55) is $\frac{1}{l-m+1}$ with the exception size at most $l-m$.

How computationally efficient the modification (7.55) of Algorithm 7.6 is depends on how easy to solve the equation defining C_i is. As we have discussed in various contexts, a standard choice of inductive conformity measure is

$$A(z_1, \ldots, z_m, (x, y)) := \frac{y - \hat{y}}{\hat{\sigma}},$$

where $\hat{y}$ is a prediction for the label y computed from x as test object and $z_1, \ldots, z_m$ as training set, and $\hat{\sigma}$ is an estimate of the quality of $\hat{y}$ computed from the same data. In this case the equation

$$A(z_1, \ldots, z_m, z_{m+i}) = A(z_1, \ldots, z_m, (x, C_i))$$

defining C_i becomes

$$\frac{y_{m+i} - \hat{y}_{m+i}}{\hat{\sigma}_{m+i}} = \frac{C_i - \hat{y}}{\hat{\sigma}},$$

where $\hat{y}_{m+i}$ (resp. $\hat{y}$) is the prediction for y_{m+i} (resp. y) computed from x_{m+i} (resp. x) as test object and $z_1, \ldots, z_m$ as training set, and $\hat{\sigma}_{m+i}$ (resp. $\hat{\sigma}$) is the estimate of the quality of $\hat{y}_{m+i}$ (resp. $\hat{y}$) computed from the same data. Therefore,

$$C_i := \hat{y} + \hat{\sigma} \frac{y_{m+i} - \hat{y}_{m+i}}{\hat{\sigma}_{m+i}}.$$

In the special case where the function $\hat{\sigma}$ is absent (or constant), we obtain (7.23), in different notation.

7.8.2 Cross-Conformal Predictive Distributions

An inductive conformity measure A is a *cross-conformity measure* if $A(z_1, \ldots, z_m, z)$ does not depend on the order of its first m arguments, in which case we return to our old notation $A(\lbag z_1, \ldots, z_m \rbag, z)$ in place of $A(z_1, \ldots, z_m, z)$; this is just a deleted conformity measure used in the context of cross-conformal prediction (in full analogy with Sect. 4.4.1).

Given a balanced isotonic cross-conformity measure A, the corresponding *cross-conformal predictive system* (CCPS) is defined following the scheme of Sect. 4.4.1. The training set $z_1, \ldots, z_l$ is randomly split into K folds z_{S_k}, $k = 1, \ldots, K$, of equal (or as equal as possible) sizes. For each $k \in \{1, \ldots, K\}$ and each potential label $y \in \mathbb{R}$ of the test object x, find the conformity scores $\alpha_{i,k}$ and α_k^y of the examples in z_{S_k} and of (x, y), respectively, by (4.22), and define the corresponding CCPS by

$$\Pi(z_1, \ldots, z_l, (x, y), \tau) := \frac{1}{l+1} \sum_{k=1}^{K} \left|\{i \in S_k : \alpha_{i,k} < \alpha_k^y\}\right|$$

Algorithm 7.7 Cross-conformal predictive system

Input: a training set $(x_i, y_i) \in \mathbf{Z}$, $i = 1, \dots, l$, and a test object $x \in \mathbf{X}$.
1: Split $z_1, \dots, z_l$ into K folds z_{S_k} as described in the main text
2: set $C := \Lbag\Rbag$, where C is a bag variable
3: **for** $k \in \{1, \dots, K\}$
4: **for** $i \in S_k$
5: define $C_{i,k}$ by the condition $A(z_{S_{-k}}, z_i) = A(z_{S_{-k}}, (x, C_{i,k}))$
6: put $C_{i,k}$ in C
7: sort C in the ascending order obtaining $C_1 \leq \dots \leq C_l$
8: set $C_0 := -\infty$ and $C_{l+1} := \infty$
9: return the predictive distribution (7.57) for the label y of x.

$$+\frac{\tau}{l+1}\sum_{k=1}^{K}\left|\left\{i \in S_k : \alpha_{i,k} = \alpha_k^y\right\}\right| + \frac{\tau}{l+1}. \tag{7.56}$$

The intuition behind (7.56) is that it becomes an ICPS when the training bags $z_{S_{-k}}$ are replaced by a single hold-out training set (one disjoint from and independent of $z_1, \dots, z_l$).

For simplicity, let us consider a balanced strictly isotonic cross-conformity measure A that is, in addition, continuous in the test label: $A(\Lbag z_1, \dots, z_m\Rbag, (x, y))$ is continuous in y. An implementation of the CCPS determined by such a cross-conformity measure is shown as Algorithm 7.7, where the predictive distribution is now defined by

$$\Pi(z_1, \dots, z_l, (x, y), \tau) := \begin{cases} \frac{i+\tau}{l+1} & \text{if } y \in (C_i, C_{i+1}) \text{ for } i \in \{0, 1, \dots, l\} \\ \frac{i'-1+(i''-i'+2)\tau}{l+1} & \text{if } y = C_i \text{ for } i \in \{1, \dots, l\} \end{cases} \tag{7.57}$$

in the notation of (7.55); the only difference from (7.55) is that we use l in place of $l - m$ (now all training examples are used for calibration). The thickness of this predictive distribution is $\frac{1}{l+1}$ with the exception size at most l.

Because of the continuity assumption, the definition of C_i is now simpler: see line 5 of Algorithm 7.7. The size of the bag C grows from 0 to l as the algorithm runs. As in the case of ICPSs, it might be easier to use (7.56) directly if the equations defining $C_{i,k}$ are difficult to solve.

7.8.3 *Practical Aspects*

As discussed in Chap. 4, the ideas of inductive and cross-conformal prediction allow us to use the standard toolkit of machine learning in conformal prediction, including standard machine-learning packages. Such packages typically include implementations of deterministic predictive systems, such as Bayesian ridge regression.

We can use any deterministic predictive system as the conformity measure for a conformal predictive system. After passing through the "conformal machine" a DPS becomes calibrated in probability. And the overall procedure is computationally feasible if the conformal machine is inductive or cross-conformal, e.g., Algorithm 7.6 or Algorithm 7.7. What is important, in this case we can also perform normalization of attributes, hyper-parameter selection, or feature selection (the first two are discussed in Sect. 4.2.4).

7.8.4 Beyond Randomness

What we discussed in the previous subsection may be regarded as a way of calibrating deterministic predictive systems: our procedures take as input a deterministic predictive system that does not have to satisfy any properties of validity and outputs a randomized (slightly fuzzy) predictive system that is automatically calibrated in probability. In this subsection we will discuss these calibration procedures outside the assumption of randomness.

Namely, we can drop the assumption of randomness and apply the procedure in (7.50)–(7.51) for calibrating a deterministic predictive system A directly instead of calibrating the prediction rule found from the proper training set. The calibrated version of A working in the online protocol is now defined as

$$\begin{aligned}\Pi(z_1, \dots, z_{n-1}, (x_n, y), \tau_n) &:= \frac{1}{n} \left|\{i = 1, \dots, n-1 : \alpha_i < \alpha^y\}\right| \\ &+ \frac{\tau}{n} \left|\{i = 1, \dots, n-1 : \alpha_i = \alpha^y\}\right| + \frac{\tau}{n}, \qquad (7.58)\end{aligned}$$

where x_n is the test object and

$$\begin{aligned}\alpha_i &:= A(z_1, \dots, z_i), \qquad i = 1, \dots, n-1, \\ \alpha^y &:= A(z_1, \dots, z_{n-1}, (x_n, y)) .\end{aligned}$$

An interesting case is where A is the deterministic predictive system corresponding to the true data-generating distribution $P \in \mathbf{P}(\mathbf{Z}^\infty)$: namely, $A(z_1, \dots, z_{n-1}, (x_n, y))$ is a version of the conditional probability that $y_n \le y$ given $z_1, \dots, z_{n-1}, x_n$, where $z_1, \dots, z_{n-1}, (x_n, y_n)$ are the first n examples generated from P. (This case is not useful from the practical point of view, but a more relevant example will be discussed at the end of this subsection.) In this case we cannot improve A, and the question is how much worse Π can become as compared with A. This is another special case of the Burnaev–Wasserman programme (see Sect. 7.3.5).

The following proposition (whose proof will be discussed in Sect. 7.10.7) is a result in this direction that is particularly easy to state.

Proposition 7.19 *Suppose a DPS A corresponds to the true data-generating distribution* $P \in \mathbf{P}(\mathbf{Z}^\infty)$ *and is strictly increasing and continuous in the label of the test object (i.e., for any n and any* $z_1, \ldots, z_{n-1}, x_n$*, the function* $y \mapsto A(z_1, \ldots, z_{n-1}, (x_n, y))$ *is strictly increasing and continuous). For any* $\epsilon > 0$ *and any n, we have*

$$(P \times \mathbf{U})\Big(\big\{((z_1, z_2, \ldots), \tau) : \\ \|\Pi(z_1, \ldots, z_{n-1}, (x_n, \cdot), \tau) - A(z_1, \ldots, z_{n-1}, (x_n, \cdot))\|_\infty \geq \epsilon\big\}\Big) \\ \leq 2\exp\left(-2\epsilon^2 n + 4\epsilon\right). \qquad (7.59)$$

According to (7.59), the uniform distance (which is defined by $\|f - g\|_\infty := \sup_y |f(y) - g(y)|$) between the true distribution function and its calibrated version will be small with a high probability for a large n.

To see an example where the calibration procedure (7.58) works very well in the absence of the randomness assumption, suppose that it is fed with $\phi(A)$ instead of the true A (i.e., A corresponding to the true P) and $\phi : [0, 1] \to [0, 1]$ is a very non-linear continuous increasing function (such as $\phi(u) := u^2$). By Proposition 7.19, the conformalized version Π (which does not depend on ϕ) of the base DPS $\phi(A)$ will quickly converge to the true A, whereas the base DPS will always remain poor.

7.9 Proofs and Calculations

7.9.1 Proofs for Sect. 7.2

Proof of Lemma 7.1

Let Π and F satisfy the conditions of the lemma. Suppose there is $y \in \mathbb{R}$ such that $\Pi(y) \neq F(y)$. Fix such a y. The probability that $\Pi(Y) \leq \Pi(y)$ is, on the one hand, $\Pi(y)$ and, on the other hand, $F(y')$, where

$$y' := \sup\{y'' : \Pi(y'') = \Pi(y)\} .$$

(The first statement follows from the distribution of Y being uniform and the second from F being a continuous distribution function.) Since $\Pi(y) \neq F(y)$, we have $y' > y$, and we know that $\Pi(y) = \Pi(y'-) = F(y') > F(y)$. We can see that Π maps the whole interval $[y, y')$ of positive probability $F(y') - F(y)$ to one point, which contradicts its distribution being uniform.

Proof of Proposition 7.2

Let us split all numbers $i \in \{1, \ldots, n\}$ into three classes: i of class I are those satisfying $\alpha_i^y > \alpha_n^y$, i of class II are those satisfying $\alpha_i^y = \alpha_n^y$, and i of class III are those satisfying $\alpha_i^y < \alpha_n^y$. Each of those numbers is assigned a *weight*: 0 for i of class I, τ/n for i of class II, and $1/n$ for i of class III; notice that the weights are larger for higher-numbered classes. According to (7.6), $\Pi_n(y,\tau)$ is the sum of the weights of all $i \in \{1,\ldots,n\}$. As y increases, each individual weight can only increase (as i can move only to a higher-numbered class), and so the total weight $\Pi_n(y,\tau)$ can also only increase.

7.9.2 *Proofs for Sect. 7.3*

Proof of Proposition 7.4

According to Proposition 7.2, $\Pi_n(y,\tau)$ will be monotonically increasing in y if $\alpha_n^y - \alpha_i^y$ is a monotonically increasing function of y. We will use the notation $e_i := y_i - \hat{y}_i$ (suppressing the dependence on y) for the ith residual, $i = 1,\ldots,n$, in the data sequence $z_1,\ldots,z_{n-1},(x_n,y)$; y_n is understood to be y. In terms of the hat matrix $\bar{H}$ (which does not depend on the labels), the difference $e_n - e_i$ can be written as

$$\begin{aligned}
\alpha_n^y - \alpha_i^y &= e_n - e_i \\
&= (y_n - \hat{y}_n) - (y_i - \hat{y}_i) \\
&= y - \hat{y}_n + \hat{y}_i + c \\
&= y - (\bar{h}_{n,1}y_1 + \cdots + \bar{h}_{n,n-1}y_{n-1} + \bar{h}_n y) \\
&\quad + (\bar{h}_{i,1}y_1 + \cdots + \bar{h}_{i,n-1}y_{n-1} + \bar{h}_{i,n}y) + c \\
&= (1 - \bar{h}_n + \bar{h}_{i,n})y + c\,, \qquad (7.60)
\end{aligned}$$

where c stands for a constant (in the sense of not depending on y), and different entries of c may stand for different constants. We can see that Π_n will be a nontrivial monotonically increasing function of y whenever

$$1 - \bar{h}_n + \bar{h}_{i,n} > 0 \qquad (7.61)$$

for all $i = 1,\ldots,n$. Since $\bar{h}_{i,n} \in [-0.5, 0.5]$ (see [46], Property 2.5(b)), we can see that it indeed suffices to assume $\bar{h}_n < 0.5$.

Proof of Proposition 7.5

We are required to show that our $c = 0.5$ is the largest c for which the assumption $\bar{h}_n < c$ is still sufficient for $\Pi_n(y,\tau)$ to be a monotonically increasing function of y. For $\epsilon \in (0,1)$, consider the data matrix

$$\bar{X} = \begin{pmatrix} -1+\epsilon \\ 1 \end{pmatrix} \qquad (7.62)$$

(so that $n = 2$; we have two examples: one training example and one test example). The hat matrix is

$$\bar{H} = \frac{1}{2 - 2\epsilon + \epsilon^2}\begin{pmatrix} (1-\epsilon)^2 & -1+\epsilon \\ -1+\epsilon & 1 \end{pmatrix}.$$

The coefficient in front of y in the last line of (7.60) (i.e., the left-hand side of (7.61)) now becomes

$$1-\frac{1}{2-2\epsilon+\epsilon^2}+\frac{-1+\epsilon}{2-2\epsilon+\epsilon^2}=\frac{\epsilon^2-\epsilon}{2-2\epsilon+\epsilon^2}<0\,.$$

Therefore, $\Pi_n(\cdot,\tau)$ is monotonically decreasing and not monotonically increasing. On the other hand,

$$\bar{h}_n=\bar{h}_2=\frac{1}{2-2\epsilon+\epsilon^2}$$

can be made as close to 0.5 as we wish by making ϵ sufficiently small.

Proof of Proposition 7.6

Let $e_{(i)}$ be the *deleted residual*, $e_{(i)} := y_i - \hat{y}_{(i)}$, as in Sect. 2.3.3. We will again use the representation

$$e_{(i)}=\frac{e_i}{1-\bar{h}_i}$$

(as in (2.46)), where e_i is the ordinary residual, as used in the proof of Proposition 7.4. Let us check when the difference $e_{(n)} - e_{(i)}$ is a monotonically increasing function of $y = y_n$. Analogously to (7.60), we have, for any $i = 1, \ldots, n-1$:

$$\begin{aligned} e_{(n)}-e_{(i)} &= \frac{e_n}{1-\bar{h}_n}-\frac{e_i}{1-\bar{h}_i}=\frac{y_n-\hat{y}_n}{1-\bar{h}_n}-\frac{y_i-\hat{y}_i}{1-\bar{h}_i} \\ &= \frac{y-\bar{h}_n y}{1-\bar{h}_n}-\frac{y_i-\bar{h}_{i,n}y}{1-\bar{h}_i}+c=y-\frac{y_i-\bar{h}_{i,n}y}{1-\bar{h}_i}+c \\ &= y\frac{1-\bar{h}_i+\bar{h}_{i,n}}{1-\bar{h}_i}+c\,. \end{aligned} \tag{7.63}$$

Therefore, it suffices to require

$$1-\bar{h}_i+\bar{h}_{i,n}>0\,, \tag{7.64}$$

which is the same condition as for the ordinary LSPM (see (7.61)) but with i and n swapped. Therefore, it suffices to assume $\bar{h}_i < 0.5$.

Proof of Proposition 7.7

The statement of the proposition is obvious from the proofs of Propositions 7.5 and 7.6: motivated by the conditions (7.61) and (7.64) being obtainable from each other by swapping i and n, we can apply the argument in the proof of Proposition 7.5 to the data matrix

$$\bar{X}=\begin{pmatrix}1\\-1+\epsilon\end{pmatrix}$$

(which is (7.62) with its rows swapped).

Proof of Proposition 7.8

Similarly to (7.60) and (7.63), we obtain:

$$\alpha_n^y - \alpha_i^y = \frac{e_n}{\sqrt{1-\bar{h}_n}} - \frac{e_i}{\sqrt{1-\bar{h}_i}} = \frac{y_n - \hat{y}_n}{\sqrt{1-\bar{h}_n}} - \frac{y_i - \hat{y}_i}{\sqrt{1-\bar{h}_i}}$$
$$= \frac{y - \bar{h}_n y}{\sqrt{1-\bar{h}_n}} - \frac{y_i - \bar{h}_{i,n} y}{\sqrt{1-\bar{h}_i}} + c = \sqrt{1-\bar{h}_n}\, y + \frac{\bar{h}_{i,n}}{\sqrt{1-\bar{h}_i}} y + c\,. \tag{7.65}$$

Therefore, we need to check the inequality

$$\sqrt{1-\bar{h}_n} + \frac{\bar{h}_{i,n}}{\sqrt{1-\bar{h}_i}} \geq 0\,.$$

This inequality can be rewritten as

$$\bar{h}_{i,n} \geq -\sqrt{(1-\bar{h}_n)(1-\bar{h}_i)}$$

and follows from [46, Property 2.6(b)].

Computations for the Studentized LSPM

Now we need the chain (7.65) with a more careful treatment of the unspecified constants c:

$$\alpha_n - \alpha_i = \frac{e_n}{\sqrt{1-\bar{h}_n}} - \frac{e_i}{\sqrt{1-\bar{h}_i}} = \frac{y_n - \hat{y}_n}{\sqrt{1-\bar{h}_n}} - \frac{y_i - \hat{y}_i}{\sqrt{1-\bar{h}_i}}$$
$$= \frac{y - \sum_{j=1}^{n-1} \bar{h}_{j,n} y_j - \bar{h}_n y}{\sqrt{1-\bar{h}_n}} - \frac{y_i - \sum_{j=1}^{n-1} \bar{h}_{i,j} y_j - \bar{h}_{i,n} y}{\sqrt{1-\bar{h}_i}}$$
$$= \left(\sqrt{1-\bar{h}_n} + \frac{\bar{h}_{i,n}}{\sqrt{1-\bar{h}_i}}\right) y - \left(\frac{\sum_{j=1}^{n-1} \bar{h}_{j,n} y_j}{\sqrt{1-\bar{h}_n}} + \frac{y_i - \sum_{j=1}^{n-1} \bar{h}_{i,j} y_j}{\sqrt{1-\bar{h}_i}}\right)$$
$$= B_i y - A_i\,, \tag{7.66}$$

where the last equality is just the definition of B_i and A_i, also given by (7.15) and (7.16) above.

The Ordinary LSPM

The analogue of the calculation (7.66) for the ordinary LSPM is

$$\alpha_n - \alpha_i = e_n - e_i = (y_n - \hat{y}_n) - (y_i - \hat{y}_i)$$
$$= \left(y - \sum_{j=1}^{n-1} \bar{h}_{j,n} y_j - \bar{h}_n y\right) - \left(y_i - \sum_{j=1}^{n-1} \bar{h}_{i,j} y_j - \bar{h}_{i,n} y\right)$$
$$= \left(1 - \bar{h}_n + \bar{h}_{i,n}\right) y - \left(\sum_{j=1}^{n-1} \bar{h}_{j,n} y_j + y_i - \sum_{j=1}^{n-1} \bar{h}_{i,j} y_j\right) = B_i y - A_i\,,$$

with the notation

$$B_i := 1 - \bar{h}_n + \bar{h}_{i,n}, \tag{7.67}$$

$$A_i := \sum_{j=1}^{n-1} \bar{h}_{j,n} y_j + y_i - \sum_{j=1}^{n-1} \bar{h}_{i,j} y_j \, . \tag{7.68}$$

Proof of (7.22)

To compute C_i, we will use the formulas (7.67)–(7.68) and (7.20)–(7.21):

$$\begin{aligned} B_i &= 1 - \bar{h}_n + \bar{h}_{i,n} \\ &= 1 - \frac{x_n'(X'X)^{-1}x_n}{1 + x_n'(X'X)^{-1}x_n} + \frac{x_i'(X'X)^{-1}x_n}{1 + x_n'(X'X)^{-1}x_n} = \frac{1 + x_i'(X'X)^{-1}x_n}{1 + x_n'(X'X)^{-1}x_n} \end{aligned}$$

and, letting $\hat{y}$ stand for the predictions computed from the first $n-1$ examples,

$$\begin{aligned} A_i &= y_i - \sum_{j=1}^{n-1} \bar{h}_{i,j} y_j + \sum_{j=1}^{n-1} \bar{h}_{j,n} y_j \\ &= y_i - \sum_{j=1}^{n-1} h_{i,j} y_j + \sum_{j=1}^{n-1} \frac{x_i'(X'X)^{-1}x_n x_n'(X'X)^{-1}x_j}{1 + x_n'(X'X)^{-1}x_n} y_j + \sum_{j=1}^{n-1} \frac{x_j'(X'X)^{-1}x_n}{1 + x_n'(X'X)^{-1}x_n} y_j \\ &= y_i - \hat{y}_i + \frac{x_i'(X'X)^{-1}x_n x_n'(X'X)^{-1}X'Y}{1 + x_n'(X'X)^{-1}x_n} + \frac{Y'X(X'X)^{-1}x_n}{1 + x_n'(X'X)^{-1}x_n} \\ &= y_i - \hat{y}_i + \frac{x_i'(X'X)^{-1}x_n \hat{y}_n}{1 + x_n'(X'X)^{-1}x_n} + \frac{\hat{y}_n}{1 + x_n'(X'X)^{-1}x_n} \\ &= y_i - \hat{y}_i + \frac{1 + x_i'(X'X)^{-1}x_n}{1 + x_n'(X'X)^{-1}x_n} \hat{y}_n \, . \end{aligned}$$

This gives

$$C_i = A_i / B_i = (y_i - \hat{y}_i) \frac{1 + x_n'(X'X)^{-1}x_n}{1 + x_i'(X'X)^{-1}x_n} + \hat{y}_n \, ,$$

i.e., (7.22).

7.9.3 *Proof of Theorem 7.16*

It suffices to prove that, for each $d \in \mathbf{D}$,

$$\int U(y, d) Q(\mathrm{d}y \mid x_n) - \int u \Pi_d^*(\mathrm{d}u) \to 0 \qquad (n \to \infty) \tag{7.69}$$

in probability, where Π_d^* is computed as in Algorithm 7.5 for the training set $z_1, \ldots, z_{n-1}$, test object x_n, and random number τ_n; indeed, under (7.69) and by (7.47), the requirement

$$R_F(z_1, \ldots, z_{n-1}, x_n, \tau_n) \to 0$$

can be rewritten as

$$\max_{d \in \mathbf{D}} \int u \Pi_d^*(\mathrm{d}u) - \int u \Pi_{F(z_1,\ldots,z_{n-1},x_n,\tau_n)}^*(\mathrm{d}u) \to 0 ,$$

which is true (even with "$\to$" replaced by "$=$") by definition for F computed by Algorithm 7.5. Fix any $d \in \mathbf{D}$ and let Q' be the image of the probability distribution Q under the mapping $(x, y) \in \mathbf{Z} \mapsto (x, U(y, d)) \in \mathbf{X} \times \mathbb{R}$. Then (7.69) can be rewritten as

$$\int u Q'(\mathrm{d}u \mid x_n) - \int u \Pi_d^*(\mathrm{d}u) \to 0 \qquad (n \to \infty) ,$$

and so the conclusion follows from the universal consistency of the randomized predictive system used in Algorithm 7.5, since the identity function $u \mapsto u$ is continuous and, under $Q'(\mathrm{d}u \mid x_n)$, bounded with probability one; the assumption that $\mathbf{X}$ is a standard Borel space is used in Theorem 7.14.

7.9.4 *Proofs for Sect. 7.8*

Proof of Proposition 7.17

We follow closely the proof of Proposition 7.2. It is clear that (7.50) is increasing in τ (and linear). To show that it is increasing in y, split, in the context of (7.50), all $i \in \{m+1, \ldots, l\}$ into three classes: the i in class I satisfy $\alpha_i > \alpha^y$, the i in class II satisfy $\alpha_i = \alpha^y$, and the i in class III satisfy $\alpha_i < \alpha^y$. Then (7.50) is the total weight of all i where the weights are 0, $\tau \in [0, 1]$, and 1 for i in classes I, II, and III, respectively. As y increases, α^y increases as well, and therefore, each i can only move to a higher-numbered class thus increasing (7.50).

Out of the remaining two conditions, let us check, e.g., (7.2). It suffices to notice that, since A is balanced, we have $\alpha^y \geq \max_{i \in \{m+1,\ldots,l\}} \alpha_i$ from some y on, for any $z_1, \ldots, z_l$ and x.

Proof of Proposition 7.18

Suppose A is not isotonic. Fix m, $z_1, \ldots, z_m$, x, y, and y' such that $y < y'$ but the consequent of (7.52) is violated. Then the putative predictive distribution function $\Pi(z_1, \ldots, z_m, (x, y), (x, \cdot), 1)$, corresponding to the proper training set $z_1, \ldots, z_m$, calibration set (x, y), test object x, and $\tau = 1$, will not be increasing: its value at y (which is 1) will be greater than its value at y' (which is 0.5).

Now suppose A is not balanced. Fix m, $z_1, \ldots, z_m$, and $x, x' \in \mathbf{X}$ such that

$$\operatorname{co} A(z_1, \ldots, z_m, (x, \mathbb{R})) \neq \operatorname{co} A(z_1, \ldots, z_m, (x', \mathbb{R}))$$

(cf. (7.53)). Suppose, for concreteness, that there is $y \in \mathbb{R}$ such that

$$\operatorname{co} A(z_1, \ldots, z_m, (x, \mathbb{R})) \ni y < \operatorname{co} A(z_1, \ldots, z_m, (x', \mathbb{R})) ,$$

where $y < S$ means $\forall s \in S : y < s$ when $S \subseteq \mathbb{R}$. (The other three possible cases can be analyzed in the same way.) Let the proper training set be $z_1, \ldots, z_m$, the calibration set be (x, y), the test object be x', and the random number be $\tau = 0$. Then we will have

$$\lim_{y' \to -\infty} \Pi(z_1, \ldots, z_m, (x, y), (x', y'), 0) > 0 \ ,$$

which contradicts requirement R2 (cf. (7.1)).

7.10 Context

7.10.1 Conformal Predictive Distributions

In Sect. 7.2 we follow [409, 410], which in turn follow, to some degree, [324, Sect. 1] and [311, Chap. 12]. The theory of predictive distributions as developed by [311] and [324] assumes that the examples are generated from a parametric statistical model. Conformal prediction extends the theory to the case of regression under unconstrained randomness, where the distribution form does not need to be specified.

For a well-known review of predictive distributions, see [216]. The more recent review [132] refers to the notion of validity used in this chapter as probabilistic calibration and describes it as critical in forecasting; [132, Sect. 2.2.3] also gives further references.

The history of the Dempster–Hill procedure, corresponding to the LSPM with one attribute that is identical 1, is discussed in Sects. 13.3.4 and 13.7.2.

The limiting distribution of the process Z in Theorem 7.10, given by (7.30), goes back to Kac et al. [183] and Durbin [95]; in a specialized related setting, Kac et al. established the convergence of finite-dimensional distributions and Durbin established the convergence in law. The key results for simple regression are due to Lyudmila Mugantseva [254]. The fact that the asymptotic variances for standard linear regression are as good as those for the location/scale model was emphasized in the pioneering paper by Pierce and Kopecky [277], and this phenomenon shows in the remarkable property that we mentioned earlier: the limiting processes for Oracles I and II do not depend on the number p of the attributes.

For proofs of Theorems 7.10 and 7.11, see [410, Sect. 6]. The proofs are based on the representation (7.22) and Mugantseva's [254] results (see also [47, Chap. 2]). Theorem 7.12 can be proved by similar methods; it is stated in [410, Theorem 4] only for the Dempster–Hill procedure (since [410] does not include the assumption of IID objects in the soft model).

7.10.2 Conformal Predictive Distributions with Kernels

Here we follow the paper [409] presented at the Braverman Readings (Boston, April 2017). Emmanuel Braverman (1931–1977) was a member of the team (headed by Mark Aizerman and also including Lev Rozonoer) at the Institute of Control Sciences promoting kernel methods in machine learning already in the 1960s. See, e.g., their monograph [1]. At the same time and at the same institution Vapnik and Chervonenkis (in Lerner's laboratory) were developing foundations of statistical learning theory. See [407], especially Chap. 5 by Vasily N. Novoseltsev.

7.10.3 Venn Prediction for Probabilistic Regression

Venn prediction, as discussed in Chap. 6, has been designed for the case of classification. It is shown by Nouretdinov et al. [266] that Venn–Abers prediction can also be used in the case of regression. It is argued in [410] that this particular implementation of Venn regression has serious limitations, but these limitations might be irrelevant in some applications, and there may be other implementations.

7.10.4 Universal Consistency and Universality

As we mentioned in Sect. 3.5.3, universal (more specifically, universally consistent) prediction algorithms are discussed mainly in two places in this book, for the cases of probabilistic classification (in Chap. 6) and probabilistic regression (in this chapter).

In Sect. 7.6 we follow [393]. It is easier to achieve universal consistency for conditional conformal predictors, and this route is also used in that paper. The main disadvantage of conditional conformal predictors as used there is that they are valid in a weaker sense: we can't say that they are strongly calibrated in probability (the values taken by the predictive distribution functions at the actual labels are no longer independent in the online mode). The construction of a universally consistent conformal predictive system adapts standard arguments for universal consistency in classification and regression [82, 144, 347].

Lei and Wasserman's [225] construction of a universal prediction algorithm in the case of regression also produces a conditional conformal predictor. However, their result is much more precise than our results since it also establishes an optimal rate of convergence; it is not just about universal consistency. See [15, Theorem 2.1] for a simplified version of Lei and Wasserman's result.

7.10.5 Various Notions of Convergence in Law

Most of this chapter is based on the most traditional approach to convergence in law of empirical processes, originated by Skorokhod and described in detail in [33]. This approach encounters severe difficulties in more general situations (such as multi-dimensional labels). Alternative approaches have been proposed by numerous authors, including Dudley (using the uniform topology and ball σ-algebra, [92, 93]) and Hoffmann-Jørgensen (dropping measurability and working with outer integrals; see, e.g., [362, Sect. 1.3] and the references in the notes to that section). Translating our results into those alternative languages might facilitate various generalizations.

Another generalization of the traditional notion of convergence in law is Belyaev's notion of weakly approaching sequences of random distributions [28]. We used it in Sect. 7.6 when discussing universal conformal predictive systems.

7.10.6 Decision Theory

The standard approach to decision making was described in von Neumann and Morgenstern's classic book [375, Chap. I] based on von Neumann's earlier work. Starting from Chap. II of their book, von Neumann and Morgenstern develop exclusively the theory of games, in which players (typically economic agents) have different, often opposite, goals, which goes very much beyond the setting of this book (which corresponds to the case of a "Robinson Crusoe" economy in the terminology of [375, Sect. 2.2]).

The crucial step made by [375] was the introduction of their notion of utility, which we referred to as von Neumann–Morgenstern utilities in Sect. 7.7. After introducing natural axioms of utilities in Sect. 3.6 of their book, they show in the appendix (added in the second edition) that the axioms determine numerical utilities up to a linear transformation. In their result they assumed the notion of probability as known, but they realized ([375, footnote 2 on p. 19]) that utilities and probabilities could be axiomatized together; this was first done by Savage [304]. After Savage this possibility has been explored in numerous papers and books, and nowadays the principle of maximizing expected utility is a cornerstone of Bayesian decision theory.

Abraham Wald [420] applied von Neumann's ideas to unify statistical hypothesis testing and point estimation. Berger [29] joked that "Statisticians seem to be pessimistic creatures who think in terms of losses", but it is also true that exposition in terms of losses often leads to simpler formulas: e.g., a natural loss function in problems of point estimation is $(y - \hat{y})^2$, the squared difference between the estimate $\hat{y}$ and the true observation y. Similar reasons prompted us to prefer nonconformity to conformity measures in the early chapters of this book (and in all of its first edition).

7.10.7 Postprocessing of Predictive Distributions

The calibration procedure of Sect. 7.8.4 is a way of postprocessing predictive distributions, namely those output by the DPS A. For a much more precise version of Proposition 7.19, see [411, Proposition 4]. According to that more precise version, the speed at which the conformalized predictive distributions approach the true predictive distributions is $O(n^{-1/2})$, which is the same as the speed of convergence that we had for the LSPM in Theorems 7.10–7.12. Proposition 7.19 itself can be deduced from [411, Proposition 4] and Massart's [240] form of the Dvoretzky–Kiefer–Wolfowitz inequality [96].

Postprocessing of predictive distributions is commonplace in meteorology. However, the underlying predictive systems typically used in meteorology output only *ensemble forecasts*, which are distribution functions taking values of the form i/N, $i = 0, \dots, N$, where N is the *size of the ensemble*. (An ensemble forecast is defined as the empirical distribution function of the point forecasts output by N simple predictors.) Ensemble forecasts are nowadays the main tool in numerical weather prediction, but in practice they tend to be biased and underdispersed [147]. See [352] for a recent review and a new proposal, quantile regression forests.

In fact, the oldest postprocessing method, introduced in [147, 148] and called the "corrected ensemble" forecast, was very similar to conformal calibration: instead of (7.50) it used, essentially,

$$\Pi(z_1, \dots, z_l, (x, y)) := \frac{1}{l - m} \left| \left\{ i = m + 1, \dots, l : \alpha_i < \alpha^y \right\} \right| \tag{7.70}$$

and was applied to ensemble forecasts. There is no difference (a.s.) whether "$<$" of "$\le$" is used in (7.70), since [147, 148] use a randomized tie-breaking procedure analogous to the one that we use. The method is called the *weighted ranks method* and studied empirically in [97]. A variation of this method was proposed by Kuleshov et al. [208, Algorithm 1].

Part III
Testing Randomness

Chapter 8
Testing Exchangeability

Abstract In Chaps. 2–7 we assumed that all examples output by Reality are exchangeable. This is a strong assumption, but it is standard in machine learning (where the even stronger assumption of randomness is usually made). We start this chapter (in Sect. 8.1) by discussing how to test this assumption in the online mode: at each point in time we would like to have a valid measure of the amount of evidence found against the hypothesis of exchangeability. Conformal prediction is a valuable tool for designing such online testing methods, and can also be adapted for detecting different kinds of deviations from exchangeability (Sect. 8.2). Such methods of *conformal testing* can be applied in multistage testing, when the task is to raise an alarm soon after the assumption of exchangeability becomes violated (Sect. 8.3), and in deciding when a machine learning algorithm depending on the exchangeability assumption should be retrained (Sect. 8.4).

Keywords Exchangeability · Statistical hypothesis testing · Anticausal classification · CUSUM · Shiryaev–Roberts procedure · Retraining

8.1 Testing Exchangeability

This section discusses online ways of monitoring the strength of evidence against the assumption of exchangeability. Such online monitoring is often a wise thing to do even if the exchangeability assumption is tentatively accepted.

In this section we define exchangeability supermartingales and exchangeability martingales. Nonnegative exchangeability supermartingales and martingales are online procedures for detecting deviations from exchangeability. Intuitively, they are betting schemes that never risk bankruptcy and do not benefit the gambler under the hypothesis of exchangeability.

Exchangeability supermartingales and martingales are easy to construct using conformal transducers, introduced in Sect. 2.7. While the exchangeability supermartingales that we construct only depend on the data, i.e., the examples produced by Reality, exchangeability martingales also require an additional source of ran-

V. Vovk et al., *Algorithmic Learning in a Random World*,
https://doi.org/10.1007/978-3-031-06649-8_8

domness. To illustrate the performance of these procedures, we again use the USPS benchmark dataset of hand-written digits (known to be somewhat heterogeneous; see Sect. B.1).

8.1.1 Exchangeability Supermartingales

In this subsection we set up our basic framework, defining the fundamental notion of exchangeability supermartingale and the closely related notion of randomized exchangeability martingale. In our standard learning protocol, Reality outputs examples $z_1, z_2, \ldots$, each of which consists of two parts, an object and its label. In the theoretical considerations of this section, however, we will not use this additional structure; therefore, the example space $\mathbf{Z}$ is not assumed to be a Cartesian product $\mathbf{X} \times \mathbf{Y}$.

We are interested in testing the hypothesis of exchangeability *online*: after observing each new example z_n we would like to have a number S_n reflecting the strength of evidence found against the hypothesis. Let us first consider testing the simple hypothesis that $z_1, z_2, \ldots$ are generated from a probability distribution P on $\mathbf{Z}^\infty$. We say that a sequence of random variables $S_0, S_1, \ldots$ is a *P-supermartingale* if, for all $n = 0, 1, \ldots$, S_n is a measurable function of $z_1, \ldots, z_n$ (in particular, S_0 is a constant) and

$$S_n \geq \mathbb{E}\left(S_{n+1} \mid S_1, \ldots, S_n\right) \quad \text{a.s.}\,, \tag{8.1}$$

where $\mathbb{E}$ refers to the expected value in the probability space $(\mathbf{Z}^\infty, P)$, in which $z_1, z_2, \ldots$ are generated from P. If $S_0 = 1$ and $\inf_n S_n \geq 0$, S_n can be regarded as the capital process of a player who starts from 1, never risks bankruptcy, at the beginning of each trial n places a fair (cf. (8.1)) bet on the z_n to be chosen by Reality, and maybe sometimes throws money away (since (8.1) is an inequality). If such a supermartingale S ever takes a large value, our belief in P is undermined; this intuition is formalized by Ville's inequality (see Sect. A.6.1), which implies

$$P\left(\{(z_1, z_2, \ldots) : \exists n : S_n \geq c\}\right) \leq 1/c\,, \tag{8.2}$$

where c is an arbitrary positive constant.

When testing a *composite hypothesis* $\mathcal{P}$ (i.e., a family of probability distributions on $\mathbf{Z}^\infty$), we will use *$\mathcal{P}$-supermartingales*, i.e., sequences of random variables $S_0, S_1, \ldots$ which are P-supermartingales for all $P \in \mathcal{P}$ simultaneously. We are primarily interested in the family $\mathcal{P}$ consisting of all exchangeable probability distributions P on $\mathbf{Z}^\infty$; in this case we will say *exchangeability supermartingales* to mean $\mathcal{P}$-supermartingales.

Remark 8.1 If $\mathcal{P}$ is the set of all power probability distributions Q^∞, Q ranging over the probability distributions on $\mathbf{Z}$, $\mathcal{P}$-supermartingales are called *randomness supermartingales*. De Finetti's theorem (see Sect. A.5) and the fact that standard Borel spaces are closed under countable products (see, e.g., [307, Lemma B.41] and Sect. A.1.1) imply that each exchangeable distribution P on $\mathbf{Z}^\infty$ is a mixture of power distributions Q^∞ provided $\mathbf{Z}$ is standard Borel. Therefore, the notions of randomness supermartingales and exchangeability supermartingales coincide in the standard Borel case. But even without the assumption that $\mathbf{Z}$ is standard Borel, all exchangeability supermartingales are randomness supermartingales.

Another useful notion is that of *randomized exchangeability martingales*; these are sequences of measurable functions $S_n(z_1, \tau_1, \ldots, z_n, \tau_n)$ (each example z_n is extended by adding a random number $\tau_n \in [0, 1]$) such that, for any exchangeable probability distribution P on $\mathbf{Z}^\infty$,

$$S_n = \mathbb{E}\left(S_{n+1} \mid S_1, \ldots, S_n\right) \quad \text{a.s.}\,, \tag{8.3}$$

$\mathbb{E}$ referring to the expected value in the probability space $((\mathbf{Z}\times[0,1])^\infty, P\times\mathbf{U}^\infty)$, in which $z_1, z_2, \ldots$ and $\tau_1, \tau_2, \ldots$ are generated from P and $\mathbf{U}^\infty$ (remember that $\mathbf{U}$ is the uniform distribution on $[0, 1]$) independently. Instead of $S_n(z_1, \tau_1, \ldots, z_n, \tau_n)$, we will also write $S(z_1, \tau_1, \ldots, z_n, \tau_n)$ or $S_n(z_1, \tau_1, z_2, \tau_2, \ldots)$ (with the dependence only on the first $2n$ elements of the infinite sequence $z_1, \tau_1, z_2, \tau_2, \ldots$).

Remark 8.2 An exchangeability martingale is natural to define as an exchangeability supermartingale such that (8.1) holds as equality for any exchangeable P, and the notion of randomized exchangeability supermartingale may be obtained by relaxing the "=" in (8.3) to "≥". We do not need these notions, however: the notion of exchangeability martingale is too restrictive and that of randomized exchangeability supermartingale is unnecessarily wide (our goals can be achieved already with randomized exchangeability martingales).

8.1.2 Conformal Test Martingales

We know from Sect. 2.7 that the p-values $p_1, p_2, \ldots$ output by a smoothed conformal transducer are independent and distributed uniformly on $[0, 1]$. They can be used for constructing exchangeability supermartingales and randomized exchangeability martingales. From now on we mostly drop "randomized" in "randomized exchangeability martingales" (as we already did in the introductory part of this section); we will never need the exchangeability martingales in the sense of Remark 8.2. We will be mostly interested in *test martingales*, i.e., martingales S that are required to be nonnegative, $S_n \geq 0$, and start from 1, $S_0 = 1$; we will also allow them to take value ∞.

The formal definition of a conformal test martingale (abbreviated to CTM) given in this paragraph will be followed by a discussion of the intuition behind it in the following paragraph. A *betting martingale* is a measurable function $F : [0, 1]^* \to$

$[0, \infty]$ such that $F(\Box) = 1$ and, for each sequence $(u_1, \ldots, u_{n-1}) \in [0, 1]^{n-1}$, $n \geq 1$, we have

$$\int_0^1 F(u_1, \ldots, u_{n-1}, u) \, \mathrm{d}u = F(u_1, \ldots, u_{n-1}) \,. \tag{8.4}$$

It can be regarded as a test martingale. A *conformal test martingale* is any sequence of functions $S_n : (\mathbf{Z} \times [0, 1])^\infty \to [0, \infty]$, $n = 0, 1, \ldots$, such that, for some nonconformity measure A and betting martingale F, for all $m \in \mathbb{N}_0$, $(z_1, z_2, \ldots) \in \mathbf{Z}^\infty$, and $(\tau_1, \tau_2, \ldots) \in [0, 1]^\infty$,

$$S_m(z_1, \tau_1, z_2, \tau_2, \ldots) = F(p_1, \ldots, p_m) \,, \tag{8.5}$$

where p_n, $n \in \mathbb{N}$, is the smoothed p-value computed by the right-hand side of (2.67) from the nonconformity measure A, the examples $z_1, z_2, \ldots$, and the nth element τ_n of the sequence $(\tau_1, \tau_2, \ldots)$. Notice that $S_m(z_1, \tau_1, z_2, \tau_2, \ldots)$ depends on $z_1, \tau_1, z_2, \tau_2, \ldots$ only via $z_1, \tau_1, \ldots, z_m, \tau_m$. We may also write $S_m(z_1, \tau_1, \ldots, z_m, \tau_m)$ or even $S(z_1, \tau_1, \ldots, z_m, \tau_m)$ for (8.5), and we will say that the CTM S *is determined by* F and A (or by F and the smoothed conformal transducer determined by A).

Each CTM is an exchangeability martingale. Of course, the definition also works for conformity measures in place of nonconformity measure, and in fact our exposition will be mostly in terms of conformity measures (and in the case of testing the usual intuitive difference between conformity and nonconformity becomes weaker).

Intuitively, a betting martingale describes the evolution of the capital of a player who gambles against the hypothesis that the p-values $p_1, p_2, \ldots$ are distributed uniformly and independently, as they should under the hypothesis of exchangeability. The requirement (8.4) expresses the fairness of the game: at step $n - 1$, the conditional expected value of the player's future capital at step n given the present situation (i.e., the first $n - 1$ p-values) is equal to his current capital. A CTM is what we get when we feed a betting martingale with the p-values produced by conformal prediction.

One way of constructing betting martingales is to use "betting functions". A *betting function* $f : [0, 1] \to [0, \infty]$ is a function satisfying $\int_0^1 f(u) \, \mathrm{d}u = 1$. A useful method of betting against the hypothesis that the p-values $p_1, p_2, \ldots$ are independent and uniformly distributed is to choose, before each step n, a betting function f_n that may depend on $p_1, \ldots, p_{n-1}$ (in a measurable manner). Then

$$F(p_1, \ldots, p_n) := f_1(p_1) \ldots f_n(p_n), \quad n = 0, 1, \ldots \,, \tag{8.6}$$

will be a betting martingale (a CTM if $p_1, p_2, \ldots$ are generated by conformal prediction).

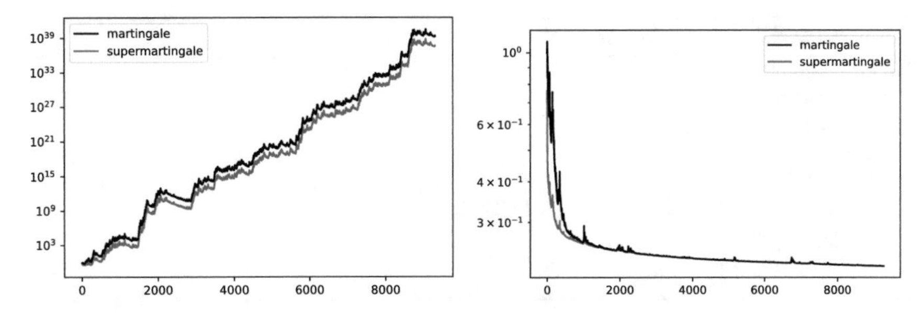

Fig. 8.1 A Composite Jumper CTM and its associated supermartingale, both based on the nearest-neighbour conformity measure and applied to the USPS dataset (left panel) and its random permutation (right panel)

All experiments described in this section are performed on the full USPS dataset with no pre-processing of images, as described in Sect. B.1. Our goal will be to detect deviations from exchangeability for this dataset. The expectation of a test martingale's final value on the USPS (or any other) dataset is 1 under the null hypothesis, and according to Jeffreys's [177, Appendix B] rule of thumb, a final value in excess of 100 provides decisive evidence against the null hypothesis.

Figure 8.1 shows paths of a CTM and its associated supermartingale. The martingale in the left panel of Fig. 8.1 is a very simple CTM, which we will define momentarily. The associated supermartingale will be defined at the end of the subsection. The final values of both are astronomically large and exceed 10^{37}, dwarfing Jeffreys's threshold of 100.

The black line in Fig. 8.1 plots the values S_n of a CTM after processing the first n examples for $n \in \{0, \dots, 9298\}$. The martingale is randomized, but its path on the USPS dataset does not depend much on the seed used in the random number generator (and this will be true for all other CTMs discussed in this chapter).

The conformity measure used in Fig. 8.1 is of the nearest-neighbour type: namely, the conformity score α_i of the ith example (x_i, y_i) in a comparison bag $\Lbag (x_1, y_1), \dots, (x_n, y_n) \Rbag$ is defined as

$$\alpha_i := \min_{j \in \{1,\dots,n\}: j \neq i} \left\| x_i - x_j \right\| , \tag{8.7}$$

where $\|\dots\|$ is Euclidean norm. Notice that the conformity measure (8.7) completely ignores the labels, although the alternative

$$\alpha_i := \min_{j \neq i : y_j = y_i} \left\| x_i - x_j \right\|$$

leads to similar (usually slightly better) results.

The CTM shown in Fig. 8.1 in black is based on the *Simple Jumper* betting martingale. The main components of the Simple Jumper are the betting functions

$$f_\epsilon(p) := 1 + \epsilon(p - 0.5), \quad p \in [0, 1] , \tag{8.8}$$

Algorithm 8.1 Simple Jumper betting martingale

Input: p-values $p_1, p_2, \ldots$.
Output: martingale $S_1, S_2, \ldots$.
1: $C_\epsilon := 1/|\mathbf{E}|$ for all $\epsilon \in \mathbf{E}$
2: $C := 1$
3: **for** $n = 1, 2, \ldots$:
4: **for** $\epsilon \in \mathbf{E}$: $C_\epsilon := (1 - J)C_\epsilon + (J/|\mathbf{E}|)C$
5: **for** $\epsilon \in \mathbf{E}$: $C_\epsilon := C_\epsilon f_\epsilon(p_n)$
6: $S_n := C := \sum_{\epsilon \in \mathbf{E}} C_\epsilon$.

where $\epsilon \in \mathbf{E} := \{-1, -0.5, 0, 0.5, 1\}$. The Simple Jumper betting martingale involves another parameter, $J \in (0, 1]$, the *jumping rate*. For any probability measure μ on $\mathbf{E}^\infty$ the function

$$F(u_1, \ldots, u_n) := \int \prod_{i=1}^{n} f_{\epsilon_i}(u_i)\mu(\mathrm{d}(\epsilon_1, \epsilon_2, \ldots)) \tag{8.9}$$

(with a variable n) is a betting martingale. The measure μ for the Simple Jumper is defined as the probability distribution of the following Markov chain with state space $\mathbf{E}$. The initial state ϵ_1 is chosen from the uniform probability measure on $\mathbf{E}$. The transition function prescribes maintaining the same state with probability $1 - J$ and, with probability J, choosing a new state from the uniform probability measure on $\mathbf{E}$. Notice that the betting martingale (8.9) is a deterministic function, even though the Markov chain is stochastic.

The experiments of this and following sections use Algorithm 8.1 for computing the Simple Jumper betting martingale. At the end of step n, C_ϵ is the integral (8.9) over the sequences $\epsilon_1, \epsilon_2, \ldots$ with $\epsilon_n = \epsilon$. The Markov nature of the probability measure μ allows us to update the array of C_ϵ efficiently.

Remark 8.3 Our choice of $\mathbf{E}$ is somewhat arbitrary, but the dependence on $\mathbf{E}$ is not heavy. Inevitably, there is some element of data snooping here, but it is limited by the use of simple definitions and round figures.

Remark 8.4 The intuition behind the betting functions (8.8) is that $\epsilon < 0$ corresponds to betting on small p-values (heavy when $\epsilon = -1$ and less heavy when $\epsilon = -0.5$), $\epsilon > 0$ corresponds to betting on large p-values, and $\epsilon = 0$ corresponds to not betting. Intuitively, the Simple Jumper tracks the best value of the parameter ϵ used to "calibrate" the p-values produced by conformal prediction into a martingale.

The dependence of the Simple Jumper betting martingale on the choice of the parameter J is also not heavy, but to weaken it further we will use the *Composite Jumper* betting martingale, defined as the arithmetic mean of the Simple Jumper over $J \in \{10^{-4}, 10^{-3}, 10^{-2}, 10^{-1}, 1\}$. The inclusion of $J = 1$ is important in that the corresponding Simple Jumper betting martingale is identical 1 (because of the symmetry of $\mathbf{E}$ around 0), and so the value of the Composite Jumper never drops below 0.2, as we can see from the right panel of Fig. 8.1. The black martingale in Fig. 8.1 is determined by the conformity measure (8.7), the conformal transducer

$$f(z_1, \dots, z_n, \tau_n) := \frac{|\{i : \alpha_i < \alpha_n\}| + \tau_n \, |\{\alpha_i = \alpha_n\}|}{n} \tag{8.10}$$

(which is (2.67) with "<" in place of ">" and in fact leads to the same randomized exchangeability martingale because of the symmetry of our family of betting functions around $p = 0.5$), and the Composite Jumper betting martingale.

The dependence of CTMs on the random numbers τ_n may be considered their disadvantage. Therefore, Fig. 8.1 also shows a path of an exchangeability supermartingale that is a deterministic version of the exchangeability martingale shown in black. Let us denote the latter by $S = (S_n)$. The least wasteful way to make S into an exchangeability supermartingale S' is to define $S'_n(z_1, z_2, \dots)$ as the infimum of $S'_n(z_1, \tau_1, z_2, \tau_2, \dots)$ over all possible values of $\tau_1, \dots, \tau_n$. This infimum will typically be a measurable function and, therefore, an exchangeability supermartingale. (In principle, the infimum might not be measurable [187, Theorem 14.2], but it will always be universally measurable [187, Theorem 21.10], assuming $\mathbf{Z}$ is standard Borel.)

A simpler way to obtain an exchangeability supermartingale from S is to replace each element $f_{\epsilon_i}(u_i)$ of each Simple Jumper component (8.9), where u_i is the p-value computed at step i, by its minimum over τ_i. This makes little difference to the exchangeability supermartingale except for its Simple Jumper component with $J = 1$. Remember that component was responsible for the inequality $S_n \geq 0.2$ since it was the identical 1, and this is no longer the case if we minimize each $f_{\epsilon_i}(u_i)$ over τ_i. So we replace that component by the identical 1. The resulting exchangeability supermartingale is shown in magenta in Fig. 8.1 and is referred to as the associated supermartingale in the caption (and in the main text). The right panel confirms that 0.2 remains a lower bound.

8.2 Testing for Concept and Label Shift in Anticausal Classification

Fawcett and Flach [100] identified an interesting and important type of classification problems. Let us say, informally, that a classification problem is *anticausal* (to use the terminology of [309]) if the objects x_n are causally dependent on the labels y_n. Despite being informal, this notion will motivate many of the definitions of this section. For example, the USPS dataset is clearly anticausal: the intended digit in the writer's mind causes the resulting matrix of pixels, not vice versa.

Under the randomness assumption, the consecutive examples (x_n, y_n) are generated from the same probability distribution Q. There is a *dataset shift* if Q in fact changes between examples. In Sect. 8.1 we discussed ways of detecting dataset shift, without using this terminology. In anticausal classification, we can also discuss natural special cases of dataset shift. Let us consider a decomposition of the data-generating distribution Q into the marginal and conditional distribution, similarly to what we did in Chap. 3 (see, e.g., (3.11)): namely, $Q_{\mathbf{Y}}$ is the marginal probability distribution $Q_{\mathbf{Y}}(y) := Q(\mathbf{X} \times \{y\})$, and $Q_{\mathbf{X}|\mathbf{Y}}$ is the conditional distribution

$$Q_{\mathbf{X}|\mathbf{Y}}(E \mid y) := \frac{Q(E \times \{y\})}{Q_{\mathbf{Y}}(y)},$$

where $E \subseteq \mathbf{X}$ is a measurable set; if $Q_{\mathbf{Y}}(y) = 0$, set $Q_{\mathbf{X}|\mathbf{Y}}(\cdot \mid y)$, e.g., to the marginal distribution of the objects x under Q. We assume $|\mathbf{Y}| < \infty$ (the case of classification). Let us say that there is a *label shift* if the marginal distribution $Q_{\mathbf{Y}}$ of the label under Q changes. There is a *concept shift* if the conditional distribution $Q_{\mathbf{X}|\mathbf{Y}}$ of the object given the label changes. (In the existing literature, e.g., in [234], the expression "label shift" is often used only in situations where there is no concept shift, but we will talk about a label shift whenever the distribution of labels changes.)

In the case of the USPS dataset, an example of label shift is where people start using new zip codes, and an example of concept shift is where people start writing the same digits differently.

Anticausal classification is particularly common in medicine. For example, suppose we are interested in the differential diagnosis between cold, flu, and Covid-19 given a set of symptoms. Under a pure label shift, the properties of the three diseases do not change (there is no concept shift), and only their prevalence changes, perhaps due to epidemics and pandemics. Under a concept shift, one or more of the diseases change leading to different symptoms. Examples are new variants of Covid-19 and new strains of flu that appear every year.

Remark 8.5 A classification problem is *causal* if, vice versa, the labels are causally dependent on the objects.

In general, exchangeability martingales may detect both label shift and concept shift. In some cases we might not be interested in label shift and only be interested in concept shift or, perhaps less commonly, vice versa. In this section we will develop exchangeability martingales targeting only concept shift or only label shift. It would be ideal to decompose the amount of evidence found by an exchangeability martingale for dataset shift into two components, one reflecting the amount of evidence found for concept shift and the other reflecting the amount of evidence found for label shift. We will see examples of such decomposable martingales.

The most obvious application of exchangeability martingales is to help us in deciding when to retrain predictors, as discussed in Sect. 8.4. We should be particularly worried about the changes in the distribution of examples that invalidate ROC analysis, which is the case of concept shift in anticausal classification [100, 425]. Our exchangeability martingales for concept shift are designed to detect such dangerous changes. In the context of conformal prediction, concept shift in anticausal classification tasks requires retraining label-conditional conformal predictors, introduced in Sect. 4.6.7. (For connection between label-conditional conformal predictors and ROC analysis, see [15, Sect. 2.7].)

The experimental results reported later in this section suggest that the exchangeability martingales constructed in the previous section are dominated (and greatly improved) by exchangeability martingales decomposable into a product of an exchangeability martingale for detecting concept shift and an exchangeability martingale for detecting label shift.

In our experiments in this section we will only use conformity measures, but for stating our theoretical result in a stronger form we will use their label-conditional version

$$A : \mathbf{Y}^* \times \left(\mathbf{Z}^{(*)}\right)^{\mathbf{Y}} \times \mathbf{Z} \to \overline{\mathbb{R}} ,$$

which is a special case of (4.30). The following definition is a special case of (4.31) corresponding to the category of each example (x, y) being its label y (and with nonconformity replaced by conformity). The (smoothed) *label-conditional conformal transducer (LCCT)* determined by the label-conditional conformity measure A is the function p producing the p-values

$$\begin{aligned} p_n &= p(z_1, \dots, z_n, \tau_n) \\ &:= \frac{|\{i : y_i = y_n \;\&\; \alpha_i < \alpha_n\}| + \tau_n \,|\{i : y_i = y_n \;\&\; \alpha_i = \alpha_n\}|}{|\{i : y_i = y_n\}|} , \end{aligned} \tag{8.11}$$

where i ranges over $\{1, \dots, n\}$, $z_i := (x_i, y_i)$, and

$$\alpha_i := A\big(y_1, \dots, y_n, \big(y \in \mathbf{Y} \mapsto \lbag z_j : j \in \{1, \dots, n\} \;\&\; y_j = y \rbag\big), z_i\big) \tag{8.12}$$

for $i = 1, \dots, n$ such that $y_i = y_n$ ((8.12) is a special case of (4.32)). If the label-conditional conformity measure A is in fact a conformity measure, we will say that the LCCT p determined by it is *simple*.

Let us say that a probability measure P on $\mathbf{Z}^\infty$ is *label-conditional exchangeable* if, for any $n \in \mathbb{N}$, any sequence $(y_1^*, \dots, y_n^*) \in \mathbf{Y}^n$, any measurable set $E \subseteq \mathbf{X}^n$, and any permutation $\pi : \{1, \dots, n\} \to \{1, \dots, n\}$,

$$\begin{aligned} &y_1^* = y_{\pi(1)}^*, \dots, y_n^* = y_{\pi(n)}^* \Longrightarrow \\ &\qquad P\left(\left\{(z_1, \dots, z_n) : y_1 = y_1^*, \dots, y_n = y_n^*, (x_1, \dots, x_n) \in E\right\}\right) \\ &\qquad = P\left(\left\{(z_1, \dots, z_n) : y_1 = y_1^*, \dots, y_n = y_n^*, (x_{\pi(1)}, \dots, x_{\pi(n)}) \in E\right\}\right) . \end{aligned}$$

In words, given the labels, the distribution of the examples does not change when we swap examples with the same label. This is an instance of de Finetti's [77] notion of partial exchangeability. Of course, exchangeability is a stronger property than label-conditional exchangeability.

The following proposition is a slight variation on Proposition 2.4, and it is also a special case of Theorem 11.1.

Proposition 8.6 *If the examples* $(z_1, z_2, \dots) \sim P$ *are label-conditional exchangeable,* $(\tau_1, \tau_2, \dots) \sim \mathbf{U}^\infty$ *is an independent sequence of random numbers, and p is a label-conditional conformal transducer, the sequence of random p-values* (8.11) *is distributed uniformly on* $[0, 1]^\infty$.

In this section we will be interested in several classes of test martingales. The *label-conditional conformal test martingales* are defined as the test martingales determined by any betting martingale F and a sequence $(p_1, p_2, \dots)$ defined by (8.11) (under the conditions of Proposition 8.6) as the input p-values.

Label-conditional CTMs are the main topic of this section. They detect concept shift. In the previous section we saw that the USPS dataset is non-exchangeable, and in this section we will explore sources of this lack of exchangeability, including concept shift.

Remark 8.7 It is important that our exchangeability martingales for detecting concept shift can be used in situations where the labels are so far from being independent that it would be unusual to talk about label shift. Discussions of label shift usually presuppose at least approximate independence of labels. Suppose a sequence of hand-written characters $x_1, x_2, \dots$ comes from a user writing a letter. The objects x_n are matrices of pixels, and the corresponding labels y_n take values in the set $\{a, b, \dots\}$. Different instances of the same character, say "a", may well be exchangeable among themselves (even conditionally on knowing the full text of the letter), whereas the text itself will be far from being exchangeable; for example, "q" will be almost invariably followed by "u" if the letter is in English. For discussions of such partial exchangeability, see, e.g., Sect. 11.3, [77], and [298].

The amount of evidence found against exchangeability will typically be larger than the amount of evidence found against concept shift. In the rest of this section we will look for possible explanations of such difference. We have already mentioned that in some situations the amount of evidence found against exchangeability decomposes into two components:

- the amount of evidence found for concept shift;
- the amount of evidence found for label shift.

In these situations the second component can be said to explain the difference.

A *label conformity measure* A is a conformity measure that satisfies, additionally, the following property: for any comparison bag $\Lbag z_1, \dots, z_n \Rbag \in \mathbf{Z}^{(*)}$ of examples of any size $n \in \mathbb{N}$ and any $i, j \in \{1, \dots, n\}$,

$$y_i = y_j \Longrightarrow A(\Lbag z_1, \dots, z_n \Rbag, z_i) = A(\Lbag z_1, \dots, z_n \Rbag, z_j) , \tag{8.13}$$

where y_i and y_j are the labels in z_i and z_j, respectively. In other words, it assigns conformity scores only to the labels rather than to the full examples. Remember that the conformal transducer f determined by a conformity measure A outputs the p-values (8.10). We will say that f is a *label conformal transducer* if A is a label conformity measure.

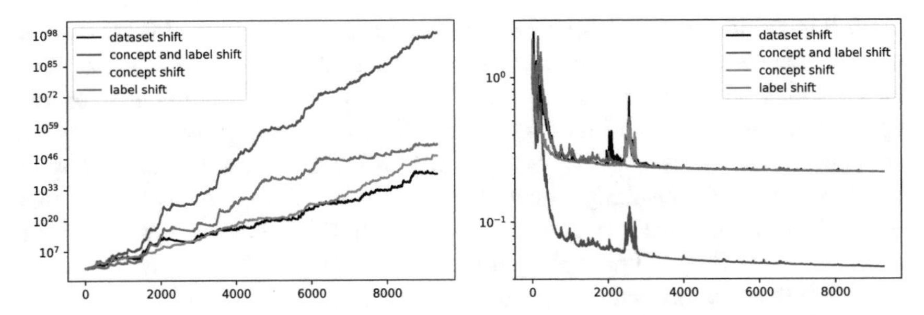

Fig. 8.2 Four exchangeability martingales determined by the Composite Jumper betting martingale and the nearest-neighbour conformity measure, applied to the USPS dataset (left panel) and its random permutation (right panel)

Our method of decomposing exchangeability martingales will be based on the following result. Its proof is given in Sect. 11.5.3.

Theorem 8.8 *If the data-generating distribution P on the examples $z_1, z_2, \dots$ is exchangeable, $(\tau_1, \tau_2, \dots)$ and $(\tau'_1, \tau'_2, \dots)$ are independent (between themselves and of the examples) sequences of random numbers distributed uniformly on $[0, 1]^\infty$, p is a simple label-conditional conformal transducer, and f is a label conformal transducer, the interleaved sequence of p-values $p_1, p'_1, p_2, p'_2, \dots$, where*

$$p_n := p(z_1, \dots, z_n, \tau_n), \quad p'_n := f(z_1, \dots, z_n, \tau'_n) ,$$

is distributed uniformly on $[0, 1]^\infty$.

If the underlying conformity measure is a label conformity measure, the CTM will be called a *label conformal test martingale*. We will say that a label-conditional CTM is *simple* if its underlying label-conditional conformal transducer is simple.

Having the stream of random p-values $p_1, p'_1, p_2, p'_2, \dots$ produced as in Theorem 8.8, we can define two derivative exchangeability martingales: a label-conditional CTM determined by $p_1, p_2, \dots$ and a label CTM determined by $p'_1, p'_2, \dots$. (There are no restrictions on the underlying betting martingales.)

Corollary 8.9 *The product of a simple label-conditional CTM and a label CTM, as described above, with independent randomizations (i.e., their sequences of random numbers τ) is an exchangeability martingale.*

Such product exchangeability martingales decompose perfectly into components for detecting concept shift and label shift. For a short derivation of this corollary from Theorem 8.8, see Sect. 8.5.1.

Figure 8.2 shows in black the CTM that is shown in Fig. 8.1 also in black. As we said in the previous section, this martingale is determined by the conformity measure (8.7), the conformal transducer (8.10), and the Composite Jumper betting martingale. The black martingale may detect any deviations from exchangeability, but in this section we are particularly interested in concept shift. In our current context, concept shift means that, for some reason, the same digit (such as "0") starts looking different; perhaps people start writing digits differently, or the digits are scanned with different equipment. To detect concept shift, we use the same conformity measure (8.7), but feed it into the label-conditional conformal transducer (8.11); the resulting sequence of p-values is fed into the same Composite Jumper betting martingale as before. The resulting CTM is shown in red in the left panel of Fig. 8.2. Its final value, of the order of magnitude 10^{47}, is even more impressive than the final value of the black martingale.

There is, of course, another reason why exchangeability may be violated: we may have label shift. To detect it, we use the label conformity measure that assigns the conformity score

$$\alpha'_i := \operatorname{med}\left(\left\{\alpha_j : j \in \{1, \dots, n\} \;\&\; y_j = y_i\right\}\right) \tag{8.14}$$

to the ith example (x_i, y_i) in the comparison bag $\wr(x_1, y_1), \dots, (x_n, y_n)\int$, where med stands for the median (the convention for $\operatorname{med}(\emptyset)$ does not affect the resulting p-values), and α_j are defined by (8.7). In other words, we average, in the sense of median, the conformity scores for each class to ensure the requirement of invariance (8.13).

Remark 8.10 In our experiments, the label conformity measure

$$\alpha'_i := \frac{\sum_{j:y_j=y_i} \alpha_j}{\left|\{j : y_j = y_i\}\right|}$$

defined in terms of arithmetic mean rather than median works slightly worse, but the difference is not substantial; this is also true for geometric and harmonic means.

The label CTM obtained by applying the same Composite Jumper to the p-values produced by the conformal transducer (8.10) applied to the conformity scores (8.14) is shown as the green line in the left panel of Fig. 8.2. It is interesting that, despite the invariance restriction, the final value (more than 10^{51}) of the green martingale, which is more volatile than the black and red ones, is even greater than the final value of the black martingale. The relatively high volatility of the green line stems from large values of the term $|\{i : \alpha_i = \alpha_n\}|$ in the numerator of (8.10) for label conformity measures, which assign the same conformity score to all images of the same class.

According to Corollary 8.9, in this context the product of a label-conditional CTM and a label CTM is still an exchangeability martingale. The product is shown as the blue line in the left panel of Fig. 8.2. By construction, the blue martingale is perfectly decomposable. Its final value (more than 10^{98}) greatly exceeds the final value achieved by the black martingale.

Remark 8.11 Corollary 8.9 has an important condition, the coin tosses τ and τ' being independent. It is satisfied in our experiments (if we ignore the fact that we use merely pseudorandom numbers) since each of our plots is produced by a single computer program that sets the seed for its random number generator only once, at the beginning.

The blue exchangeability martingale, on the one hand, dominates the black martingale over the USPS dataset and, on the other hand, decomposes into a product of exchangeability martingales for detecting concept shift and for detecting label shift. Therefore, the red and green pair in the left panel of Fig. 8.2 is a significant improvement over the black martingale.

Of course, when the USPS dataset is permuted, as in the right panel of Fig. 8.2, these successful exchangeability martingales start losing capital.

8.3 Multistage Nonrandomness Detection

In this and next sections it will be more convenient to talk about testing the hypothesis of randomness (i.e., the data being IID), but in view of Remark 8.1, what we say will be also applicable to testing exchangeability.

It is often argued that the kind of validity enjoyed by test martingales (and given by Ville's inequality) is too strong, and we should instead be looking for a testing procedure that is valid only in the sense of not raising false alarms too often. In the context of testing randomness, the interpretation of the data-generating process adopted in this section is that at first the data conforms to the assumption of randomness, but starting from some moment T the assumption breaks down; the special case $T = 0$ describes the situation where the randomness assumption is never satisfied (and so our interpretation does not restrict generality). We want our procedures to be efficient in the sense of raising an alarm soon after the null hypothesis (such as the assumption of randomness) becomes violated; both validity and efficiency can be required to hold with high probability or on average. In this section CTMs are adapted to such less demanding requirements of validity using the standard CUSUM and Shiryaev–Roberts procedures for online change detection.

A typical example of online change detection is where we observe attacks, which we assume to be IID, on a computer system. When a new kind of attacks appears, the process of attacks ceases to be IID, and we would like to raise an alarm soon afterwards.

In the previous sections we did not insist on having an explicit rule for raising an alarm, and simply regarded the value of a CTM as the amount of evidence found against the hypothesis of randomness, but in this section it will be more convenient to couch our discussion in terms of such rules. First we discuss why the kind of guarantees enjoyed by the policy of raising an alarm when $S_n \geq c$ for a CTM S and for a given threshold $c > 1$ (cf. (8.2)) may be too strong to be really useful. The reason is that, under the null hypothesis of randomness, the CTM S is trying to gamble against an exchangeable sequence of examples, which is futile, and so its value decreases. If a change occurs at some point in the distant future, it might take

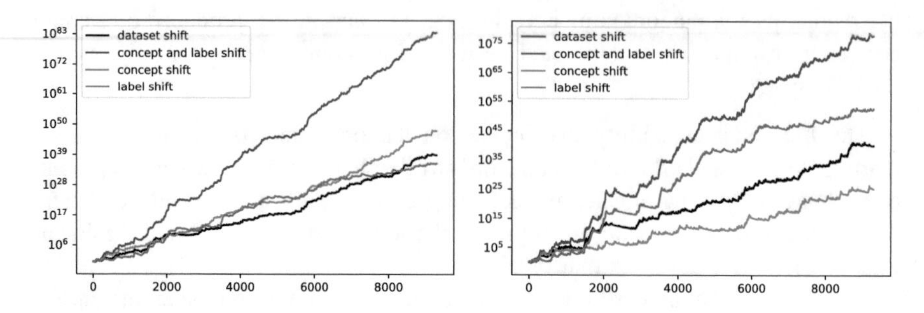

Fig. 8.3 Four exchangeability martingales determined by the nearest-neighbour conformity measure and the Simple Jumper betting martingale with the jumping rate 0.1 (left panel) and 0.01 (right panel), applied to the USPS dataset

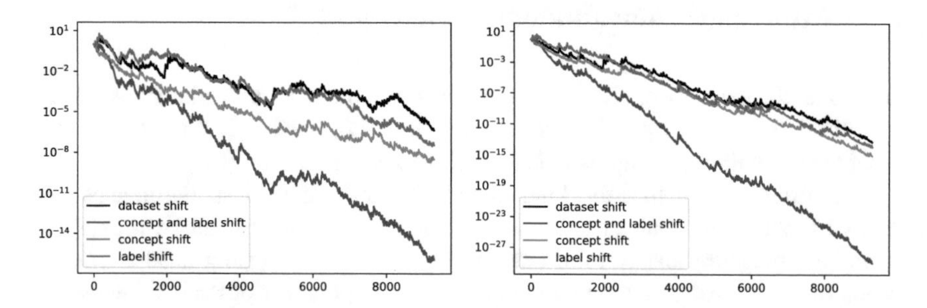

Fig. 8.4 Four exchangeability martingales determined by the nearest-neighbour conformity measure and the Simple Jumper betting martingale with the jumping rate 0.1 (left panel) and 0.01 (right panel), applied to the randomly permuted USPS dataset

a long time for the martingale to recover its value. This is a general phenomenon; we must pay for giving ourselves the chance to detect lack of randomness by losing capital in the situation of randomness.

However, a look at the right panels of Figs. 8.1 and 8.2 apparently contradicts this intuition, since, e.g., the Composite Jumper martingale shown in black never drops below 0.2. Why should it take a long time for this martingale to recover its value when randomness becomes violated? The reason for the lower bound $S_n \geq 0.2$ is that the Composite Jumper martingale consists of 5 components, and one of those components is passive, identical 1. The value of the active components goes down exponentially fast, and the more active a component is the faster its value tends to go down under randomness. Figures 8.3 and 8.4 show two of such component Simple Jumper martingales for the USPS dataset, those that contribute most to the Composite Jumper martingale in the left panel of Fig. 8.2.

Remark 8.12 It is interesting that the product exchangeability martingale in Fig. 8.2 (left panel, blue line) takes much higher values than those in Fig. 8.3, despite the contribution of the other component Simple Jumper martingales being negligible. The explanation is that the concept shift is detected better for $J = 0.1$, whereas the label shift is detected better for $J = 0.01$, and it is essentially those components that get combined.

Weaker validity guarantees are provided by multistage procedures originated, in a basic form, by Shewhart in his control chart techniques [325] and perfected by Page [268] and Kolmogorov and Shiryaev [206]. As Shiryaev mentions in his fascinating historical account [328, Sect. 1], he and Kolmogorov rejected the policy of raising an alarm when $S_n \geq c$ in favour of a multistage procedure, which was "the *correct formulation of the problem*" (the emphasis is Shiryaev's), in January 1959 or soon afterwards, after talking to a practitioner, Yuri B. Kobzarev, the founder of the Soviet school of radiolocation. In their multistage scheme, we reset the detection system after it raises an alarm instead of stopping it [328, Sect. 1.2].

8.3.1 CUSUM-Type Change Detection

A standard multistage procedure of raising alarms is the CUSUM procedure proposed by Page [268] (see also [281, Sect. 6.2]). According to this procedure, we raise the kth alarm at the time

$$\tau_k := \min \left\{ n > \tau_{k-1} : \max_{i=\tau_{k-1},\dots,n-1} \frac{S_n}{S_i} \geq c \right\}, \quad k \in \mathbb{N}, \tag{8.15}$$

where the threshold $c > 1$ is a parameter of the algorithm, $\tau_0 := 0$, and $\min \emptyset := \infty$. If $\tau_k = \infty$ for some k, an alarm is raised only finitely often; otherwise it is raised infinitely often. The procedure is usually applied to the likelihood ratio process between two power distributions, but it can be applied to any positive martingale, and we are interested in the case where S is a CTM, which is now additionally assumed to be positive, ensuring that the denominator in (8.15) is always non-zero. CUSUM is often interpreted as a repeated sequential probability ratio test [268, Sect. 4.2]. The *conformal CUSUM procedure* (i.e., CUSUM applied to a positive CTM) was introduced in [374]; however, a basic and approximate version of this procedure has been known since 1990: see [244].

Properties of validity for the conformal CUSUM procedure will be obtained in this section as corollaries of the corresponding properties of validity for the Shiryaev–Roberts procedure, which we consider next.

8.3.2 Shiryaev–Roberts Change Detection

A popular alternative to the CUSUM procedure is the Shiryaev–Roberts procedure [291, 327], which modifies (8.15) as follows:

$$\tau_k := \min\left\{ n > \tau_{k-1} : \sum_{i=\tau_{k-1}}^{n-1} \frac{S_n}{S_i} \geq c \right\}, \quad k \in \mathbb{N} \tag{8.16}$$

(i.e., we just replace the max in (8.15) by $\sum$). We will again apply it to a CTM S, still assumed to be positive, obtaining the *conformal Shiryaev–Roberts procedure*.

The procedure defining τ_1 is based on the statistics

$$R_n := \sum_{i=0}^{n-1} \frac{S_n}{S_i}, \tag{8.17}$$

which are called the *Shiryaev–Roberts statistics* and which admit the recursive representation

$$R_n = \frac{S_n}{S_{n-1}} (R_{n-1} + 1), \quad n \in \mathbb{N}, \tag{8.18}$$

with $R_0 := 0$. An interesting finance-theoretic interpretation of this representation is that R_n is the value at time n of a portfolio that starts from \$0 at time 0 and invests \$1 into the martingale S at each time $i = 1, 2, \ldots$ [90, Sect. 2]. If and when an alarm is raised at time n, we apply the same procedure to the remaining examples $z_{n+1}, z_{n+2}, \ldots$.

The following proposition gives a non-asymptotic property of validity of the Shiryaev–Roberts procedure. Roughly, it says that we do not expect the first alarm to be raised too soon under the hypothesis of randomness. For brevity, by the *standard assumptions* we will mean the assumptions that the examples $z_1, z_2, \ldots$ are IID (i.e., the assumption of randomness), $\tau_1, \tau_2, \ldots$ are independent and distributed uniformly on $[0, 1]$, and the sequences $z_1, z_2, \ldots$ and $\tau_1, \tau_2, \ldots$ are independent.

Proposition 8.13 *The conformal Shiryaev–Roberts procedure* (8.16) *satisfies* $\mathbb{E}(\tau_1) \geq c$, *for any* $c > 1$, *under the standard assumptions.*

For the proof, see Sect. 8.5.2. Of course, we can apply the proposition to other alarm times as well obtaining $\mathbb{E}(\tau_k - \tau_{k-1}) \geq c$ for all $k \in \mathbb{N}$. Therefore, the time interval between raising successive alarms is not too short in expectation under the hypothesis of randomness.

All results of this subsection (from Proposition 8.13 to Corollary 8.16) are general and applicable to any positive martingale S. As we already mentioned, they are usually stated for S being the likelihood ratio between two power distributions (pre-change and post-change). To simplify exposition, we will state them only for S being a positive CTM with the underlying filtration $(\mathcal{G}_n)_{n=0,1,\ldots}$, where $\mathcal{G}_n$ is generated by the first n p-values $p_1, \ldots, p_n$. However, our arguments (which are standard in literature on change detection) will be applicable to any filtration and any positive martingale with respect to that filtration.

Corollary 8.14 *The conformal CUSUM procedure* (8.15) *also satisfies* $\mathbb{E}(\tau_1) \geq c$ *under the standard assumptions.*

Proof All our properties of validity for the CUSUM procedure will be deduced from the corresponding properties for Shiryaev–Roberts and the fact that Shiryaev–Roberts raises alarms more often than CUSUM does, in the following sense. Let τ_k (resp. τ'_k) be the time of the kth alarm raised by Shiryaev–Roberts (resp. CUSUM). Then $\tau_k \leq \tau'_k$ for all k; this can be checked by induction on k. □

The next proposition is an asymptotic counterpart of Proposition 8.13 given in terms of frequencies.

Proposition 8.15 *Let* A_n *be the number of alarms*

$$A_n := \max\{k : \tau_k \leq n\}$$

raised by the conformal Shiryaev–Roberts procedure (8.16) *after seeing the first* n *examples* $z_1, \ldots, z_n$. *Then, under the standard assumptions,*

$$\limsup_{n\to\infty} \frac{A_n}{n} \leq \frac{1}{c} \quad a.s. \tag{8.19}$$

Under the standard assumptions, all alarms are false, and so (8.19) limits the frequency of false alarms. Of course, the statement of Proposition 8.15 (to be proved in Sect. 8.5.2) also holds for the CUSUM procedure.

Corollary 8.16 *Let* A_n *be the number of alarms raised by the conformal CUSUM procedure* (8.15) *after seeing the examples* $z_1, \ldots, z_n$. *Then* (8.19) *holds under the standard assumptions.*

Proof As in the proof of Corollary 8.14, combine Proposition 8.15 with the fact that Shiryaev–Roberts raises alarms more often than CUSUM does. □

8.4 When Do We Retrain?

In its simplest form, conformal prediction works under the assumption of randomness, which is the main assumption of mainstream machine learning. Specifically, the data is assumed to consist of independent examples generated from the same unknown probability measure. The main idea behind conformal prediction is to test the hypothesis of exchangeability for each possible value of the unknown label of a future example. The assumption of randomness is both a great strength and great weakness of mainstream machine learning in general and basic conformal prediction in particular.

Suppose a company engages in a prediction task (*big prediction task*) for many years or even decades. An example can be a pharmaceutical company predicting the toxicity and other properties of chemical compounds in the process of drug discovery [99]. Another example is predicting the number of passengers by a ferry operator [426]. The company can use various prediction algorithms, and it can adapt and retrain them repeatedly. The process is akin to (but of course happens on a much smaller scale than) the process of adapting, discarding, and creating theories in science, as discussed by, e.g., Popper [282] and Kuhn [207]. Popper represented his picture of the latter process by formulas ("evolutionary schemas") similar to

$$\mathrm{PS}_1 \to \mathrm{TT}_1 \to \mathrm{EE}_1 \to \mathrm{PS}_2 \to \dots \tag{8.20}$$

(introduced in his 1965 talk on which [282, Chap. 6] is based). In response to a problem situation PS, scientists create a tentative theory TT and then subject it to attempts at error elimination EE, whose success leads to a new problem situation PS, after which scientists come up with a new tentative theory TT, etc.

In the machine-learning version of the process (8.20), tentative theories are prediction rules, problem situations are situations in which our current prediction rule ceases to work, and error elimination may correspond to detecting the failure of the current prediction rule's assumptions. In Chaps. 2–7 we concentrated on the TT entries: designing good prediction rules and their counterparts in conformal prediction. What we have been doing in this chapter can be used for the EE entries: prediction rules typically depend on the assumption of randomness, and when we detect that this assumption is violated (and especially when detecting a concept shift, as discussed in Sect. 8.2), we may want to retrain the prediction rule. In any case, as a result we are in a problem situation, a PS entry.

In this section we concentrate on EE entries, namely on monitoring the validity of the assumption of randomness and deciding when to retrain the current prediction rule. However, there are lots of other interesting questions arising in connection with the scheme (8.20), some of which will be discussed in later chapters. For example:

- Can we test the current prediction rule directly, instead of testing the assumption of randomness? (We will discuss this in Chap. 10.)
- Can we make our prediction rules more adaptive and less sensitive to violations of various assumptions? (Again we will touch on this in Chap. 10.)
- How do we predict after we decide to retrain the current prediction rule (but before the retraining is over)?

At this time we are far from having an overall picture of the prediction process for a given big prediction task.

8.4.1 The Ville Procedure for the Wine Quality Dataset

First we discuss a possible informal scheme for deciding when to retrain a prediction algorithm. It may be called the *Ville procedure*, and consists in monitoring a test

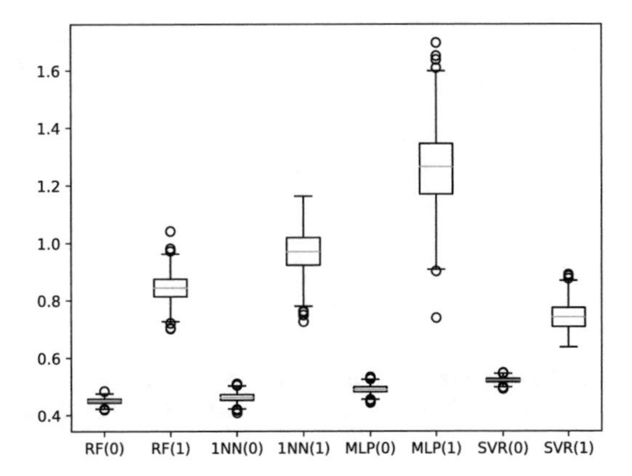

Fig. 8.5 The accuracies of various prediction algorithms on the Wine Quality dataset, as described in the main text

martingale and retraining when it takes a large value (perhaps without deciding in advance on the threshold for triggering an alarm).

As an example, we consider the Wine Quality dataset (see Sect. B.2). The label is the score between 0 and 10 reflecting the wine quality. We consider the problem of predicting the label given the object (consisting of eleven attributes) as regression problem.

The dataset consists of two parts, 4898 white wines and 1599 red wines. We randomly choose a subset of 1599 white wines and refer to it as *test set 0*, and the remaining white wines (randomly permuted) will be our *training set*. All 1599 red wines form our *test set 1*; therefore we have two test sets of equal sizes.

We will be interested in two scenarios. In scenario 0 we train our prediction algorithm on the full training set and test the resulting prediction rule on test set 0. We can expect the quality of prediction to be good, since the training and test sets are coming from the same distribution. After normalizing the attributes (see Sect. B.4), we can achieve the test MAD (mean absolute deviation) of about 0.45. The best values achieved by the algorithms implemented in a standard machine-learning package (see Sect. B.2 for details) are given in Fig. 8.5, where RF stands for random forest, 1NN for 1-nearest neighbour, MLP for multilayer perceptron (a neural network), and SVR for support vector regression. The relevant boxplots are those marked with 0 in parentheses; the boxplots are over 1000 simulations and shown to give an idea of the dependence on the seed used for the random number generator (the seed affects the split into the training and test sets and may be used internally by the prediction algorithm, e.g., by random forest). The algorithms are ordered by their performance in scenario 0. In scenario 1 we test the same prediction rule on test set 1; since its distribution is different (from the very start of the test set), the resulting test MAD will be significantly worse, as indicated in Fig. 8.5 by the boxplots marked with 1 in parentheses.

To detect a possible changepoint in the test set (which does not exist in test set 0 and is the very start in test set 1), the training set of 3299 white wines is randomly split into three *folds* of nearly equal sizes, 1100, 1100, and 1099. We use each fold

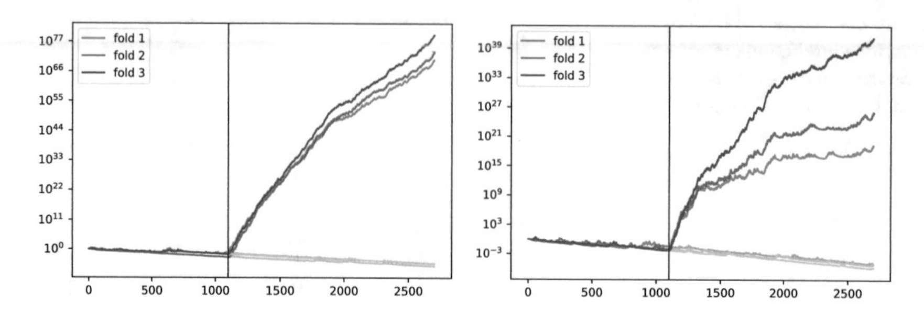

Fig. 8.6 Paths of the three conformal test martingales based on random forest, conformity measure $y - \hat{y}$ (left panel) or $|y - \hat{y}|$ (right panel), and Simple Jumper applied to the Wine Quality dataset

in turn as the *calibration set* and the remaining folds as the *proper training set*. For each fold $k \in \{1, 2, 3\}$ we train a prediction algorithm on the proper training set and run an exchangeability martingale (based, in some way, on the resulting prediction rule) on the $1100 + 1599 = 2699$ examples $z'_1, \dots, z'_{1100}, z''_1, \dots, z''_{1599}$, where $z'_1, \dots, z'_{1100}$ is the calibration set and $z''_1, \dots, z''_{1599}$ is the test set (for one of the folds, 1100 should be replaced by 1099, but we will ignore this in our discussion). This way we obtain three paths, graphs of the values of the exchangeability martingales vs time. We still have two scenarios: scenario 0 uses test set 0, and scenario 1 uses test set 1; thus we have 6 paths overall.

As we noticed in the previous section, the performance of the Simple Jumper betting martingale does not depend crucially on the jumping rate J, and $J = 0.01$ is a reasonable default value. Therefore, in this section we will only consider this betting martingale; its advantage over the Composite Jumper is that it is easier to interpret.

Results for specific conformity measures (based on random forest) and the Simple Jumper betting martingale are shown in Fig. 8.6; let us first concentrate on the left panel. For each fold we have a conformal test martingale, and paths of these three martingales are shown using different colours, as indicated in the legend. We have two paths for each of the martingales: the one over the calibration set and test set 0 (with the part over test set 0 shown in lighter colours), and the other over the calibration set and test set 1. The behaviour of the three martingales in scenario 1 is similar, all achieve a high value, of the order of magnitude about 10^{70}, and start rising sharply soon after the changepoint (shown as the thin vertical line); the presence of the changepoint becomes obvious shortly after it happens.

The conformity measure used in the left panel of Fig. 8.6 is

$$\alpha_i := y_i - \hat{y}_i , \tag{8.21}$$

where $\hat{y}_i$ is the prediction for the label y_i of the object x_i produced by the prediction rule found from the proper training set. (Intuitively, it is not obvious why (8.21) should be interpreted as a measure of conformity, but we already mentioned that in conformal testing the difference between conformity and nonconformity often disappears; in particular, the Simple Jumper works equally well with $-\alpha_i$ in place of α_i.)

Replacing the conformity measure (8.21) by its absolute value,

$$\alpha_i := \left| y_i - \hat{y}_i \right| , \tag{8.22}$$

leads to a slower growth, as illustrated in the right panel of Fig. 8.6. In the case of fold 1 (the red line), we can see a pronounced phenomenon of "decay" setting in around example 2000; as it were, the new distribution (corresponding to red wines) becomes a new normal, and the growth of the martingale stops.

The choice between the signed version (8.21) vs the unsigned version (8.22) depends on the kind of changes we regard as requiring retraining. If we are only interested in changes in the accuracy of the underlying prediction algorithm, and do not care whether our predictions are typically an undershoot, $\hat{y}_i < y_i$, or overshoot, $\hat{y}_i > y_i$, (8.22) might be preferable.

The conformity measures (8.21) and (8.22) can be applied to any regression algorithm, including multilayer perceptron, support vector regression, and 1-nearest neighbour. We do not present all analogues of Fig. 8.6 for those algorithms, but the behaviour of the CTMs based on them is as expected. Our martingales detect lack of exchangeability best in situations where it matters most: e.g., according to Fig. 8.5, the accuracy of multilayer perceptron suffers most of the dataset shift, and the conformal test martingales based on this algorithm achieve the fastest growth. In some cases, the phenomenon of decay is even more pronounced than in the right panel of Fig. 8.6.

The procedure described in this section may be used for deciding when to retrain: e.g., we may decide to retrain when one of the three martingales exceeds the threshold 100. In this case, the probability of ever raising a false alarm never exceeds 3%.

Detecting Covariate Shift

The conformity measures that we have used so far in this section were functions of the true labels and predictions. If the underlying prediction algorithm is robust to moderate *covariate shift*, i.e., a change in the marginal distribution $Q_{\mathbf{X}}$ of the objects, the resulting testing procedures will not detect deviations from exchangeability under such covariate shift. It makes sense since no retraining is required in this case.

The right panel of Fig. 8.7 shows the results for the conformity score α of an example (x, y) (in the calibration or test set) computed as the Euclidean distance from x to the nearest object in the proper training set. This *nearest-distance* conformity measure completely ignores the labels but still achieves spectacular values for conformal test martingales based on it. The two panels of Fig. 8.7 show the paths of conformal test martingales based on the nearest neighbour idea, but

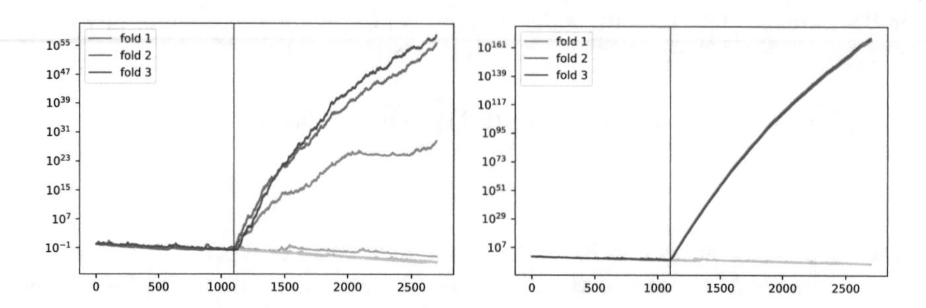

Fig. 8.7 Left panel: analogue of the left panel of Fig. 8.6 with $\hat{y}_i$ produced by 1-nearest neighbour. Right panel: paths of the three conformal test martingales based on the nearest-distance conformity measure (ignoring the labels and predictions)

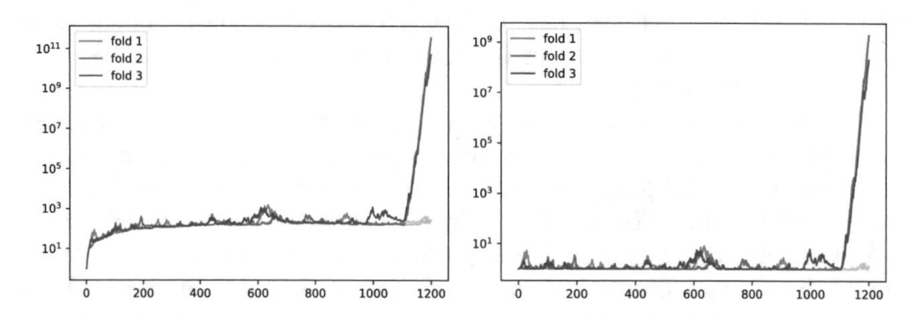

Fig. 8.8 The Shiryaev–Roberts (left panel) and CUSUM (right panel) statistics over the first 1200 examples of the combined calibration and test sets for the multilayer perceptron and the conformity measure $y - \hat{y}$

the left one is restricted to evaluating the quality of predictions, and the restriction shows in a slower growth.

8.4.2 CUSUM and Shiryaev–Roberts for Wine Quality

We have already discussed the well-known disadvantage of the Ville procedure that it becomes less and less efficient as time passes and the value of the test martingale goes down, which inevitably happens in the absence of changepoints: cf. scenario 0 (lighter colours) in Figs. 8.6 and 8.7. It may take a long time to recover the lost capital and so to detect the changepoint. We can say that, whereas the Ville procedure may be suitable during the first stages of the testing process (while our capital is still not negligible), it is less suitable for later stages. Therefore, in this subsection we experiment with the CUSUM and Shiryaev–Roberts procedures.

Figure 8.8 shows the evolution of the Shiryaev–Roberts statistics (8.17) and CUSUM statistics $\gamma_n := \max_{i<n} S_n/S_i$ over the calibration set and the first 100

Table 8.1 The median delay and the interquartile intervals for the delay for different conformity measures and different procedures for change detection, as described in the main text

c. measure	Ville	CUSUM	SR
$y - \hat{y}$, RF	70 [56, 87]	68 [57, 85]	64 [52, 80]
$y - \hat{y}$, 1NN	92 [69, 138]	93 [70, 137]	86 [65, 126]
$y - \hat{y}$, MLP	55 [44, 72]	55 [45, 72]	51 [42, 67]
$y - \hat{y}$, SVR	156 [100, 294]	156 [102, 297]	142 [94, 270]
$\lvert y - \hat{y} \rvert$, RF	222 [131, 538]	221 [130, 501]	203 [119, 450]
$\lvert y - \hat{y} \rvert$, 1NN	192 [114, 416]	188 [115, 412]	172 [106, 342]
$\lvert y - \hat{y} \rvert$, MLP	88 [62, 143]	88 [62, 142]	81 [58, 130]
$\lvert y - \hat{y} \rvert$, SVR	720 [300, ∞]	721 [303, ∞]	562 [272, ∞]
ND	29 [26, 32]	29 [27, 31]	27 [25, 29]

examples of the test set (in scenarios 1 and 0) for the conformity measure $y - \hat{y}$, $\hat{y}$ being computed by multilayer perceptron. The CUSUM statistic is shown in the form $\max(\gamma_n, 1)$ (since the values $\gamma_n < 1$ are less interesting). Both statistics will raise an alarm in scenario 1 for c up to 10^9 (we refer both to the individual R_n and γ_n and to their sequences as “statistics”).

Let us now compare the performance of different prediction algorithms and conformity measures in change detection more systematically, still concentrating on the Wine Quality dataset. The betting martingale is, as before, the Simple Jumper (with $J := 0.01$). Our results will be summarized in Table 8.1.

We randomly choose a subset of 1000 white wines as training set, a disjoint subset of 1000 white wines as calibration set, and a subset of 1000 red wines as test set, all three subsets randomly ordered. We train various prediction algorithms, labelled with the same abbreviations as in Fig. 8.5, on the training set, use the conformity measure given in the column “c. measure” to obtain a conformal test martingale, as described earlier, and run the Ville, CUSUM, and Shiryaev–Roberts (SR) procedures with the thresholds $c = 10^2, 10^4, 10^6$, respectively, on the calibration set continued by the test set. The alarm is raised (i.e., the threshold is exceeded) on the test set in the vast majority of cases, and we define the *delay* as the ordinal number of the example in the test set at which the alarm happens. Table 8.1 reports the median delay accompanied by the interquartile intervals for the delays (i.e., the intervals whose end-points are the lower and upper quartiles) for nine different conformity measures (already described earlier) and 1000 simulations; ND stands for the nearest distance conformity measure discussed earlier (and used in the right panel of Fig. 8.7). A delay of ∞ refers to the case where no alarm is ever raised (on the test set). One striking feature is how less sensitive the conformity measures $\lvert y - \hat{y} \rvert$ are as compared with $y - \hat{y}$. The nearest distance conformity measure (detecting covariate shift) is quickest in raising alarms.

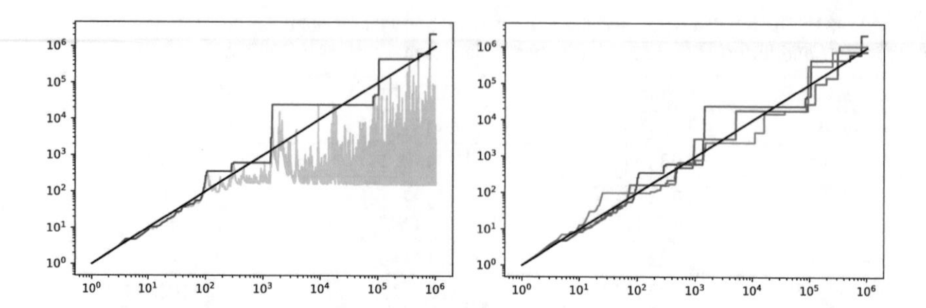

Fig. 8.9 Left panel: behaviour of the Shiryaev–Roberts statistic in the ideal setting (of simulated p-values). The path of the statistic itself is shown in light blue and the path of its maximum process in dark blue. Right panel: behaviour of the maximum process of the Shiryaev–Roberts statistic for three seeds of the random number generator in the ideal setting

8.4.3 Fixed and Variable Schedules

In this subsection we discuss two procedures for deciding when to retrain, which we call the fixed training schedule and variable training schedule. The former is our preferred procedure, but we start from the latter, which we describe in less detail.

Variable Training Schedule

Unlike the CUSUM procedure, the Shiryaev–Roberts procedure has an interesting property of validity, Proposition 8.13, which holds under the randomness assumption (in the ideal setting, as we will say later). This suggests the idea of what we call a variable training schedule: we retrain when the Shiryaev–Roberts statistic R_n exceeds a parameter C interpreted as the *target lifespan* of the prediction rule. The corresponding property of validity for the CUSUM procedure, Corollary 8.14, is very conservative, but in the case of the Shiryaev–Roberts procedure we can say that not only $\mathbb{E}(\tau_1) \geq C$ (where τ_1 is defined in (8.16)), but also $\mathbb{E}(\tau_1) \approx C$ when overshoots do not play a big role (see the proof in Sect. 8.5.2). Suppose the target lifespan is $C = 10^6$. The left panel of Fig. 8.9 shows the behaviour of the Shiryaev–Roberts statistic R_n (see (8.16) and (8.17)) and its *maximum process*

$$R_n^* := \max_{i \in \{0,\dots,n\}} R_i \tag{8.23}$$

in the *ideal setting*, where the p-values are independent and uniformly distributed on $[0, 1]$ (as they are under the standard assumptions). The black line is the compensator n of the submartingale R_n, in the sense of $R_n - n$ being a martingale (as defined in Sect. A.6). The typical behaviour of R_n is illustrated by the light blue line in Fig. 8.9 (left panel), which is very different from its expectation, the black line. The right panel of Fig. 8.9 shows three paths of the maximum process R_n^*.

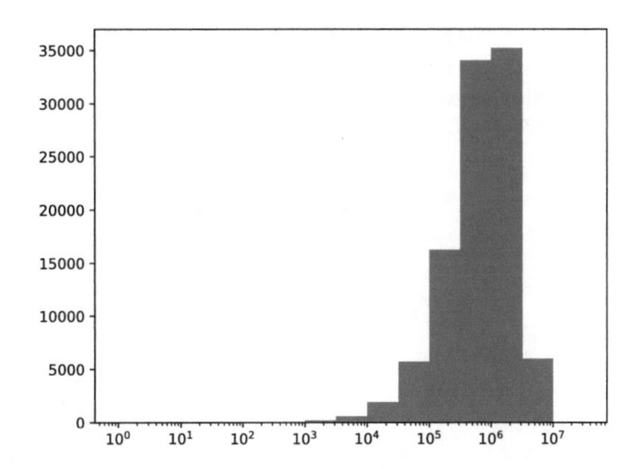

Fig. 8.10 The histogram for the time of alarm for the Shiryaev–Roberts statistic for the target lifespan 10^6 and 10^5 simulations in the ideal setting (of simulated p-values)

Figure 8.10 gives the histogram for the times of the alarm (points at which we decide to retrain), $\min\{n : R_n \geq 10^6\}$, over 10^5 simulations in the ideal setting. In numbers, rounded to the nearest 10^3: the mean time to the alarm is 1.125×10^6, which exceeds 10^6 because of overshoots, with the large standard deviation of 1.123×10^6; the median is 0.781×10^6, and the interquartile interval is $[0.320 \times 10^6, 1.561 \times 10^6]$. These figure and numbers illustrate the property of validity of the variable training schedule: the expected lifespan of the trained predictor (prediction rule) is indeed around $C = 10^6$. However, the variability of the lifespan, even in the ideal setting, will be a problem in many applications, where retraining is a complicated process that needs to be planned in advance.

Fixed Training Schedule

In the fixed training schedule we decide in advance when we would like to retrain our predictor, and change our plans only when we have significant evidence that the distribution of the data has changed (presumably, this will be a rare event, and so retraining will typically happen at the prespecified time). Our property of validity for the fixed training schedule will be computational (i.e., ensured by computer simulations).

For simplicity, instead of free parameters we will use specific numbers (such as splitting the training set into three folds; we do not claim that 3 is always the best possible number of folds). Fix a prediction algorithm (such as random forest) and a conformity measure (such as $y - \hat{y}$). Training the algorithm on a dataset then gives both a prediction rule and an exchangeability martingale (conformal test martingale). The fixed training schedule also needs the target lifespan, C, of the prediction rule. It proceeds as follows, starting from a training set and then processing a stream of test examples (arriving sequentially).

1. Split the training set into 3 approximately equal folds, 1, 2, and 3.
2. For each $k \in \{1, 2, 3\}$:

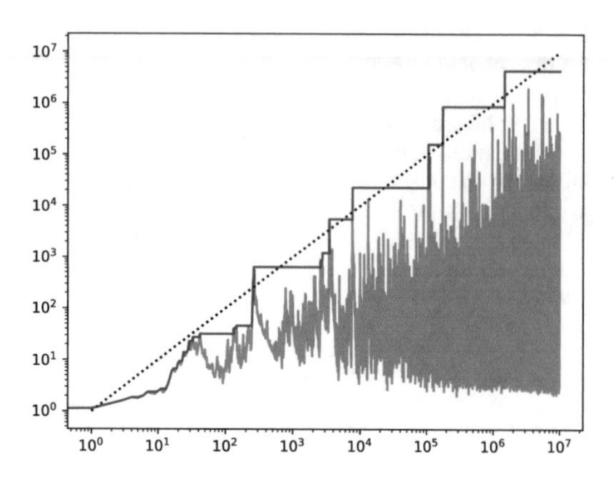

Fig. 8.11 The maximum of 100 CUSUM paths in red and its maximum process in blue in the ideal setting

- Train the prediction algorithm on the folds different from k getting a CTM S^k.
- Start running the CTM S^k on fold k (randomly permuted) and then on the stream of test examples in chronological order. Run the CUSUM statistic $\gamma_n^k := \max_{i<n} S_n^k / S_i^k$ on top of each S^k.

3. Retrain after processing C test examples unless one of the following two events happens earlier when processing those test examples:
   (a) If and when two out of the three martingales S^k raise an alarm at level 100, $S_n^k \geq 100$, retrain straight away.
   (b) Alternatively, if and when two out of the three CUSUM statistics γ^k raise an alarm at level $f_n = f_n(C)$, $\gamma_n^k \geq f_n$, retrain straight away.

The remaining parameters (after fixing, e.g., the number of folds, as discussed earlier) are the lifespan C and the sequence f_n, $n = 1, \ldots, C$. We consider, for concreteness, $C = 10^6$, and we will discuss a suitable choice of f_n momentarily.

The exploitation stage 3 of the fixed training schedule consists of two components, which we will refer to as the *Ville component* (item 3(a)) and the *CUSUM component* (item 3(b)). The validity guarantees for the latter will be computational, to be discussed shortly, and those for the former will be theoretical (although we could have replaced them with tighter computational guarantees, similar to those that we use for the CUSUM component).

The values $f_n = f_n(C)$ should be chosen in such a way that the probability of the CUSUM statistic reaching level f_n before time C should not exceed 1% in the ideal setting. The overall probability of the fixed training schedule raising a false alarm is then at most 3%. This follows from the Ville component raising a false alarm with probability at most 1.5% and the CUSUM component raising a false alarm with probability at most 1.5%. See Lemma 8.18 below for the bound 1.5%.

Figure 8.11 shows in red the maximum of 100 simulated paths of the CUSUM statistic, and it suggests that a reasonable barrier f_n is a straight line with slope 1 in the loglog representation; in the original (x, y)-axes the barrier has the equation

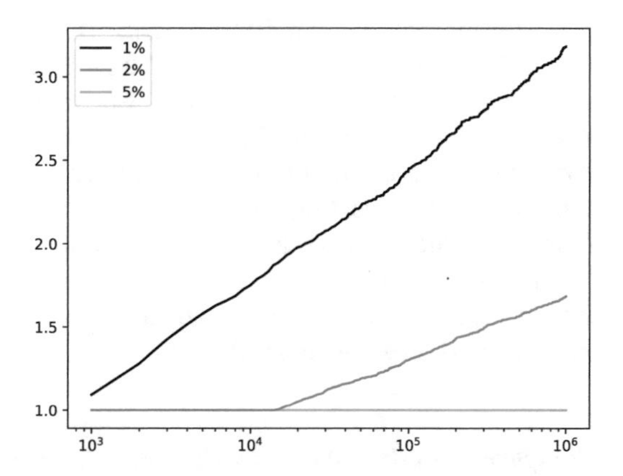

Fig. 8.12 The slopes for the CUSUM stage in the ideal setting, as described in the main text

Table 8.2 The confidence intervals, based on 10^5 simulations, for the probability of false alarm raised by the CUSUM component in the fixed training schedule for lifespan $C = 10^6$ and various choices of the coefficients for the threshold $f_n = cn$

c	Alarms	99.9% confidence interval
3.2	988	[0.89%, 1.10%]
3.3	958	[0.86%, 1.06%]
3.4	930	[0.83%, 1.03%]
3.5	901	[0.81%, 1.00%]
4	793	[0.70%, 0.89%]
5	622	[0.54%, 0.71%]

$y = cx$ (a straight line passing through the origin). The blue line in Fig. 8.11 is the maximum process, defined by (8.23).

Figure 8.12 summarizes the empirical performance of various slopes for such barriers. A path of the CUSUM statistic γ_n over $n = 1, \ldots, N$ will trigger an alarm for a barrier $y = cx$ if $\gamma_n \geq cn$ for some $n \leq N$. Let us generate a large number K ($K = 10^5$ in the case of Fig. 8.12 and Table 8.2) of paths of the CUSUM statistic in the ideal setting. For each N, let c_N be the number such that 1% of the K paths trigger an alarm (a false one) for the barrier $y = c_N x$. (The definite article "the" in "the number" is almost justified for large K.) The black line in Fig. 8.12 plots c_N vs N for $N \in \{1000, 2000, \ldots, 10^6\}$; it looks like a straight line. The blue line, corresponding to the 2% frequency of false alarms, also looks straight, except that it cannot go under $y = 1$. This allows us to estimate the right slope c for a given target lifespan of our prediction rule. For example, if the target lifespan is $N = 10^6$, we can see from Fig. 8.12 that $c_N \approx 3.2$ for the 1% frequency of false alarms (more precise values are 3.183 for 1% and 1.685 for 2%).

To find a suitable barrier, we need a confidence interval for the probability of a false alarm at a suitable confidence level that is completely inside [0, 1%]. To ensure the validity of the barrier $f_n = cn$, we have computed the exact Clopper–Pearson confidence intervals [52] for several round values for c at confidence level 99.9%

(to allow for multiple hypothesis testing, as we are looking at several candidates for c). For each of those c, we computed the number of the paths of γ_n that trigger an alarm (which is a false alarm, since we are in the ideal setting) at level cn over $n = 1, \ldots, 10^6$; these numbers are given in the column "alarms" in Table 8.2. For example, according to Table 8.2, we can set $c := 4$, since the corresponding confidence interval is a subset of $[0, 1\%]$. (It is sometimes argued that the Clopper–Pearson confidence intervals are too conservative and less conservative approximate intervals are desirable, but in our current context there is no need to sacrifice the exact validity of the Clopper–Pearson confidence intervals since the number of simulations is under our control.)

Comparison of the Ville and CUSUM Components

To compare the rules used in the Ville and CUSUM components, we need to understand how the tester's capital S_n evolves during these stages of the testing process. The left panel of Fig. 8.13 gives three typical paths of the Simple Jumper's capital S_n in the ideal setting. On the log-scale for the capital, they look linear and very close. The right panel of Fig. 8.13 is the histogram of the final values S_{10^6} of the Simple Jumper over 10^5 simulations. In numbers, the median final capital is $10^{-1720.0}$, and the interquartile interval is $[10^{-1731.6}, 10^{-1708.1}]$. Since the Simple Jumper is a martingale, the true expectation of its final value is 1, but this fact is not visible in the histogram at all (we need many more simulations for it to become visible). If the changepoint is at step n, after which the Simple Jumper S starts a quick growth, the Ville component triggers an alarm when $S_n \geq 100$, whereas the CUSUM component triggers an alarm around the time when $S_n / 10^{-0.00172n} \geq 4n$. Solving numerically the equation $4n \times 10^{-0.00172n} = 100$, we obtain 906.7 as its second solution (see Fig. 8.14, in which the blue line is the left-hand side of the equation as function of n, and the red line is the right-hand side). Therefore, we

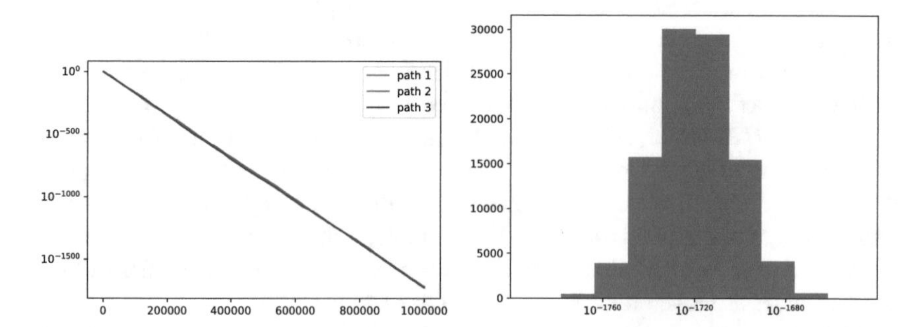

Fig. 8.13 Left panel: three typical paths of the Simple Jumper's capital in the ideal setting. Right panel: the histogram of the final values of the Simple Jumper's capital based on 10^5 simulations in the ideal setting

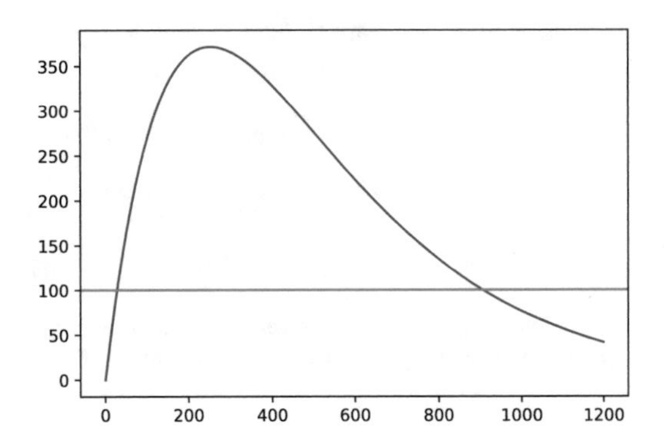

Fig. 8.14 The Ville vs CUSUM components

can regard the boundary between the natural domains for the Ville and CUSUM components to be approximately 10^3; after that boundary, the CUSUM rule (of triggering an alarm when $\gamma_n \geq 4n$) can be said to dominate the Ville rule (of triggering an alarm when $S_n \geq 100$).

Remarks

In practice, the predictions provided to the users of our prediction algorithm should be computed from the full training set, of course. The predictions computed from two out of the three folds should only be used for monitoring the validity of the randomness assumption.

We could marginally improve the sensitivity of our procedures by resetting the conformal test martingales S^k, $k \in \{1, 2, 3\}$, to 1 when they leave fold k and enter the test set, since we know that the exchangeability assumption holds for fold k.

Not all examples on which the main prediction rule (trained on the full training set) is run are necessarily included in the test stream (used for testing the assumption of randomness). In general, we have a training set, a test stream, and a possibly wider exploitation stream.

We should also be careful about including examples in the test stream in order not to violate the assumption of randomness for irrelevant reasons. For example, new test examples can be added in randomly shuffled batches of reasonable sizes.

8.5 Proofs

8.5.1 *Proof of Corollary 8.9*

Let the simple label-conditional CTM be $S_n = F(p_1, \ldots, p_n)$ and the label CTM be $S'_n = F'(p'_1, \ldots, p'_n)$, $n = 0, 1, \ldots$, where F and F' are betting martingales and $p_1, p'_1, p_2, p'_2, \ldots$ is a stream of p-values as in Theorem 8.8. Let us check that $S_n S'_n$, $n = 0, 1, \ldots$, is a martingale w.r.t. to the filtration generated by the p-values: for any $n \in \{1, 2, \dots\}$,

$$\begin{aligned}
\mathbb{E}_{p_1,p'_1,\dots,p_{n-1},p'_{n-1}}(S_n S'_n) &= \mathbb{E}_{p_1,p'_1,\dots,p_{n-1},p'_{n-1}}\left(\mathbb{E}_{p_1,p'_1,\dots,p_{n-1},p'_{n-1},p_n}(S_n S'_n)\right) \\
&= \mathbb{E}_{p_1,p'_1,\dots,p_{n-1},p'_{n-1}}\left(S_n \mathbb{E}_{p_1,p'_1,\dots,p_{n-1},p'_{n-1},p_n}(S'_n)\right) \\
&= \mathbb{E}_{p_1,p'_1,\dots,p_{n-1},p'_{n-1}}(S_n S'_{n-1}) = \mathbb{E}_{p_1,p'_1,\dots,p_{n-1},p'_{n-1}}(S_n) S'_{n-1} = S_{n-1} S'_{n-1} ,
\end{aligned}$$

where each lower index for $\mathbb{E}$ signifies the conditioning σ-algebra (namely, the conditioning σ-algebra is generated by the listed random variables). The third and last equalities follow from (8.4).

8.5.2 *Proofs for Sect. 8.3*

Proof of Proposition 8.13

The proof will follow from the fact that $R_n - n$ is a martingale; this fact (noticed, in a slightly different context, in [280, Theorem 1]) follows from (8.18): since S is a martingale,

$$\mathbb{E}(R_n \mid \mathcal{G}_{n-1}) = \frac{\mathbb{E}(S_n \mid \mathcal{G}_{n-1})}{S_{n-1}} (R_{n-1} + 1) = R_{n-1} + 1 .$$

Another condition for $R_n - n$ being a martingale requires the integrability of R_n, which follows from the integrability of each addend in (8.17):

$$\mathbb{E}\left(\frac{S_n}{S_i}\right) = \mathbb{E}\left(\mathbb{E}\left(\frac{S_n}{S_i} \mid \mathcal{G}_i\right)\right) = \mathbb{E}(1) = 1 < \infty .$$

Fix the threshold $c > 1$. By Doob's optional sampling theorem (see, e.g., [330, Theorem 7.2.1]) applied to the martingale $R_n - n$,

$$\mathbb{E}(\tau_1) = \mathbb{E}(R_{\tau_1}) \geq c .$$

Applying this theorem, however, requires some regularity conditions, and the rest of this proof is devoted to checking technical details.

If $\tau_1 = \infty$ with a positive probability, we have $\mathbb{E}(\tau_1) = \infty \geq c$, and so we assume that $\tau_1 < \infty$ a.s. Doob's optional sampling theorem is definitely applicable to the stopping time $\tau_1 \wedge L$, where L is a positive constant (see, e.g., [330, Corollary 7.2.1]), and so the nonnegativity of R implies

$$\mathbb{E}(\tau_1) \geq \mathbb{E}(\tau_1 \wedge L) = \mathbb{E}(R_{\tau_1 \wedge L}) \geq \mathbb{E}(R_{\tau_1} 1_{\{\tau_1 \leq L\}}) \geq c\mathbb{P}(\tau_1 \leq L) \to c \quad (L \to \infty) .$$

Proof of Proposition 8.15

Fix a positive CTM S and a threshold $c > 0$. We can rewrite (8.16) as

$$\tau_k := \min\left\{n > \tau_{k-1} : R^k_n \geq c\right\} , \tag{8.24}$$

where

$$R_n^k := \sum_{i=\tau_{k-1}}^{n-1} \frac{S_n}{S_i} .$$

It will be convenient to modify (8.24) by forcing an alarm L steps after the last one:

$$\tau_k' := (\tau_{k-1}' + L) \wedge \min \left\{ n > \tau_{k-1}' : R_n'^k \geq c \right\} ,$$

where $\tau_0' := 0$ and

$$R_n'^k := \sum_{i=\tau_{k-1}'}^{n-1} \frac{S_n}{S_i} .$$

(The value of L will be chosen later.) Similarly to the proof of Corollary 8.14, by induction on k we can check that, for all k, $\tau_k' \leq \tau_k$.

We still have a recursive representation similar to (8.18) for (R^k and) R'^k. Notice that $R_n'^k$, $n \geq \tau_{k-1}'$, is a nonnegative submartingale with $n - \tau_{k-1}'$ as its compensator (and we can set $R_n'^k$ and its compensator to 0 for $n < \tau_{k-1}'$).

Remember that $\mathcal{G}_n$ is the σ-algebra generated by the p-values $p_1, \ldots, p_n$, and let $\mathcal{G}_{\tau_k'}$ be the σ-algebra of events E such that $E \cap \{\tau_k' \leq n\} \in \mathcal{G}_n$ for all n (informally, $\mathcal{G}_{\tau_k'}$ consists of the events E expressible in terms of the p-values and settled at time τ_k').

Let us say that $k \in \mathbb{N}$ is *slow* if

$$\mathbb{P}\left(\tau_k' - \tau_{k-1}' = L \mid \mathcal{G}_{\tau_{k-1}'}\right) \geq c/L ;$$

otherwise, k is *fast*. Notice that the event that k is fast (or slow) is $\mathcal{G}_{\tau_{k-1}'}$-measurable. By Doob's optional sampling theorem and the nonnegativity of $R_n'^k$, where $n \geq \tau_{k-1}'$, for a fast k we obtain, similarly to the proof of Proposition 8.13,

$$\begin{aligned}
\mathbb{E}\left(\tau_k' - \tau_{k-1}' \mid \mathcal{G}_{\tau_{k-1}'}\right) &= \mathbb{E}\left(R_{\tau_k'}'^k \mid \mathcal{G}_{\tau_{k-1}'}\right) \\
&= \mathbb{E}\left(R_{\tau_k'}'^k 1_{\tau_k' - \tau_{k-1}' = L} \mid \mathcal{G}_{\tau_{k-1}'}\right) + \mathbb{E}\left(R_{\tau_k'}'^k 1_{\tau_k' - \tau_{k-1}' < L} \mid \mathcal{G}_{\tau_{k-1}'}\right) \\
&\geq 0 + c\mathbb{P}\left(\tau_k' - \tau_{k-1}' < L \mid \mathcal{G}_{\tau_{k-1}'}\right) \geq c(1 - c/L) = c - c^2/L .
\end{aligned}$$

Let $F \subseteq \mathbb{N}$ be the random set of all fast k, $S := \mathbb{N} \setminus F$ be the random set of all slow k, and F_K (resp. S_K) be the set consisting of the K smallest elements of F (resp. S). The strong law of large numbers for bounded martingale differences (see Sect. A.6.2) now implies

$$\liminf_{K \to \infty} \frac{1}{K} \sum_{k \in S_K} (\tau_k' - \tau_{k-1}') \geq L(c/L) = c \quad \text{a.s.} \tag{8.25}$$

and

$$\liminf_{K \to \infty} \frac{1}{K} \sum_{k \in F_K} (\tau_k' - \tau_{k-1}') \geq c - c^2/L \quad \text{a.s.} ; \tag{8.26}$$

the inequality in (8.25) (resp. (8.26)) is interpreted as true when $|S| < \infty$ (resp. $|F| < \infty$). Combining (8.25) and (8.26), we obtain

$$\liminf_{K\to\infty} \frac{\tau'_K}{K} = \liminf_{K\to\infty} \frac{1}{K} \sum_{k=1}^{K} (\tau'_k - \tau'_{k-1}) \geq c - c^2/L \quad \text{a.s.}$$

Therefore, setting

$$A'_n := \max\{k : \tau'_k \leq n\} ,$$

we have

$$\limsup_{n\to\infty} \frac{A_n}{n} \leq \limsup_{n\to\infty} \frac{A'_n}{n} \leq \frac{1}{c - c^2/L} ,$$

and it remains to let $L \to \infty$.

Remark 8.17 It might be tempting to deduce (8.19) from Proposition 8.13 directly using a suitable law of large numbers. However, a simple application of the Borel–Cantelli–Lévy lemma (Sect. A.6.1) shows that we cannot do so without using the specifics of our stopping times τ_k. Indeed, assuming $c \in \{2, 3, \dots\}$, we can define a filtered probability space and stopping times τ_k, $k = 0, 1, \dots$, with $\tau_0 := 0$, in such a way that

$$\tau_k - \tau_{k-1} = \begin{cases} 1 & \text{with probability} 1 - k^{-2} \\ (c-1)k^2 + 1 & \text{with probability} k^{-2} \end{cases}$$

for all $k \in \mathbb{N}$ (where the probabilities may be conditional on a suitable σ-algebra $\mathcal{G}_{\tau_{k-1}}$). Then $\mathbb{E}(\tau_k - \tau_{k-1}) = c$ (and $\mathbb{E}(\tau_k - \tau_{k-1} \mid \mathcal{G}_{\tau_{k-1}}) = c$) for all k but, almost surely, $\tau_k - \tau_{k-1} = 1$ from some k on.

8.5.3 A Proof for Sect. 8.4

When discussing the fixed training schedule in Sect. 8.4.3, we used the following Lemma (with $k = 2$, $n = 3$, and $p = 1\%$).

Lemma 8.18 *Let $E_1, \dots, E_n$ be events of probability at most p, and let $n, k \in \mathbb{N}$ with $n \geq k$. Then the probability of the occurrence of at least k of those events is at most $\frac{n}{k}p$. This bound is tight, provided $\frac{n}{k}p \leq 1$.*

Proof The bound can be derived from Markov's inequality:

$$\mathbb{P}(1_{E_1} + \cdots + 1_{E_n} \geq k) \leq \frac{\mathbb{E}(1_{E_1} + \cdots + 1_{E_n})}{k} \leq \frac{n}{k}p .$$

Now suppose $\frac{n}{k}p \leq 1$. To show that the bound is tight, consider a probability space consisting of all $\binom{n}{k}$ subsets of $\{1, \dots, n\}$ of size k and another element, which is not a subset of $\{1, \dots, n\}$. The probability of each such subset of $\{1, \dots, n\}$ is $\epsilon \in [0, 1/\binom{n}{k}]$. The event E_i, $i \in \{1, \dots, n\}$, is defined as the family of all subsets of $\{1, \dots, n\}$ of size k that contain i, and we will choose ϵ to ensure $\mathbb{P}(E_i) = p$. Then we have

$$\mathbb{P}(E_i) = \binom{n-1}{k-1}\epsilon = p\,,$$

which implies

$$\mathbb{P}(1_{E_1} + \cdots + 1_{E_n} \geq k) = \binom{n}{k}\epsilon = \frac{\binom{n}{k}}{\binom{n-1}{k-1}}p = \frac{n}{k}p\,.$$

□

8.6 Context

8.6.1 Conformal Test Martingales

In Sect. 8.1 we follow mainly [403]. Before that paper, it was not even clear that nontrivial exchangeability supermartingales exist; we saw that they not only exist, but can attain huge final values on a benchmark (USPS) dataset starting from 1 and never risking bankruptcy.

Jean Ville (1910–1988).
Used with permission from Mary Paige Snell. The photo appeared previously in [337] and [317, Sect. 8.6]

Martingales were introduced and studied as a powerful tool in the foundations of probability by Jean Ville [373] (see Sect. A.7.2 for further details). In our definitions of martingale (8.3) and supermartingale (8.1) we follow Doob [87], Sect. II.7 and the beginning of Sect. VII.1. (Doob, however, did not use the term "supermartingale"; for details, see [338, Section "Martingales"].) A more standard approach (described in Sect. A.6 and introduced already in [87]) replaces the condition " $\mid S_1, \ldots, S_n$" in (8.3) and (8.1) by " $\mid \mathcal{F}_n$" for a filtration of σ-algebras $\mathcal{F}_n$. In (8.3) and (8.1), $\mathcal{F}_n$ is generated by the process itself, namely by $S_1, \ldots, S_n$, but the most natural σ-algebra $\mathcal{F}_n$ is generated by the examples $z_1, \ldots, z_n$ (i.e., $\mathcal{F}_n$ represents all information available by the end of trial n). To see how restrictive such modifications of (8.3) and (8.1) are, notice that such exchangeability martingales and exchangeability supermartingales become trivial when this apparently small change is made: the former are constant, and the latter are decreasing processes ($S_0 \geq S_1 \geq \ldots$). This is formally stated and proved in the binary case by Ramdas et al. [287, Theorem 17].

There are two reasons why nontrivial randomized exchangeability martingales exist:

- Our underlying filtration is poorer than $\mathcal{F}_n$. A CTM S is a martingale in its own filtration. Moreover, it is a martingale in the filtration $(\mathcal{G}_n)_{n=0,1,\ldots}$ where $\mathcal{G}_n$ is generated by the first n p-values $p_1, \ldots, p_n$.
- Randomization is essential. CTMs are randomized in that they also depend on the random numbers $\tau_1, \tau_2, \ldots$.

The first reason alone seems to be insufficient for getting really useful exchangeability martingales: e.g., in the binary case $\mathbf{Z} = \{0, 1\}$, the examples $z_1, \ldots, z_n$ are determined by the values $S_0, S_1, \ldots, S_n$, unless $S_i = S_{i-1}$ for some $i \in \{1, \ldots, n\}$ (let us check this for $n = 1$: depending on the value of z_1, we have either $S_1(z_1) > S_0$ or $S_1(z_1) < S_0$, and knowing which inequality is true determines z_1; for general n, use induction on n). In many practically interesting cases there is not much randomness in CTMs; it is only used for tie-breaking. However, even a tiny amount of randomness can be conceptually important (other fields where this phenomenon has been observed are differential privacy and defensive forecasting [320, Sect. 12.7]).

In the first edition of this book (Sect. 7.1) we only considered decreasing betting functions, the intuition being that lack of randomness should lead to strange new examples and thus abnormally low p-values. However, later it turned out [101] that this intuition is deceptive, and non-monotonic betting functions often lead to better results. For example, such betting functions, along with betting on small p-values, are also allowed to bet on large p-values. In general, they gamble against the non-uniformity of p-values. Considering non-monotonic betting functions allowed us to improve the results reported in the first edition greatly and simultaneously simplify definitions; in particular, the Simple Jumper martingale is simpler than the "Sleepy Jumper" described in the first edition [402, Sect. 7.1].

The idea of tracking the best betting function goes back to [157] ("tracking the best expert"). A general "Aggregating Algorithm" (AA) for merging expert advice was introduced in [377], and Herbster and Warmuth [157] showed how to extend the AA to "track the best expert", in order to try and outperform even the best static expert. Vovk [380] noticed that Herbster and Warmuth's algorithm ("Fixed Share") is in fact a special case of the AA, when it is applied not to the original experts but to "superexperts". In the context of Sect. 8.1, the experts are the betting functions f_ϵ, while the superexperts are the betting martingales

$$F_{\epsilon_1,\epsilon_2,\ldots}(u_1, \ldots, u_n) := \prod_{i=1}^{n} f_{\epsilon_i}(u_i) .$$

Statically mixing the superexperts in (8.9) corresponds to tracking the best ϵ.

8.6.2 The Other Conformal Martingales

It is natural to refer to CTMs S as conformal martingales if we drop the assumptions $S_n \geq 0$ and $S_0 = 1$. There is an unfortunate terminological clash between such conformal martingales and the older (and classical) notion of conformal martingales introduced by Getoor and Sharpe in 1972 [129] and discussed later in, e.g., [422] and [289, Sect. V.2]. One of the equivalent definitions of a conformal local martingale Z is that Z is a continuous-time stochastic process taking values in the complex plane $\mathbb{C}$ such that both Z and Z^2 are local martingales [289, Proposition V.2.1]. This older notion of conformal martingales is never used in this book outside this subsection, but we still prefer to couch our discussions in terms of "conformal test martingales", or "conformal nonnegative martingales" in Chap. 9, which are unambiguous expressions.

8.6.3 Optimality of Sequential Testing Procedures

The efficiency of our procedures for testing randomness will be the topic of the following chapter, Chap. 9, and in this subsection we only discuss the nature of the problem. Perhaps the most satisfactory results about the efficiency of sequential testing procedures are optimality results such as that obtained by Wald and Wolfowitz [421] for Wald's [418, 419] sequential probability ratio test. The goal of establishing such optimality results for our procedures for testing randomness would be, however, too ambitious.

Wald's sequential probability ratio test was designed by him in April 1943 [418, Sect. B] for the problem of testing a simple null hypothesis against a simple alternative, both being power distributions. Let S_n be the likelihood ratio of the alternative hypothesis to the null hypothesis after seeing n examples. The test consists in fixing two positive constants A and B such that $A > 1 > B$ and stopping as soon as S_n leaves the interval (B, A). If $S_n \geq A$ at that time n, we reject the null hypothesis; otherwise, we accept it.

A sequential test can make errors of two kinds: reject the null hypothesis when it is true (error of the first kind) or accept it when it is false (error of the second kind). Any sequential probability ratio test T is efficient in a strong sense: if another sequential test T' has errors of the first and second kind that are not worse than those for T, the expected time of reaching a decision is as good for T as it is for T' (or better), under both null and alternative hypotheses. In other words, sequential probability ratio tests optimize the number of examples needed to arrive at a decision, under natural constraints.

Wald showed the efficiency of his test in the sections "Efficiency of the Sequential Probability Ratio Test" in [418, 419] ignoring the possibility of S_n overshooting A or undershooting B. In [421] he and Wolfowitz provided a full proof.

The strength of this result is made possible by the restricted nature of the testing problem. Both null and alternative hypotheses are known probability distributions. The test is specified by two numbers, A and B. The situation with testing randomness using CTMs is very different. A CTM is determined by the underlying nonconformity measure, which can even involve an element of intelligence. In this book we have defined numerous nonconformity measures based on powerful algorithms of machine learning, including neural networks. We cannot expect to be able to *prove* that such a procedure is successful (as argued by philosophers of science in other contexts; see, e.g., [283, Sect. 20]). The task of designing a CTM is too open-ended for that.

Our approach to establishing the efficiency of our procedures will not be based on optimality. The idea is to show that our procedures do not constrain us: whatever a procedure for testing randomness can achieve, can be achieved with CTMs. Notice that even the Wald–Wolfowitz result can be interpreted in this way. However, our efficiency results in Chap. 9 will be much cruder.

8.6.4 Standard Procedures for Change Detection and Their Optimality

There is vast literature on online change detection; see, e.g., [281, 331] for reviews. However, the standard case is where the pre-change and post-change distributions are known, and the only unknown is the time of change. Generalizations of this picture usually stay fairly close to it (see, e.g., [281, Sect. 7.3]). Conformal change detection relaxes the standard assumptions radically.

Remark 8.19 The literature on offline change detection is also vast; see, e.g., [32] for an early review. Here the problem is to detect changes in a data sequence all of which is given to us in a batch rather than sequentially. The importance of this problem has grown in recent decades because of its applications in bioinformatics; see, e.g., [332]. However, we concentrate on online problems.

It is remarkable that both CUSUM and Shiryaev–Roberts procedures are optimal under some natural conditions and for some natural criteria of optimality. In the standard setting of detecting a change from one known power distribution for the incoming data to another known power distribution, the CUSUM and Shiryaev–Roberts procedures are applied to a specific martingale, the likelihood ratio of the post-change distribution to the pre-change distribution. Therefore, they depend on just one parameter, the threshold c (for given pre-change and post-change distributions), whereas in the context of testing randomness we have a wide class of CUSUM and Shiryaev–Roberts procedures, built on top of different CTMs.

The five standard criteria for the quality of such specific procedures have been referred to by the letters A–E; see, e.g., Shiryaev [331]. Under natural conditions, Shiryaev–Roberts is optimal under two of the criteria, and CUSUM is optimal under one of them. Such statements of optimality are very satisfactory results about the efficiency of the corresponding procedures.

In Sect. 8.3 we mainly follow [391, Sect. 4]. In that section, and in this book in general, we only discuss validity results for CUSUM and Shiryaev–Roberts in the context of randomness testing and do not claim their optimality. As discussed at the end of the previous subsection, this is a difficult task already for the basic Ville-type testing procedure using CTMs. The null hypothesis (that of randomness) is composite and, moreover, very large (for large $\mathbf{Z}$), and we do not specify any alternatives; we simply do not have enough structure to specify a meaningful optimization problem.

8.6.5 Deciding When to Retrain

In Sect. 8.4, we mostly follow [414], where additional experimental illustrations can be found. Our version of the fixed schedule is streamlined and combines the opening and the middle game, in the terminology of [414]. The Clopper–Pearson confidence intervals are computed using the R package `binom` [88].

8.6.6 Avoiding Numerical Problems

If the Shiryaev–Roberts statistic R_n is implemented on a modern computer directly using the formula in (8.17), we will obtain Fig. 8.15 instead of the left panel of Fig. 8.9. The behaviour of the Shiryaev–Roberts statistic in Fig. 8.15 abruptly changes shortly before the 200,000th example. This happens because of a numerical underflow. Up to that point the value S_n of the martingale has been

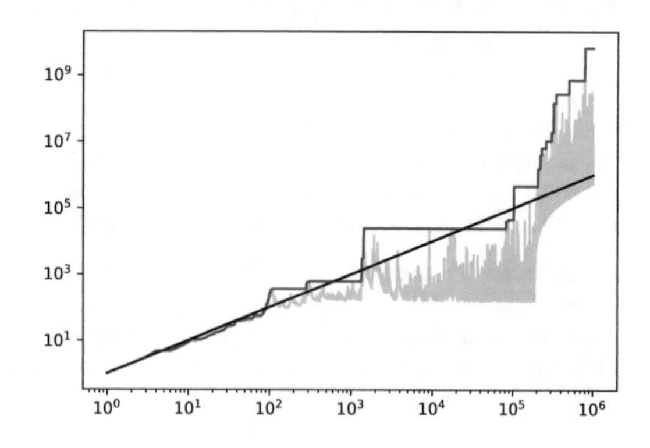

Fig. 8.15 The Shiryaev–Roberts statistic, implemented naively, over the first 10^6 examples

exponentially decreasing, down to a small multiple of $2^{-1074} \approx 5 \times 10^{-324}$, the smallest positive number representable as double-precision floating-point number [36] (the underflow happening shortly before the 200,000th example roughly agrees with Fig. 8.13). After that point the value S_n of the martingale cannot decrease further substantially and keeps fluctuating in the region of small multiples.

To get rid of the underflow, we can use the recursion (8.18) with occasional rescaling of S_n (in our code we rescale S_n, setting it to 1, every 10,000th example). This results in Fig. 8.9. Similar precautions need to be taken for the CUSUM statistic as well, in which case the recursion is

$$\gamma_n = \frac{S_n}{S_{n-1}} \max(\gamma_{n-1}, 1) .$$

8.6.7 *Low-Dimensional Dynamic Models*

In the first edition of this book [402, Sect. 7.2] we combined the ideas of conformal testing and conformal prediction to get rid of (or at least weaken) the assumption of randomness. We assumed, instead, that there is a static "random core" on which a less complicated dynamic structure is superimposed. The idea is to fix a family of "detrending transformations" $(F_\theta : \theta \in \Theta)$ (playing the role of a statistical model) such that, for some θ, applying F_θ to the true data sequence is believed to lead to an exchangeable sequence. After observing a data sequence we can reject some θ (as not leading to an exchangeable sequence) and use the remaining θ for prediction. This approach to prediction was motivated by Barnard's [22] pivotal inference.

8.6.8 *Islands of Randomness*

In the first edition of this book [402, Sect. 7.3] we also considered another relaxation of the assumption of randomness, assuming that only certain subsequences of the full data sequence output by Reality satisfy the assumption of exchangeability. In this case instead of a comprehensive theory explaining all examples we only have a patchwork of theories each explaining only a relatively small piece of the observed data sequence; we discussed only the simplest case where each of the "local theories" was just the hypothesis of randomness applied to a subsequence of the full data sequence. We proved a simple mathematical result about the conservative validity of conformal predictors in this case.

Chapter 9
Efficiency of Conformal Testing

Abstract Our question in this chapter is how much we can potentially lose when relying on conformal test martingales as compared with unrestricted testing of either exchangeability or statistical randomness. We will see that at a crude scale we do not lose much when restricting our attention to testing statistical randomness with conformal test martingales. Following the old tradition of the algorithmic theory of randomness, in this chapter we consider the binary case only; this makes our results less relevant in practical applications, but they are still reassuring.

We consider two very different settings of the problem. In the first, we are given an arbitrary critical region for testing exchangeability of finite binary sequences of a given length, and the question is whether there exists a conformal test martingale emulating this critical region. In the second, we are given a natural alternative, simple or composite, to the hypothesis of exchangeability, and the question is whether there exists a conformal test martingale competitive with the likelihood ratio of the alternative to the null hypothesis (suitably defined).

Keywords Kelly gambling · Conformal probability · Exchangeability probability · Randomness probability · Changepoint alternative · Markov alternative

9.1 Conformal Testing with a Known Critical Region

In this section we consider finite binary sequences of a known length N. First we discuss the relation between exchangeability and statistical randomness for such sequences, and then explore the power of CTMs for detecting lack of exchangeability.

V. Vovk et al., *Algorithmic Learning in a Random World*,
https://doi.org/10.1007/978-3-031-06649-8_9

9.1.1 Randomness Probability vs Exchangeability Probability

We are interested in two related interpretations of the efficiency of conformal test martingales. On the one hand, we can think of them as detecting deviations from statistical randomness, and on the other, we can think of them as detecting deviations from exchangeability, and CTMs can have different efficiency for these two tasks. While for infinite sequences randomness and exchangeability are connected very closely by de Finetti's theorem (see Sects. 2.9.1 and A.5.1), for finite sequences their relation becomes much less close. In this subsection we will discuss the relation between exchangeability and statistical randomness in our simple binary case.

Let $\Omega := \{0, 1\}^N$ be the set of all binary sequences of a given length N, interpreted as sequences of examples. The time horizon $N \in \mathbb{N}$ can be regarded as fixed in the rest of this section (apart from (9.4)).

Let $\mathbf{B}_\pi$ be the Bernoulli probability measure on $\{0, 1\}$ with the probability of 1 equal to $\pi \in [0, 1]$: $\mathbf{B}_\pi(\{1\}) := \pi$. The *upper randomness probability* of a set $E \subseteq \Omega$ is defined to be

$$\mathbb{P}^{\text{rand}}(E) := \sup_{\pi \in [0,1]} \mathbf{B}_\pi^N(E) , \tag{9.1}$$

and the *upper exchangeability probability* of $E \subseteq \Omega$ is

$$\mathbb{P}^{\text{exch}}(E) := \sup_P P(E) , \tag{9.2}$$

P ranging over the exchangeable probability measures on Ω (in the current binary case we can say that a probability measure P on Ω is *exchangeable* if $P(\{\omega\})$ depends on $\omega \in \Omega$ only via the number of 1s in ω).

Remark 9.1 The *lower probabilities* corresponding to (9.1) and (9.2) are $1 - \mathbb{P}^{\text{rand}}(\Omega \setminus E)$ and $1 - \mathbb{P}^{\text{exch}}(\Omega \setminus E)$, respectively. In this book we never need lower probabilities.

The function $\mathbb{P}^{\text{rand}}$ can be used when testing the hypothesis of randomness: if $\mathbb{P}^{\text{rand}}(E)$ is small (say, 1%), E is chosen in advance, and the observed sequence ω is in E, we can reject the hypothesis of randomness for ω. Similarly, $\mathbb{P}^{\text{exch}}$ can be used when testing the hypothesis of exchangeability. This is an instance of application of *Cournot's principle*, often regarded to be the only bridge between probability theory and its applications and discussed further in Sect. 9.5.2. It is the cornerstone of statistical hypothesis testing, although statisticians rarely mention Cournot's principle by name. In the standard statistical terminology [222, Sect. 3.1, (3.3)], E is a *critical region for testing randomness* of *size* $\mathbb{P}^{\text{rand}}(E)$, and it is a *critical region for testing exchangeability* of *size* $\mathbb{P}^{\text{exch}}(E)$.

Cournot's principle suggests the following understanding of the efficiency of a method of testing the hypothesis of randomness or exchangeability: given any event E such that $\mathbb{P}^{\text{rand}}(E)$ or $\mathbb{P}^{\text{exch}}(E)$ is very small, the method should allow us to reject the hypothesis of randomness after observing E. The following proposition, whose proof (an easy calculation based on Stirling's formula) is given in Sect. 9.4.1, establishes a connection between the requirements of being small for $\mathbb{P}^{\text{rand}}(E)$ and for $\mathbb{P}^{\text{exch}}(E)$.

Proposition 9.2 *For any $E \subseteq \Omega$,*

$$\mathbb{P}^{\text{rand}}(E) \leq \mathbb{P}^{\text{exch}}(E) \leq 1.5\sqrt{N}\mathbb{P}^{\text{rand}}(E) . \tag{9.3}$$

Remark 9.3 The constant 1.5 in the right-hand inequality of (9.3) is not too far from being optimal: when $N = 2$ and $E = \{(0, 1)\}$, it can be improved only to $\sqrt{2} \approx 1.414$. Our argument in Sect. 9.4.1 in fact gives $\sqrt{2\pi}e^{1/6}/2 \approx 1.481$ instead of 1.5.

Kolmogorov's [204, 205] implicit interpretation of (9.3) was that $\mathbb{P}^{\text{rand}}$ and $\mathbb{P}^{\text{exch}}$ are close; on the log scale we have

$$-\log \mathbb{P}^{\text{rand}}(E) = -\log \mathbb{P}^{\text{exch}}(E) + O(\log N) , \tag{9.4}$$

whereas typical values of $-\log \mathbb{P}^{\text{rand}}(E)$ and $-\log \mathbb{P}^{\text{exch}}(E)$ have the order of magnitude N for small (but non-zero) $|E|$. See Sect. 9.5.1 for details.

From the point of view of Cournot's principle, Proposition 9.2 may be interpreted as saying that there is not much difference between testing randomness and testing exchangeability. If we have a test with critical region E of size ϵ for testing exchangeability, we can use it for testing randomness and its size will not increase; in the opposite direction, if we have a test with critical region E of size ϵ for testing randomness, we can use it for testing exchangeability, and its size will increase to at most $1.5\sqrt{N}\epsilon$.

9.1.2 Conformal Probability

First we will define upper conformal probability $\mathbb{P}^{\text{conf}}$, an analogue of $\mathbb{P}^{\text{rand}}$ and $\mathbb{P}^{\text{exch}}$ for testing randomness using conformal test martingales. A *conformal nonnegative martingale* is the product cS for a constant $c \geq 0$ and a CTM S; in other words, it is a CTM S with the requirement $S_0 = 1$ dropped. Our simple version of upper conformal probability will be sufficient for our current purpose; there are other natural definitions. The *upper conformal probability* of $E \subseteq \Omega$ is

$$\mathbb{P}^{\text{conf}}(E) := \inf\{S_0 : \forall(z_1, \ldots, z_N) \in E : S_N(z_1, \tau_1, z_2, \tau_2, \ldots) \geq 1 \; \tau\text{-a.s.}\} , \tag{9.5}$$

where S ranges over the conformal nonnegative martingales, "τ-a.s." refers to the uniform probability measure over $(\tau_1, \tau_2, \ldots) \in [0, 1]^\infty$, and S_0 stands for the constant initial value of S. The definition (9.5) is in the spirit of [320, Sect. 2.1];

$\mathbb{P}^{\mathrm{conf}}(E) < \epsilon$ for a small $\epsilon > 0$ means that there exists a conformal nonnegative martingale with a small initial value, below ϵ, that almost surely increases its value manyfold, to at least 1, if the event E happens. Therefore, we do not expect this event to happen under the hypothesis of randomness. This is spelled out in the following lemma.

Lemma 9.4 *For any event E, $\mathbb{P}^{\mathrm{rand}}(E) \le \mathbb{P}^{\mathrm{conf}}(E)$.*

Proof Let P be a power distribution on Ω and S be a conformal nonnegative martingale satisfying the condition in (9.5). It suffices to prove

$$P(E) \le S_0 ; \tag{9.6}$$

indeed, we can then obtain $\mathbb{P}^{\mathrm{rand}}(E) \le \mathbb{P}^{\mathrm{conf}}(E)$ by taking sup of the left-hand side of (9.6) over P and taking inf of the right-hand side over S.

To check (9.6), remember that $S_N \ge 1_E$ a.s., where 1_E is the indicator of E. Since S is a nonnegative martingale under P (like any conformal nonnegative martingale), we have

$$P(E) = \int 1_E \, \mathrm{d}P \le \int S_N \, \mathrm{d}P = S_0 .$$

□

Upper conformal probability can be used to formalize the notion of efficiency for conformal testing. Namely, if $\mathbb{P}^{\mathrm{rand}}$ and $\mathbb{P}^{\mathrm{conf}}$ are shown to be close, this could be interpreted as CTMs being able to detect any deviations from randomness. By Cournot's principle, any deviations from randomness are demonstrated by indicating in advance an event E of small probability under any power distribution, i.e., such that $\mathbb{P}^{\mathrm{rand}}(E)$ is small, which then happens. If $\mathbb{P}^{\mathrm{rand}}$ and $\mathbb{P}^{\mathrm{conf}}$ are close, $\mathbb{P}^{\mathrm{conf}}(E)$ will also be small. This means that there exists a conformal nonnegative martingale (equivalently, CTM) S that increases its value manyfold when E happens. We can choose S in advance since E is chosen in advance. And such an S will be successful in detecting deviations from randomness whenever E is.

The following proposition shows that $\mathbb{P}^{\mathrm{exch}}$ and $\mathbb{P}^{\mathrm{conf}}$ are close, in the sense similar to the closeness of $\mathbb{P}^{\mathrm{rand}}$ and $\mathbb{P}^{\mathrm{exch}}$ asserted in Proposition 9.2 (see also (9.4)).

Proposition 9.5 *For any $E \subseteq \Omega$,*

$$\mathbb{P}^{\mathrm{exch}}(E) \le \mathbb{P}^{\mathrm{conf}}(E) \le N\mathbb{P}^{\mathrm{exch}}(E) . \tag{9.7}$$

Proposition 9.5 is a statement of efficiency for CTMs. It says that, at our crude scale, lack of exchangeability can be detected using CTMs. Namely, given a critical region E of a very small size $\mathbb{P}^{\mathrm{exch}}(E) < \epsilon \ll 1$, we can construct a conformal nonnegative martingale with initial capital $N\epsilon$ that attains capital of at least 1 when E happens (and we can scale it up to a CTM attaining a large value, $1/(N\epsilon)$, on E).

Combining the right-hand inequalities in (9.3) and (9.7) we obtain

$$\mathbb{P}^{\mathrm{conf}}(E) \le 1.5N^{1.5}\mathbb{P}^{\mathrm{rand}}(E) . \tag{9.8}$$

This inequality says that CTMs are efficient at detecting deviations not only from exchangeability but also from randomness. Given a critical region E of a very small size $\mathbb{P}^{\mathrm{rand}}(E) < \epsilon \ll 1$, there exists a conformal nonnegative martingale that increases an initial capital of $1.5N^{1.5}\epsilon$ to at least 1 when E happens.

The proof of Proposition 9.5 given in Sect. 9.4.2 is very intuitive: to show that $\mathbb{P}^{\mathrm{conf}}(E) \le N\mathbb{P}^{\mathrm{exch}}(E)$ (this is the nontrivial inequality), we consider the trivial nonconformity measure assigning nonconformity score z to each example $z \in \{0, 1\}$ and for each element $\omega \in E$ consider a conformal nonnegative martingale S^{ω} (based on that nonconformity measure) that knows ω and gambles recklessly on it; a linear combination of such S^{ω} will witness that $\mathbb{P}^{\mathrm{conf}}(E) \le N\mathbb{P}^{\mathrm{exch}}(E)$.

9.2 Conformal Testing with a Known Alternative

In this section we consider an infinite sequence of binary examples, $z_n \in \{0, 1\}$, $n = 1, 2, \ldots$, and our task is to test the hypothesis that z_n are IID.

Let us start from an auxiliary problem. Suppose that, in the context of conformal testing, we know the true distribution of the nth p-value p_n (conditional on knowing the first $n - 1$ p-values), and suppose it is continuous with density ρ. What betting function f should we choose at step n in order for the resulting betting martingale to grow as fast as possible? This is a continuous version of the standard problem of horse race betting [60, 188]. If the assumption of randomness is violated, the distribution of the conformal p-values can be expected to be non-uniform, and the following simple lemma sheds some light on ways of exploiting such non-uniformity.

Lemma 9.6 *For any probability density functions ρ and f on* $[0, 1]$ *(so that* $\int_0^1 \rho(p)\,\mathrm{d}p = 1$ *and* $\int_0^1 f(p)\,\mathrm{d}p = 1$*),*

$$\int_0^1 \Bigl(\log \rho(p)\Bigr)\rho(p)\,\mathrm{d}p \ge \int_0^1 \Bigl(\log f(p)\Bigr)\rho(p)\,\mathrm{d}p \tag{9.9}$$

and

$$\int_0^1 \Bigl(\log \rho(p)\Bigr)\rho(p)\,\mathrm{d}p \ge 0 . \tag{9.10}$$

(We define the right-hand side of (9.9) to be $-\infty$ if $f = 0$ with a positive ρ-probability.)

Proof It is well known (and immediately follows from the inequality $\log x \le x-1$) that the Kullback–Leibler divergence is always nonnegative:

$$\int_0^1 \log\left(\frac{\rho(p)}{f(p)}\right)\rho(p)\,\mathrm{d}p \ge 0\,. \tag{9.11}$$

This is equivalent to (9.9). And (9.10) is a special case of (9.9) corresponding to the probability density function $f := 1$. □

If we choose a betting function f, the logarithm of our capital will increase by the right-hand side of (9.9) in expectation. Therefore, according to (9.9), the largest increase in expectation is achieved when we use ρ as the betting function. (Increasing the log capital as much as possible in expectation is a natural objective since such increases add to give the log of the final capital, and so we can apply the law of large numbers, as in horse racing [60, Sect. 6.1] or log-optimal portfolios [60, Chap. 16].) The discrete version of this strategy is known as *Kelly gambling* [60, Theorem 6.1.2].

How efficient the betting function ρ is depends on the left-hand side of (9.9), which is the minus (differential) entropy of ρ. The maximum entropy distribution on $[0, 1]$ is the uniform distribution, as asserted by (9.10), whose right-hand side is equal to the minus entropy of the uniform distribution. (This is a very special case of standard maximum entropy results, such as [60, Theorem 12.1.1].) The uniform true distribution for the p-values gives zero expected increase in the log capital; otherwise, it is positive.

9.2.1 *Bayes–Kelly Algorithm*

Now let us fix a probability measure P on the example space $\{0, 1\}^\infty$; it will play the role of the alternative hypothesis in the Neyman–Pearson setting of statistical hypothesis testing. To some degree the alternative hypothesis P also plays the role of the soft statistical model in Chap. 2 (see Sect. 2.5.1): the validity of our methods depends only on the null hypothesis of randomness, but we will try to make them as efficient as possible under the alternative hypothesis P. In this subsection the alternative hypothesis is simple, i.e., is a probability measure rather than a set of probability measures.

Our construction of a CTM efficient under the alternative P will use a Bayesian method. The conformal p-values $p_1, p_2, \dots$ will always be computed using the *identity nonconformity measure* (assigning conformity score z to each example $z \in \{0, 1\}$ regardless of the comparison bag):

$$p_n := \frac{|\{i \in \{1,\dots,n\} : z_i > z_n\}| + \tau_n\, |\{i \in \{1,\dots,n\} : z_i = z_n\}|}{n} \tag{9.12}$$

(this is a special case of (2.67)).

Under the alternative hypothesis, the p-values are generated by a completely specified stochastic mechanism. According to Lemma 9.6, the optimal (in the sense of the Kelly criterion) betting function f_n is given by the density of the predictive distribution of p_n conditional on knowing $p_1, \ldots, p_{n-1}$. Let us find these predictive distributions. We will use the notation $\mathbf{U}[a, b]$, where $a < b$, for the uniform probability distribution on the interval $[a, b]$ (so that its density is $1/(b-a)$); remember that $\mathbf{U} := \mathbf{U}[0, 1]$.

We are in a typical situation of Bayesian statistics. The Bayesian parameter is a binary sequence $(z_1, z_2, \ldots) \in \{0, 1\}^\infty$ of examples, and the prior distribution on it is the alternative P. The Bayesian observations are the conformal p-values $p_1, p_2, \ldots$. Given the Bayesian parameter, the distribution of p_n is

$$p_n \sim \begin{cases} \mathbf{U}[0, k/n] & \text{if } z_n = 1 \\ \mathbf{U}[k/n, 1] & \text{if } z_n = 0 , \end{cases}$$

where $k := z_1 + \cdots + z_n$ is the number of 1s among the first n examples.

For any examples $z_1, \ldots, z_n$ (not necessarily those output by Reality), the *cylinder* $[z_1, \ldots, z_n]$ is defined to be the set of all infinite sequences of binary examples starting from $z_1, \ldots, z_n$. Let $w[z_1, \ldots, z_n] := w([z_1, \ldots, z_n])$, where $(z_1, \ldots, z_n) \in \mathbf{Z}^*$, be the total posterior probability at the end of step n of the parameter values $z'_1, z'_2, \ldots$ for which $z'_1 = z_1, \ldots, z'_n = z_n$; we will use them as the weights when computing the predictive distributions for the p-values. We can compute the weights $w[z'_1, \ldots, z'_n]$ recursively in the length n of the argument as follows. We start from

$$w[\Box] := 1 . \tag{9.13}$$

At each step $n \geq 1$, first we compute the unnormalized weights

$$\tilde{w}[z_1, \ldots, z_n] := w[z_1, \ldots, z_{n-1}] P(z_n \mid z_1, \ldots, z_{n-1}) f^n_{z_1+\cdots+z_n, z_n}(p_n) , \tag{9.14}$$

where f is the likelihood defined by

$$f^n_{k,L}(p) := \begin{cases} \frac{n}{k} & \text{if } L = 1 \text{ and } p \leq \frac{k}{n} \\ \frac{n}{n-k} & \text{if } L = 0 \text{ and } p \geq \frac{k}{n} \\ 0 & \text{otherwise} , \end{cases} \tag{9.15}$$

and then we normalize them:

$$w[z_1, \ldots, z_n] := \frac{\tilde{w}[z_1, \ldots, z_n]}{\sum_{z'_1, \ldots, z'_n} \tilde{w}[z'_1, \ldots, z'_n]} . \tag{9.16}$$

Given the posterior weights for the previous step, we can find the predictive distribution for p_n as

$$p_n \sim \sum_{(z_1,\ldots,z_{n-1})\in\{0,1\}^{n-1}} w[z_1,\ldots,z_{n-1}] \times \left(P(1 \mid z_1,\ldots,z_{n-1}) \mathbf{U}\left[0, \frac{z_1+\cdots+z_{n-1}+1}{n}\right] + P(0 \mid z_1,\ldots,z_{n-1}) \mathbf{U}\left[\frac{z_1+\cdots+z_{n-1}}{n}, 1\right]\right).$$

Therefore, the betting functions for the resulting *Bayes–Kelly conformal test martingale* are

$$f_n(p) = \sum_{(z_1,\ldots,z_{n-1})\in\{0,1\}^{n-1}} w[z_1,\ldots,z_{n-1}] \times \left(P(1 \mid z_1,\ldots,z_{n-1}) \frac{n}{z_1+\cdots+z_{n-1}+1} 1_{p\le\frac{z_1+\cdots+z_{n-1}+1}{n}} + P(0 \mid z_1,\ldots,z_{n-1}) \frac{n}{n-z_1-\cdots-z_{n-1}} 1_{p\ge\frac{z_1+\cdots+z_{n-1}}{n}}\right). \tag{9.17}$$

Notice that $p_1 \sim \mathbf{U}$ and, therefore, $f_1 = 1$.

The procedure is summarized as Algorithm 9.1, which takes as input the p-values computed by (9.12) from the sequence of examples output by Reality. However, it is not an efficient algorithm since it requires exponential computation time and memory. For specific alternatives P we will be able to make it polynomial-time.

How can we tell if an algorithm such as Algorithm 9.1 is any good? For that we will introduce a number of benchmarks. Our notation for the power probability measure $\mathbf{B}_\pi^\infty$ on $\{0, 1\}^\infty$ will be $\text{Ber}(\pi)$ ("Ber" standing for "Bernoulli"). Our null hypothesis is the *randomness model* $(\text{Ber}(\pi) : \pi \in [0, 1])$, which we will also call the *Bernoulli model* in this binary context.

Let us assume that the alternative probability measure P on $\{0, 1\}^\infty$ assigns a non-zero measure to each cylinder $[z_1, \ldots, z_n]$; we will call such probability measures *positive*. The corresponding *lower benchmark* is defined as

Algorithm 9.1 Bayes–Kelly $((p_1, p_2, \ldots) \mapsto (S_1, S_2, \ldots))$

1: $S_1 := 1$
2: set the initial weights as per (9.13)
3: **for** $n = 2, 3, \ldots$
4: $\quad S_n := f_n(p_n)S_{n-1}$, with f_n defined by (9.17)
5: $\quad$ update the weights as per (9.14)–(9.16).

$$\mathrm{LB}_n^P(z_1, z_2, \dots) := \frac{P([z_1, \dots, z_n])}{\mathrm{Ber}(\hat{\pi})([z_1, \dots, z_n])}, \tag{9.18}$$

where $\hat{\pi} := k/n$ (the maximum likelihood estimate) and $k = k(n)$ is the number of 1s among $z_1, \dots, z_n$. By definition, $\mathrm{LB}_0^P := 1$. The notion of validity for the lower benchmark LB_n^P is that, for any power distribution $\mathrm{Ber}(\pi)$, it is dominated by a test martingale $S_n^{(\pi)}$ w.r. to $\mathrm{Ber}(\pi)$: $\mathrm{LB}_n^P \le S_n^{(\pi)}$ for all n and π. (To obtain $S_n^{(\pi)}$, simply replace $\hat{\pi}$ by π on the right-hand side of (9.18).)

Remark 9.7 We interpret the lower benchmark (9.18) as the likelihood ratio of the alternative probability measure to the randomness model. Since in the binary case the randomness model is tiny, we can get rid of the uncertainty in the choice of the power distribution by replacing the parameter π by its maximum likelihood estimate.

Remark 9.8 The expression (9.18) is the infimum over the power distributions of the likelihood ratios that are individually optimal (for each power distribution) in Wald's sense (see Sect. 8.6.3). However, this does not mean that the infimum (9.18) itself is optimal. The extreme case for binary examples is where the null hypothesis consists of all probability measures on $\{0, 1\}^\infty$. The analogue of the lower benchmark will quickly tend to 0 for a typical alternative P, and so its performance will be much worse than that of the identical 1 (which is a test martingale under any null hypothesis). For more general example spaces, such as in the case of real numbers changing their distribution (e.g., with $N(0, 1)$ as pre-change distribution and $N(1, 1)$ as post-change distribution), the randomness model becomes too large, and we are in a situation that is even worse: the analogues of the ratios in (9.18) become zero. (Remember that such analogues have the supremum over all power distributions in the denominator, not the supremum over some parametric model containing both pre-change and post-change distributions.) Our current binary case, however, is very far from such difficult cases.

In the following proposition, which is somewhat analogous to Proposition 9.5, S_n^P stands for the Bayes–Kelly algorithm (Algorithm 9.1) applied to the alternative distribution P.

Proposition 9.9 *For any (positive) alternative P, the Bayes–Kelly martingale S^P satisfies, for any binary sequence of examples $z_1, z_2, \dots$ and any step $n \ge 1$,*

$$S_n^P(z_1, z_2, \dots) \ge \mathrm{LB}_n^P(z_1, z_2, \dots) \frac{n!}{n^n} \frac{k^k}{k!} \frac{(n-k)^{n-k}}{(n-k)!} \quad a.s., \tag{9.19}$$

where $k = z_1 + \cdots + z_n$ is the number of 1s in $z_1, \dots, z_n$.

The qualification "a.s." in (9.19) refers to the internal randomness in the definition of a CTM. We will prove Proposition 9.9 in Sect. 9.4.3.

By Stirling's formula, (9.19) implies

$$\begin{aligned} S_n^P(z_1, z_2, \ldots) &\geq \mathrm{LB}_n^P(z_1, z_2, \ldots) \frac{n!}{n^n} \frac{k^k}{k!} \frac{(n-k)^{n-k}}{(n-k)!} \\ &\approx \mathrm{LB}_n^P(z_1, z_2, \ldots)(2\pi)^{-1/2} \left(\frac{n}{k(n-k)} \right)^{1/2} \\ &\geq \mathrm{LB}_n^P(z_1, z_2, \ldots)(2/\pi)^{1/2} n^{-1/2} . \end{aligned} \tag{9.20}$$

Roughly, (9.20) says that the CTM S^P is competitive with the lower benchmark, and the term $n^{-1/2}$ at the end of (9.20) gives us an idea of how competitive it is. (It is instructive to compare that term with the $\sqrt{N} = N^{1/2}$ in (9.3).)

A finite-sample version of (9.20) that is true for all $n \geq 1$ is

$$S_n^P(z_1, z_2, \ldots) \geq \mathrm{LB}_n^P(z_1, z_2, \ldots) 2^{-1/2} n^{-1/2}$$

(the constant $2^{-1/2}$ is optimal given (9.19) and attained at $n = 2$). In our plots later in this section we will use this inequality on the log scale:

$$\log S_n^P(z_1, z_2, \ldots) \geq \log \mathrm{LB}_n^P(z_1, z_2, \ldots) - 0.5 \log 2 - 0.5 \log n . \tag{9.21}$$

9.2.2 Bayes–Kelly Algorithm with a Composite Alternative

An interesting and common case is where the alternative model P depends on some natural parameters. For example, later in this section we will consider the alternative model consisting of Markov distributions P; if we fix the probability that the first bit is 1 to 0.5, each such distribution is characterized by two parameters, the probability of observing a 1 after a 0, $\pi_{1|0}$, and the probability of observing a 1 after a 1, $\pi_{1|1}$.

We will usually concentrate on the case of a simple alternative, assuming that the parameters of the alternative distribution are known. This will be often sufficient even in the situation when the parameter is not known, and so the alternative hypothesis that we are really interested in is composite: we can consider a grid of parameter values and average the conformal test martingales constructed for those parameter values. We will apply this simple strategy, which we will call *martingale averaging*, to both of our specific alternative models (changepoint, discussed in Sects. 9.2.3–9.2.4, and Markov, discussed in Sects. 9.2.5–9.2.7). In general, we can average the conformal test martingales over any distribution on the parameter values, which we will refer to as the *prior distribution*, but the simplest choice is the uniform distribution on the grid.

A more advanced solution is to average the alternative distributions for various values of the parameters obtaining a single simple alternative distribution P. We do this only for the second of our specific alternative models (Markov in Sect. 9.2.7). In the Markov case we can average over the distribution of the parameter values known as Jeffreys's prior, which can be done in a computationally efficient manner. After that we can apply the Bayes–Kelly martingale directly to the average distribution P. This procedure of *distribution averaging* may be computationally more efficient (if a dense grid is used in martingale averaging) but requires extra work.

Conceptually there is not much difference between martingale averaging and distribution averaging; for the same prior distribution on the parameter values, both methods will give the same result. The difference is computational: in martingale averaging, we implement a simple alternative, and then average the resulting conformal test martingales over a grid. In distribution averaging, we work out the details of the Bayes–Kelly algorithm for the average distribution P on paper, and then implement it.

9.2.3 Changepoint Alternatives

An interesting alternative hypothesis is the changepoint hypothesis, already discussed in Sects. 8.3 and 8.6.4. Now our data consist of binary examples generated independently from Bernoulli distributions $\mathbf{B}_\pi$. We assume that the examples are IID except that the value of the parameter π changes at some point (called *changepoint*, as before). Let π_0 be the pre-change parameter and π_1 be the post-change parameter. The total number of examples is N, of which the first N_0 come from the pre-change distribution $\mathbf{B}_{\pi_0}$ and the remaining $N_1 := N - N_0$ from the post-change distribution $\mathbf{B}_{\pi_1}$.

Our main model situation is where $\pi_0 = 0.4$, $\pi_1 = 0.6$, and $N_0 = N_1 = 1000$. Figure 9.1 shows the path of the Simple Jumper CTM based on the identity nonconformity measure; the martingale's parameter (jumping rate) is $J = 0.01$, and it uses the same 5 betting functions (8.8) as before. This process can serve as measure of the amount of evidence found against the null hypothesis, and it performs very well finding decisive evidence against it.

The Simple Jumper CTM was not designed for the change detection problem, but now we will take the nature of the problem more seriously. Our goal will be to explore final values of test martingales that are attainable (with high probability) under the alternative hypothesis. Our null hypothesis is the randomness model.

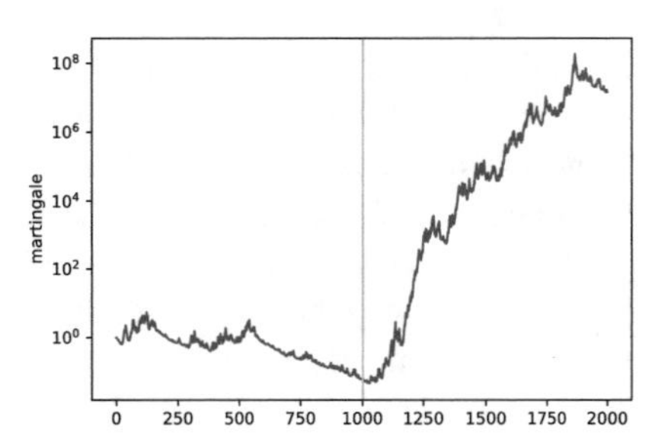

Fig. 9.1 The Simple Jumper CTM in the change detection problem (model situation)

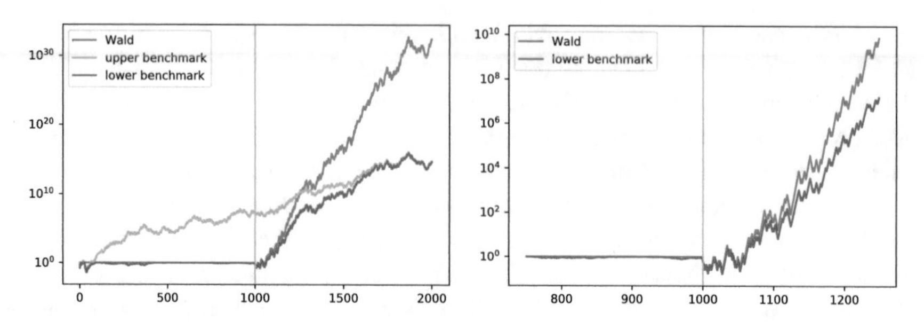

Fig. 9.2 Left panel: Wald's martingale (red line), the upper benchmark (yellow line), and the lower benchmark (green line) over the whole dataset. Right panel (close-up of the left panel): Wald's martingale and the lower benchmark over the middle 500 examples

In addition to the lower benchmark, we will define several notions of an upper benchmark. The first one is utterly unrealistic. It is the likelihood ratio of the true (i.e., alternative) distribution P to the pre-change distribution:

$$W_n := \begin{cases} 1 & \text{if } n \le N_0 \\ \left(\frac{\pi_1}{\pi_0}\right)^{k(n)-k(N_0)} \left(\frac{1-\pi_1}{1-\pi_0}\right)^{(n-N_0)-(k(n)-k(N_0))} & \text{otherwise ,} \end{cases}$$

where $k(n)$ is the number of 1s among the first n examples (we will use this notation repeatedly). This is the optimal test martingale in Wald's sense (Sect. 8.6.3), and we will call it *Wald's martingale*. This process, however, is a test martingale only with respect to the null hypothesis $\mathbf{B}_{\pi_0}^{\infty} = \mathbf{B}_{0.4}^{\infty}$, whereas our null hypothesis is the randomness model. Therefore, it is not a reasonable benchmark. Its path is shown in red in Fig. 9.2 (over the full dataset in the left panel, and over its middle part in the right panel).

Figure 9.2 shows in green the lower benchmark

$$L_n := \begin{cases} \dfrac{\pi_0^{k(n)}(1-\pi_0)^{n-k(n)}}{\left(\frac{k(n)}{n}\right)^{k(n)}\left(1-\frac{k(n)}{n}\right)^{n-k(n)}} & \text{if } n \le N_0 \\ \dfrac{\pi_0^{k(N_0)}(1-\pi_0)^{N_0-k(N_0)}\pi_1^{k(n)-k(N_0)}(1-\pi_1)^{(n-N_0)-(k(n)-k(N_0))}}{\left(\frac{k(n)}{n}\right)^{k(n)}\left(1-\frac{k(n)}{n}\right)^{n-k(n)}} & \text{otherwise} \end{cases} \quad (9.22)$$

(where $0^0 := 1$). Its final value $\mathrm{LB}_N := L_N$ is indicative of the best result that can be attained in our testing problem.

In order to develop an alternative to the lower benchmark (9.22) that would also work outside the binary case, let us replace the denominator of (9.22), which is the maximum likelihood chosen *a posteriori*, by the likelihood at a parameter value chosen a priori but with the knowledge of the stochastic mechanism generating the data. Let us generalize our setting slightly, assuming that the examples take values in a finite set and take value i with probability $\pi_{0,i}$ before the changepoint and $\pi_{1,i}$ after the changepoint (so that $\sum_i \pi_{0,i} = \sum_i \pi_{1,i} = 1$). Our goal is to find a probability measure (u_i) for one example such that the (random) likelihood ratio of the true data-generating distribution to the Nth power of (u_i) is as small as possible. By the Kelly criterion, the corresponding optimization problem for the optimal probability measure (u_i) in the denominator is

$$N_0 \sum_i \pi_{0,i} \log \frac{\pi_{0,i}}{u_i} + N_1 \sum_i \pi_{1,i} \log \frac{\pi_{1,i}}{u_i} \to \min ,$$

which simplifies to

$$\sum_i \frac{N_0 \pi_{0,i} + N_1 \pi_{1,i}}{N} \log u_i \to \max .$$

By the nonnegativity of Kullback–Leibler divergence (cf. (9.11)), the optimal solution is

$$u_i := \frac{N_0 \pi_{0,i} + N_1 \pi_{1,i}}{N} ,$$

i.e., the weighted average of π_0 and π_1.

In the binary case, the *upper benchmark* is

$$\mathrm{UB}_N := \frac{\pi_0^{k(N_0)} (1 - \pi_0)^{N_0 - k(N_0)} \pi_1^{k(N) - k(N_0)} (1 - \pi_1)^{N - N_0 - (k(N) - k(N_0))}}{\pi^{k(N)} (1 - \pi)^{N - k(N)}} , \tag{9.23}$$

where

$$\pi := \frac{N_0}{N} \pi_0 + \frac{N_1}{N} \pi_1 . \tag{9.24}$$

The upper benchmark is the final value $\mathrm{UB}_N = U_N$ of the likelihood ratio martingale

$$U_n := \begin{cases} \frac{\pi_0^{k(n)} (1-\pi_0)^{n-k(n)}}{\pi^{k(n)} (1-\pi)^{n-k(n)}} & \text{if } n \le N_0 \\ \frac{\pi_0^{k(N_0)} (1-\pi_0)^{N_0-k(N_0)} \pi_1^{k(n)-k(N_0)} (1-\pi_1)^{(n-N_0)-(k(n)-k(N_0))}}{\pi^{k(n)} (1-\pi)^{n-k(n)}} & \text{otherwise} , \end{cases} \tag{9.25}$$

where $n = 0, \ldots, N$. Unlike (9.22), (9.25) easily extends to change detection for other statistical models. Some of the standard statistical models are closed under convex mixture (as in (9.24)), and for them the upper benchmark has a particularly simple expression.

The path of the likelihood ratio martingale (9.25) is shown as the yellow line in Fig. 9.2. It is close to a straight line, which makes it look very different from the

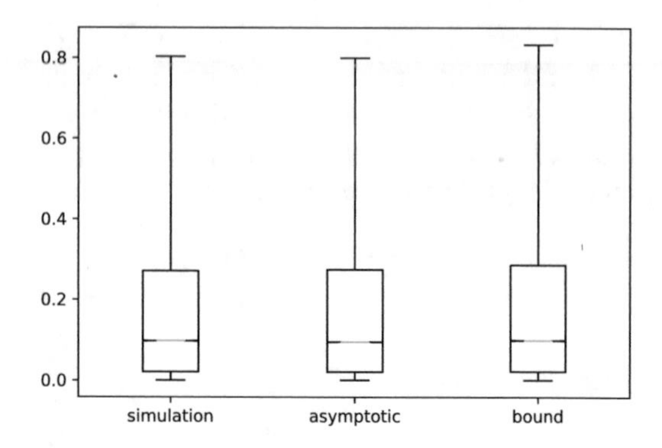

Fig. 9.3 The decimal logarithm of $\mathrm{UB}_N / \mathrm{LB}_N$ in our model situation, its asymptotic approximation, and an upper bound for it, as described in the main text, based on 10^6 simulations

lower benchmark. If, instead, we showed UB_n (as defined in (9.23) with n in place of N) vs $n \geq N_0$, the graphs for the two benchmarks would be indistinguishable. Figure 9.2 only shows that the final values are close. However, the line $n \mapsto \mathrm{UB}_n$ is slightly more difficult to interpret, since it shows the values not of one martingale but many different ones. We will call it the *adaptive upper benchmark* (and it will be used in Fig. 9.4 below).

The following proposition, whose proof we omit, says that the final values of the upper and lower benchmarks are fairly close to each other asymptotically.

Proposition 9.10 *Let* $\xi \sim N(0, 1)$. *As* $N_0 \to \infty$ *and* $N_1 \to \infty$,

$$\frac{2N\pi(1-\pi)}{N_0\pi_0(1-\pi_0) + N_1\pi_1(1-\pi_1)} \ln \frac{\mathrm{UB}_N}{\mathrm{LB}_N} \xrightarrow{\text{law}} \xi^2 . \tag{9.26}$$

Informally, (9.26) implies

$$\log_{10} \frac{\mathrm{UB}_N}{\mathrm{LB}_N} \approx \frac{N_0\pi_0(1-\pi_0) + N_1\pi_1(1-\pi_1)}{2N\pi(1-\pi)\ln 10}\xi^2 \leq \frac{\xi^2}{2\ln 10}, \tag{9.27}$$

where $\approx$ is used to signify the approximate equality of distributions, and the inequality follows from Jensen's inequality applied to the concave function $\pi \in [0, 1] \mapsto \pi(1-\pi)$. Figure 9.3 shows the distributions of $\log_{10}(\mathrm{UB}_N / \mathrm{LB}_N)$, its approximation as given by the expression following $\approx$ in (9.27), and its upper bound as given by the expression following $\leq$ in (9.27). We can see that the number of examples $N = 2000$ (split in half by the changepoint) is sufficient for the asymptotic approximation to work. The median in the column "simulation" is approximately 0.1, and so the difference between the two benchmarks will not be very noticeable in our plots.

Let us specialize the Bayes–Kelly algorithm (Algorithm 9.1) to the alternative hypothesis of a known changepoint and known pre- and post-change distributions: the examples are independent, $z_n = 1$ with probability π_0 if $n \leq N_0$, and $z_n = 1$ with probability π_1 if $n > N_0$. The specialized algorithm will run in polynomial time. We set

$$\pi_n^* := \begin{cases} \pi_0 & \text{if } n \le N_0 \\ \pi_1 & \text{if } n > N_0 \ ; \end{cases}$$

this is the probability of $z_n = 1$.

Let w_k^n, where $n = 0, 1, \ldots$ and $k = 0, \ldots, n$, be the total posterior probability at the end of step n of the parameter values $z_1, z_2, \ldots$ for which $z_1 + \cdots + z_n = k$; we will use them as the weights when computing the predictive distributions for the p-values. In terms of the weights used in the general Bayes–Kelly algorithm,

$$w_k^n := \sum_{(z_1,\ldots,z_n):z_1+\cdots+z_n=k} w[z_1, \ldots, z_n] \ .$$

We only need to keep track of the aggregate weights w_k^n. We can compute those aggregate weights recursively in n, starting from $w_0^0 := 1$, as follows. First we compute the unnormalized weights

$$\tilde{w}_k^n := w_{k-1}^{n-1} \pi_n^* f_{k,1}^n(p_n) + w_k^{n-1}(1 - \pi_n^*) f_{k,0}^n(p_n) \tag{9.28}$$

where f is the likelihood defined by (9.15) and $w_{-1}^{n-1} := 0$. After that we normalize them:

$$w_k^n := \frac{\tilde{w}_k^n}{\tilde{w}_0^n + \cdots + \tilde{w}_n^n} \ . \tag{9.29}$$

Given the posterior weights for the previous step, we can find the predictive distribution for p_n as

$$p_n \sim \sum_{k=0}^{n-1} w_k^{n-1} \left(\pi_n^* \mathbf{U}\left[0, \frac{k+1}{n}\right] + (1 - \pi_n^*) \mathbf{U}\left[\frac{k}{n}, 1\right] \right) .$$

Therefore, the betting functions for the Bayes–Kelly CTM in this case are

$$f_n(p) = \sum_{k=0}^{n-1} w_k^{n-1} \left(\frac{n}{k+1} \pi_n^* 1_{p \le \frac{k+1}{n}} + \frac{n}{n-k} (1 - \pi_n^*) 1_{p \ge \frac{k}{n}} \right) \tag{9.30}$$

(and so we again have $f_1 = 1$).

The specialization of the Bayes–Kelly CTM to the changepoint alternative is shown as Algorithm 9.2. At step n it requires $O(n)$ memory for storing the array w_k^n and $O(n)$ time for the computations. To process the first N examples, it requires $O(N^2)$ time and $O(N)$ memory.

The left panel of Fig. 9.4 shows the Bayes–Kelly CTM and two benchmarks, the lower benchmark and the adaptive upper benchmark. The Bayes–Kelly CTM

Algorithm 9.2 Bayes–Kelly for change detection

1: $S_0 := 1$
2: set the initial weight as $w_0^0 := 1$
3: **for** $n = 1, 2, \ldots$
4: $\quad S_n := f_n(p_n)S_{n-1}$, with f_n defined by (9.30)
5: $\quad$ update the weights as per (9.28)–(9.29).

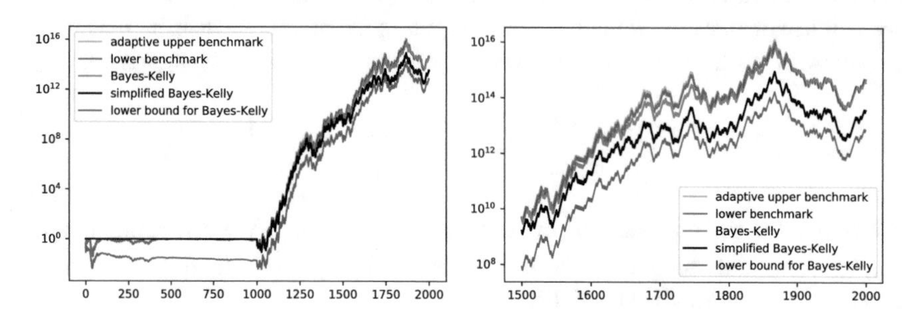

Fig. 9.4 Two benchmarks, the Bayes–Kelly CTM, its simplified version, and a bound, as described in the main text. Left panel: over all examples. Right panel: over the last quarter

is close to the two benchmarks. The left panel also shows the bound (9.21) and a simplified Bayes–Kelly CTM, to be discussed momentarily. The right panel magnifies the part of the left panel over the last 500 examples (the most informative part of the plot).

The specialized Bayes–Kelly CTM is polynomial-time, but it still requires significant resources. If we assume that $n \geq N_0$ and the weights w_k^n, $k = 0, \ldots, n$, are concentrated at

$$k \approx k + 1 \approx N_0\pi_0 + (n - N_0)\pi_1 \ ,$$

(9.30) will simplify to

$$f_n(p) := \begin{cases} \frac{n\pi_1}{N_0\pi_0+(n-N_0)\pi_1} & \text{if } p \leq \frac{N_0\pi_0+(n-N_0)\pi_1}{n} \\ \frac{n(1-\pi_1)}{N_0(1-\pi_0)+(n-N_0)(1-\pi_1)} & \text{otherwise .} \end{cases} \tag{9.31}$$

Figure 9.5 shows the weights for the last step of the Bayes–Kelly CTM. They are indeed concentrated in the neighbourhood of $N_0\pi_0 + N_1\pi_1 = 1000$.

The simplified Bayes–Kelly martingale for change detection is shown as Algorithm 9.3. It does not gamble over the first N_0 examples (before the changepoint). It is plotted in Fig. 9.4 alongside the Bayes–Kelly and its lower bound; in this case it is sandwiched between them.

Figure 9.6 gives some boxplots for 10^3 simulations of our model scenario; namely, 1000 examples from $\mathbf{B}_{0.4}$ are followed by 1000 examples from $\mathbf{B}_{0.6}$. The boxplots show the median and the quartiles of the empirical distributions over the

Algorithm 9.3 Simplified Bayes–Kelly for change detection

1: $S_n := 1$ for $n = 0, \dots, N_0$
2: **for** $n = N_0 + 1, N_0 + 2, \dots$
3: $\quad S_n := f_n(p_n)S_{n-1}$, with f_n defined by (9.31).

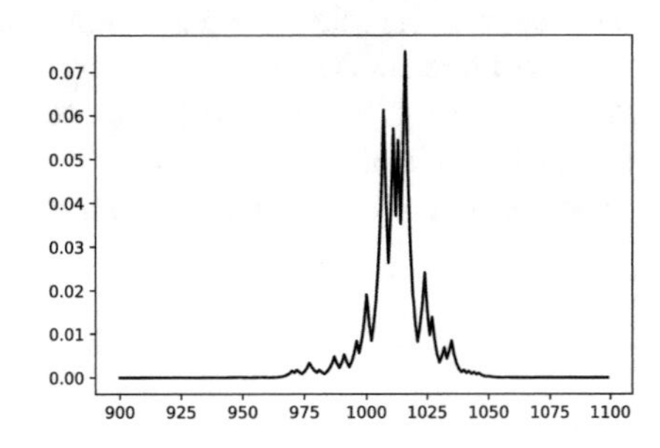

Fig. 9.5 The weights $w_k^{10^4}$, for a range of k (the weights are virtually zero outside this range) at the last step for the Bayes–Kelly CTM

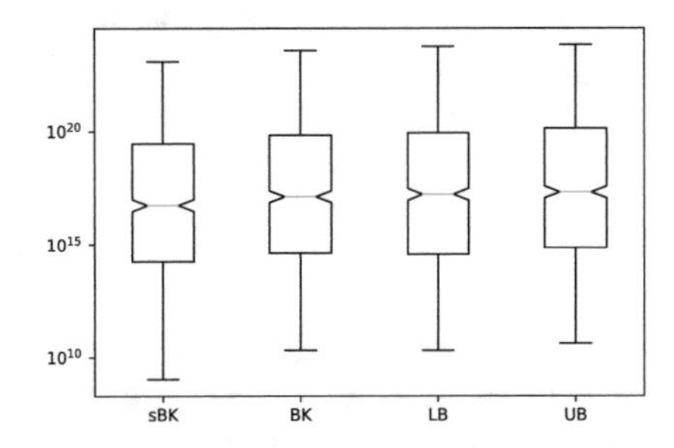

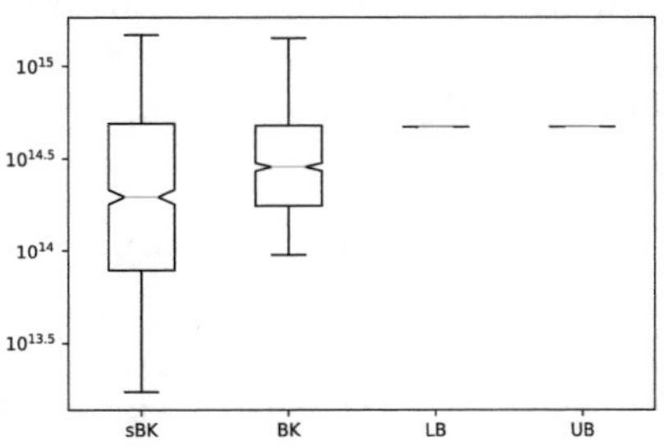

Fig. 9.6 Boxplots for several test martingales and related processes ("sBK" standing for "simplified Bayes–Kelly", "BK" for "Bayes–Kelly", "LB" for "lower benchmark", and "UB" for "upper benchmark", with the parameters described in the main text. Left panel: for a randomly generated dataset. Right panel: for a fixed dataset

10^3 simulations for the final values of the Bayes–Kelly, simplified Bayes–Kelly, and the lower and upper benchmarks, and their whiskers show the 5 and 95% quantiles. The boxplots are notched, with the notches indicating confidence intervals for the median. (We used the same conventions in Fig. 9.3, but the details were less important there.) The left panel shows that the Bayes–Kelly, and even simplified Bayes–Kelly, martingales are competitive with the two benchmarks. The right panel is for a fixed dataset and suggests that Bayes–Kelly martingale is less volatile than its simplified version. Interestingly, the final value of the Bayes–Kelly martingale, and even of the simplified Bayes–Kelly martingale, may well exceed the final value of the upper benchmark.

9.2.4 More Natural Conformal Test Martingales for Changepoint Alternatives

The Bayes–Kelly and simplified Bayes–Kelly martingales depend very much on the knowledge of the true data-generating mechanism. We can get rid of the knowledge of the parameters N_0, π_0, and π_1 by mixing over those parameters. But can we obtain comparable results with even less knowledge of the stochastic mechanism? This is the topic of this subsection.

Let us generalize the betting function (9.31) to

$$f_{(a,b)}(p) := \begin{cases} \frac{b}{a} & \text{if } p \le a \\ \frac{1-b}{1-a} & \text{otherwise} \end{cases}, \qquad (9.32)$$

where $a, b \in (0, 1)$. It is easy to see that $\int f_{(a,b)} = 1$. Apart from the betting functions (9.32) we will use the trivial function $f_\square$, $f_\square(p) := 1$ for all p.

Let S_n be the conformal test martingale

$$S_n := \int f_{s_1}(p_1) \dots f_{s_n}(p_n) \mu(\mathrm{d}(s_1, s_2, \dots)), \qquad (9.33)$$

where $p_1, p_2, \dots$ is the underlying sequence of conformal p-values and μ is the distribution of the following Markov chain with states $s_1, s_2, \dots$ defined in the spirit of tracking the best expert in prediction with expert advice (cf. Sects. 8.1.2 and 8.6.1). The state space is $\{\square\} \cup (0, 1)^2$, and $R \in (0, 1)$ is the parameter (typically a small number). The initial state is $s_1 := \square$ (the *sleeping* state). The transition function is:

- if the current state is $\square$, with probability $1 - R$ the state remains $\square$, and with probability R a new state (a, b) is chosen from the uniform distribution on $(0, 1)^2$;
- the states $(a, b) \in (0, 1)^2$ are absorbing: if the current state is $(a, b) \in (0, 1)^2$, it will stay (a, b).

In our implementation of the procedure (9.33), we replace the square $(0, 1)^2$ by the grid $\mathbf{G}$, where

$$\mathbf{G} := \left\{ \frac{1}{G}, \frac{2}{G}, \dots, \frac{G-1}{G} \right\}^2 \qquad (9.34)$$

and G (positive integer) is another parameter. The resulting procedure is shown as Algorithm 9.4.

The intuition behind Algorithm 9.4 is that, in order to gamble against the uniformity of $(p_1, p_2, \dots)$, we distribute our initial capital of 1 among accounts $S_{a,b}$ indexed by $(a, b) \in \mathbf{G}$, and there is also a sleeping account $S_\square$. We start from

Algorithm 9.4 Sleeper/Stayer $((p_1, p_2, \dots) \mapsto (S_1, S_2, \dots))$

1: $S_\square := 1$
2: **for** $(a, b) \in \mathbf{G}$: $S_{a,b} := 0$
3: **for** $n = 1, 2, \dots$
4: **for** $(a, b) \in \mathbf{G}$: $S_{a,b} := S_{a,b} f_{(a,b)}(p_n)$
5: $S_n := S_\square + \sum_{(a,b)\in\mathbf{G}} S_{a,b}$
6: **for** $(a, b) \in \mathbf{G}$: $S_{a,b} := S_{a,b} + RS_\square/(G-1)^2$
7: $S_\square := (1 - R)S_\square$.

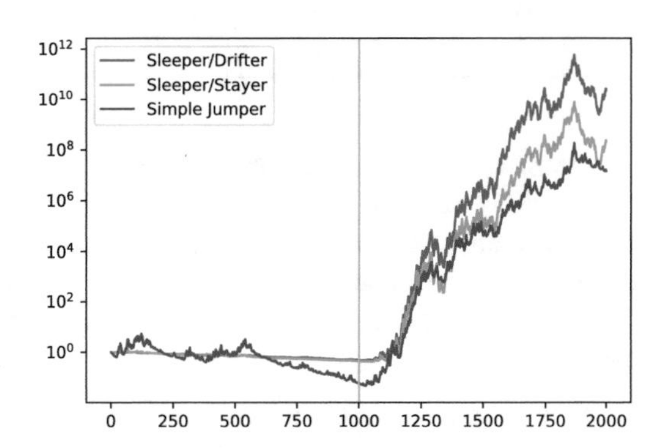

Fig. 9.7 Various conformal test martingales, as described in the main text

all money invested in the sleeping account, but at the end of each step a fraction R of that money is moved to the active accounts $S_{a,b}$ and divided between them equally (see lines 6 and 7). On account $S_{a,b}$ we gamble against the uniformity of the input p-values using the betting function $f_{(a,b)}$.

Figure 9.7 (the line in cyan) suggests that we can improve on the result of Fig. 9.1 using a natural, and in fact very basic, conformal test martingale (the improvement can be very significant for larger datasets). In Fig. 9.7 we use the identity nonconformity measure and the Sleeper/Stayer betting martingale of Algorithm 9.4, and the parameters are $R := 0.001$ and $G := 10$; therefore, a and b are chosen from the grid $\{0.1, 0.2, \dots, 0.9\}$.

To improve further the performance of a natural conformal test martingale, let us make another step towards the simplified Bayes–Kelly martingale (9.31) following the idea of martingale averaging described in Sect. 9.2.2. The new martingale will be defined as an average of the following "expert martingales". An expert martingale is characterized by a vector parameter $(N_0, \pi_0, \pi_1) \in \{1, 2, \dots\} \times (0, 1)^2$ and is the simplified Bayes–Kelly martingale for these postulated (N_0, π_0, π_1), rather than the unknown real ones. (In this and the next paragraphs, we will use N_0, π_0, π_1 as local variables; in the end they will be integrated out, and we will again be able to use them in the global sense, as before.) The expert sleeps (does not gamble) until time N_0, and at each time $n > N_0$ it uses the betting function (9.31). This betting function is of the form (9.32) with $b := \pi_1$ and $a = a_n$ being the weighted average of π_0 and π_1 with the weights N_0/n and $1 - N_0/n$, respectively. Therefore, a_n gradually drifts from π_0 towards π_1.

Algorithm 9.5 Sleeper/Drifter $((p_1, p_2, \dots) \mapsto (S_1, S_2, \dots))$

$S_0 := S_\square := 1$
for $i = 1, 2, \dots$ and $(a, b) \in \mathbf{G}$: $S_{i,a,b} := 0$
for $n = 1, 2, \dots$
  for $i < n/M$ and $(a, b) \in \mathbf{G}$
    $a' := \frac{iM}{n} a + \left(1 - \frac{iM}{n}\right) b$
    $S_{i,a,b} := S_{i,a,b} f_{(a',b)}(p_n)$
  $S_n := S_\square + \sum_{(i,a,b) \in \mathbb{N} \times \mathbf{G}} S_{i,a,b}$
  if n is divisible by M
    for $(a, b) \in \mathbf{G}$: $S_{n/M,a,b} := RMS_\square/(G-1)^2$
    $S_\square := (1 - RM)S_\square$.

The *Sleeper/Drifter martingale* depends on three hyperparameters: G, determining the grid (9.34), M ($M := 1$ is a good value, but larger values of M improve computational efficiency), and R (the rate at which the experts, who are originally sleeping, wake up). (We say "hyperparameters" to distinguish them from N_0, π_0, π_1.) It is the average of the experts w.r. to the following probability measure:

- all three parameters N_0, π_0, π_1 are independent;
- $N_0 = iM$, where $i \in \{1, 2, \dots\}$ is generated according to the geometric distribution with parameter RM;
- π_0 and π_1 are generated from the uniform distribution on the grid (9.34).

The overall procedure is given as Algorithm 9.5. The key array in this algorithm is $(S_{i,a,b})$, where $S_{i,a,b}$ is the total capital of the experts drifting from a towards b who woke up at time iM. Besides, $S_\square$ is the total capital of the experts who are still asleep; as an expert wakes up, its capital moves from $S_\square$ to one of the $S_{i,a,b}$.

The performance of Algorithm 9.5 is shown as the magenta line in Fig. 9.7. The parameters used there are $G = 10$, $M = 100$, and $R = 0.001$. (There is not much sensitivity to the values of the parameters; e.g., if we decrease R to 10^{-4} or 10^{-5}, we will get final values of about the same order of magnitude.)

In general, custom-made conformal test martingales, such as the simplified Bayes–Kelly, provide clear goals for more natural conformal test martingales, and even give ideas of how these goals can be attained. These ideas, in turn, add to the toolbox that we can use for dealing with practical problems, where we often have only a vague notion of the true data-generating distribution.

9.2.5 Symmetric Markov Alternatives

Now we consider two new model situations. Our data consist of binary examples generated from a Markov model. We will use the notation Markov$(\pi_{1|0}, \pi_{1|1})$ for the probability distribution of a Markov chain with the transition probabilities $\pi_{1|0}$

for transitions $0 \to 1$ and $\pi_{1|1}$ for transitions $1 \to 1$; the probability that the first example is 1 will always be assumed 0.5. The probability of the first observation plays a very minor role, and our null hypothesis is, essentially, that $\pi_{1|0} = \pi_{1|1}$. In the *hard scenario*, the model is Markov(0.4, 0.6), and in the *easy scenario*, the model is Markov(0.1, 0.9). (The hard case is harder to distinguish from the null hypothesis than the easy case.) The number of examples is $N := 10^3$.

As before, $\mathbf{B}_\pi$ is the Bernoulli distribution on $\{0, 1\}$ with parameter $\pi \in [0, 1]$, and $\text{Ber}(\pi) = \mathbf{B}_\pi^\infty$. Formally, our null hypothesis is still the randomness model (and the alternative hypothesis is a Markov chain).

Figure 9.8 shows paths of the Simple Jumper CTM in the two scenarios, hard and easy, for various values of the jumping rate; it performs poorly in this context (unlike in the case of change detection). In this subsection we will design better martingales.

To evaluate the quality of our CTMs, we can still use the *lower benchmark*

$$\text{LB}_n := \frac{\text{Markov}(\pi_{1|0}, \pi_{1|1})([z_1, \ldots, z_n])}{\text{Ber}(\hat{\pi})([z_1, \ldots, z_n])} \tag{9.35}$$

(cf. (9.18)), where $\hat{\pi} := k/n$ (the maximum likelihood estimate) and $k = k(n)$ is the number of 1s among $z_1, \ldots, z_n$. The *upper benchmark* is now defined as

$$\text{UB}_n := \frac{\text{Markov}(\pi_{1|0}, \pi_{1|1})([z_1, \ldots, z_n])}{\text{Ber}(0.5)([z_1, \ldots, z_n])} . \tag{9.36}$$

By definition, $\text{UB}_0 = \text{LB}_0 := 1$. Now we do not need the notion of an adaptive upper benchmark that we used for the changepoint alternative; (9.36) is automatically adaptive.

The paths of the upper and lower benchmarks are shown in Fig. 9.9 in yellow and green; the figure also shows the paths of the Bayes–Kelly and simplified Bayes–Kelly martingales specialized to the Markov case, to be discussed momentarily, and the lower bound (9.21). The two benchmarks coincide or almost coincide. Notice

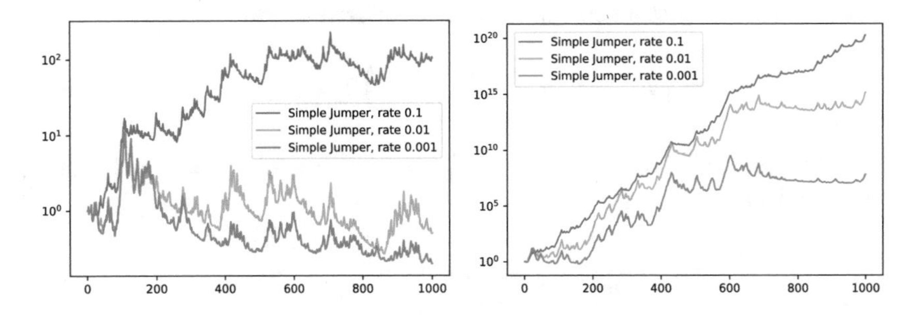

Fig. 9.8 The Simple Jumper martingale for various values of the jumping rate. Left panel: hard scenario. Right panel: easy scenario

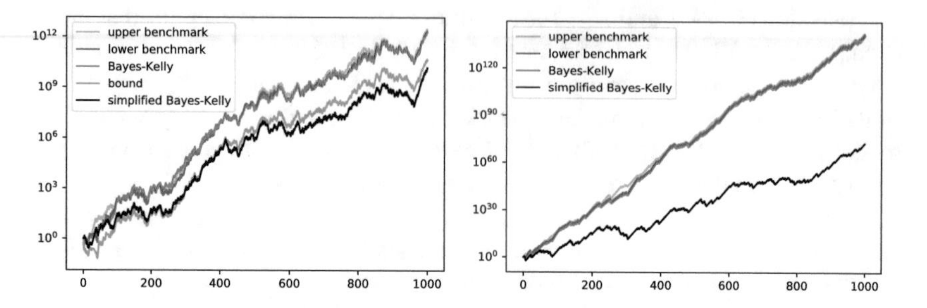

Fig. 9.9 The two benchmarks, Bayes–Kelly CTM, its simplified version, and its lower bound (9.21). Left panel: hard case. Right panel: easy case (with the lower bound not shown as it virtually coincides with the Bayes–Kelly CTM)

that the upper benchmark can never be less than the lower benchmark; this is as true in the Markovian case as it was in the change detection case.

Let us now spell out the Bayes–Kelly martingale in the Markovian case. Now we can represent the weights more efficiently. Namely, let $w^n_{k,L}$, where $n = 1, 2, \dots$, $k = 0, \dots, n$, and $L \in \{0, 1\}$, be the total posterior probability given the p-values $p_1, \dots, p_n$, of the parameter values $z_1, z_2, \dots$ for which $z_1 + \dots + z_n = k$ and $z_n = L$. In other words,

$$w^n_{k,L} := \sum_{(z_1,\dots,z_n):z_1+\dots+z_n=k,z_n=L} w[z_1, \dots, z_n] ;$$

we will use them as the weights when computing the predictive distributions for the p-values.

We can compute the weights $w^n_{k,L}$ recursively in n as follows, using the shorthand $\pi_{0|j} := 1 - \pi_{1|j}$. We start from

$$w^1_{0,0} := w^1_{1,1} := 0.5, \quad w^1_{0,1} := w^1_{1,0} := 0 . \tag{9.37}$$

At each step $n \geq 2$, first we compute the unnormalized weights

$$\tilde{w}^n_{k,L} := \left(w^{n-1}_{k-L,0} \pi_{L|0} + w^{n-1}_{k-L,1} \pi_{L|1} \right) f^n_{k,L}(p_n) , \tag{9.38}$$

where f is the likelihood (9.15) and $w^{n-1}_{-1,0} := w^{n-1}_{-1,1} := w^{n-1}_{0,1} := 0$, and then we normalize them:

$$w^n_{k,L} := \tilde{w}^n_{k,L} / \sum_{k=0}^{n} \sum_{L=0}^{1} \tilde{w}^n_{k,L} . \tag{9.39}$$

Algorithm 9.6 Bayes–Kelly for the Markov alternative

1: $S_0 := S_1 := 1$
2: set the initial weights as per (9.37)
3: **for** $n = 2, 3, \ldots$:
4: $S_n := f_n(p_n)S_{n-1}$, with f_n defined by (9.40)
5: update the weights as per (9.38)–(9.39).

Given the posterior weights for the previous step, we can find the predictive distribution for p_n as

$$p_n \sim \sum_{k=0}^{n-1}\sum_{L=0}^{1} w_{k,L}^{n-1}\left(\pi_{1|L}\mathbf{U}\left[0, \frac{k+1}{n}\right] + \pi_{0|L}\mathbf{U}\left[\frac{k}{n}, 1\right]\right).$$

Therefore, the betting functions for the specialized Bayes–Kelly CTM are

$$f_n(p) = \sum_{k=0}^{n-1}\sum_{L=0}^{1} w_{k,L}^{n-1}\left(\frac{n}{k+1}\pi_{1|L}1_{p\le\frac{k+1}{n}} + \frac{n}{n-k}\pi_{0|L}1_{p\ge\frac{k}{n}}\right). \tag{9.40}$$

The procedure is summarized as Algorithm 9.6. Alternatively, in line 4 we could have summed the unnormalized weights for step n to obtain S_n.

For experimental results, see Fig. 9.9. The Bayes–Kelly CTM appears to be very close to the two benchmarks.

As in the case of the changepoint alternative, we now consider a radical simplification of the Bayes–Kelly CTM (9.40). We still assume that the Markov chain is symmetric, as in our model situations. (This assumption will be relaxed in the next subsection.) If we assume that the weights $w_{k,L}^n$, $k = 0, \ldots, n$, are concentrated at

$$k \approx k + 1 \approx n/2$$

and set

$$L := \begin{cases} 1 & \text{if } p_{n-1} \le 0.5 \\ 0 & \text{if not}, \end{cases} \tag{9.41}$$

(9.40) will simplify to

$$f_n(p) = 2\pi_{1|L}1_{p\le 0.5} + 2\pi_{0|L}1_{p>0.5}, \tag{9.42}$$

with L defined by (9.41). Figure 9.10 shows the weights (averaged over $L \in \{0, 1\}$) for the last step of the Bayes–Kelly conformal test martingale for 1000 examples. They are indeed concentrated around values of k not so different from $0.5N = 500$;

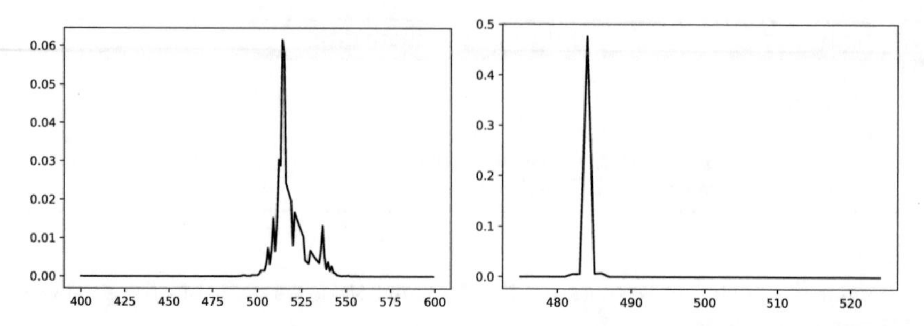

Fig. 9.10 The weights $w_k^{1000} := (w_{k,0}^{1000} + w_{k,1}^{1000})/2$, with the range of k indicated on the horizontal axis (the weights are virtually zero outside this range) at the last step for the Bayes–Kelly CTM (the hard case on the left and easy on the right)

Algorithm 9.7 Simplified Bayes–Kelly $((p_1, p_2, \dots) \mapsto (S_1, S_2, \dots))$

1: $S_0 := S_1 := 1$
2: **for** $n = 2, 3, \dots$
3: **if** $p_{n-1} \leq 0.5$: $L := 1$
4: **else**: $L := 0$
5: **if** $p_n \leq 0.5$: $S_n := 2\pi_{1|L} S_{n-1}$
6: **else**: $S_n := 2\pi_{0|L} S_{n-1}$.

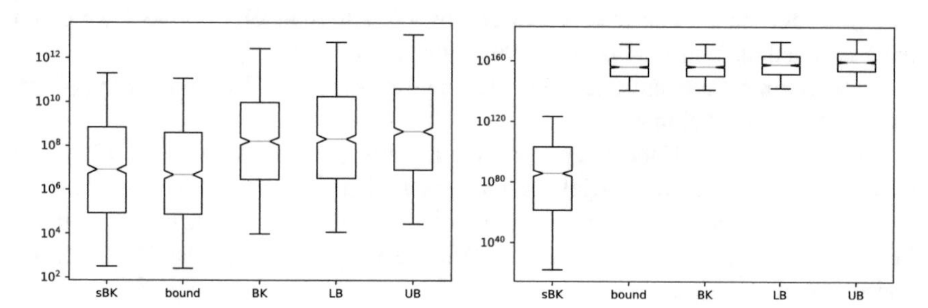

Fig. 9.11 Boxplots based on 10^3 runs for the final values of the two benchmarks (upper, UB, and lower, LB), the Bayes–Kelly conformal test martingale (BK), its lower bound (9.21), and its simplified version (sBK) in the Markov case. Left panel: hard case. Right panel: easy case

this is because $B_{0.5}$ is the stationary distribution in both hard and easy cases, both being symmetric. If $k(n-1) := z_1 + \cdots + z_{n-1} \approx (n-1)/2$, then the expression on the right-hand side of (9.41) is equal to z_{n-1} with high probability, which justifies setting (9.41). The procedure is summarized as Algorithm 9.7. The performance of the simplified version is also shown in Fig. 9.9. It is significantly worse than that of the Bayes–Kelly CTM and the two benchmarks.

Figure 9.11 once again shows that the statistical performance of the simplified Bayes–Kelly CTM particularly suffers in the easy case. It should be mentioned, however, that the simplified version catches up for a larger number of examples.

Figure 9.12 illustrates the work of the simplified Bayes–Kelly CTM: we can see that each p-value carries a lot of information about the following p-value in the easy

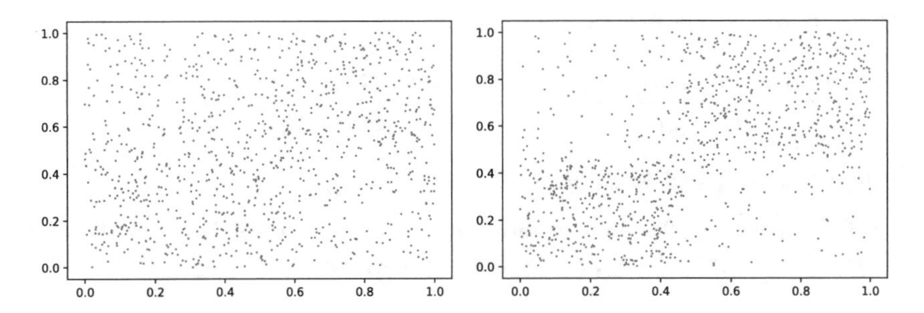

Fig. 9.12 The scatterplot of $10^3 - 1$ pairs (p_{n-1}, p_n) of consecutive p-values, $n = 2, \ldots, 10^3$, in the hard (left panel) and easy (right panel) cases

case. In the hard case, predicting the next p-value given the current one is much trickier: the density of the points in the four squares $[0, 0.5]^2$, $[0, 0.5] \times [0.5, 1]$, $[0.5, 1] \times [0, 0.5]$, $[0.5, 1]^2$ is not noticeably different.

9.2.6 Asymmetric Markov Alternatives

So far we have considered symmetric Markov chains as alternatives, i.e., the case $\forall i, j : \pi_{i|j} = \pi_{j|i}$. Now we do not assume symmetry and only assume $\min_{i,j} \pi_{i|j} > 0$; in particular, the Markov chain is aperiodic and irreducible [261]. We still assume that the initial distribution of the Markov chain is uniform (although Proposition 9.11 below only needs the initial distribution to be positive, i.e., both probabilities, for 0 and for 1, to be positive).

The definition of the lower benchmark (9.35) still works in the asymmetric case, but in the definition of the upper benchmark (9.36) we replace Ber(0.5) in the denominator by Ber(π_1), where π_1 is the probability of 1 under the stationary distribution for the Markov chain. By definition, the stationary distribution (π_0, π_1), where π_0 is the probability of 0, satisfies

$$\begin{cases} \pi_{0|0}\pi_0 + \pi_{0|1}\pi_1 = \pi_0 \\ \pi_{1|0}\pi_0 + \pi_{1|1}\pi_1 = \pi_1 \ . \end{cases} \tag{9.43}$$

By the ergodic theorem [261, Theorem 1.10.2], this choice of the denominator for the likelihood ratio process makes the upper benchmark as close to the lower benchmark as possible asymptotically. The following proposition says that this choice of the denominator is asymptotically optimal.

Proposition 9.11 *For any* $x \in (0, 1) \setminus \{\pi_1\}$,

$$\frac{\text{Markov}(\pi_{1|0}, \pi_{1|1})([z_1, \ldots, z_n])}{\text{Ber}(x)([z_1, \ldots, z_n])} > \frac{\text{Markov}(\pi_{1|0}, \pi_{1|1})([z_1, \ldots, z_n])}{\text{Ber}(\pi_1)([z_1, \ldots, z_n])}$$

from some n *on almost surely under* $\text{Markov}(\pi_{1|0}, \pi_{1|1})$.

Proof We have, by the ergodic theorem and strong law of large numbers for martingales (see Sect. A.6.2), almost surely as $n \to \infty$,

$$\begin{aligned}
&\frac{1}{n} \log \frac{\text{Markov}(\pi_{1|0}, \pi_{1|1})([z_1, \ldots, z_n])}{\text{Ber}(x)([z_1, \ldots, z_n])} \\
&= \pi_{0|0}\pi_0 \log \frac{\pi_{0|0}}{1-x} + \pi_{1|0}\pi_0 \log \frac{\pi_{1|0}}{x} + \pi_{0|1}\pi_1 \log \frac{\pi_{0|1}}{1-x} + \pi_{1|1}\pi_1 \log \frac{\pi_{1|1}}{x} + o(1) \\
&= \pi_{0|0}\pi_0 \log \frac{1}{1-x} + \pi_{1|0}\pi_0 \log \frac{1}{x} + \pi_{0|1}\pi_1 \log \frac{1}{1-x} + \pi_{1|1}\pi_1 \log \frac{1}{x} + c + o(1) \\
&= \pi_0 \log \frac{1}{1-x} + \pi_1 \log \frac{1}{x} + c + o(1) > \pi_0 \log \frac{1}{\pi_0} + \pi_1 \log \frac{1}{\pi_1} + c + o(1) \\
&= \frac{1}{n} \log \frac{\text{Markov}(\pi_{1|0}, \pi_{1|1})([z_1, \ldots, z_n])}{\text{Ber}(\pi_1)([z_1, \ldots, z_n])} + o(1) ,
\end{aligned}$$

where c is a constant (depending only on the πs), the penultimate "=" follows from (9.43), and the last inequality, ">", disregards the $o(1)$ terms and follows from the positivity of the Kullback–Leibler distance in this context. □

The Bayes–Kelly CTM (Algorithm 9.6) also works for asymmetric Markov chains. Let us derive the simplified Bayes–Kelly CTM (Algorithm 9.7) in the non-symmetric case. The solution to (9.43) is

$$\pi_1 = \frac{\pi_{1|0}}{\pi_{1|0} + \pi_{0|1}} .$$

When $k \approx k + 1 \approx n\pi_1$ and

$$L := \begin{cases} 1 & \text{if } p_{n-1} \leq \pi_1 \\ 0 & \text{if not ,} \end{cases}$$

(9.40) will lead to

$$f_n(p) = \frac{\pi_{1|L}}{\pi_1} 1_{p \leq \pi_1} + \frac{\pi_{0|L}}{\pi_0} 1_{p > \pi_1}$$

in place of (9.42).

9.2.7 Mixing Over All Markov Alternatives

So far we have assumed the parameters $(\pi_{1|0}, \pi_{1|1})$ of the alternative distribution known. As discussed earlier, a simple solution to the problem of unknown parameters is to mix Algorithm 9.6 over a grid of parameters such as (9.34). This is what we called martingale averaging in Sect. 9.2.2.

In this subsection, we will also consider distribution averaging, also discussed in Sect. 9.2.2. Let us mix all Markov models Markov$(\pi_{1|0}, \pi_{1|1})$ with initial distribution $(0.5, 0.5)$ over $\pi_{1|0}$ and $\pi_{1|1}$ independent and each having beta distribution with parameters $(0.5, 0.5)$. The explicit formula for this mixture P, which we call the *Jeffreys mixture* following [287], is

$$P([z_1, \ldots, z_n]) := \frac{\Gamma(n_{0|0} + 1/2)\Gamma(n_{0|1} + 1/2)\Gamma(n_{1|0} + 1/2)\Gamma(n_{1|1} + 1/2)}{2\Gamma(1/2)^4\Gamma(n_{0|0} + n_{1|0} + 1)\Gamma(n_{0|1} + n_{1|1} + 1)}, \tag{9.44}$$

where $n_{i|j}$ is the number of times i follows j in the binary sequence $z_1, \ldots, z_n$. It should be said, however, that the prior distribution over $(\pi_{1|0}, \pi_{1|1})$ that we are using is not the true Jeffreys distribution for the Markov model: see [354]. Our prior is the product of two independent Jeffreys priors for the Bernoulli model (see, e.g., [141, Example 8.3]); its advantage is the simple expression (9.44) for the mixture.

Applying the Bayes–Kelly algorithm to Jeffreys mixture (9.44), we will obtain what we call the Bayes–Kelly–Jeffreys (BKJ) algorithm. It will not use (9.44) directly, and we will only need the conditional probabilities that (9.44) implies:

$$P(0 \mid \ldots 0) = \frac{n_{0|0} + 1/2}{n_{0|0} + n_{1|0} + 1} \tag{9.45}$$

$$P(0 \mid \ldots 1) = \frac{n_{0|1} + 1/2}{n_{0|1} + n_{1|1} + 1} \tag{9.46}$$

$$P(1 \mid \ldots 0) = \frac{n_{1|0} + 1/2}{n_{0|0} + n_{1|0} + 1} \tag{9.47}$$

$$P(1 \mid \ldots 1) = \frac{n_{1|1} + 1/2}{n_{0|1} + n_{1|1} + 1} \tag{9.48}$$

where $P(i \mid \ldots j)$ stands form the conditional probability that $z_{n+1} = i$ given the first n examples $z_1, \ldots, z_n$ with $z_n = j$. These expressions are very natural and in the spirit of Laplace's rule of succession (cf. Sect. 6.2.3).

An explicit algorithm is shown as Algorithm 9.8. The algorithm uses the likelihood $f^n_{k,L}(p)$ given by (9.15). Its main data structure is an infinite array $w_{F,L,C,\nu}$ of weights assigned to types of binary sequences. In line 42 we update the weights, in line 43 we use the unnormalized weights to compute the relative increase of the martingale at the nth step, and in line 44 we normalize the weights.

Algorithm 9.8 Bayes–Kelly–Jeffreys (BKJ) martingale

1: $S_0 := S_1 := 1$ ▷ no nontrivial gambling on step 1
2: **for** $n = 2, 3, \dots$
3: $\tilde{w} :=$ zeros$(2, 2, \lceil n/2 \rceil, n)$ ▷ unnormalized weights $\tilde{w}_{F,L,C,\nu}$
4: **if** $n = 2$
5: $\tilde{w}_{0,0,0,1} := 0.25 f^n_{0,0}(p_n)$ ▷ defining $\tilde{w}_{F,L,C,\nu}$ in terms of the prior weights
6: $\tilde{w}_{0,1,0,0} := 0.25 f^n_{1,1}(p_n)$ ▷ and the likelihood $f^n_{k,L}(p)$
7: $\tilde{w}_{1,0,0,0} := 0.25 f^n_{1,0}(p_n)$
8: $\tilde{w}_{1,1,0,0} := 0.25 f^n_{2,1}(p_n)$
9: **else**
10: **for** $F \in \{0, 1\}$ ▷ F is the first bit
11: **for** $L \in \{0, 1\}$ ▷ L is the last bit
12: **if** n is even
13: $C_{\max} := n/2 - 1$
14: **else**
15: $C_{\max} := (n-1)/2 - |F - L|$
16: **for** $C \in \{0, \dots, C_{\max}\}$ ▷ C is the complexity
17: $\nu_{\min} := 0$
18: **if** $C = 0$ and $F = 0$ and $L = 0$
19: $\nu_{\min} := \nu_{\max} := n - 1$
20: **elif** $C = 0$ and $F = 1$ and $L = 1$
21: $\nu_{\max} := 0$
22: **else**
23: $\nu_{\max} := n - 1 - 2C - |F - L|$
24: **for** $\nu \in \{\nu_{\min}, \dots, \nu_{\max}\}$ ▷ ν is the number of transitions $n_{0|0}$
25: $n_{0|0} := \nu, \quad n_{0|1} := n_{1|0} := C$
26: **if** $F = 0$ and $L = 1$
27: $n_{1|0} := C + 1$
28: **if** $F = 1$ and $L = 0$
29: $n_{0|1} := C + 1$
30: $n_{1|1} := n - 1 - n_{0|0} - n_{0|1} - n_{1|0}$
31: $k = F + n_{1|0} + n_{1|1}$ ▷ k is the number of 1s
32: **for** $L^* \in \{0, 1\}$ ▷ L^* is the penultimate bit
33: **if** $n_{L|L^*} = 0$ ▷ such L^* is impossible
34: continue ▷ go to the next iteration
35: $C^* := C$ and $\nu^* := \nu$
36: **if** $L^* \neq L$
37: **if** $C^* = n_{L|L^*}$
38: $C^* -= 1$
39: **else**
40: **if** $L^* = L = 0$
41: $\nu^* := \nu - 1$
42: $\tilde{w}_{F,L,C,\nu} \mathrel{+}= w_{F,L^*,C^*,\nu^*} \frac{n_{L|L^*} - 0.5}{n_{0|L^*} + n_{1|L^*}} f^n_{k,L}(p_n)$
43: $S_n := \text{sum}(\tilde{w}) S_{n-1}$ ▷ the new value of the martingale
44: $w := \tilde{w}/\text{sum}(\tilde{w})$. ▷ normalizing the weights

In the following several paragraphs we will go over the pseudocode of Algorithm 9.8. The type of a binary sequence $z_1, \dots, z_n$ (intuitively, a path of a Markov chain) of a known length $n \in \{1, 2, \dots\}$ is defined as the quadruple of the following numbers:

- The first example $F := z_1$.
- The last example $L := z_n$.
- The "complexity" $C := \min(n_{0|1}, n_{1|0})$.
- Finally, $\nu := n_{0|0}$.

The weight $w_{F,L,C,\nu} = w^n_{F,L,C,\nu}$ is defined to be the posterior probability after observing $p_1, \dots, p_n$ of the type of the first n examples $z_1, \dots, z_n$ chosen by Reality being (F, L, C, ν). (It is also instructive to think of $w_{F,L,C,\nu} = w^n_{F,L,C,\nu}$ as the total posterior normalized capital after observing $p_1, \dots, p_n$ of the martingales $S^{z_1,z_2,\dots}_n$, as defined in Sect. 9.4.3 below, for $z_1, z_2, \dots$ such that the type of $z_1, \dots, z_n$ is (F, L, C, ν).)

In lines 10–31 we go over the valid types (F, L, C, ν) (i.e., over the quadruples (F, L, C, ν) that are the type of at least one binary sequence of length n). First of all, the complexity $C := \min(n_{0|1}, n_{1|0})$ takes values $0, \dots, C_{\max}$, where:

- when n is even, $C_{\max} := n/2 - 1$;
- when n is odd and $F = L$, $C_{\max} := (n - 1)/2$ (which can also be written as $C_{\max} := (n - 1)/2 - |F - L|$);
- when n is odd and $F \neq L$, $C_{\max} := (n - 1)/2 - 1$ (which can also be written as $C_{\max} := (n - 1)/2 - |F - L|$).

The range of $\nu = n_{0|0}$ is determined as follows:

- If $F = L = C = 0$, then $n_{0|0} = n - 1$.
- If $F = L = 1$ and $C = 0$, then $n_{0|0} = 0$.
- Otherwise, $n_{0|0}$ ranges between 0 and $\nu_{\max}$, where

$$\nu_{\max} := n - 1 - 2C - |F - L| \ .$$

To find the numbers of transitions $n_{0|1}$ and $n_{1|0}$:

- if $F = L$, $n_{0|1} = n_{1|0} = C$,
- if $F = 0$ and $L = 1$, $n_{0|1} = C$ and $n_{1|0} = C + 1$,
- if $F = 1$ and $L = 0$, $n_{0|1} = C + 1$ and $n_{1|0} = C$.

After that, we can find

$$n_{1|1} = n - 1 - n_{0|0} - n_{0|1} - n_{1|0}$$

since the total number of all transitions $n_{i|j}$ is $n - 1$. And the number of 1s

$$k = F + n_{1|0} + n_{1|1}$$

is essentially composed of the 1s following a 0 and the 1s following a 1.

In lines 5–8 we initialize the unnormalized weights $\tilde{w}$ globally; $n = 2$ is the first nontrivial step. At each step n we also have local initialization. Namely, in line 3 we initialize the unnormalized weights $\tilde{w}$ for this step. The Python-type notation "zeros" stands for an array of the given dimensions filled with all 0 and with the indices starting from 0; e.g., zeros(a, b) is an $a \times b$ all-0 array with indices $(i, j) \in \{0, \dots, a-1\} \times \{0, \dots, b-1\}$. As we can see from lines 12–23, the tight upper bounds for the last two indices of the weights $w_{F,L,C,\nu}$ of the valid types of sequences of examples at step n are

$$C_{\max} \le \lceil n/2 \rceil - 1$$

$$\nu_{\max} \le n - 1 .$$

Therefore, at this step, the weights may be non-zero only for $C \le \lceil n/2 \rceil - 1$ and $\nu \le n - 1$, so that the corresponding dimensions are $\lceil n/2 \rceil$ and n, respectively, as in line 3.

In line 42, $a \mathrel{+}= b$ means $a := a + b$; we also use $a \mathrel{-}= b$ to mean $a := a - b$ in line 38.

In line 42 we compute the unnormalized weights for step n. These weights are coming from the weights from the previous step, and we have to compute the possible types (F, L^*, C^*, ν^*) of the path at the previous step. The key element of this computation is the loop in line 32 over the possible penultimate bits L^* of the current path. Once we know L^*, we can compute C^* and ν^*. The formulas for Jeffreys's update in line 42 look different from those in (9.45)–(9.48) because the latter should be applied to the path on the previous step.

In line 44, we normalize the weights. The notation sum$(\tilde{w})$ stands for the sum of all elements of the array $\tilde{w}$. In line 44 the elements of the finite array $\tilde{w}$ are normalized and written to the locations with the same indices of the infinite array w. In line 43 we use sum$(\tilde{w})$ as the value of the betting function at step n.

At step n, Algorithm 9.8 requires $O(n^2)$ memory, mainly for storing the weights, and its running time is $O(n^2)$, since each weight needs to be processed. If we want to run it for N steps, we need $O(N^2)$ memory, and the running time is $O(N^3)$.

Figure 9.13 demonstrates the performance of the Bayes–Kelly–Jeffreys martingale and three other processes, as described in the caption. In the easy case, all four graphs are very close to each other, and for the hard case the Bayes–Kelly–Jeffreys martingale performs even better than the lower benchmark for our default seed for the random number generator.

9.3 Testing the Validity of Putative Test Martingales

In this section we will discuss a potential pitfall of conformal testing and a useful sanity check. As we mentioned when discussing the right panel of Fig. 9.6, one

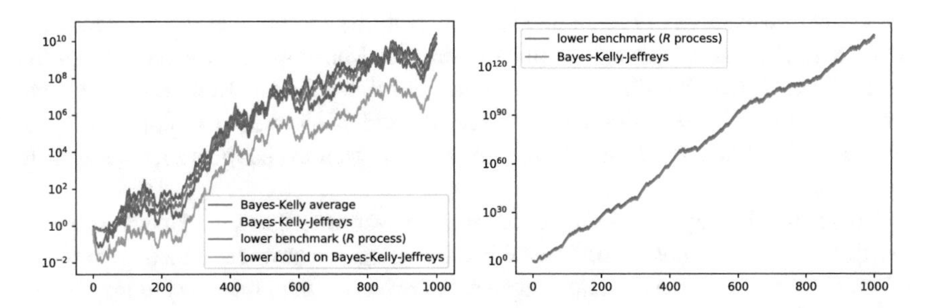

Fig. 9.13 The BKJ martingale, the lower benchmark (which now coincides with the process R_n introduced in [287]), the lower bound (9.21) on the BKJ martingale in terms of the lower benchmark, and the mixture (indicated as "Bayes–Kelly average" in the legend) of Algorithm 9.6 over the grid (9.34) with $G = 10$. The hard case is on the left and the easy case is on the right. In the easy case the bound and the mixture are not shown as they are indistinguishable from the BKJ martingale

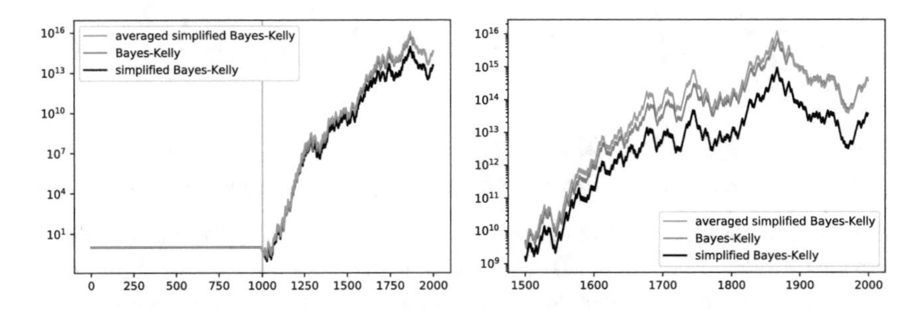

Fig. 9.14 The averaged simplified Bayes–Kelly martingale in the change detection problem. Left panel: over all examples. Right panel: over the last 500

disadvantage of the simplified Bayes–Kelly martingale is its high volatility; its dependence on the random numbers τ used in computing the p-values is significant. An easy way of reducing volatility is averaging (familiar in machine learning from the method of bagging), and a natural idea is to average several realizations of the simplified Bayes–Kelly martingale for a given dataset. An additional boon of averaging is that different realizations often lead to vastly different results (on the linear rather than log scale), and in this situation averaging is akin to taking the maximum.

Figure 9.14 reproduces the paths, given in Fig. 9.4, of the Bayes–Kelly martingale and its simplified version complementing it with the averaged (over 1000 simulations) simplified Bayes–Kelly martingale. All three processes use the same dataset but independent streams of random numbers τ used in computing the p-values. The simplified Bayes–Kelly martingale improves significantly as result of averaging and performs similarly to the Bayes–Kelly martingale. This looks too good to be true for a valid method.

In fact, the average of randomized exchangeability martingales is not guaranteed to be a randomized exchangeability martingale. Martingales in a fixed filtration form a linear space, but different runs of the simplified Bayes–Kelly martingale are martingales in different filtrations, as discussed earlier. Figure 9.14 suggests that the averaged simplified Bayes–Kelly martingale is not an exchangeability martingale any more.

Therefore, it may be useful to be able to test whether a given process is a martingale. It is also potentially useful in debugging: we have theoretical guarantees of validity for conformal test martingales, but even for them mistakes in implementation are always possible. The testing method of this section will use the following large deviations inequality based on Doléans's supermartingale of [320, Sect. 3.2].

Proposition 9.12 *Let M be a positive integer and $F_1, \ldots, F_K$, $K \geq 2^{2M}$, be independent nonnegative random variables with expected value 1. Then the expected value of*

$$E := \frac{1}{M} \sum_{m=1}^{M} \exp\left(K^{1-m/2M}(\bar{F} - 1) - K^{-m/M} \sum_{k=1}^{K} (F_k - 1)^2 \right), \tag{9.49}$$

where $\bar{F} := \frac{1}{K}\sum_{k=1}^{K} F_k$ is the average of the F_k, is at most 1. Besides, $\frac{1}{E} \wedge 1$ is a p-value, in the sense of the probability of $\frac{1}{E} \wedge 1 \leq \epsilon$ being at most ϵ for any $\epsilon > 0$.

Proof The statement about the expected value of (9.49) being at most 1 follows from the right-hand side of (9.49) being the final value of a test supermartingale (i.e., a nonnegative supermartingale with initial value 1), namely an average of Doléans supermartingales. See [320, Proposition 3.4], whose applicability depends on our assumption $K \geq 2^{2M}$, which implies $K^{-m/2M} \leq 0.5$. The statement about $\frac{1}{E} \wedge 1$ being a p-value follows from Markov's inequality. □

Values taken by nonnegative random variables whose expectation does not exceed 1 are known as *e-values*. A large e-value is interpreted as evidence against our postulated stochastic mechanism (the null hypothesis). According to Jeffreys [177, Appendix B] (who used different terminology, of course), the evidence is strong when the e-value exceeds 10 and, as already mentioned, decisive when it exceeds 100. We will often use Proposition 9.12 in the form of the following inequality.

Corollary 9.13 *Let M be a positive integer, let $F_1, \ldots, F_K$, $K \geq 2^{2M}$, be independent nonnegative random variables with expected value 1, and let $\epsilon > 0$. Define $X > 0$ as the only solution to*

$$\sum_{m=1}^{M} \exp\left(K^{1-m/2M} X - K^{-m/M} \sum_{k=1}^{K} (F_k - 1)^2 \right) = \frac{M}{\epsilon} \tag{9.50}$$

(the left-hand side is strictly increasing in X). Then

$$\mathbb{P}\left(\frac{1}{K}\sum_{k=1}^{K} F_k < 1 + X\right) \geq 1 - \epsilon \,. \tag{9.51}$$

Proof If the inner inequality in (9.51) is violated, we will have

$$\frac{1}{M}\sum_{m=1}^{M} \exp\left(K^{-m/2M}\sum_{k=1}^{K}(F_k - 1) - K^{-m/M}\sum_{k=1}^{K}(F_k - 1)^2\right) \geq \frac{1}{\epsilon}$$

instead of (9.50). The probability of this event is at most ϵ since the reciprocal to (9.49) is a p-value. □

Let us use $M := 5$ (which requires $K \geq 1024$). For a few sets of values for (N_0, N_1) and (π_0, π_1), Table 9.1 gives some statistics for the final values F_k of the simplified Bayes–Kelly CTM (fed with independent sequences of examples and random numbers τ) based on the betting functions (9.31) designed for the pre-/post-change parameters (π_0, π_1) but run on the IID data with parameter π_0; the numbers of pre- and post-change examples are N_0 and N_1, respectively. The bound is as given by Corollary 9.13 with $\epsilon := 0.01$ (this value is fixed throughout the rest of this section). The closeness of the means and bounds to 1 suggests that the processes are really test martingales. Of course, the bound is never exceeded by the actual mean.

Table 9.2 is analogous to Table 9.1 but gives statistics for the average over 10^3 simulations (using independent streams of random numbers τ) of the simplified Bayes–Kelly CTM. The means are still close to 1 and do not exceed the bounds. Unfortunately, this kind of statistics does not allow us to check deviations of the average CTM from being a martingale, since the expectation of the final value of the average is still 1.

The method that we have used so far can be modified to enable it to check the martingale property, and it will show that the average CTM is not a martingale itself (under the null hypothesis of IID data). Let S_n be an average CTM; it will be

Table 9.1 The mean $\frac{1}{K}\sum_k F_k$, its upper bound in (9.51) (with $\epsilon := 0.01$), and the median and interquartile range of $F_1, \ldots, F_K$

(π_0, π_1)	(N_0, N_1)	K	Mean	Bound	Median	Quartiles
(0.1, 0.4)	(10, 10)	10^9	0.99993	1.00054	0.33016	[0.13964, 0.84562]
(0.4, 0.5)	(10, 10)	10^9	1.00000	1.00008	0.89615	[0.66667, 1.21212]
(0.4, 0.5)	(100, 100)	10^9	0.99985	1.00040	0.36630	[0.14232, 0.94952]

Table 9.2 The analogue of Table 9.1 for the average CTM over 10^3 simulations

(π_0, π_1)	(N_0, N_1)	K	Mean	Bound	Median	Quartiles
(0.1, 0.4)	(10, 10)	10^6	0.99894	1.00570	0.67879	[0.38007, 1.37617]
(0.4, 0.5)	(10, 10)	10^6	1.00007	1.00207	0.94866	[0.74567, 1.15930]
(0.4, 0.5)	(100, 100)	10^6	0.99972	1.00994	0.43602	[0.17872, 1.06452]

Table 9.3 Statistics for the conditional validity of the average CTM with $(\pi_0, \pi_1) = (0.1, 0.4)$, as described in the main text

K^*	K	A	Mean	Bound	Median	Quartiles
10^6	482,311	10^3	1.00426	1.00101	0.99580	[0.89924, 1.00682]
10^9	400,000,071	1	1.00001	1.00007	0.83333	[0.83333, 1.42857]
10^9	447,299,138	10	1.00266	1.00005	0.96172	[0.88585, 1.06718]
10^9	470,992,540	10^2	1.00353	1.00005	0.98566	[0.91111, 1.02118]
10^9	482,226,950	10^3	1.00452	1.00004	0.99589	[0.89931, 1.00684]

assumed positive. The defining property of a martingale is $\mathbb{E}(S_n \mid S_1, \ldots, S_{n-1}) = S_{n-1}$ (almost surely; see (8.3)). The method that we have used tests the crude implication $\mathbb{E}(S_n) = 1$ of the defining property, which we know to hold for an average of martingales; the modification will test $\mathbb{E}(S_n \mid S_{n-1}) = S_{n-1}$, i.e., $\mathbb{E}(S_n/S_{n-1} \mid S_{n-1}) = 1$.

Table 9.3 summarizes a case where $\mathbb{E}(S_n/S_{n-1} \mid S_{n-1} \geq 1) > 1$ (so that S possesses a momentum: a rise in the value of S creates a tendency to a further rise). The CTM is the one with the betting functions (9.31), where $N_0 := 2$ and $(\pi_0, \pi_1) = (0.1, 0.4)$; it is averaged over A simulations (with independent streams of random numbers τ). The value of K is the number of runs (using independent sequences of examples) of the average CTM with $S_{n-1} \geq 1$, where $n := 5$. These runs are selected from K^* runs by discarding the runs leading to $S_{n-1} < 1$. The mean, median, and quartiles are those of S_n/S_{n-1} over the K selected runs. We can see that the bound is exceeded by the actual mean except for the case where $A = 1$ (and so there is no averaging). The mean mostly depends on A, and the bound on K.

To get an idea of how serious the violation of the bounds in Table 9.3 is, we can apply Proposition 9.12 directly. The p-values computed using Proposition 9.12 from Table 9.3 are tiny, except, of course, for the second row, where the e-value (9.49) is 0.25 and the p-value is 1. Even for the top row, the p-value is below 10^{-44}.

9.4 Proofs and Discussions

9.4.1 Proof of Proposition 9.2

The first inequality in (9.3) follows from each power distribution on Ω being exchangeable. If E contains either the all-0 sequence $0 \ldots 0$ or the all-1 sequence $1 \ldots 1$, the second inequality in (9.3) is obvious ($\mathbb{P}^{\text{rand}}(E) = \mathbb{P}^{\text{exch}}(E) = 1$). If E is empty, it is also obvious ($\mathbb{P}^{\text{rand}}(E) = \mathbb{P}^{\text{exch}}(E) = 0$). Finally, if E is nonempty and contains neither sequence, we have, for some $k \in \{1, \ldots, N-1\}$,

$$\mathbb{P}^{\text{exch}}(E) = \mathbb{P}^{\text{exch}}(E \cap \Omega_k) = \frac{1/\binom{N}{k}}{(k/N)^k (1-k/N)^{N-k}} \mathbb{P}^{\text{rand}}(E \cap \Omega_k) \tag{9.52}$$

$$\leq \frac{k!(N-k)!N^N}{N!k^k(N-k)^{N-k}}\mathbb{P}^{\text{rand}}(E) \leq \sqrt{2\pi}e^{1/6}\sqrt{\frac{k(N-k)}{N}}\mathbb{P}^{\text{rand}}(E) \tag{9.53}$$

$$\leq (\sqrt{2\pi}e^{1/6}/2)\sqrt{N}\mathbb{P}^{\text{rand}}(E) \leq 1.5\sqrt{N}\mathbb{P}^{\text{rand}}(E) , \tag{9.54}$$

where Ω_k is the set of all sequences in Ω containing k 1s (and, in this context, π is the well-known mathematical constant). The first equality in (9.52) follows from each exchangeable probability measure on Ω being a convex mixture of the uniform probability measures on Ω_k, $k = 0, \dots, N$. The second equality in (9.52) follows from the maximum of $\mathbf{B}_p(\{\omega\})$, $\omega \in \Omega_k$, over $p \in [0, 1]$ being attained at $p = k/N$. The first inequality in (9.53) is equivalent to the obvious $\mathbb{P}^{\text{rand}}(E \cap \Omega_k) \leq \mathbb{P}^{\text{rand}}(E)$. The second inequality in (9.53) follows from Stirling's formula

$$n! = \sqrt{2\pi}n^{n+1/2}e^{-n}e^{r_n}, \quad 0 < r_n < \frac{1}{12n} ,$$

valid for all $n \in \mathbb{N}$; see, e.g., [290], where it is also shown that $r_n > \frac{1}{12n+1}$. The first inequality in (9.54) follows from $\max_{p\in[0,1]} p(1-p) = 1/4$.

9.4.2 Proof of Proposition 9.5

First we check the left inequality in (9.7) (which strengthens Lemma 9.4, in view of the first inequality in (9.3)). We will do even more: we will check that it remains true even if the right-hand side of (9.5) is replaced by

$$\inf\{S_0 : \forall(z_1, \dots, z_N) \in E : \mathbb{E}_\tau S_N(z_1, \tau_1, z_2, \tau_2, \dots) \geq 1\} ,$$

where $\mathbb{E}_\tau$ refers to the uniform probability measure $\mathbf{U}^\infty$ over $(\tau_1, \tau_2, \dots) \in [0, 1]^\infty$. Notice that $S_0, \dots, S_N$ in (9.5) is a martingale in the filtration $(\mathcal{G}_n)$ generated by the p-values $p_1, \dots, p_N$ under any exchangeable probability measure on Ω; this follows from the fact that $p_1, \dots, p_N$ are independent and uniform on [0, 1] under any exchangeable probability measure (see Proposition 2.11). Therefore, for each $E \subseteq \Omega$ and each CTM S such that $\mathbb{E}_\tau S_N \geq 1_E$, we have

$$\mathbb{P}_z(E) \leq \mathbb{P}_z(\mathbb{E}_\tau S_N \geq 1) \leq \mathbb{E}_z(\mathbb{E}_\tau S_N) = \mathbb{E}_{z,\tau} S_N = S_0 , \tag{9.55}$$

where $\mathbb{P}_z$ refers to $(z_1, \dots, z_N) \sim P$, P is an exchangeable probability measure on Ω, $\mathbb{E}_\tau$ refers to $(\tau_1, \dots, \tau_N) \sim \mathbf{U}^N$, $\mathbf{U}$ is the uniform probability measure on [0, 1], and $\mathbb{E}_{z,\tau}$ refers to $(z_1, \dots, z_N) \sim P$ and $(\tau_1, \dots, \tau_N) \sim \mathbf{U}^N$ independently. Taking the sup of the leftmost expression in (9.55) over P and the inf of the rightmost expression in (9.55) over S, we obtain the left-hand inequality in (9.7).

It remains to check the right-hand inequality in (9.7). Let us first check the part "$\leq$" of the first equality in

$$\mathbb{P}^{\text{conf}}(\{\omega\}) = \frac{k!(N-k)!}{N!} = \mathbb{P}^{\text{exch}}(\{\omega\}) ,$$

where $k \in \{0, \dots, N\}$ and $\omega \in \Omega$ contains k 1s (the part "$\geq$" was established in the previous paragraph; it will not be used in the rest of this proof).

Let $\omega = (z_1, \dots, z_N)$ be the representation of ω as a sequence of bits. Consider the CTM S^ω obtained from the identity nonconformity measure and the scaled betting martingale F such that $F(\Box) = 1/\binom{N}{k}$ (where $\Box$ is the empty sequence) and

$$\frac{F(p_1,\dots,p_{n-1},p_n)}{F(p_1,\dots,p_{n-1})} := \begin{cases} \frac{n}{k_n} & \text{if } p_n \le k_n/n \text{ and } z_n = 1 \\ \frac{n}{n-k_n} & \text{if } p_n \ge k_n/n \text{ and } z_n = 0 \\ 0 & \text{otherwise ,} \end{cases} \tag{9.56}$$

where $n = 1, \dots, N$ and k_n is the number of 1s in ω observed so far,

$$k_n := \left|\{j \in \{1,\dots,n\} : z_j = 1\}\right| ;$$

in particular, $k_N = k$. The numerator of the left-hand side of (9.56) is defined to be 0 when the denominator is 0. Intuitively, S^ω gambles recklessly on the nth example being z_n. If the actual sequence of examples chosen by Reality happens to be ω, on step n the value of the martingale S^ω is multiplied, a.s., by the fraction whose numerator is n and whose denominator is the number of bits z_n observed in ω so far. The product of all these fractions over $n = 1, \dots, N$ will have $N!$ as its numerator and $k!(N-k)!$ as its denominator. This CTM is almost deterministic, in the sense of not depending on τ_n provided $\tau_n \notin \{0, 1\}$, and its final value on ω is, a.s.,

$$\frac{1}{\binom{N}{k}} \frac{N!}{k!(N-k)!} = 1 .$$

To move from singletons to arbitrary $E \subseteq \Omega$, notice that a finite linear combination of CTMs S^ω with positive coefficients is again a CTM, since the component CTMs involve the same nonconformity measure, and betting martingales can be combined. Fix $E \subseteq \Omega$ and remember that Ω_k is the set of all sequences in Ω containing k 1s. Represent E as the disjoint union

$$E = \bigcup_{k=0}^{N} E_k, \quad E_k \subseteq \Omega_k ,$$

and let $\mathbf{U}_k$ be the uniform probability measure on Ω_k. We then have

$$\begin{aligned} \mathbb{P}^{\text{conf}}(E) &\le \sum_{\omega\in E} \mathbb{P}^{\text{conf}}(\{\omega\}) = \sum_{k=0}^{N} \sum_{\omega\in E_k} \mathbb{P}^{\text{conf}}(\{\omega\}) = \sum_{k=0}^{N} \sum_{\omega\in E_k} \mathbb{P}^{\text{exch}}(\{\omega\}) \\ &= \sum_{k=0}^{N} \mathbf{U}_k(E_k) \le N \max_{k=0,\dots,N} \mathbf{U}_k(E_k) = N\mathbb{P}^{\text{exch}}(E) , \end{aligned}$$

where the last inequality holds when, e.g., E does not contain the all-0 sequence $0\dots0 \in \Omega$. If E does contain the all-0 sequence, it is still true that

$$\mathbb{P}^{\text{conf}}(E) \le 1 \le N = N\mathbb{P}^{\text{exch}}(E) .$$

9.4.3 An Alternative Interpretation of the Bayes–Kelly Algorithm and the Proof of Proposition 9.9

In this subsection we discuss a useful alternative interpretation of the Bayes–Kelly algorithm, which will immediately imply Proposition 9.9. In the proof of Proposition 9.5 in the previous subsection, the key role is played by what we called reckless gambling, which was performed by the CTMs with betting functions (9.56). These betting functions are identical to the likelihood (9.15). Since (9.17) implies

$$f_n(p_n) = \sum_{(z_1,\dots,z_n)\in\{0,1\}^n} \tilde{w}[z_1,\dots,z_n], \tag{9.57}$$

the CTM produced by the Bayes–Kelly algorithm is exactly the mixture of the martingales defined by (9.56) over $(z_1, z_2, \dots)$ distributed according to the probability measure P.

Notice that (9.57) provides a simpler alternative way of computing $f_n(p_n)$ for use in line 4 of Algorithm 9.1: we can just sum the unnormalized weights $\tilde{w}$.

Let us summarize the alternative form of the Bayes–Kelly algorithm. For each infinite sequence of examples $\omega = (z_1, z_2, \dots)$, let S_n^ω be the CTM that gambles recklessly on the examples being $z_1, z_2, \dots$: $S_0^\omega := 1$ and $S_n^\omega := S_{n-1}^\omega f_n$, where

$$f_n := \begin{cases} \frac{n}{k_n} & \text{if } p_n \le k_n/n \text{ and } z_n = 1 \\ \frac{n}{n-k_n} & \text{if } p_n \ge k_n/n \text{ and } z_n = 0 \\ 0 & \text{otherwise}, \end{cases}$$

and $k_n := z_1 + \dots + z_n$. The "distributed reckless representation" of the Bayes–Kelly CTM is

$$S_n = S_n^P := \int_\Omega S_n^\omega P(\mathrm{d}\omega). \tag{9.58}$$

In terms of this representation, the weights $w[z_1,\dots,z_n]$ after observing p-values $p_1,\dots,p_n$ can be written as

$$w[z_1,\dots,z_n] = P([z_1,\dots,z_n] \mid p_1,\dots,p_n) \propto P([z_1,\dots,z_n]) S^{(z_1,z_2,\dots)}(p_1,\dots,p_n)$$

(in the last expression, there is no dependence on $z_{n+1}, z_{n+2}, \dots$).

Proof of Proposition 9.9

Using (9.58),

$$\begin{aligned} S_n^P(z_1,z_2,\dots) &\ge S_n^{(z_1,z_2,\dots)} P([z_1,\dots,z_n]) = \frac{n!}{k!(n-k)!} P([z_1,\dots,z_n]) \\ &= \frac{n!}{k!(n-k)!} \frac{k^k (n-k)^{n-k}}{n^n} \mathrm{LB}_n^P(z_1,z_2,\dots) \quad \text{a.s.} \end{aligned}$$

9.5 Context

9.5.1 *Randomness and Exchangeability: Algorithmic Perspective*

In the early 1960s Kolmogorov started revival of von Mises's approach to the foundations of probability in terms of random sequences, considering it important for understanding the applications of probability theory and mathematical statistics. (See Sect. 11.6.1 for further historical information.) He concentrated on binary sequences (as a simple starting point), in which context he often referred to random sequences as *Bernoulli sequences*. His first imperfect publication on this topic was the 1963 paper [202] (Kolmogorov refers to it as "incomplete discussion", according to the English translation of [203]). In the same year he conceived using the notion of computability for formalizing randomness. Kolmogorov's main publications on the algorithmic theory of randomness were [203–205].

Kolmogorov did not emphasize the difference between statistical randomness and exchangeability (his preferred definition was a formalization of exchangeability), but the algorithmic framework that he created is a useful tool for discussing relations between them. In this subsection we will discuss Kolmogorov's framework expressed in Martin-Löf's [238] terms, which are closer to the traditional statistical language. (Kolmogorov's original definitions, equivalent but given in terms of algorithmic complexity, are discussed in the online supplement to [391].) In our terminology we will follow [396].

A *measure of randomness* is an upper semicomputable function $f : \{0,1\}^* \to [0,1]$ such that, for any $N \in \mathbb{N}$, any power distribution $P = Q^N$ on $\{0,1\}^N$, the restriction of f to $\{0,1\}^N$ is a *p-variable* (meaning that $P(f \le \epsilon) \le \epsilon$ for any $\epsilon > 0$). The upper semicomputability of f means that there exists an algorithm that, when fed with a rational number r and sequence $\omega \in \{0,1\}^*$, eventually stops if $f(\omega) < r$ and never stops otherwise.

In other words, a measure of randomness is a family of p-variables for testing randomness in $\{0,1\}^N$. The requirement of upper semicomputability is natural: e.g., if $f(\omega) < 0.01$ (the p-value, i.e., the value taken by the p-variable, is highly statistically significant), we should learn this eventually.

Analogously, a *measure of exchangeability* is an upper semicomputable function $f : \{0,1\}^* \to [0,1]$ such that, for any $N \in \mathbb{N}$, any exchangeable measure P on $\{0,1\}^N$, and any $\epsilon > 0$, the restriction of f to $\{0,1\}^N$ is a p-variable.

Lemma 9.14 *There exists a measure of randomness f (called* universal*) such that any other measure of randomness f' satisfies $f = O(f')$. There exists a measure of exchangeability f (called* universal*) such that any other measure of exchangeability f' satisfies $f = O(f')$.*

The proof of Lemma 9.14 is standard; see, e.g., [238] or [396, Lemma 4].

In the algorithmic theory of randomness, it is customary to measure lack of randomness or exchangeability on the log scale. Therefore, we fix a universal measure of randomness f, set $d^{\text{rand}} := -\log f$, and refer to $d^{\text{rand}}(\omega)$ as the *deficiency of randomness* of the sequence $\omega \in \{0,1\}^*$. Similarly, we fix a universal measure of exchangeability f, set $d^{\text{exch}} := -\log f$, and refer to $d^{\text{exch}}(\omega)$ as the *deficiency of exchangeability* of ω. (Traditionally, the log is binary.)

Proposition 9.2 immediately implies

$$d^{\text{exch}}(\omega) - O(1) \le d^{\text{rand}}(\omega) \le d^{\text{exch}}(\omega) + \frac{1}{2}\log N + O(1), \tag{9.59}$$

where ω ranges over $\{0,1\}^*$ and N is the length of ω. In fact, we can interpret (9.59) as the algorithmic version of Proposition 9.2. Kolmogorov regarded the coincidence to within an additive $O(\log N)$ as being sufficient, at least for some purposes: cf. the last two paragraphs of [204]; therefore, he preferred the simpler definition $d^{\text{exch}}(\omega) \approx 0$ of ω being a Bernoulli sequence.

Proposition 9.2 is very crude, but more accurate results are known in the context of the algorithmic theory of randomness; some were obtained in the paper [376] written under Kolmogorov's supervision. To clarify relations between algorithmic randomness and exchangeability, we will need another notion, binomiality. The binomial probability distribution $\text{bin}_{N,p}$ on $\{0,\ldots,N\}$ with parameter p is defined by

$$\text{bin}_{N,p}(\{k\}) := \binom{N}{k} p^k (1-p)^{N-k}, \quad k \in \{0,\ldots,N\}.$$

A *measure of binomiality* is an upper semicomputable function $f : \{(N,k) : N \in \mathbb{N}, k \in \{0,\ldots,N\}\} \to [0,1]$ such that, for any $N \in \mathbb{N}$, any $p \in [0,1]$, and any $\epsilon > 0$,

$$\mathrm{bin}_{N,p}(\{k : f(N,k) \le \epsilon\}) \le \epsilon .$$

Lemma 9.15 *There exists a measure of binomiality f (called* universal*) such that any other measure of binomiality f' satisfies $f = O(f')$.*

We fix a universal measure of binomiality f, set $d^{\mathrm{bin}}(k;N) := -\log f(N,k)$, and refer to $d^{\mathrm{bin}}(k;N)$ as the *deficiency of binomiality* of k (in $\{0,\ldots,N\}$).

Proposition 9.16 *For any constant $\epsilon > 0$,*

$$(1-\epsilon)\left(d^{\mathrm{exch}}(\omega) + d^{\mathrm{bin}}(k;N)\right) - O(1) \le d^{\mathrm{rand}}(\omega)$$
$$\le (1+\epsilon)\left(d^{\mathrm{exch}}(\omega) + d^{\mathrm{bin}}(k;N)\right) + O(1) ,$$

N ranging over $\mathbb{N}$, ω over $\{0,1\}^N$, and k being the number of 1s in ω.

Proposition 9.16 follows immediately from (and is stated, in a more precise form, after) [376, Theorem 1]. It says, informally, that the randomness of ω is equivalent to the conjunction of two conditions: ω should be exchangeable, and the number of 1s in it should be binomial. For example, suppose that N is a large even number and the number of 1s in $\omega \in \{0,1\}^N$ is $k = N/2$. Then ω might be perfectly exchangeable whereas it will not be random since it belongs to the set of all binary sequences with the number of 1s precisely $N/2$, whose probability

$$\binom{N}{N/2} 2^{-N} \sim \sqrt{2/\pi} N^{-1/2} < N^{-1/2}$$

is small under any power distribution.

9.5.2 Testing Exchangeability and Randomness

The theory of testing statistical hypotheses based on Cournot's principle has a very long history, going back at least to Arbuthnott (1712, [5]), and the literature devoted to this topic is vast. The principle was widely discussed at the beginning of the twentieth century and defended by, e.g., Borel, Lévy, and Kolmogorov [318, Sect. 2.2]. Kolmogorov's statement of Cournot's principle in his *Grundbegriffe* [199, Chap. I, Sect. 2] is

> If $\mathsf{P}(A)$ is very small, then one can be practically certain that the event A will not occur on a single realization of the conditions $\mathfrak{S}$.

(The conditions $\mathfrak{S}$ in this quote refer to the probability trial under discussion.) In the form stated by Kolmogorov, the principle goes back to Jacob Bernoulli [31] (see, e.g., [318, Sect. 2.2]). It establishes a bridge between probability theory and our expectations about reality; observing an event A (assumed to be chosen in advance) of a small probability casts doubt on P. Cournot's [57, p. 78] contribution was to state that this is the only bridge between probability theory and reality.

Testing using test martingales is also popular: for example, test martingales (in the form of probability ratios) are widely used in sequential analysis for this purpose. However, even when testing is done using test martingales, the basic principle is usually still Cournot's (see, e.g., [419, 421]); the value taken by a test martingale is rarely interpreted as measuring the weight

Harold Jeffreys (1891–1989).

Photo by Walter Stoneman. Used with permission of The Royal Society. ©Godfrey Argent Studio

of evidence found against the statistical hypothesis. The martingale approach to testing free of Cournot's principle is discussed in [320, 379].

The exposition of Sect. 9.1 is based on [391]. While the exponent $1/2$ in $\sqrt{N} = N^{1/2}$ in (9.2) is clearly optimal, the optimality of the exponents 1 and 1.5 in (9.5) and (9.8) is an open question.

When the alternative hypothesis is composite (but the null hypothesis is still simple), it is natural still to use the likelihood ratio of a simple alternative hypothesis to the null but to integrate it over some "prior" probability measure on the composite alternative hypothesis. A popular choice is Jeffreys's prior; one of its appealing features is that it does not depend on the parametrization of the alternative hypothesis (provided it is smooth, in a natural sense).

Ramdas et al. [287, Sect. 2] construct a process $R = R_n$ (already mentioned in Sect. 9.2.7) that is the lower benchmark for the Markov alternative integrated w.r. to the product Jeffreys prior, but Ramdas et al. also apply it to the changepoint alternative. This work motivated [389] and [412], which served as the basis of Sect. 9.2.

Implicitly, we met Jeffreys's prior in Chap. 6: the Krichevsky–Trofimov estimate $(k+1/2)/(l+1)$ estimate discussed in Sect. 6.2.3 is the conditional probability computed from the mixture of the power distributions $\mathbf{B}_\pi^\infty$ w.r. to Jeffreys's prior over π.

In Sect. 9.3 we follow [389]. The notion of e-value has been promoted as an alternative to the standard statistical notion of p-value in statistical hypothesis testing in, e.g., [316] (under the name of "betting score") and [398] (which introduced the term "e-value").

Chapter 10
Non-conformal Shortcut

Abstract In Chap. 8 we argued for complementing a prediction rule, i.e., a trained prediction algorithm, with a system for testing deviations from exchangeability. As soon as serious violations of exchangeability are detected, we start retraining the prediction algorithm or take other appropriate measures. In this short chapter we suggest a shortcut: we test directly the predictions output by the prediction rule, and a successful way of testing then automatically translates to improved ("protected") predictions. In this way we apply methods developed in the first two chapters of this part outside conformal prediction. Our procedures of protected prediction can be used from the very beginning, or after detecting lack of exchangeability (but before retraining is complete). We consider separately the case of regression, in which we combine the probability integral transformation and betting martingales, and classification, in which we move even further from conformal prediction.

Keywords Protected prediction · Price of protection · Prediction with expert advice · Robustness

10.1 General Picture

It is a common knowledge, and we have repeatedly observed it in our collaboration with industrial partners, that in practical applications of machine learning the distribution of the data usually changes, often drastically, after a prediction algorithm has been trained and a prediction rule is ready to be deployed. In our experience the change in distribution was easily detectable by conformal test martingales. Such a change is likely to lead to a significant deterioration in the performance of the prediction rule.

Despite the existence of efficient ways of online detection of a change in distribution, there are inevitably awkward gaps between the change in distribution and its detection and between the detection of the change and the deployment of a retrained prediction algorithm.

The idea of this chapter is to turn successful betting accomplished by a test martingale into improvement in the quality of a prediction rule. In the case of

V. Vovk et al., *Algorithmic Learning in a Random World*,
https://doi.org/10.1007/978-3-031-06649-8_10

conformal testing, we are betting against the hypothesis of exchangeability. In the case of our new procedures, we will be betting against the prediction rule computed from the training set (the *base prediction rule*, as we will call it). It seems plausible that in the absence of exchangeability the prediction rule will work poorly on the test data, and a suitable test martingale will attain high values when betting against the base prediction rule. This automatically leads to an improved prediction rule.

Formally, the methods proposed in this chapter are independent of conformal prediction. In particular, they do not depend on the exchangeability assumption. However, conformal prediction and testing not only motivate them but also provide required technical tools.

This chapter proposes to use testing procedures (including betting martingales, such as the Simple Jumper and Composite Jumper) for developing better and more robust prediction algorithms. In this respect it is reminiscent of the method of defensive forecasting [320, Chap. 12], which starts from a test martingale (more generally, a strategy for Sceptic) and then develops a prediction algorithm that prevents the test martingale (more generally, Sceptic's capital) from growing. An advantage of our current procedure is that in typical cases it is computationally more efficient (in particular, it never requires finding fixed points, as in defensive forecasting, and when it requires solving equations, this can be done efficiently).

The prediction rules that we are interested in this chapter are deterministic predictive systems, as defined in Sect. 7.2.1, which we will abbreviate to *predictive systems* and both use in the case of regression and adapt to the case of classification. Our goal is to prevent a catastrophic drop in the quality of a base predictive system when the data distribution changes. Given a base predictive system, our procedures give a protected predictive system that is more robust to changes in the data distribution. To use Anscombe's [4] insurance metaphor (repeatedly used already in [172]), our procedure provides an insurance policy against such changes. The main desiderata for such a policy are its low price and its efficiency.

We are interested in two seemingly different questions about the base predictive system:

Online testing Can we gamble successfully against the base predictive system (at the odds determined by its predicted probabilities)? We are interested in online testing, as in the previous two chapters, i.e., in constructing test martingales with respect to the base predictive system that take large values on the sequence of examples chosen by Reality.

Online prediction Can we improve the base predictive system, modifying its predictions p_n to better predictions $p_n^\dagger$?

If the quality of online prediction is measured using the log loss function [135], the difference between the two questions almost disappears, as we will see later (see, e.g., (10.5)).

We discuss online testing and online prediction in Sect. 10.2 in the case of regression and in Sect. 10.3 in the case of classification. We illustrate our procedures for regression in simulation studies (Sect. 10.2.3) and our procedures for classification in empirical studies (Sect. 10.3.3).

In principle, it is possible that the protected predictive system will perform worse than the base predictive system on the actual data chosen by Reality. We will introduce the notion of the price of protection, which is the worst possible drop in the performance of the protected predictive system as compared with the base one. We are particularly interested in procedures with a low (definitely finite) price of protection. In Sect. 10.3.5 we will give an example of a theoretical performance guarantee (Theorem 10.3, an application of the technique of prediction with expert advice) for our protection procedure, including the price of protection and efficiency of protection.

10.2 Case of Regression

We first explain the idea of turning online testing into protected prediction and then illustrate it in simulation studies.

10.2.1 *Probability Integral Transformation in Place of Conformal Prediction*

Suppose we are given a deterministic predictive system Π, as defined in Sect. 7.2.1, processing examples $z_n = (x_n, y_n)$; for simplicity, let us assume that the predictive distribution functions $\Pi(z_1, \ldots, z_{n-1}, (x_n, \cdot))$ that it outputs are always differentiable with a positive derivative. This is our *base predictive system.*

Conformal testing consists of two steps: first we turn an exchangeable sequence of examples into a sequence of p-values distributed uniformly on $[0, 1]^\infty$ (thus turning the composite null hypothesis of exchangeability into a simple one), and then we gamble against the p-values being independent and uniformly distributed. A very similar strategy works for betting against the base predictive system. Lévy's [231, Sect. 39] probability integral transformation turns a predictive system into a source of independent random variables uniformly distributed on [0, 1]. Namely, for $y_n \sim \Pi(z_1, \ldots, z_{n-1}, (x_n, \cdot))$, $n = 1, 2, \ldots$, the probability integral transforms

$$p_n := \Pi(z_1, \ldots, z_{n-1}, z_n) \sim \mathbf{U} \tag{10.1}$$

will be independent. This plays the role of the first step of conformal testing. We can then apply the numerous betting martingales developed in conformal testing. Such betting martingales translate into test martingales that gamble against the base predictive system.

Let F be a betting martingale; we combine it with the base predictive system Π to get a *protected predictive system* $\Pi^\dagger$ as follows. Suppose we are given examples $z_1, \ldots, z_{n-1}$ and an object x_n. Let f_n be the betting function for F at step n (defined by the equality (8.6), where $p_1, \dots, p_{n-1}$ are the values (10.1) computed in the first $n-1$ steps, and $p_n \in [0,1]$ is a variable). If Q is the probability measure on $\mathbb{R}$ with the distribution function $\Pi(z_1, \dots, z_{n-1}, (x_n, \cdot))$, $\Pi^\dagger(z_1, \dots, z_{n-1}, (x_n, \cdot))$ will be the distribution function of the probability measure

$$Q_n^\dagger := f_n(p_n) Q_n \tag{10.2}$$

(i.e., of the probability measure $Q_n^\dagger$ defined by $Q_n^\dagger(E) := \int_E f_n(p_n) \mathrm{d}Q_n$ for Borel $E \subseteq \mathbb{R}$, where p_n is defined by (10.1) and considered a function of the label in z_n).

In this section we evaluate the performance of the base and protected predictive systems using the *log loss function*, which is the analogue of (6.7) in the case of regression. The loss of a predictive density q (the derivative of the predictive distribution, which we assumed to exist and be positive at the beginning of this subsection) for a realized label y is defined to be

$$-\log q(y)\,. \tag{10.3}$$

Let $L_n(\Pi)$ (resp. $L_n(\Pi^\dagger)$) be the cumulative log loss of the base (resp. protected) predictive system over the first n example. Since

$$q_n^\dagger(y_n) = f_n(p_n) q_n(y_n)\,, \tag{10.4}$$

where $q_n^\dagger$ and q_n are the densities of the probability measures $Q_n^\dagger$ and Q_n, respectively, in (10.2), and y_n is the label of the nth example output by Reality, we have

$$L_n(\Pi^\dagger) = L_n(\Pi) - \log S_n\,, \tag{10.5}$$

where S_n is the value taken at the end of trial n by the betting martingale F fed with (10.1). According to (10.5), successful gambling means improved prediction.

10.2.2 The Simple Jumper Protection Procedure

To turn the probability integral transforms $p_1, p_2, \dots$ into a test martingale we will use the Simple Jumper betting martingale S given by the betting functions (8.8), where we take $\epsilon \in \{-E, 0, E\}$ for a constant E. In the next subsection we set $J := 0.01$. A safer option would be to use the Composite Jumper betting martingale including a component Simple Jumper that never gambles, but in order to make our simulation studies easy to interpret we prefer simple choices. In the next section we will obtain a finite price of protection in the context of classification, but it will be clear that the same method also works in the case of regression.

Algorithm 10.1 Simple Jumper protection

Input: base predictive distribution functions $G_1, G_2, \dots$.
Output: protected predictive distribution functions $G_1^\dagger, G_2^\dagger, \dots$.
1: $C_{-E} := C_0 := C_E := 1/3$
2: $C := 1$
3: **for** $n = 1, 2, \dots$
4: **for** $\epsilon \in \{-E, 0, E\}$: $C_\epsilon := (1 - J)C_\epsilon + (J/3)C$
5: $\epsilon^\dagger := E(C_E - C_{-E})/C$
6: output $G_n^\dagger := F_{\epsilon^\dagger}(G_n)$
7: **for** $\epsilon \in \{-E, 0, E\}$: $C_\epsilon := C_\epsilon f_\epsilon(G_n(y_n))$
8: $S_n := C := C_{-E} + C_0 + C_E$.

Suppose the predictive distribution function output by the base predictive system Π at step n is $G_n = G_n(y)$, and so the corresponding predictive density is $g_n = g_n(y) = G_n'(y)$, and the betting function output by S at that step is f. By (10.4), the protected predictive density is $f(G_n)g_n$. It integrates to 1 since $f(G_n)g_n = (F(G_n))'$, where F is the indefinite integral $F(v) := \int_0^v f$ of f, so that $F' = f$. We can see that the distribution function for the protected algorithm is $F(G_n)$.

The procedure of protection is given as Algorithm 10.1, where

$$F_\epsilon(v) := \int_0^v f_\epsilon(u)\,\mathrm{d}u = \frac{\epsilon}{2}v^2 + \left(1 - \frac{\epsilon}{2}\right)v \tag{10.6}$$

(cf. (8.8)). Algorithm 10.1 uses the fact that the Simple Jumper outputs betting functions (8.8) for $\epsilon = \epsilon^\dagger$. Let us check it. The value of the Simple Jumper martingale at the end of trial n is, in the notation of Algorithms 8.1 and 10.1,

$$\begin{aligned}
\sum_\epsilon C_\epsilon f_\epsilon(p_n) &= \sum_\epsilon C_\epsilon(1 + \epsilon(p_n - 0.5)) \\
&= C_E(1 + E(p_n - 0.5)) + C_0 + C_{-E}(1 - E(p_n - 0.5)) \\
&= (C_E + C_0 + C_{-E}) + (C_E - C_{-E})E(p_n - 0.5) \\
&= \left(1 + \frac{C_E - C_{-E}}{C}E(p_n - 0.5)\right)C,
\end{aligned}$$

where $C := C_E + C_0 + C_{-E}$ is the value of the martingale at the beginning of the trial. We can see that the slope of the resulting straight line is indeed as given in line 5 of Algorithm 10.1.

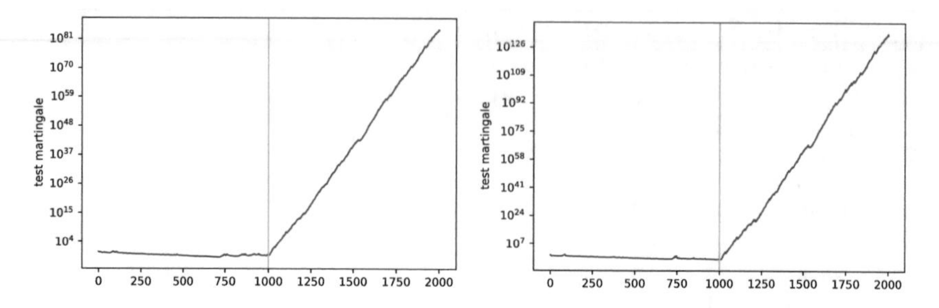

Fig. 10.1 The Simple Jumper test martingale. Left panel: $\epsilon \in \{-1, 0, 1\}$. Right panel: $\epsilon \in \{-2, 0, 2\}$

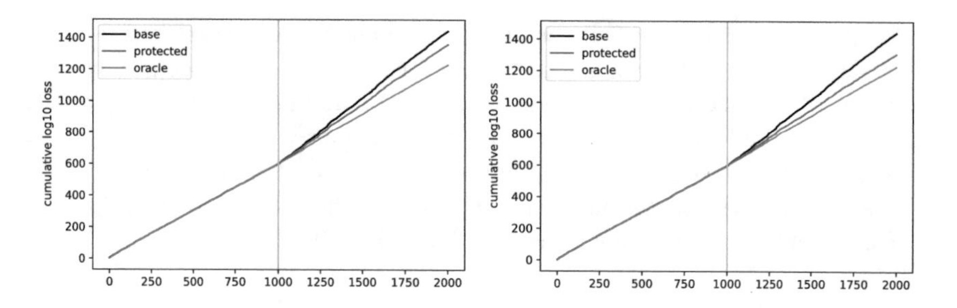

Fig. 10.2 The cumulative log losses of three predictive systems (to the left of the changepoint the three lines coincide or are visually indistinguishable). Left panel: $\epsilon \in \{-1, 0, 1\}$. Right panel: $\epsilon \in \{-2, 0, 2\}$

10.2.3 A Simulation Study

We consider a dataset that consists of independent Gaussian examples: the first 1000 are generated from $\mathbf{N}_{0,1}$, and another 1000 from $\mathbf{N}_{1,1}$. Our base predictive system does not know that there is a changepoint at time 1000 and always predicts $\mathbf{N}_{0,1}$.

First we run the Simple Jumper martingale (with $J = 0.01$) on our dataset. The left panel of Fig. 10.1 shows its path for $E := 1$; it loses capital before the changepoint, but quickly regains it afterwards. Its final value is 5.55×10^{84}.

The cumulative log loss of the protected version of the base predictive system is shown as the green line in the left panel of Fig. 10.2. The black line corresponds to the base predictive system, and the red line to the unrealistic *oracle predictive system*, which knows the truth and predicts with $\mathbf{N}_{0,1}$ before the changepoint and $\mathbf{N}_{1,1}$ afterwards. According to Fig. 10.1 (left panel) and (10.5), the difference between the final values of the black and green lines is about 85.

To understand better the mechanism of protection in this case, notice that Algorithm 10.1 with the simplest choice of $E := 1$ outputs betting functions f_ϵ of the form (8.8), where $\epsilon \in [-1, 1]$. The corresponding predictive distributions $f_\epsilon(\Phi)\phi$ (where ϕ is the standard normal density and Φ its distribution function) are shown in the left panel of Fig. 10.3 for five values of ϵ. We can see that our range

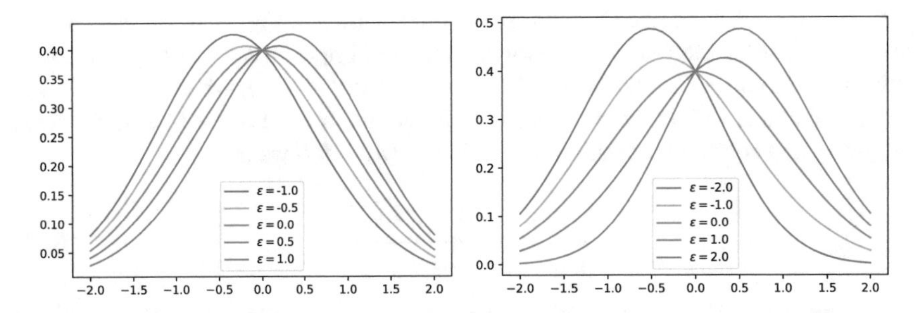

Fig. 10.3 The protected predictive distributions. Left panel: the range of ϵ is $\{-1, -0.5, 0, 0.5, 1\}$. Right panel: $\epsilon \in \{-2, -1, 0, 1, 2\}$

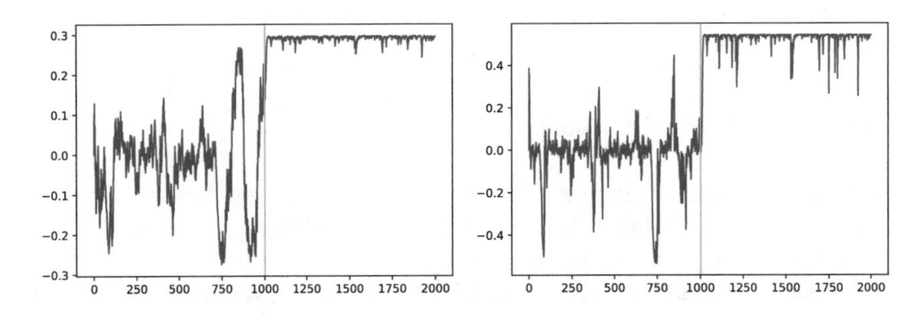

Fig. 10.4 The protected point predictions (medians of the protected predictive distributions). Left panel: $\epsilon \in \{-1, 0, 1\}$. Right panel: $\epsilon \in \{-2, 0, 2\}$

of ϵ, $\epsilon \in [-1, 1]$, is not sufficiently ambitious and does not allow us to approximate $\mathbf{N}_{1,1}$ well.

Replacing the range $[-1, 1]$ for ϵ by $[-2, 2]$, we obtain the right panel of Fig. 10.3. The right-most graph in that panel now looks closer to the density of $\mathbf{N}_{1,1}$. We cannot extend the range of ϵ further without (8.8) ceasing to be a betting function (which is required to be nonnegative). (Of course, the betting function does not have to be linear, but let us stick to the simplest choices for now.)

Using the range $\{-2, 0, 2\}$ for ϵ leads to the right panels of Figs. 10.1 and 10.2. We can see that in the right panel of Fig. 10.2 the performance of the protected algorithm is much closer to that of the oracle predictive system than in the left panel.

Figure 10.2 provides useful and precise information, but it is not very intuitive. A cruder approach is to convert the probabilistic predictions into point predictions. Figure 10.4 uses the medians of the predictive distributions as point predictions. In the case of the base predictive system, the prediction is always 0 (the median of $\mathbf{N}_{0,1}$), for the oracle predictive system it is 0 before the changepoint and 1 afterwards, and for the protected predictive systems the predictions are shown in the figure. We can see that the right panel of Fig. 10.4, corresponding to a wider range of ϵ, is a better approximation to the oracle predictions.

Remark 10.1 To compute the point predictions shown in Fig. 10.4, we can use the representation of the betting function f_ϵ for the Simple Jumper in the form (8.8) with $\epsilon = E(C_E - C_{-E})/C$ (as in line 5 of Algorithm 10.1), where $E = 1$ in the left panel and $E = 2$ in the right one. The indefinite integral of the betting function is (10.6), and solving the quadratic equation $F_\epsilon(v) = 0.5$ we get

$$v = \frac{\epsilon - 2 + \sqrt{\epsilon^2 + 4}}{2\epsilon} . \tag{10.7}$$

Since the distribution function of the protected probability forecast is $F_\epsilon(\Phi)$, where Φ is the distribution function of the base probabilistic prediction $\mathbf{N}_{0,1}$, we obtain the median of the protected prediction as the v quantile of $\mathbf{N}_{0,1}$, with v defined by (10.7).

Custom-Made Betting Functions

The betting functions we have used so far are fairly poor, even those shown in the right panel of Fig. 10.3. Let us find better custom-made betting functions, as we did in Sect. 9.2 in a somewhat different context.

According to Lemma 9.6, the optimal, in a natural sense, betting function for the last 1000 examples z_n (distributed as $\mathbf{N}_{1,1}$ but predicted with $\mathbf{N}_{0,1}$) is the probability density function of p_n. The probability distribution of those p_n is the image of $\mathbf{N}_{1,1}$, whose density we will denote g, under the mapping Φ (which is the distribution function of $\mathbf{N}_{0,1}$). The probability density function of the image of g under Φ is

$$u \mapsto \frac{g(\Phi^{-1}(u))}{\phi(\Phi^{-1}(u))} ,$$

where $\phi = \Phi'$. In terms of distribution functions the statement is simpler: if G is the distribution function corresponding to g, so that $G' = g$, the image of G under the mapping Φ is the distribution function $u \mapsto G(\Phi^{-1}(u))$. Since

$$\frac{g(y)}{\phi(y)} = \frac{\exp(-(y-1)^2/2)}{\exp(-y^2/2)} = \exp(y - 1/2) ,$$

the optimal betting function, which we will refer to as *ideal*, is

$$f(p) = \frac{g(\Phi^{-1}(p))}{\phi(\Phi^{-1}(p))} = \exp(\Phi^{-1}(p) - 1/2) . \tag{10.8}$$

The ideal betting function is shown in Fig. 10.5 in blue. For comparison, the polynomial betting functions $(d + 1)u^d$ are also shown for $d = 1$ (linear), $d = 2$

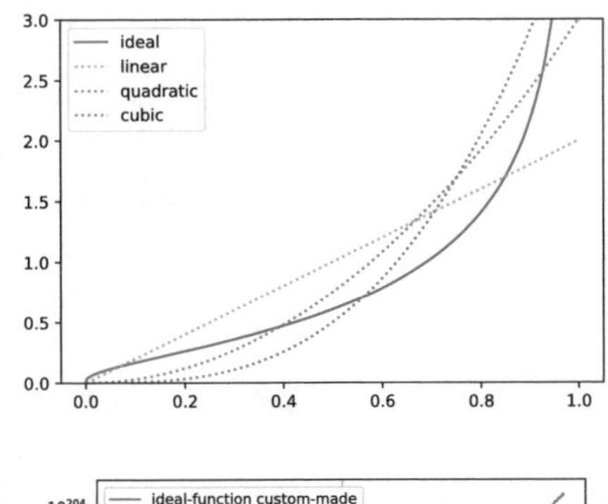

Fig. 10.5 The ideal betting function after the changepoint (solid blue line)

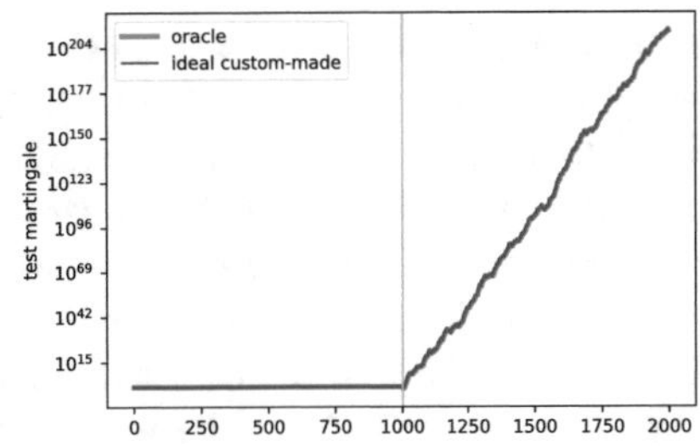

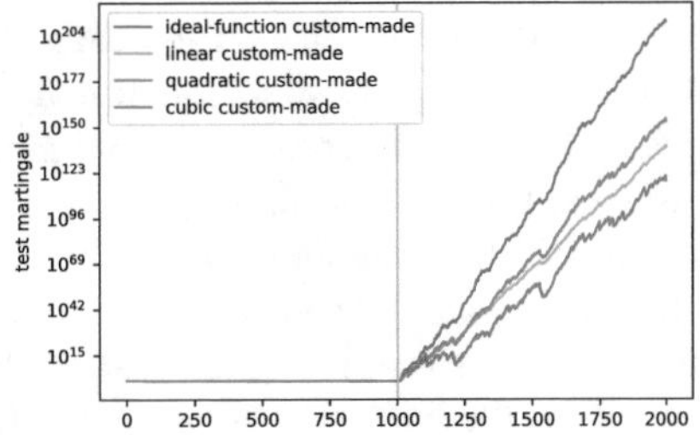

Fig. 10.6 Left panel: the oracle martingale and the ideal custom-made martingale. Right panel: the ideal-function, linear, quadratic, and cubic custom-made sleeping martingales

Algorithm 10.2 Sleeping betting martingale $((p_1, p_2, \dots) \mapsto (S_1, S_2, \dots))$

1: $S := 1$
2: $A := 0$
3: **for** $n = 1, 2, \dots$
4: $\quad S := (1 - R)S$
5: $\quad A := A + RS$
6: $\quad A := Af(p_n)$
7: $\quad$ output $S_n := S + A$.

(quadratic), and $d = 3$ (cubic); the linear betting function will give results very similar to those produced by the Simple Jumper protection with $E = 2$.

The left panel of Fig. 10.6 shows the oracle martingale and ideal custom-made martingale (not betting over the first 1000 examples and using the ideal betting function afterwards); their paths are visually indistinguishable (unless drawn in different colours and with different widths, as in the figure) since they are merely different implementations of the same martingale.

The ideal custom-made martingale is unrealistic since it knows the precise data-generating distribution; in particular, it knows the location of the changepoint. Algorithm 10.2 represents a test martingale that assumes less. At the beginning it does not gamble, or "sleeps". The parameter R (0.001 in our experiment) is the rate at which it wakes up; S is the total sleeping capital, and A is the total active capital; f is the betting function. The right panel of Fig. 10.6 shows the results of applying

Algorithm 10.2 to four betting functions: the ideal (10.8), shown in Fig. 10.5 as solid blue line, and the linear, quadratic, and cubic ones (dotted in Fig. 10.5). Of the polynomial functions, the best one is quadratic; linear and cubic are worse (and the higher degrees become even worse). We can see that the performance of the Simple Jumper protection can be greatly improved, at least as measured by the log loss function.

10.3 Case of Classification

This section concentrates on the case of classification and at first pays particular attention to binary classification, usually assuming that the label space is $\{0, 1\}$. In general, a *predictive system*, as used in this section, maps past data and an object x with an unknown label $y \in \mathbf{Y}$ to a probability measure $p \in \mathbf{P}(\mathbf{Y})$, regarded as probabilistic prediction for y. But in the binary case $\mathbf{Y} = \{0, 1\}$ we replace p by $p\{1\}$, which carries the same information. We are given a *base predictive system*, and our goal is to protect it. In this section we are mostly interested in the case where the base predictive system is a prediction rule obtained by training a prediction algorithm, and so the probabilistic prediction depends only on the object x, but allowing the dependence on the past data does not complicate the exposition.

10.3.1 Testing Predictions by Betting

We consider a potentially infinite sequence of examples $z_1, z_2, \ldots$ output by Reality, each consisting of two components $z_n = (x_n, y_n)$, where $x_n \in \mathbf{X}$ is an object chosen from an object space $\mathbf{X}$, and $y_n \in \{0, 1\}$ is a binary label. A *predictive system* is a function that maps any object x and any finite sequence of examples $z_1, \ldots, z_i$ (intuitively, the past data) for any $i \in \{0, 1, \ldots\}$ to a number $p \in [0, 1]$ (intuitively, the probability that the label of x is 1). Fix a *base predictive system*, and let $p_1, p_2, \ldots$ be its predictions for the examples output by Reality: p_n is the prediction output by the base predictive system on x_n and $z_1, \ldots, z_{n-1}$; it is interpreted as the predicted probability that $y_n = 1$. (We will not need any measurability assumptions; in particular, $\mathbf{X}$ is not supposed to be a measurable space.)

Our first online testing procedure, familiar by now but applied in a new context, is given as Algorithm 10.3. One of its two parameters is a finite non-empty family $f_\theta : [0, 1] \to [0, 1]$, $\theta \in \Theta$, of *calibration functions*. The intuition behind f_θ is that we are trying to improve the base predictions p_n, or *calibrate* them; the idea is to use a new prediction $f_\theta(p_n)$ instead of p_n. We assume that Θ contains a distinguished element $\mathbf{0} \in \Theta$ (used in line 1 of Algorithm 10.3) such that $f_\mathbf{0}$ is the identity function.

Perhaps the simplest choice is to use a finite subset of the family

Algorithm 10.3 Simple Jumper martingale for testing a predictive system

Input: base probabilistic predictions $p_1, p_2, \ldots$.
Output: test martingale $S_1, S_2, \ldots$.
1: $C_\theta := 1_{\theta=\mathbf{0}}$ for all $\theta \in \Theta$
2: $C := 1$
3: **for** $n = 1, 2, \ldots$
4: **for** $\theta \in \Theta$: $C_\theta := (1 - J)C_\theta + (J/|\Theta|)C$
5: $p_n^\dagger := \sum_\theta f_\theta(p_n)C_\theta / \sum_{\theta'} C_{\theta'}$ ▷ protected prediction (not needed for testing)
6: **for** $\theta \in \Theta$: $C_\theta := C_\theta \mathbf{B}_{f_\theta(p_n)}\{y_n\}/\mathbf{B}_{p_n}\{y_n\}$
7: output $S_n := C := \sum_{\theta\in\Theta} C_\theta$.

$$f_\theta(p) := p + \theta p(1 - p) , \tag{10.9}$$

where $\theta \in [-1, 1]$, and $\theta = 0$ is the distinguished element $\mathbf{0}$. For $\theta > 0$ we are correcting for the forecasts p being underestimates of the true probability of 1, while for $\theta < 0$ we are correcting for p being overestimates; f_0 is the identity function (no correction).

We do not know in advance which f_θ will work best, and moreover, it seems plausible that suitable values of θ will change over time, and we proceed as in Sect. 8.1 averaging the *elementary test martingales*

$$\prod_{i=1}^{n} \frac{\mathbf{B}_{f_{\theta_i}(p_i)}\{y_i\}}{\mathbf{B}_{p_i}\{y_i\}}, \qquad n = 0, 1, \ldots , \tag{10.10}$$

where $(\theta_1, \theta_2, \ldots) = \boldsymbol{\theta}$ is a sequence of elements of Θ and, as in Algorithm 10.3, $\mathbf{B}_p$ is the Bernoulli distribution on $\{0, 1\}$ with parameter p, $\mathbf{B}_p\{1\} = p$. The Simple Jumper Markov chain over which we perform the averaging is slightly different in that its initial state θ_0 is $\mathbf{0}$ (line 1 of Algorithm 10.3) instead of being chosen from the uniform probability measure on Θ (as we did before). This expresses the "presumption of innocence" towards the base predictive system: we do not calibrate it unless calibration is needed (and we definitely do not calibrate it at the very beginning of the test set if the base predictive system is a prediction rule obtained by training a prediction algorithm). Otherwise the Markov chain is as before: the transition function prescribes maintaining the same state with probability $1 - J$ and, with probability J, choosing a new state from the uniform probability measure on Θ (line 4), where $J \in [0, 1]$ is the parameter called the *jumping rate*.

Algorithm 10.3 gives both the average

$$S_n := \int \prod_{i=1}^{n} \frac{\mathbf{B}_{f_{\theta_i}(p_i)}\{y_i\}}{\mathbf{B}_{p_i}\{y_i\}} \mu(\mathrm{d}\theta)$$

of (10.10), where μ is the distribution of the Simple Jumper Markov chain, and the corresponding protected predictions $p_n^\dagger$ (line 5). In most of this section we measure

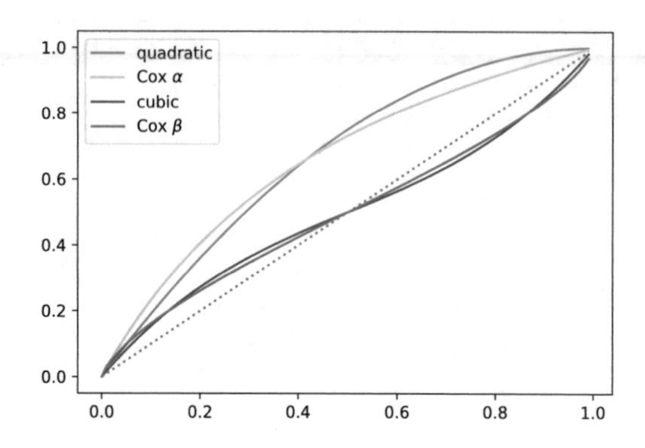

Fig. 10.7 Examples of calibration functions

the quality of predictions using the *log loss function*

$$\lambda(y, p) := \begin{cases} -\log p & \text{if } y = 1 \\ -\log(1-p) & \text{if } y = 0 \,, \end{cases} \tag{10.11}$$

where $y \in \{0, 1\}$ is the true label and $p \in [0, 1]$ is its prediction (the logarithm is typically natural, but in our experiments we consider decimal logarithms); in Sect. 6.4.3 (see (6.7)) we denoted it $\lambda_{\log}$. The relation between testing and prediction in our procedures (apart from the last one, Algorithm 10.5) is that, at each step n,

$$\lambda(y, p_n^\dagger) = \lambda(y, p_n) - \log b_n(y) \tag{10.12}$$

for all $y \in \mathbf{Y}$, where $b_n = S_n/S_{n-1}$ is the betting function used at step n. Summing (10.12) applied to the labels y_n chosen by Reality over the first n trials, we obtain (10.5), where L_n now stands for the cumulative loss in the sense of (10.11) over the first n trials.

In one respect the Simple Jumper martingale is not adaptive enough: we assume that a suitable jumping rate J is known in advance. Instead, we will use the *Composite Jumper* martingale that depends on two parameters, $\pi \in (0, 1)$ and a finite set $\mathbf{J}$ of non-zero jumping rates. It is defined to be the weighted average

$$S_n := \pi + \frac{1-\pi}{|\mathbf{J}|} \sum_{J \in \mathbf{J}} S_n^J \,, \tag{10.13}$$

where S^J is computed by Algorithm 10.3 fed with J as its parameter.

In Sect. 10.5.2 we will see that already the simple choice (10.9) leads to very successful betting for popular benchmark datasets. However, there are numerous other natural calibration functions, some of which are shown in Fig. 10.7. The function in red is in the quadratic family (10.9) (and its parameter is $\theta = 1$); the functions in this family are fully above, fully below, or (for $\theta = 0$) situated on the bisector of the first quadrant (shown as the dotted line). In many situations other calibration functions will be more suitable. For example, it is well known that untrained humans tend to be overconfident [184, Part VI, especially Chap. 22]. A possible calibration function correcting for overconfidence is the cubic function

$$f_{a,b}(p) := p + ap(p-b)(p-1) ,$$

where $(a, b) = \theta \in [0, 1]^2$. An example of such a function is shown in Fig. 10.7 in blue (with parameters $a = 1.5$ and $b = 0.5$). The meaning of the parameters is that b is the value of p (such as 0.5) that we believe does not need correction, and that a indicates how aggressively we want to correct for overconfidence ($a < 0$ meaning that in fact we are correcting for underconfidence). If the predictor predicts a p that is close to 0 or 1, we correct for his overconfidence (assuming $a > 0$) by moving p towards the neutral value b.

In the experimental subsection (Sect. 10.3.3) we will use Cox's [63, Sect. 3] calibration functions

$$f_{\alpha,\beta}(p) := \frac{p^\beta \exp(\alpha)}{p^\beta \exp(\alpha) + (1-p)^\beta} , \tag{10.14}$$

where $(\alpha, \beta) = \theta \in \mathbb{R}^2$. We obtain the identity function $f_{\alpha,\beta}(p) = p$ for $\alpha = 0$ and $\beta = 1$; these are the "neutral" values, $\mathbf{0} = (0, 1) \in \Theta$. Cox starts his exposition in [63, Sect. 3] from the one-parameter subfamily

$$f_\beta(p) := \frac{p^\beta}{p^\beta + (1-p)^\beta} , \tag{10.15}$$

where $\beta \in \mathbb{R}$, obtained by fixing $\alpha := 0$. We are mostly interested in $\beta > 0$, although $\beta = 0$ (when every p is transformed to 0.5) and $\beta < 0$ (which reverses the order of the label probabilities) are also possible. Similarly, we can set β to its neutral value 1 obtaining another one-parameter subfamily,

$$f_\alpha(p) := \frac{p \exp(\alpha)}{p \exp(\alpha) + (1-p)} . \tag{10.16}$$

An example of a function in the class (10.15) is shown in Fig. 10.7 in green (with $\beta = 0.75$); the graph in orange shows a function in the class (10.16) (with $\alpha = 1$).

Remark 10.2 Cox's calibration functions look particularly natural (are linear functions) in terms of the log odds ratios (as presented in Cox [63, Sect. 3]). Namely, (10.14) can be rewritten as

$$\log \frac{f_{\alpha,\beta}(p)}{1 - f_{\alpha,\beta}(p)} := \alpha + \beta \log \frac{p}{1-p} . \tag{10.17}$$

Algorithm 10.4 Composite Jumper predictive system

Input: base probabilistic predictions $p_1, p_2, \ldots$.
Output: protected probabilistic predictions $p_1^\dagger, p_2^\dagger, \ldots$.
1: $S_0 := 1$ ▷ Composite Jumper martingale (optional)
2: $P := \pi$
3: $A_\theta^J := \frac{1-\pi}{|\mathbf{J}|} 1_{\theta=\mathbf{0}}$ for all $J \in \mathbf{J}$ and $\theta \in \Theta$
4: **for** $n = 1, 2, \ldots$
5: **for** $J \in \mathbf{J}$
6: $A := \sum_\theta A_\theta^J$
7: **for** $\theta \in \Theta$: $A_\theta^J := (1 - J)A_\theta^J + AJ/|\Theta|$
8: output $p_n^\dagger := p_n P + \sum_{J,\theta} f_\theta(p_n) A_\theta^J$
9: $S_n := S_{n-1} \mathbf{B}_{p_n^\dagger}\{y_n\}/\mathbf{B}_{p_n}\{y_n\}$ ▷ Composite Jumper martingale (optional)
10: $P := P\mathbf{B}_{p_n}\{y_n\}$
11: **for** $J \in \mathbf{J}$ and $\theta \in \Theta$: $A_\theta^J := A_\theta^J \mathbf{B}_{f_\theta(p_n)}\{y_n\}$
12: $C := P + \sum_{J,\theta} A_\theta^J$
13: $P := P/C$
14: **for** $J \in \mathbf{J}$ and $\theta \in \Theta$: $A_\theta^J := A_\theta^J/C$.

10.3.2 Protecting Prediction Algorithms

Algorithm 10.4 implements the protected predictive system corresponding to the Composite Jumper martingale, i.e., the test martingale of Algorithm 10.3 averaged as in (10.13). The martingale determines the protected prediction $p_n^\dagger$ at each step n by the requirement (10.12).

The prior distribution on the elementary martingales is built on top of the Simple Jumper Markov chain (described in the previous subsection) and taking the averaging (10.13) into account:

- With probability π, we choose the *passive* elementary martingale, which is the identical 1.
- Otherwise, we choose the jumping rate J from the uniform probability measure on $\mathbf{J}$.
- Having chosen J, we generate $\boldsymbol{\theta} := (\theta_1, \theta_2, \ldots)$ as in the previous section, which gives us the *active* elementary martingale (10.10) indexed by $(J, \boldsymbol{\theta})$ (to have the component J will be useful in Algorithm 10.4, even though (10.10) does not depend on it).

This determines the prior distribution on the elementary martingales.

Algorithm 10.4 can be considered an implementation of the Bayesian merging rule. The variable P in Algorithm 10.4 holds the posterior weight (proportional to the capital) of the passive elementary martingale (and the corresponding predictor, which is the base predictive system), and A_θ^J holds the total posterior weight (proportional to the capital) of the elementary martingales $(J, \boldsymbol{\theta})$ (and the corresponding predictors) that are in the state θ (i.e., $\theta_n = \theta$, where $\boldsymbol{\theta} = (\theta_1, \theta_2, \ldots)$ and n is the current step). We initialize them to their prior weights in lines 2–3. Unlike in

Algorithm 10.3, now we normalize the weights P and A^J_θ at the end of each step (lines 12–14). Line 7 corresponds to the transition function of the Simple Jumper Markov chain with the jumping rate J, and lines 10–11 reflect the evolution of the elementary martingales' capital (and can be considered to be Bayesian weight updates). In line 12 we compute the total posterior weight of the elementary martingales (and the corresponding predictors), and in lines 13–14 we normalize the weights. In line 8 we compute the protected prediction to ensure (10.12) (but it can also be interpreted as weighted average of the predictions produced by the predictors corresponding to the elementary martingales).

By construction, we still have (10.5) for Algorithm 10.4.

The protection provided by Algorithm 10.4, and similar procedures, has its price, since the loss suffered by the protected predictive system can be greater than the loss suffered by the base predictive system. The *price of protection* for Algorithm 10.4 or a similar procedure is defined to be

$$\sup \left(L_n(\Pi^\dagger) - L_n(\Pi) \right) = \sup \left(-\log S_n \right) , \tag{10.18}$$

where the sup is over all n, all object spaces $\mathbf{X}$, all base predictive systems Π, and all sequences of examples. While the definition given by the left-hand side of (10.18) is applicable to any procedure of protection, for Algorithm 10.4 we have an equivalent definition given by the right-hand side of (10.18), S being the Composite Jumper martingale. We are particularly interested in the case of a finite price of protection, and in Sect. 10.3.5 (Theorem 10.3) we will see that it is at most $\log \frac{1}{\pi}$ for Algorithm 10.4, where π is the parameter in line 2 (we will set it to 0.5 in our experiments).

10.3.3 Experimental Results

Not many of the popular benchmark datasets are suitable for our experiments. First, they should be ordered chronologically (to fit the scenario discussed in Sect. 10.1) or contain timestamps for all examples. And second, they should not be of the time-series type, so that applying typical prediction algorithms implemented in standard machine-learning packages makes sense.

Our main dataset will be Bank Marketing (the only dataset in the top twelve most popular datasets in the UCI Machine Learning Repository that satisfies our requirements). The dataset consists of 45,211 examples listed in chronological order. We take the first 10,000 examples as the training set and train random forest on it. (See Sect. B.3 for details.) Random forest often outputs probabilities of success that are equal to 0 or 1, and when such a prediction turns out to be wrong (which happens repeatedly), the log loss is infinite. It is natural, therefore, to truncate a probability $p \in [0, 1]$ of 1 to the interval $[\epsilon, 1 - \epsilon]$ replacing p by

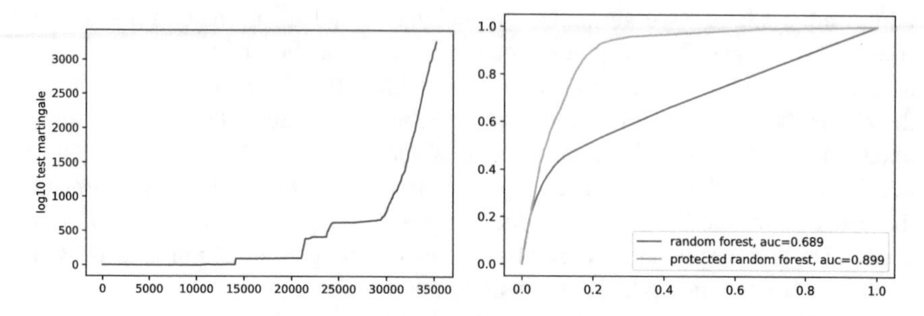

Fig. 10.8 Left panel: the Composite Jumper test martingale for the Bank Marketing dataset and random forest. Right panel: the ROC curve for the Composite Jumper protection

$$p^* := \begin{cases} \epsilon & \text{if } p \le \epsilon \\ p & \text{if } p \in (\epsilon, 1-\epsilon) \\ 1-\epsilon & \text{if } p \ge 1-\epsilon \ , \end{cases} \tag{10.19}$$

where we set $\epsilon := 0.01$. The resulting prediction rule is our base predictive system. After we find it, we never use the training set again, and the numbering of examples starts from the first element of the test set (i.e., the dataset in the chronological order without the training set).

The left panel of Fig. 10.8 shows the path of $\log_{10} S_n$, $n = 1, \ldots, 35211$, where S_n is the value of the Composite Jumper test martingale over the test set with the jumping rates $\mathbf{J} := \{10^{-2}, 10^{-3}, 10^{-4}\}$, $\pi := 0.5$, and Cox's family (10.14) with $\alpha \in \{-1, 0, 1\}$ and $\beta \in \{0, 1, 2\}$. With this choice of the ranges of α and β, there are $|\Theta| = 9$ of parameter vectors $\theta := (\alpha, \beta)$. The final value of the test martingale in the left panel of Fig. 10.8 is approximately 10^{3260}.

The right panel of Fig. 10.8 gives the ROC curve for random forest and random forest protected by Algorithm 10.4. We can see that the improvement is substantial. In terms of the log loss function and decimal logarithms, the loss goes down from approximately 7210 to 3950 (the difference between these two numbers being, predictably, the exponent 3260 in the final value of the test martingale in the left panel; cf. (10.5)).

Table 10.1 gives the AUC (area under curve) for Algorithm 10.4 and several base prediction algorithms as implemented in a standard machine-learning package; see Sect. B.3 for details. Since the dataset is imbalanced (only 12% of the labels are positive), AUC is a more suitable measure of quality than error rate.

In our experiments so far we have assumed that every test example is used for calibrating the prediction rule. From the practical point of view this may be unrealistic, and in reality we can only hope to get feedback on a fraction of the test examples. Figure 10.9 is the counterpart of the right panel of Fig. 10.8 for the case where the vast majority of examples are predicted by the protected predictive system without getting any feedback, and the weight updates in Algorithm 10.4 are

Table 10.1 The AUC for the Bank Marketing dataset and key prediction algorithms together with their protected versions

Prediction algorithm	Base	Protected
Random forest	0.689	0.899
Gradient boosting	0.733	0.901
Decision trees	0.564	0.815
Multilayer perceptron	0.650	0.877
SVM	0.685	0.843
Naive Bayes	0.646	0.807
Logistic regression	0.609	0.838

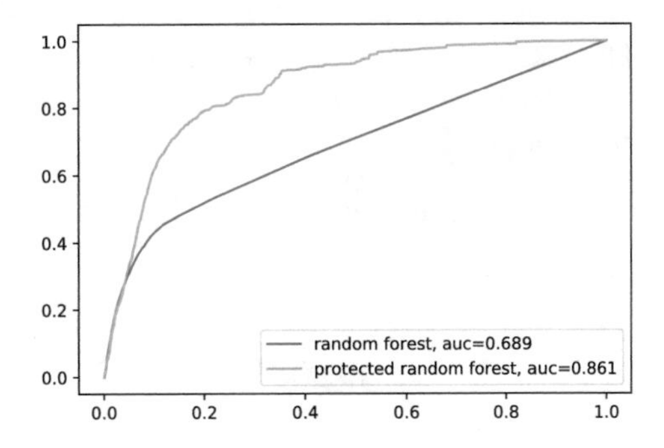

Fig. 10.9 The ROC curve for the Bank Marketing dataset and random forest with the Composite Jumper protection and feedback provided for every 100th test example

run only on every 100th test example (so that the same weights P and A_θ^J are used for the examples $100k + 1$, $100k + 2$,..., $100k + 100$ for $k = 0, 1, \ldots$). Now the improvement is more modest.

10.3.4 *Classification with the Brier Loss Function*

So far in this chapter we have only used the log loss function (often implicitly), and in this section we have only considered the binary case $\mathbf{Y} := \{0, 1\}$. We will now remove both these restrictions, arriving at Algorithm 10.5, which is our preferred procedure of protection (now we write ln in place of log to emphasize that the logarithm in natural). We will go over it in the rest of this subsection, and derive it from the general Aggregating Algorithm described in Sect. 10.4.1.

Now we have an arbitrary finite label space $\mathbf{Y}$ (equipped with the discrete σ-algebra, as usual); the predictions are chosen from $\mathbf{P}(\mathbf{Y})$, i.e., are probability measures on $\mathbf{Y}$. Algorithm 10.5 uses a finite family $(f_\theta : \theta \in \Theta)$ of *calibration functions* $f_\theta : \mathbf{P}(\mathbf{Y}) \to \mathbf{P}(\mathbf{Y})$ containing the identity function $f_\mathbf{0}$. The prime example is Cox's calibration functions (10.14), properly generalized.

The advantage of the representations (10.14)–(10.16) over (10.17) is that they are easy to extend to the multiclass case. The calibration functions (10.14) become

Algorithm 10.5 Composite Jumper Brier predictor

Input: base predictions $p_1, p_2, \ldots$ from $\mathbf{P}(\mathbf{Y})$.
Output: protected predictions $p_1^\dagger, p_2^\dagger, \ldots$.
1: $P := \pi$
2: **for** $J \in \mathbf{J}$ and $\theta \in \Theta$: $A_\theta^J := \frac{1-\pi}{|\mathbf{J}|} 1_{\theta=\mathbf{0}}$
3: **for** $n = 1, 2, \ldots$
4: **for** $J \in \mathbf{J}$
5: $A := \sum_\theta A_\theta^J$
6: **for** $\theta \in \Theta$: $A_\theta^J := (1-J)A_\theta^J + AJ/|\Theta|$
7: **for** $y \in \mathbf{Y}$: $G(y) := -\ln\left(P \exp(-\lambda(y, p_n)) + \sum_{J,\theta} A_\theta^J \exp(-\lambda(y, f_\theta(p_n)))\right)$
8: solve $\sum_{y \in \mathbf{Y}} (s - G(y))^+ = 2$ in $s \in \mathbb{R}$
9: **for** $y \in \mathbf{Y}$: $p_n^\dagger\{y\} := (s - G(y))^+/2$
10: output $p_n^\dagger$
11: $P := P \exp(-\lambda(y_n, p_n))$
12: **for** $J \in \mathbf{J}$ and $\theta \in \Theta$: $A_\theta^J := A_\theta^J \exp(-\lambda(y_n, f_\theta(p_n)))$.

$$f_{\alpha,\beta}(p)\{y\} := \frac{p\{y\}^\beta \exp(\alpha(y))}{\sum_{y' \in \mathbf{Y}} p\{y'\}^\beta \exp(\alpha(y'))}, \qquad (10.20)$$

where $\alpha \in \mathbb{R}^{\mathbf{Y}}$ and $\beta \in \mathbb{R}$. Formally, the number of parameters in (10.20) is $|\mathbf{Y}|+1$, but the effective number of parameters is $|\mathbf{Y}|$, since the transformation (10.20) does not change when the same constant (positive or negative) is added to all $\alpha(y)$, $y \in \mathbf{Y}$. The neutral value of the parameter $\theta := (\alpha, \beta)$ in (10.20) is $\mathbf{0} := (0, \ldots, 0, 1)$; for this value, f_θ is the identity function, $f_{\mathbf{0}}(p) = p$.

Arguably the log loss function is the most fundamental loss function (see, e.g., [387]), but its disadvantage is that, for typical base prediction algorithms, we need truncation, as in (10.19), to prevent an infinite loss. A popular alternative to the log loss function is the *Brier loss function*

$$\lambda(y, p) := \sum_{y' \in \mathbf{Y}} \left(p\{y'\} - 1_{y'=y}\right)^2 . \qquad (10.21)$$

It is much more forgiving and does not require truncation. In the binary case, the Brier loss is twice as large as the square loss (6.8) (so these are equivalent for us).

Algorithm 10.5 is closely connected with the Brier loss function. We still use the notation L_n for the cumulative loss for this loss function, and the *price of protection* for Algorithm 10.5 is then defined by the left-hand side of (10.18) (the right-hand side is not applicable any longer). We will see in Theorem 10.3 below that the price of protection for Algorithm 10.5 is at most $\log \frac{1}{\pi}$.

In one respect Algorithm 10.5 is simpler than Algorithm 10.4: the former does not have any analogue of lines 12–14, which perform normalization of the weights P and A_θ^J, in the latter. The reason is that the procedure for computing the protected prediction in lines 7–9 of Algorithm 10.5 is invariant with respect to multiplying the weights by a positive constant. One potential danger in the absence of weight normalization at each step is numerical underflow as result of the weight updates 11–12 in Algorithm 10.5 (which can only reduce the weights). To prevent it, it is sufficient to perform occasional weight normalization, as we did in Sect. 8.6.6.

For the Bank Marketing dataset, Algorithm 10.5 leads to similar results to Algorithm 10.4 (we will say about this more in Sect. 10.5.2). In practice, before solving the equation in line 8 of Algorithm 10.5, we can first solve it ignoring the $^+$, which gives $s := (2 + \sum_y G(y))/|\mathbf{Y}|$, and then check that the probabilities we get in line 9 are still nonnegative if we ignore the $^+$. (If not, we will have to solve the equation in line 8 in a different way, but this never happens in the binary case, such as for the Bank Marketing dataset.)

10.3.5 Theoretical Guarantees

A simple performance guarantee for Algorithms 10.4 and 10.5 is given by the following result (proved in Sect. 10.4), which uses the notation (more specific than the $L_n(\Pi)$ that we used earlier)

$$\mathrm{Loss}(p_1, \dots, p_n \mid y_1, \dots, y_n) := \sum_{i=1}^{n} \lambda(y_i, p_i)$$

for the cumulative log loss (in the case of Algorithm 10.4) or Brier loss (in the case of Algorithm 10.5) of predictions p_i on labels y_i.

Theorem 10.3 *The price of protection for Algorithms 10.4 and 10.5 is at most* $\log \frac{1}{\pi}$. *Besides, for any n, any jumping rate* $J \in \mathbf{J}$, *any sequence of examples, and any sequence* $f_{\theta_1}, \dots, f_{\theta_n}$ *of calibration functions (from the family* $(f_\theta \mid \theta \in \Theta)$*),*

$$\mathrm{Loss}(p_1^\dagger, \dots, p_n^\dagger \mid y_1, \dots, y_n) \le \mathrm{Loss}(f_{\theta_1}(p_1), \dots, f_{\theta_n}(p_n) \mid y_1, \dots, y_n) + \log \frac{1}{1-\pi} + \log |\mathbf{J}| + k \log \frac{|\Theta| - 1}{J'} + (n-k-1) \log \frac{1}{1-J'}, \tag{10.22}$$

where

$$J' := \frac{|\Theta| - 1}{|\Theta|} J \tag{10.23}$$

and $k = k(\theta_1, \dots, \theta_n)$ *is the number of switches,*

$$k := |\{i \in \{1, \dots, n\} : \theta_i \ne \theta_{i-1}\}| ,$$

with $\theta_0 = \mathbf{0}$ *being the given initial value.*

The price of protection $\log \frac{1}{\pi}$ in Theorem 10.22 is small unless π is very close to 0, even for test sets of a moderate size. For example, setting $\pi := 0.5$ appears a reasonable compromise between the two terms involving π in the price of protection and in (10.22), and this is what we used in our experiments in Sect. 10.3.3.

The value J' introduced in (10.23) is an alternative parametrization of the jumping rate (and it is used in, e.g., [380] as the main one); it is usually close to (or at least has the same order of magnitude as) J. We may call J' the effective jumping rate: it is the probability that the state of the underlying Markov chain actually changes at a given step.

The *regret term*, given by the last four addends of (10.22), can be interpreted as follows. First, it gives the degree to which we are competitive with an "oracular" predictive system that knows in advance which calibration function should be used at each step. We have already discussed the addend $\log \frac{1}{1-\pi}$, and $\log |\mathbf{J}|$ is the price, typically very moderate ($\log 3$ in our experiments), that we pay for using several jumping rates. The following addend in the regret term gives us the price, in terms of the log or Brier loss, for each switch. Namely, each switch costs us $\log(|\Theta| - 1)$ (which is $\log 8$ in our experiments) plus an amount that depends on the switching rate, namely $\log \frac{1}{J'}$. The last term in (10.22) is close, assuming $k \ll n$ and $J' \ll 1$, to nJ'. A reasonable choice of J' is the inverse $1/N$ of the expected number N of examples in the test set (but remember that Algorithms 10.4 and 10.5 cover a range of J, which is motivated by N typically not being known in advance). With this choice the price $\log \frac{1}{J'}$ to pay per switch becomes $\log N$.

Theorem 10.3 is also applicable to the case of regression (with the log loss function (10.3)), namely the counterpart of Algorithm 10.1 with Composite Jumper in place of Simple Jumper. The price of protection in this case is also defined by the left-hand side of (10.18).

10.4 Complements and Proofs

We start from the description of the general Aggregating Algorithm (AA) containing Algorithms 10.4 and 10.5 as special cases and then derive Theorem 10.3 from its properties.

10.4.1 Aggregating Algorithm

The AA has a loss function $\lambda : \mathbf{Y} \times \mathbf{P}(\mathbf{Y}) \to [0, \infty]$ as its parameter, which in our current context is either the log loss function

$$\lambda(y, p) := -\log p\{y\} \tag{10.24}$$

(the multiclass version of (10.3) and (10.11)) or the Brier loss function (10.21). Another parameter is a probability measure P_0 on a measurable space $\boldsymbol{\Theta}$; this is the *prior distribution* specifying the initial weights assigned to the *experts* $\boldsymbol{\theta} \in \boldsymbol{\Theta}$. Let $\xi_n : \boldsymbol{\Theta} \to \mathbf{P}(\mathbf{Y})$ be the function, assumed to be measurable, mapping each expert $\boldsymbol{\theta} \in \boldsymbol{\Theta}$ to the prediction that he makes at trial n.

First we describe an algorithm (the *Aggregating Pseudo-Algorithm*, or APA) that is allowed to make not permitted predictions $p \in \mathbf{P}(\mathbf{Y})$ but "mixtures", in some sense, of permitted predictions. A *generalized prediction* is defined to be any function of the type $\mathbf{Y} \to [0, \infty]$; a permitted prediction $p \in \mathbf{P}(\mathbf{Y})$ is identified with the generalized prediction g defined by $g(y) := \lambda(y, p)$. The APA suffers loss $g_n(y_n)$ after choosing generalized prediction g_n when the actual outcome (the label chosen by Reality) is y_n.

The APA proceeds as follows while processing the labels $y_1, y_2, \ldots$. At every trial $n = 1, 2, \ldots$ we update the experts' weights,

$$P_n(\mathrm{d}\boldsymbol{\theta}) := \exp(-\lambda(y_n, \xi_n(\boldsymbol{\theta})))P_{n-1}(\mathrm{d}\boldsymbol{\theta}), \quad \boldsymbol{\theta} \in \boldsymbol{\Theta} , \tag{10.25}$$

where P_0 is the prior distribution. (So the weight of an expert $\boldsymbol{\theta}$ whose prediction $\xi_t(\boldsymbol{\theta})$ leads to a large loss $\lambda(y_n, \xi_n(\boldsymbol{\theta}))$ gets slashed.) The generalized prediction chosen by the APA at trial n is the weighted average [198] of the experts' predictions:

$$g_n(y) := -\ln \int_{\boldsymbol{\Theta}} \exp(-\lambda(y, \xi_n(\boldsymbol{\theta})))P^*_{n-1}(\mathrm{d}\boldsymbol{\theta}) , \tag{10.26}$$

where P^*_{n-1} are the normalized weights,

$$P^*_{n-1}(\mathrm{d}\boldsymbol{\theta}) := \frac{P_{n-1}(\mathrm{d}\boldsymbol{\theta})}{P_{n-1}(\boldsymbol{\Theta})}$$

(assuming that the denominator is positive; if it is 0, P_0-almost all experts have suffered infinite loss, in which case the APA and AA are allowed to choose any (generalized) prediction).

The AA is obtained from the APA by replacing each generalized prediction g_n by a permitted prediction $p_n \in \mathbf{P}(\mathbf{Y})$ that dominates g_n, in the sense that, for any $y \in \mathbf{Y}$, $\lambda(y, p_n) \le g_n(y)$. This is possible for both the log loss function (10.24) and the Brier loss function (10.21). For the log loss function, this statement is obvious, since any mixture (10.26) of permitted predictions is again a permitted prediction. For the Brier loss function, it can be checked by computing the Gauss–Kronecker curvature of the boundary of the set of permitted predictions in $[0, \infty)^{\mathbf{Y}}$; for details, see [399, Sect. 4].

In the log loss case, the AA coincides with the APA. For the AA in the Brier case, we need to choose a specific permitted prediction that dominates a given generalized prediction g_n (for which such a permitted prediction exists). The generalized prediction g corresponding to $p \in \mathbf{P}(\mathbf{Y})$ satisfies $\forall y : g(y) = a - 2p\{y\}$ for some constant a. Assuming that g_n is given up to an additive constant (so that we can use unnormalized weights, as we do in Algorithm 10.5), it seems obvious that we should choose p such that $2p\{y\} \ge a - g_n(y)$ for all y and for an a that is as large as possible. In other words, we should choose

$$p\{y\} := (a - g_n(y))^+/2 , \tag{10.27}$$

where a is chosen to make p a probability measure. This is checked in detail in [399, Sect. 5], and this choice is used in Algorithm 10.5 (lines 7–9).

The proof of Theorem 10.3 will be based on the following properties of the APA and AA, in which we use the notation

$$L_n(\boldsymbol{\theta}) := \sum_{i=1}^{n} \lambda(y_i, \xi_i(\boldsymbol{\theta})), \quad L_n(\mathrm{APA}) := \sum_{i=1}^{n} g_i(y_i), \quad L_n(\mathrm{AA}) := \sum_{i=1}^{n} \lambda(y_i, p_i)$$

for the cumulative loss of expert $\boldsymbol{\theta}$, the APA (outputting generalized predictions g_i), and the AA (outputting predictions p_i) over the first n trials, respectively.

Lemma 10.4 *For any prior distribution P_0 and any $n = 1, 2, \dots$,*

$$L_n(\mathrm{APA}) = -\ln \int_{\boldsymbol{\Theta}} \exp(-L_n(\boldsymbol{\theta})) P_0(\mathrm{d}\boldsymbol{\theta}) . \tag{10.28}$$

Proof We are required to deduce (10.28) from the formula (10.26) for computing generalized predictions. It is more natural, however, to move in the opposite direction, from (10.28) to (10.26). If we want to achieve goal (10.28) (which is the generalization of the standard formula for the Bayesian mixture), we have little choice but to compute generalized predictions using (10.26). Indeed, noticing that

$$P_{n-1}(\mathrm{d}\boldsymbol{\theta}) = \exp(-L_{n-1}(\boldsymbol{\theta})) P_0(\mathrm{d}\boldsymbol{\theta})$$

(this immediately follows from (10.25)) and assuming (10.28), we obtain:

$$g_n(y_n) = L_n(\mathrm{APA}) - L_{n-1}(\mathrm{APA}) = -\ln \frac{\int_{\boldsymbol{\Theta}} \exp(-L_n(\boldsymbol{\theta})) P_0(\mathrm{d}\boldsymbol{\theta})}{\int_{\boldsymbol{\Theta}} \exp(-L_{n-1}(\boldsymbol{\theta})) P_0(\mathrm{d}\boldsymbol{\theta})}$$

$$= -\ln \frac{\int_{\boldsymbol{\Theta}} \exp(-L_{n-1}(\boldsymbol{\theta}) - \lambda(y_n, \xi_n(\boldsymbol{\theta}))) P_0(\mathrm{d}\boldsymbol{\theta})}{\int_{\boldsymbol{\Theta}} \exp(-L_{n-1}(\boldsymbol{\theta})) P_0(\mathrm{d}\boldsymbol{\theta})} = -\ln \frac{\int_{\boldsymbol{\Theta}} \exp(-\lambda(y_n, \xi_n(\boldsymbol{\theta}))) P_{n-1}(\mathrm{d}\boldsymbol{\theta})}{P_{n-1}(\boldsymbol{\Theta})} ,$$

which coincides with (10.26). Reversing this argument, we can see that (10.26) is not only necessary but also sufficient for (10.28). □

Lemma 10.4 implies

$$L_n(\mathrm{AA}) \le -\ln \int_{\boldsymbol{\Theta}} \exp(-L_n(\boldsymbol{\theta})) P_0(\mathrm{d}\boldsymbol{\theta}) . \tag{10.29}$$

Algorithm 10.4 is the AA applied to the log loss function (10.24), the prior distribution P_0, as described at the beginning of Sect. 10.3.2, on the sequences $\boldsymbol{\theta} = (\theta_1, \theta_2, \dots)$ of indices of calibration functions f_θ, and with $\xi_n(\boldsymbol{\theta})$ defined as $f_{\theta_n}(p_n)$. This is an explicit self-contained description of how P_0 generates $\boldsymbol{\theta} := (\theta_1, \theta_2, \dots)$: set $\theta_0 := \mathbf{0}$; with probability π, set $\theta_1 = \theta_2 = \dots = \mathbf{0}$; otherwise, choose the jumping rate J from the uniform probability measure on $\mathbf{J}$ and generate $\theta_1, \theta_2, \dots$ from the Markov chain whose transition function prescribes maintaining the same state with probability $1 - J$ and, with probability J, choosing a new state from the uniform probability measure on Θ. This description of the procedure of protection assumes $p_1, p_2, \dots$ given; we are interested in them being produced by the base predictive system. Algorithm 10.5 is analogous but uses the Brier loss function and the rule (10.27) for replacing generalized predictions by permitted predictions.

10.4.2 Proof of Theorem 10.3

The price of protection being at most $\log \frac{1}{\pi}$ is an immediate implication of (10.29): with P_0-probability at least π, $\boldsymbol{\theta} = (\mathbf{0}, \mathbf{0}, \dots)$, and so $p_n^\dagger = p_n$ for all n.

To prove (10.22), notice that the probability that the Simple Jumper Markov chain with a given jumping rate $J \in \mathbf{J}$ and starting from $\theta_0 := \mathbf{0}$ produces $\theta_1, \ldots, \theta_n$ with k switches as its first n states is

$$\left(\frac{J'}{|\Theta| - 1}\right)^k (1 - J')^{n-k-1} .$$

Therefore, the P_0-probability of obtaining $\theta_1, \ldots, \theta_n$ as the first n states is at least

$$\frac{1-\pi}{|\mathbf{J}|} \left(\frac{J'}{|\Theta| - 1}\right)^k (1 - J')^{n-k-1} ,$$

and it remains to remember that the cumulative loss of any expert whose sequence of states begins with $\theta_1, \ldots, \theta_n$ is $\mathrm{Loss}(f_{\theta_1}(p_1), \ldots, f_{\theta_n}(p_n) \mid y_1, \ldots, y_n)$.

10.5 Context

In our collaboration with Stena Line in 2018–2021, we analyzed company data for ferry bookings, in particular its Gothenburg–Kiel route. We found that standard conformal test martingales, such as the Simple Jumper, were extremely sensitive. This suggested that it is very common for the distribution of the data to change, right after a prediction algorithm has been trained.

Unfortunately, standard benchmark datasets are ill-suited for experiments with protected prediction. The data with timestamps tend to be time series data, not the data of the kind processed by standard machine-learning packages. For the latter, the timestamps are usually regarded as superfluous and discarded when the final dataset is created. And usually there is no guarantee that the examples are reported in their chronological order.

10.5.1 Regression

This section is based on [390].

The uniformity of the distribution of the probability integral transforms was used by Lévy [231, Sect. 39] as the foundation of his theory of denumerable probabilities (which allowed him to avoid using the then recent axiomatic foundation suggested by Kolmogorov in his *Grundbegriffe* [199]). Modern papers, including [74], usually refer to Rosenblatt [295], who disentangled Lévy's argument from his concern with the foundations of probability; Rosenblatt, however, refers to Lévy's [231] book in his paper.

It is well known that the probability integral transformation can be used for testing predictive systems considered as data-generating distributions. See, e.g., [74, Sects. 3.8 and 4.7]. According to [134, Sect. 3.1], it forms the cornerstone of checking calibration of probability forecasts.

Remark 10.5 The probability integral transformation can be regarded as a way of normalizing the labels, so that standard test martingales become more likely to work. In principle, all the methods proposed in this chapter can be applied directly to the labels rather than to their transforms.

10.5.2 *Classification*

Section 10.3 is based on [413]. In that paper we use two more datasets with the examples listed in the chronological order. One is the electricity dataset [124, 150], and the other is the UJIIndoorLoc dataset (with three classes) [356]. For both datasets the improvement resulting from protection is significant.

The Brier loss function (10.21) was introduced by Glenn Brier [37] in 1950. It is especially popular in meteorology.

A useful feature of our procedure of protection is that the insurance it provides is cheap, which is achieved by mixing Simple Jumper martingales with constant 1: see (10.13). The role of mixing with 1 is to insure against a catastrophic loss of evidence against the null hypothesis (given by the base predictive system) found by those test martingales. There are much more sophisticated ways of insuring against loss of evidence [320, Chap. 11], and they will provide further protection.

Notice that calibration functions may depend not only on the current predicted probability p but also on the current object x. (So that "calibration" may be understood in a very wide sense, as in [71], and include elements of "resolution" [72].)

Further Experimental Results for Classification

First we report results for the family (10.9) of quadratic calibration functions with θ restricted to a finite set Θ. We choose a minimal Θ, namely $\Theta := \{-1, 0, 1\}$. The left panel of Fig. 10.10 is the counterpart of the right panel of Fig. 10.8 for this family.

In the rest of this subsection we only consider Cox's calibration functions. The inclusion of the value $\beta = 0$ (or β close to 0) is essential for the impressive performance we are getting. The right panel of Fig. 10.10 is the counterpart of the right panel of Fig. 10.8 for Algorithm 10.5.

Next we discuss some of the reasons for a good performance of our protection procedures on the Bank Marketing and electricity datasets. It is revealing that protection still gives good results for both datasets when the objects are randomly shuffled (so that they become uninformative). See Fig. 10.11. Therefore, already the order of the test labels is informative. Of course, the ROC curves for the unprotected procedures become trivial (close to the main diagonal) when the objects are shuffled.

To understand the reasons for the ROC curves being nontrivial after protection in Fig. 10.11, we compute the moving averages of the labels (including both training and test sets). Figure 10.12 shows in red the path of the moving average of the labels: the value of the path at time n is the

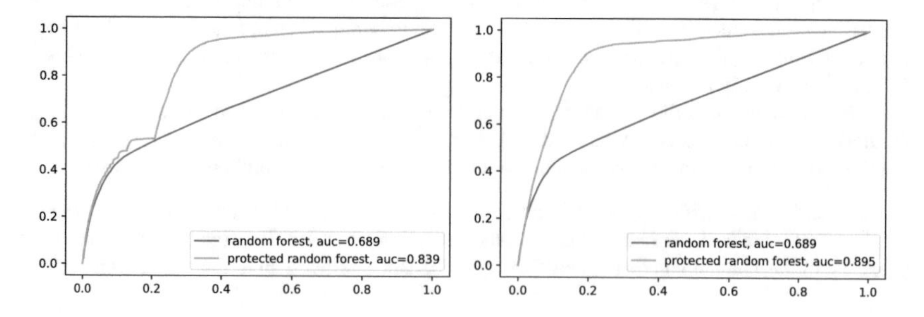

Fig. 10.10 The ROC curves for the Bank Marketing dataset and random forest with the Composite Jumper protection. Left panel: based on quadratic rather than Cox calibration. Right panel: using the Brier loss function (Algorithm 10.5)

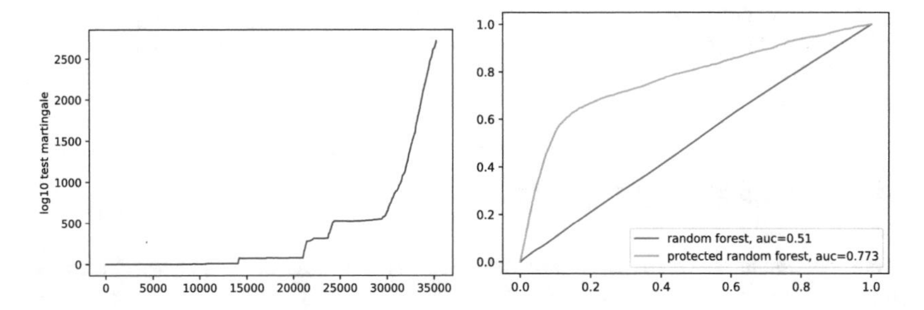

Fig. 10.11 The analogue of Fig. 10.8 for the Bank Marketing dataset with randomly permuted objects

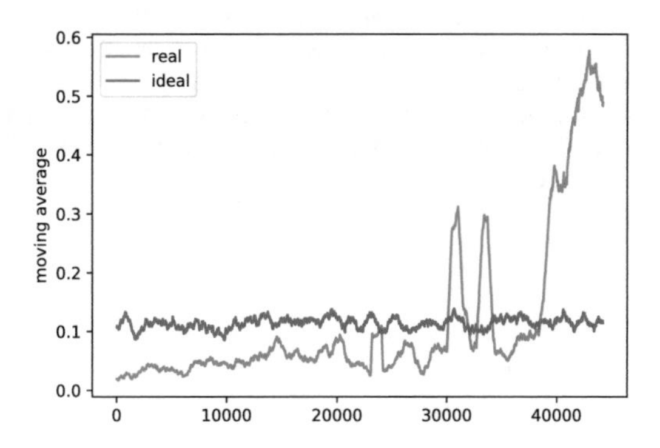

Fig. 10.12 The moving averages of the labels for the Bank Marketing dataset, as described in the main text

arithmetic mean of the 1000 consecutive labels starting from y_n (namely, the arithmetic mean of $y_n, \ldots, y_{n+999}$). For comparison, the analogous moving average for a simulated Bernoulli sequence with the right percentage of 1s (12% for Bank Marketing) is shown in green. The behaviour of the moving average is clearly anomalous: the proportion of successful calls increases drastically towards the end of the dataset, which explains the quick growth of the Composite Jumper martingale in Figs. 10.8 and 10.11 starting from approximately the 30,000th example in the test set, which corresponds to the 40,000th example in the full dataset. The percentage of successful calls is 3.5% for the training set and 14.0% for the test set. The behaviour for the electricity dataset, mentioned at the beginning of this subsection, is less anomalous, but there are still clear non-random patterns in it.

Generally, in hindsight, we can usually explain the reason why the exchangeability assumption fails when it does. But of course, this does not mean that such failures can be prevented, even though standard machine-learning methods depend on this assumption.

10.5.3 Standard Loss Functions

Starting from Chap. 6 ((6.7) and (6.8)), we have been discussing two main loss functions used for evaluating predictions, log loss and square/Brier loss, in their various guises, including the cases of regression (10.3), binary classification (10.11), and multiclass classification ((10.24) and (10.21)). They are strictly proper loss functions, meaning that the optimal expected loss is only attained by the true predictive distribution (or density, in the case of (10.3)); see, e.g., [72] or [133, Sect. 4].

In Sect. 6.4.3 we discussed a weaker property, propriety. In Chap. 3 the notion of propriety for loss functions served for us as motivation for introduction of conditionally proper criteria of efficiency (Sect. 3.1.7).

10.5.4 Aggregating Algorithm

In our description of the AA we follow [380, Sect. 2] and [381]. In its simplest form it was proposed in [377] (as mentioned in Sect. 8.6.1) as a common generalization of the Weighted Majority Algorithm [236, 378] and the Bayesian merging scheme. The case of Brier loss was analyzed in detail in [399].

As we mentioned in Sect. 8.6.1, the AA also covers tracking the best expert in prediction with expert advice. In designing betting martingales we can apply all arsenal of the tools developed in this area, such as switching between different experts, averaging over parameters, sleeping, etc. See, e.g., [380] for a review. We have used a small subset of these methods (including a basic version of sleeping in Sect. 10.2.3). In the case where $\pi = 0$ and $|\mathbf{J}| = 1$, Algorithms 10.4 and 10.5 can be considered special cases of the Fixed Share algorithm of [157], already mentioned in Sect. 8.6.1.

Part IV
Online Compression Modelling

Chapter 11
Generalized Conformal Prediction

Abstract In the previous chapters we assumed that the data are generated from an exchangeable probability measure. In this chapter we generalize the method of conformal prediction to cover arbitrary statistical models that belong to the class of, as we call them, online compression models. Interesting online compression models include, e.g., partial exchangeability models, Gaussian models, and causal networks.

Keywords Online compression model · Repetitive structure · One-off structure · Exchangeability model · Partial exchangeability model · Gaussian model · Gauss linear model · Multivariate Gaussian model

11.1 Introduction

We know that each conformal predictor is automatically valid, in particular has correct long-run frequencies of errors at different significance levels, when used in the online mode and provided that the data sequence is generated by an exchangeable distribution. In this chapter we state a more general result replacing the assumption of exchangeability of the data-generating distribution by the assumption that the data agrees with a given "online compression model"; the exchangeability model is just one of many interesting models of this type. This chapter's result is a step towards implementation of Kolmogorov's programme for applications of probability; in particular, the concept of online compression model is an online version of the concept considered by Kolmogorov (and is closely connected to Martin-Löf's repetitive structures and Freedman's summarizing statistics).

As in the previous chapters, we also state explicitly simpler versions of our definitions and results that are more relevant for the one-off and offline modes of prediction.

We start Sect. 11.2 from defining the online compression models, which include, besides the exchangeability model and its extensions, the Gaussian model, the hypergraphical model, and other interesting models. An online compression model (OCM) is an automaton (usually infinite) for summarizing statistical information efficiently. It is usually impossible to restore the statistical information from the

V. Vovk et al., *Algorithmic Learning in a Random World*,
https://doi.org/10.1007/978-3-031-06649-8_11

OCM's summary (so the OCM performs lossy compression), but it can be argued that the only information lost is noise, since one of our requirements is that the summary should be a "sufficient statistic". In Sect. 11.2.2 we construct conformal transducers for an arbitrary OCM and state a simple theorem (proved in Sect. 11.5.1) showing that the confidence information provided by conformal transducers is valid; this theorem generalizes the validity result of Chap. 2. We then briefly remind the reader how conformal transducers are used for confidence prediction. In Sect. 11.2.4 we describe an alternative language for online compression modelling; it is based on Martin-Löf's notion of repetitive structure and is very convenient when specific models are discussed. In the following Sect. 11.2.5, we simplify the notions of online compression models and repetitive structures to the notion of a one-off structure (OOS). One-off structures are easier to define, are more general, but satisfy a weaker property of validity. In the following two sections, 11.3 and 11.4, we consider several interesting examples of online compression models: exchangeability and partial exchangeability, Gaussian, Gauss linear, and multivariate Gaussian models. Another interesting example, hypergraphical models, will be discussed and used in conformal prediction starting from Sect. 12.5 in the next chapter; these models are closely related to causal networks. In Sect. 11.6 we discuss the origins of the idea of online compression modelling and of specific online compression models.

11.2 Online Compression Models and Their Modifications

We are interested in making predictions about a sequence of examples $z_1, z_2, \dots$ output by Reality. Typically we will want to say something about example z_n, $n = 1, 2, \dots$, given the previous examples $z_1, \dots, z_{n-1}$. In this section we will discuss assumptions that we might be willing to make about the examples and, in Sect. 11.2.2, prediction algorithms based on those assumptions.

11.2.1 Online Compression Models

An *online compression model* (OCM) is a quintuple

$$M = (\Sigma, \Box, \mathbf{Z}, F, B) ,$$

where:

1. Σ is a measurable space called the *summary space*; its elements are called *summaries*; $\Box \in \Sigma$ is a summary called the *empty summary*;
2. $\mathbf{Z}$ is a measurable space from which the examples z_i are drawn;
3. F is a measurable function of the type $\Sigma \times \mathbf{Z} \to \Sigma$ called the *forward function*;

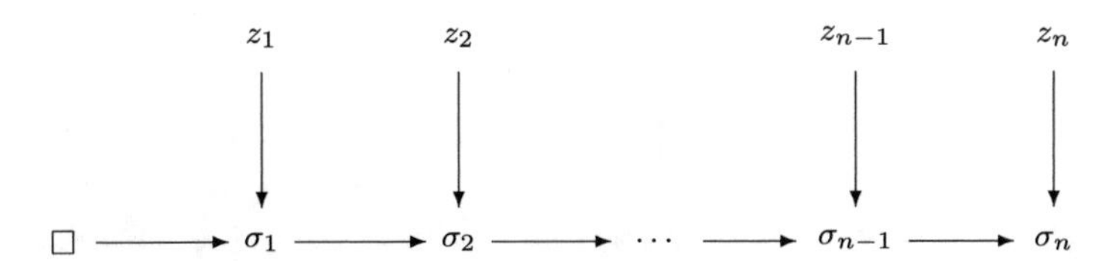

Fig. 11.1 Using the forward function F to compute σ_n from $z_1, \ldots, z_n$

4. B is a Markov kernel (see Sect. A.4) of the type $\Sigma \hookrightarrow \Sigma \times \mathbf{Z}$ called the *backward kernel*; it is required that B be an inverse to F in the sense that

$$B\left(F^{-1}(\sigma) \mid \sigma\right) = 1$$

for each $\sigma \in F(\Sigma \times \mathbf{Z})$.

Next we explain briefly the intuition behind this definition, introduce some further notation, and impose another condition.

An OCM is a way of summarizing statistical information. At the beginning we do not have any information, which is represented by the empty summary $\sigma_0 := \Box$. When the first example z_1 arrives, we update our summary to $\sigma_1 := F(\sigma_0, z_1)$, etc.; when example z_n arrives, we update the summary to $\sigma_n := F(\sigma_{n-1}, z_n)$. This process is represented in Fig. 11.1. Let t be the *summarizing statistic* in the OCM, which maps any finite sequence of examples $z_1, \ldots, z_n$ (of any length $n \geq 0$) to σ_n:

$$\begin{aligned} t() &:= \sigma_0 = \Box; \\ t(z_1, \ldots, z_n) &:= F(t(z_1, \ldots, z_{n-1}), z_n), \quad n = 1, 2, \ldots . \end{aligned} \tag{11.1}$$

The value $t(z_1, \ldots, z_n)$ is a summary of the full data sequence $z_1, \ldots, z_n$ available at the end of trial n; this includes the summary $t()$ of the empty sequence. Our definition requires that the summaries should be computable online: the function F updates σ_{n-1} to σ_n for any n.

Condition 3 in the definition of OCM reflects its online character, as explained in the previous paragraph. We want, however, the system of summarizing statistical information represented by the OCM to be accurate, so that no useful information is lost. This is reflected in Condition 4: the distribution B of the more detailed description (σ_{n-1}, z_n) given the less detailed σ_n is known, and so (σ_{n-1}, z_n) does not carry any additional information about the distribution generating the examples $z_1, z_2, \ldots$; in other words, σ_n contains the same useful information as (σ_{n-1}, z_n), and the extra information in (σ_{n-1}, z_n) is noise. This intuition would be captured in statistical terminology (see, e.g., [64, Sect. 2.2]) by saying that σ_n is a "sufficient statistic" of (σ_{n-1}, z_n) and, eventually, of $z_1, \ldots, z_n$ (although this expression does not have a formal meaning in our present context, since we do not have a full statistical model $(P_\theta : \theta \in \Theta)$ at this point).

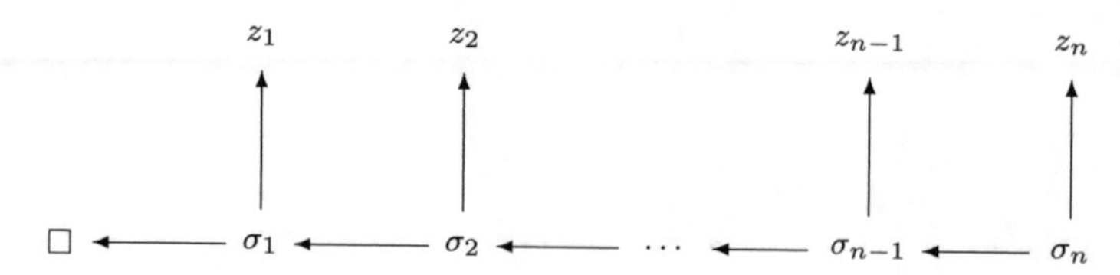

Fig. 11.2 Using the backward kernel B to extract the distribution of $z_1, \ldots, z_n$ from σ_n

Analogously to Fig. 11.1, we can find the distribution of the data sequence $z_1, \ldots, z_n$ from σ_n (see Fig. 11.2): given σ_n, we generate the pair (σ_{n-1}, z_n) from the distribution $B(\sigma_n)$, then we generate (σ_{n-2}, z_{n-1}) from $B(\sigma_{n-1})$, etc. Formally, using the Markov kernels $B(\mathrm{d}\sigma_{n-1}, \mathrm{d}z_n|\sigma_n)$, we can define the conditional distribution P_n of $z_1, \ldots, z_n$ given σ_n by the requirement that, for all bounded measurable functions $f : \mathbf{Z}^n \to \mathbb{R}$,

$$\int f(z_1, \ldots, z_n) P_n(\mathrm{d}z_1, \ldots, \mathrm{d}z_n \mid \sigma_n) := \int \cdots \int f(z_1, \ldots, z_n) B(\mathrm{d}\sigma_0, \mathrm{d}z_1 \mid \sigma_1) B(\mathrm{d}\sigma_1, \mathrm{d}z_2 \mid \sigma_2) \ldots B(\mathrm{d}\sigma_{n-2}, \mathrm{d}z_{n-1} \mid \sigma_{n-1}) B(\mathrm{d}\sigma_{n-1}, \mathrm{d}z_n \mid \sigma_n) \; ; \quad (11.2)$$

the existence of such a probability distribution P_n immediately follows from the Stone–Daniell theorem (see, e.g., [94, Theorem 4.5.2]). A shorter way to write (11.2) is

$$P_n(\mathrm{d}z_1, \ldots, \mathrm{d}z_n \mid \sigma_n) := B(\mathrm{d}\sigma_0, \mathrm{d}z_1 \mid \sigma_1) B(\mathrm{d}\sigma_1, \mathrm{d}z_2 \mid \sigma_2) \ldots B(\mathrm{d}\sigma_{n-2}, \mathrm{d}z_{n-1} \mid \sigma_{n-1}) B(\mathrm{d}\sigma_{n-1}, \mathrm{d}z_n \mid \sigma_n) \, .$$

It is sometimes convenient to use a slightly modified online compression model $(\Sigma, \Box, \mathbf{Z}, F, B)$ removing the "unused" parts of Σ, F, and B (e.g., $B(\Box)$ is never used). We will write Σ_n for $t(\mathbf{Z}^n)$, $n \geq 0$, including $\Sigma_0 = \{\Box\}$; the elements of Σ_n will be called *n-summaries*. The final condition that we impose on OCMs is that the sets Σ_n are required to be disjoint. Let

$$F_n := F|_{\Sigma_{n-1} \times \mathbf{Z}}, \quad B_n := B|_{\Sigma_n} \, ,$$

$f|_E$ standing for the restriction of f to E. All our definitions and results will involve only these summaries and restrictions. The *reduced online compression model* is $((\Sigma_n), \Box, \mathbf{Z}, (F_n), (B_n))$ (n ranging over $\mathbb{N}$ except that Σ_0 is also defined).

We say that a probability distribution P on $\mathbf{Z}^\infty$ *agrees* with the OCM $(\Sigma, \Box, \mathbf{Z}, F, B)$ if, for each n and each event $A \subseteq \Sigma \times \mathbf{Z}$, $B(A \mid \sigma)$ is a version of the conditional probability, w.r.t. to P, that $(t(z_1, \ldots, z_{n-1}), z_n) \in A$ given $t(z_1, \ldots, z_n) = \sigma$ and given the values of $z_{n+1}, z_{n+2}, \ldots$.

11.2.2 Conformal Transducers and Validity of OCM

In Chap. 2, any function f of the type $(\mathbf{Z} \times [0, 1])^* \to [0, 1]$ was called a *randomized transducer*; it is regarded as mapping each input sequence $(z_1, \tau_1, z_2, \tau_2, \ldots)$ in $(\mathbf{Z} \times [0, 1])^\infty$ into the output sequence of *p-values* $(p_1, p_2, \ldots)$ defined by $p_n := f(z_1, \tau_1, \ldots, z_n, \tau_n)$, $n = 1, 2, \ldots$. We say that the randomized transducer f is *exactly valid* w.r.t. to an OCM M if the output p-values $p_1 p_2 \ldots$ are always distributed according to the uniform distribution $\mathbf{U}^\infty$ on $[0, 1]^\infty$ provided the input examples $z_1 z_2 \ldots$ are generated by a probability distribution that agrees with M and $\tau_1 \tau_2 \ldots$ are generated, independently of $z_1 z_2 \ldots$, from $\mathbf{U}^\infty$. If we drop the dependence on the random numbers τ_n, we obtain the notion of *deterministic transducer*.

Any measurable function $A : \Sigma \times \mathbf{Z} \to \overline{\mathbb{R}}$ is called a *nonconformity measure* w.r.t. to the OCM $M = (\Sigma, \Box, \mathbf{Z}, F, B)$. Let $B_\mathbf{Z}$ be the marginal distribution of the backward kernel B on $\mathbf{Z}$:

$$B_\mathbf{Z}(E \mid \sigma) := B(\Sigma \times E \mid \sigma) \tag{11.3}$$

for any event $E \subseteq \mathbf{Z}$. The *conformal transducer* determined by A is the deterministic transducer where p_n are defined as

$$p_n := B_\mathbf{Z}\left(\{z \in \mathbf{Z} : A(\sigma_n, z) \geq A(\sigma_n, z_n)\} \mid \sigma_n\right) \tag{11.4}$$

and $\sigma_n := t(z_1, \ldots, z_n)$. The randomized version, called the *smoothed conformal transducer* determined by A, is obtained by replacing (11.4) with

$$\begin{aligned} p_n := {} & B_\mathbf{Z}\left(\{z \in \mathbf{Z} : A(\sigma_n, z) > A(\sigma_n, z_n)\} \mid \sigma_n\right) \\ & + \tau_n B_\mathbf{Z}\left(\{z \in \mathbf{Z} : A(\sigma_n, z) = A(\sigma_n, z_n)\} \mid \sigma_n\right) . \end{aligned} \tag{11.5}$$

A *conformal transducer in* an OCM M is a conformal transducer (deterministic or smoothed) determined by some nonconformity measure w.r.t. to M.

Theorem 11.1 *Suppose the examples $z_n \in \mathbf{Z}$ are generated from a probability distribution P that agrees with an online compression model and $\tau_n \sim \mathbf{U}$, $n = 1, 2, \ldots$, all independent. Any smoothed conformal transducer in that model is exactly valid (will produce independent p-values p_n distributed uniformly on* $[0, 1]$*).*

As discussed in Chap. 2, conformal transducers can be used for hedged prediction; we will briefly summarize how this is done. Suppose each example z_n consists of two components, x_n (the object) and y_n (the label); at trial n we are given x_n

and the goal is to predict y_n. Suppose we are given a significance level $\epsilon > 0$ (the maximum probability of error we are prepared to tolerate). When given x_n, we can output as the prediction set $\Gamma_n^\epsilon \subseteq \mathbf{Y}$ the set of labels y such that $y_n = y$ would lead to a p-value $p_n > \epsilon$. (When a conformal transducer is applied in this mode, it is referred to as a *conformal predictor*.) If an error at trial n is defined as $y_n \notin \Gamma_n^\epsilon$, then by Theorem 11.1 errors at different trials are independent and the probability of error at each trial is ϵ, assuming the p_n are produced by a smoothed conformal transducer. In particular, such confidence predictors are asymptotically exact, in the sense that the number Err_n^ϵ of errors made in the first n trials satisfies

$$\lim_{n\to\infty} \frac{\mathrm{Err}_n^\epsilon}{n} = \epsilon \quad \text{a.s.}$$

This implies that if the p_n are produced by a deterministic conformal transducer, we will still have the conservative version of this property,

$$\limsup_{n\to\infty} \frac{\mathrm{Err}_n^\epsilon}{n} \leq \epsilon \quad \text{a.s.}$$

11.2.3 Finite-Horizon Result

In this subsection we will state a modification of Theorem 11.1 which, although slightly less elegant than Theorem 11.1 itself, is mathematically stronger and more readily applicable (in particular, it explains why shuffling finite datasets, as described in Sect. B.5, makes smoothed conformal transducers produce independent p-values distributed uniformly on [0, 1]).

Theorem 11.1 is so general that it implies almost all other validity results in this book, but its generality also gives rise to some problems. The statement of the theorem is given in terms of probability distributions P that agree with the given OCM, but even the simplest questions about the class $\mathcal{P}$ of such P are very difficult and can at present be answered only in some special cases (see, e.g., [214]; the exposition in that book is in terms of "repetitive structures", which, as explained in Sect. 11.2.4, provide an equivalent language for talking about online compression modelling). Even the question whether $\mathcal{P}$ is nonempty is difficult (although the answer is known to be positive if $\mathbf{Z}$ is finite; see [214, Proposition 7.1 in Chap. II]). The question whether $\mathcal{P}$ is nontrivial (i.e., whether $|\mathcal{P}| > 1$) is even more difficult; in statistical mechanics this question appears as the question of existence of phase transition. Moreover, the mathematical techniques used to answer these questions are heavily asymptotic and seem very remote from the practice of machine learning.

Fix a positive integer N, the *horizon*. We say that a probability distribution P on $\mathbf{Z}^N$ *agrees* with an OCM $(\Sigma, \Box, \mathbf{Z}, F, B)$ if, for each $n = 1, \dots, N$ and each event $A \subseteq \Sigma \times \mathbf{Z}$, $B(A \mid \sigma)$ is a version of the conditional probability, w.r. to P, that $(t(z_1, \dots, z_{n-1}), z_n) \in A$ given $t(z_1, \dots, z_n) = \sigma$ and given the values of

$z_{n+1}, \ldots, z_N$. The description of the family of all $P \in \mathbf{P}(\mathbf{Z}^N)$ that agree with the OCM is trivial: these are the mixtures of $P_N(\cdot \mid \sigma)$ over $\sigma \in \Sigma_N = t(\mathbf{Z}^N)$. (Indeed, each $P_N(\cdot \mid \sigma)$ agrees with the OCM, and each $P \in \mathbf{P}(\mathbf{Z}^N)$ that agrees with the OCM is $\int_\Sigma P_N(\cdot \mid \sigma) P'(\mathrm{d}\sigma)$, where P' is the image of P under the mapping t.)

Nonconformity measures and conformal transducers for OCM are defined as before, with the only difference that, for the latter, n now ranges over $\{1, \ldots, N\}$.

Theorem 11.2 *Let $N \in \mathbb{N}$, the examples $z_n \in \mathbf{Z}$, $n = 1, \ldots, N$, be generated from a probability distribution P on $\mathbf{Z}^N$ that agrees with an online compression model, and $\tau_n \sim \mathbf{U}$, all independently. Any smoothed conformal transducer in that model is exactly valid, i.e., will produce independent p-values p_n, $n = 1, \ldots, N$, distributed uniformly on $[0, 1]$, even when conditioned on $\sigma_N = t(z_1, \ldots, z_N)$.*

In particular, any smoothed conformal predictor will produce p-values distributed as $\mathbf{U}^N$ if the examples are generated from any conditional distribution $P_N(\sigma)$, $\sigma \in \Sigma_N$.

It is clear that Theorem 11.2 implies Theorem 11.1: if a probability distribution P on $\mathbf{Z}^\infty$ agrees with an OCM, the restriction of P to the first N examples will still agree with the OCM, and therefore, the first N p-values $p_1, \ldots, p_N$ will be distributed according to the uniform distribution on $[0, 1]^N$; now standard results (such as [428, Lemma 1.6]) imply that the infinite sequence $p_1, p_2, \ldots$ has the uniform distribution on $[0, 1]^\infty$.

11.2.4 Repetitive Structures

There are two equivalent languages for discussing online compression modelling: our online compression models and Martin-Löf's repetitive structures. The former is more convenient in the general theory, considered so far, and the latter is better suited to discussing specific models, which we do in the following sections. In this subsection we define repetitive structures and compare the two languages.

Let Σ and $\mathbf{Z}$ be measurable spaces (of "summaries" and "examples", respectively). A *repetitive structure* $(\Sigma, \mathbf{Z}, t, (P_n))$ contains, additionally, the following two elements:

- a *summarizing statistic* (i.e., measurable function) $t : \mathbf{Z}^* \to \Sigma$, with the sets $t(\mathbf{Z}^n)$ disjoint for different n;
- a system of Markov kernels $P_n : \Sigma \hookrightarrow \mathbf{Z}^n$, $n = 0, 1, 2, \ldots$.

These two elements are required to satisfy the following consistency requirements:

Agreement between P_n and t: for each $\sigma \in t(\mathbf{Z}^n)$, the probability distribution $P_n(\cdot \mid \sigma)$ is concentrated on the set $t^{-1}(\sigma) \subseteq \mathbf{Z}^n$;

Online character of t: for all integers $n \geq 1$, $t(z_1, \ldots, z_n)$ is determined by $t(z_1, \ldots, z_{n-1})$ and z_n, in the sense that the function $t(z_1, \ldots, z_n)$ is measurable w.r.t. the σ-algebra generated by the function $t(z_1, \ldots, z_{n-1})$ and z_n;

Consistency of P_n: for all integers $n > 1$ and all $\sigma_n \in t(\mathbf{Z}^n)$, $P_{n-1}(\cdot \mid \sigma_{n-1})$ is a version of the conditional distribution of $z_1, \dots, z_{n-1}$ when $z_1, \dots, z_n$ is generated from $P_n(dz_1, \dots, dz_n | \sigma_n)$ and it is known that $t(z_1, \dots, z_{n-1}) = \sigma_{n-1}$ and $z_n = z$ (σ_{n-1} ranging over $t(\mathbf{Z}^{n-1})$ and z over $\mathbf{Z}$).

The *reduced version* of a repetitive structure $(\Sigma, \mathbf{Z}, t, (P_n))$ is defined to be $((\Sigma_n), \mathbf{Z}, (t_n), (P'_n))$, where $\Sigma_n := t(\mathbf{Z}^n)$, $t_n := t|_{\mathbf{Z}^n}$ (we will also refer to t_n as *summarizing statistics*), and $P'_n := P_n|_{\Sigma_n}$. In our examples of repetitive structures, the Markov kernels $P'_n(\cdot \mid \sigma_n)$, where $\sigma_n \in \Sigma_n$, will often be the uniform probability measures on $t_n^{-1}(\sigma_n)$.

The online character of t can be restated as follows: there exists a sequence of measurable functions $F_n : \Sigma_{n-1} \times \mathbf{Z} \to \Sigma_n$, $n = 1, 2, \dots$, such that

$$t_n(z_1, \dots, z_n) = F_n\left(t_{n-1}(z_1, \dots, z_{n-1}), z_n\right) \tag{11.6}$$

for all n and $z_1, \dots, z_n \in \mathbf{Z}$. This makes repetitive structures more similar to online compression models; the full equivalence is established in the following proposition, whose proof we omit.

Proposition 11.3 *If $M = ((\Sigma_n), \Box, \mathbf{Z}, (F_n), (B_n))$ is a reduced online compression model, then $M' := ((\Sigma_n), \mathbf{Z}, (t_n), (P_n))$ (see* (11.1)*, supplying t with the index indicating the number of arguments, and* (11.2)*) is a reduced repetitive structure. If $M = ((\Sigma_n), \mathbf{Z}, (t_n), (P_n))$ is a reduced repetitive structure, a reduced online compression model $M' = ((\Sigma_n), \Box, \mathbf{Z}, (F_n), (B_n))$ can be defined as follows:*

- *$\Box$ is, e.g., the empty set;*
- *F_n are the functions from* (11.6)*, $n = 1, 2, \dots$.*
- *$B_n(d\sigma_{n-1}, dz_n \mid \sigma_n)$ is the image of the distribution $P_n(dz_1, \dots, dz_n \mid \sigma_n)$ under the mapping $(z_1, \dots, z_n) \mapsto (\sigma_{n-1}, z_n)$, where $\sigma_{n-1} := t_{n-1}(z_1, \dots, z_{n-1})$.*

If M is a reduced online compression model, $M'' = M$. If M is a reduced repetitive structure, $M'' = M$.

11.2.5 One-Off Structures (OOS)

So far in this chapter we have been interested in the online mode of prediction. Now we introduce a model that combines features of online compression models and repetitive structures. Let us fix $n \in \mathbb{N}$, the size of the augmented training set.

A *one-off n-structure* (OOS) is a quadruple $(\Sigma, \mathbf{Z}, t, B)$, where, as before, Σ and $\mathbf{Z}$ are measurable spaces, the *summary space* and *example space*, respectively,

- $t : \mathbf{Z}^n \to \Sigma$ is a measurable function (the *summarizing statistic*);
- $B : \Sigma \hookrightarrow \mathbf{Z}$ is a Markov kernel such that, for each $\sigma \in t(\mathbf{Z}^n)$,

$$B\left(\left\{z : \sigma \in t(\mathbf{Z}^{n-1} \times \{z\})\right\} \mid \sigma\right) = 1 . \tag{11.7}$$

It is clear that every online compression model and every repetitive structure uniquely determine a one-off structure (with the B in (11.7) corresponding to the $B_{\mathbf{Z}}$ in (11.3)).

A *nonconformity n-measure* is a measurable function $A : \Sigma \times \mathbf{Z} \to \mathbb{R}$. Given a training set $z_1, \ldots, z_{n-1}$ and a test object x_n in the one-off setting, with each potential label $y \in \mathbf{Y}$ for x_n we associate the *p-value*

$$\begin{aligned} p^y = p^y(z_1, \ldots, z_{n-1}, x_n) &= p(z_1, \ldots, z_{n-1}, (x_n, y)) \\ &:= B\left(\left\{z \in \mathbf{Z} : A(\sigma^y, z) \geq A(\sigma^y, (x_n, y))\right\} \mid \sigma^y\right) , \end{aligned}$$

where $\sigma^y := t(z_1, \ldots, z_{n-1}, (x_n, y))$. In particular, the prediction set at significance level ϵ is

$$\Gamma^\epsilon = \Gamma^\epsilon(z_1, \ldots, z_{n-1}, x_n) := \{y \in \mathbf{Y} : p^y(z_1, \ldots, z_{n-1}, x_n) > \epsilon\} . \tag{11.8}$$

Let us say that a probability distribution P on $\mathbf{Z}^n$ *agrees* with an OOS $(\Sigma, \mathbf{Z}, t, B)$ if, for each $A \subseteq \mathbf{Z}$, $B(A \mid \sigma)$ is a version of the conditional probability w.r.t. to P that $z_n \in A$ given $t(z_1, \ldots, z_n) = \sigma$. The following proposition is a basic version of Theorem 11.2.

Proposition 11.4 *Let $n \in \mathbb{N}$ and the examples $z_1, \ldots, z_n$ be generated from a probability distribution P on $\mathbf{Z}^n$ that agrees with an OOS $(\Sigma, \mathbf{Z}, t, B)$. Then, for any $\epsilon > 0$, $p(z_1, \ldots, z_n) \leq \epsilon$ with probability at most ϵ. In particular, the prediction set* (11.8) *makes an error, $y_n \notin \Gamma^\epsilon$, with probability at most ϵ.*

Each repetitive structure $(\Sigma, \mathbf{Z}, t, (P_n))$ determines an OOS once we fix n: namely, $(t(\mathbf{Z}^n), \mathbf{Z}, t|_{\mathbf{Z}^n}, B)$ is an OOS, where $B(\sigma)$, $\sigma \in t(\mathbf{Z}^n)$, is the marginal distribution of the nth example generated by $P_n(\sigma)$. Similarly, each OCM determines an OOS. Therefore, notions that we define for OOSs (such as Venn predictors) become applicable to repetitive structures and OCMs (at each n).

Remark 11.5 In the general results about OOSs stated in this chapter, such as Proposition 11.4, the sequence structure of the first $n-1$ examples, $z_1, \ldots, z_{n-1}$, is not essential, and we can replace those $n-1$ examples by one "superexample" $z' \in \mathbf{Z}'$ for an unstructured measurable space $\mathbf{Z}'$. The summarizing statistic will then map $\mathbf{Z}' \times \mathbf{Z}$ to Σ. Of course, the structure of z' will be essential in our specific examples.

11.3 Exchangeability Model and Its Modifications

In this section we discuss some familiar (although not defined formally so far) one-off structures and online compression models, viz., the exchangeability model and its modifications. In Sect. 11.4 we will consider new models, viz., the Gaussian model and its modifications. All specific OCMs will be defined through their summarizing statistic t and conditional distributions P_n (i.e., through the corresponding repetitive structure, which will be called the "repetitive-structure representation" of the OCM). For prediction, however, it will be important to move to the representation as an online compression model (the "online compression representation", as we will say).

11.3.1 Exchangeability OOS

The *exchangeability one-off structure* has the summarizing statistic

$$t(z_1, \ldots, z_n) := \lbag z_1, \ldots, z_n \rbag \,, \tag{11.9}$$

and given the value of the statistic σ, $B(\cdot \mid \sigma)$ assigns the same probability $1/n$ to each element of the bag σ; formally, $B(\{z\} \mid \sigma)$ is the number of times $z \in \mathbf{Z}$ occurs in σ. The notion of a nonconformity measure A in the exchangeability OOS is as in (2.18); as we are only interested in bags of size n, we can also take $A : \mathbf{Z}^{(n)} \times \mathbf{Z} \to \overline{\mathbb{R}}$; $A(\lbag z_1, \ldots, z_n \rbag, z_i)$ is the nonconformity score of z_i in $\lbag z_1, \ldots, z_n \rbag$.

11.3.2 Conditional-Exchangeability OOS

For a given taxonomy $K : \mathbf{Z}^{(*)} \times \mathbf{Z} \to \mathbf{K}$, the summary space of the *conditional-exchangeability OOS* is $\Sigma := \mathbf{K}^n \times (\mathbf{Z}^{(*)})^{\mathbf{K}}$ (the Cartesian product of $\mathbf{K}^n$ and the family of all mappings of the type $\mathbf{K} \to \mathbf{Z}^{(*)}$), the summarizing statistic is

$$\begin{aligned} t(z_1, \ldots, z_n) := \Bigl(& K(\lbag z_1, \ldots, z_n \rbag, z_1), \ldots K(\lbag z_1, \ldots, z_n \rbag, z_n), \\ & \bigl(k \in \mathbf{K} \mapsto \lbag z_i : i \in \{1, \ldots, n\},\ K(\lbag z_1, \ldots, z_n \rbag, z_i) = k \rbag\bigr)\Bigr) \,, \end{aligned} \tag{11.10}$$

and the backward kernel B chooses randomly (with equal probabilities) an element of the bag of examples of the same category as the last example. Formally, if $\sigma = (\kappa_1, \ldots, \kappa_n, S)$ and $z \in \mathbf{Z}$, $B(\{z\} \mid \sigma) := a/b$ where b is the size of the bag $S(\kappa_n)$ and a is the number of times z occurs in $S(\kappa_n)$.

In Chap. 8 we already saw an example illustrating that the conditional-exchangeability model may be less restrictive in an important way than the exchangeability model. In Remark 8.7 we pointed out that the label-conditional conformal predictor may be applicable to a user writing a letter, despite the exchangeability model for the resulting character string being grossly wrong.

Conditional-exchangeability models add another dimension to the usual question "what is exchangeable with what?" As in the case of Venn predictors, we want, on one hand, our predictions to be as specific as possible (which creates pressure on the categories to become smaller) and, on the other hand, we need enough statistics for each category (which resists the pressure). The pressure is increased by smaller categories meaning a weaker assumption about Reality.

11.3.3 Exchangeability OCM

The *exchangeability model* has the summarizing statistic (11.9); given the value of the summarizing statistic, all orderings have the same probability $1/n!$. It is easy to see what the online compression representation of the exchangeability model is. The function $F_n : (\sigma_{n-1}, z_n) \mapsto \sigma_n$ puts another example z_n in the bag σ_{n-1} producing a bigger bag σ_n. The probability distribution $B_n(\cdot\,|\sigma_n)$ can be implemented as follows: draw an example z_n from the bag σ_n at random and output the pair (σ_{n-1}, z_n), where σ_{n-1} is σ_n with z_n removed.

The set of probability distributions on $\mathbf{Z}^\infty$ that agree with the exchangeability OCM is exactly the exchangeability statistical model (Lemma A.3 in Appendix A shows that each exchangeable probability distribution agrees with the exchangeability OCM, and Lemma A.2 immediately implies that each probability distribution that agrees with the exchangeability OCM is exchangeable).

It is clear that the notion of a nonconformity measure in the exchangeability model is identical with that of a nonconformity measure as defined in Chap. 2, and so Proposition 2.4 is a special case of Theorem 11.1.

11.3.4 Generality and Specificity

Let us say that a (reduced) repetitive structure $M_2 = (\Sigma^2, \mathbf{Z}, (t_n^2), (P_n^2))$ is *more specific* than a repetitive structure $M_1 = (\Sigma^1, \mathbf{Z}, (t_n^1), (P_n^1))$ if there exists a sequence of measurable functions $f_n : \Sigma^1 \to \Sigma^2$ such that

- $t_n^2(z_1, \ldots, z_n) = f_n(t_n^1(z_1, \ldots, z_n))$ for all n and all data sequences $(z_1, \ldots, z_n) \in \mathbf{Z}^n$;
- for each n and each $\sigma^2 \in t_n^2(\mathbf{Z}^n)$, the function $P_n^1(\cdot \mid \sigma^1)$, $\sigma^1 \in f_n^{-1}(\sigma^2)$, is a version of the conditional probability given that $t_n^1(z_1, \ldots, z_n) = \sigma^1$ in the probability space $(\mathbf{Z}^n, P_n^2(\cdot \mid \sigma^2))$.

The first condition says that the statistics t_n^1 are more complete summaries of the data sequence $z_1, \ldots, z_n$ than the statistics t_n^2 are (since a summary preserves all useful information in the data sequence, this means that t_n^1 contains more noise, assuming both models are accepted). The second condition says that the probability distributions P_n^1 can be obtained from P_n^2 by conditioning on the more complete information.

The fact that M_2 is more specific than M_1 will be denoted $M_1 \preceq M_2$, and will also be expressed by saying that M_1 is *more general* than M_2.

It is clear that if a probability distribution on $\mathbf{Z}^\infty$ agrees with a repetitive structure, it will agree with a more general repetitive structure. As we know, a repetitive structure formalizes the assumption we are willing to make about Reality, and this assumption weakens as we replace a repetitive structure by a more general structure.

For simplicity, we did not state some results of Chap. 2 in their full generality. It is true that inductive and Mondrian conformal predictors are valid in the exchangeability model. But more than this is true: they are valid under weaker models, which will be considered in the following subsections.

11.3.5 Inductive-Exchangeability Models

As in Sect. 4.2.1, let $m_1, m_2, \ldots$ be a strictly increasing sequence, finite or infinite, of positive integers; if the sequence is finite, say $m_1, \ldots, m_r$, we set $m_{r+1} := \infty$. For each such sequence $m_1, m_2, \ldots$ we can define the corresponding *inductive-exchangeability model* $(\Sigma, \mathbf{Z}, t, (P_n))$, where:

- Σ is the set of finite sequences of bags of elements of $\mathbf{Z}$;
- the summary is defined as the sequence

$$t(z_1, \ldots, z_n) := \Bigl(\langle z_1, \ldots, z_{m_1} \rangle, \langle z_{m_1+1}, \ldots, z_{m_2} \rangle, \ldots,$$
$$\langle z_{m_{k-1}+1}, \ldots, z_{m_k} \rangle, \langle z_{m_k+1}, \ldots, z_n \rangle \Bigr),$$

  where k is such that $m_k < n \le m_{k+1}$;
- for each $\sigma \in \Sigma_n$, $P_n(\cdot \mid \sigma)$ is defined as the uniform probability distribution on the set of $m_1! \ldots m_k!(n - m_k)!$ (with k defined as in the previous item) sequences obtained from σ by ordering its bags in different ways.

It is clear that inductive conformal predictors are conformal predictors in the inductive-exchangeability model (although not all conformal predictors in the inductive-exchangeability model are inductive conformal predictors); therefore, Proposition 4.1 is a special case of Theorem 11.1.

It is also clear that each inductive-exchangeability model is more general than the exchangeability model with the same example space. (The role of the hyphen is to emphasize that inductive-exchangeability models are not instances of exchangeability models, unless the sequence $m_1, m_2, \dots$ is empty.) It is easy to see that an inductive-exchangeability model is strictly more general if $m_1, m_2, \dots$ is not empty: if, e.g., $m_1, m_2, \dots$ has only one term m, any product $Q_1^m \times Q_2^\infty$, where $Q_1, Q_2 \in \mathbf{P}(\mathbf{Z})$, will agree with the inductive-exchangeability model, whereas it will agree with the exchangeability model only if $Q_1 = Q_2$. The inductive-exchangeability model does not appear to be an interesting generalization of the exchangeability model (indeed, if $Q_1 \neq Q_2$, the inductive conformal predictor will be still valid, but its efficiency is likely to suffer), but in the next subsection we will discuss Mondrian-exchangeability models, which can be quite useful (cf. Remark 8.7).

11.3.6 Mondrian-Exchangeability Models

Following Sect. 4.6, fix a Mondrian taxonomy $\kappa : \mathbb{N} \times \mathbf{Z} \to \mathbf{K}$. The corresponding *Mondrian-exchangeability model* $(\Sigma, \mathbf{Z}, t, (P_n))$ is defined as follows:

- $\Sigma := \mathbf{K}^* \times (\mathbf{Z}^{(*)})^{\mathbf{K}}$;
- similarly to (11.10), the summary $t(z_1, \dots, z_n)$ is defined as the pair

$$\Bigl((\kappa(1, z_1), \dots, \kappa(n, z_n)), \bigl(k \in \mathbf{K} \mapsto \Lbag z_i : i \in \{1, \dots, n\}, \kappa(i, z_i) = k \Rbag \bigr) \Bigr) ;$$

- let $\sigma \in \Sigma_n$ consist of a sequence of categories $k_1, \dots, k_n$ and a family of bags $(B_k : k \in \mathbf{K})$ of examples, such that each $k \in \mathbf{K}$ occurs $|B_k|$ times in the sequence $k_1, \dots, k_n$; $P_n(\cdot \mid \sigma)$ is then defined as the uniform probability distribution on the set of $\prod_{k \in K} |B_k|!$ sequences $z_1, \dots, z_n$ obtained from σ by ordering its bags in different ways and putting the elements of each ordered bag B_k consecutively in the places occupied by k in the sequence $k_1, \dots, k_n$.

Proposition 4.6 is a special case of Theorem 11.2: indeed, the latter shows that $p_1, \dots, p_N$ are distributed as $\mathbf{U}^N$ given the observed categories $k_1, \dots, k_N$.

It is easy to see that the notion of generality for repetitive structures (Sect. 11.3.4) as applied to Mondrian-exchangeability models agrees with the notion of generality for Mondrian taxonomies (Sect. 4.6.4).

11.3.7 Exchangeability with a Known Dataset Shift

In conclusion of this section, we will define an exceptional repetitive structure with non-uniform Markov kernels. The notion of a dataset shift was introduced in Sect. 8.2.

The *exchangeability model with a known dataset shift* has as its parameter a sequence of measurable functions $f_n : \mathbf{Z} \to (0, \infty)$ (intuitively, these specify the dataset shift). The summarizing statistic is the same as for the exchangeability model, (11.9), but given a bag $\sigma_n \in \Sigma_n$, the Markov kernel assigns the probability

$$P_n(\{(z_1, \ldots, z_n)\} \mid \sigma_n) := \frac{f_1(z_1) \ldots f_n(z_n)}{\sum_\pi f_1(z_{\pi(1)}) \ldots f_n(z_{\pi(n)})} \tag{11.11}$$

to each ordering $(z_1, \ldots, z_n)$ of σ_n, where π ranges over all permutations of the set $\{1, \ldots, n\}$. The exchangeability model is a special case corresponding to all f_n being equal positive constants.

An example of a probability measure that agrees with the exchangeability model with the dataset shift given by $f_1, f_2, \ldots$ is $P := (f_1 Q) \times (f_2 Q) \times \ldots$, where $Q \in \mathbf{P}(\mathbf{Z})$ and $f Q$ is the probability measure that has density f w.r.t. to Q. The intuition behind P is that we are in a situation of a dataset shift: the nth example z_n is chosen by Reality from $f_n Q$ independently of the other examples, where Q is an unknown probability measure and f_n reflects the dataset shift at step n. We only need to know the functions f_n up to a constant factor (which does not affect (11.11)). See Sect. 11.6.3 for a further discussion.

11.4 Gaussian Model and Its Extensions

11.4.1 Gaussian Model

In the *Gaussian model*, $\mathbf{Z} := \mathbb{R}$, the summarizing statistic is

$$t(z_1, \ldots, z_n) := \left(n, \overline{z}_n, \hat{\sigma}_n\right),$$

$$\overline{z}_n := \frac{1}{n} \sum_{i=1}^n z_i, \quad \hat{\sigma}_n^2 := \frac{1}{n-1} \sum_{i=1}^n (z_i - \overline{z}_n)^2 \tag{11.12}$$

(except that $\hat{\sigma}_1 := 0$), and $P_n(\mathrm{d}z_1, \ldots, \mathrm{d}z_n \mid \sigma)$, $\sigma \in \Sigma_n$, is the uniform distribution on $t^{-1}(\sigma)$ (in other words, it is the uniform distribution on the $(n-2)$-dimensional sphere in $\mathbb{R}^n$ with centre $(\overline{z}_n, \ldots, \overline{z}_n) \in \mathbb{R}^n$ of radius $\sqrt{n-1}\hat{\sigma}_n$ lying inside the hyperplane $\frac{1}{n}(z_1 + \cdots + z_n) = \overline{z}_n$).

It is clear that there are many possible representations of essentially the same model; for example, we obtain an equivalent model if we replace (11.12) by

$$t(z_1, \ldots, z_n) := \left(n, \sum_{i=1}^n z_i, \sum_{i=1}^n z_i^2\right). \tag{11.13}$$

Let us first find the forward functions in the online compression representation of the Gaussian model. It is easy to check that the updating formulae for $\overline{z}_n$ and $\hat{\sigma}_n$ are

$$\overline{z}_n = \frac{n-1}{n}\overline{z}_{n-1} + \frac{1}{n}z_n ,$$

$$\hat{\sigma}_n^2 = \frac{n-2}{n-1}\hat{\sigma}_{n-1}^2 + \frac{1}{n}(z_n - \overline{z}_{n-1})^2 ;$$

this defines the forward function F. The expression for the backward kernel is much more complicated, and we do not give it explicitly; it can be derived from the fact that (11.15) below has Student's t-distribution.

Let us now explicitly find the prediction set for the Gaussian model and nonconformity measure

$$A(\sigma, z_n) = A((n, \overline{z}_n, \hat{\sigma}_n), z_n) := |z_n - \overline{z}_n| , \quad \sigma \in \Sigma_n \tag{11.14}$$

(it is easy to check that this nonconformity measure is equivalent, in the sense of leading to the same p-values, to $|z_n - \overline{z}_{n-1}|$, as well as to several other natural expressions, including (11.15) in absolute value). Under $P_n(\mathrm{d}z_1, \ldots, \mathrm{d}z_n \mid (n, \overline{z}_n, \hat{\sigma}_n))$ and assuming $n > 2$, the expression

$$\sqrt{\frac{n-1}{n}}\frac{z_n - \overline{z}_{n-1}}{\hat{\sigma}_{n-1}} \tag{11.15}$$

has the t-distribution with $n-2$ degrees of freedom. (This fact is proved in, e.g., [65, Sect. 29.4], where it is assumed, however, that $z_1, \ldots, z_n$ are independent and have the same normal distribution. The latter assumption may be replaced by our assumption of the uniform distribution; for a general argument, see the proof of Proposition 11.6 in Sect. 11.5.4.) Let $\mathbf{t}_{\delta,k}$ be the value defined by $\mathbb{P}\{\xi \geq \mathbf{t}_{\delta,k}\} = \delta$ with ξ having the t-distribution with k degrees of freedom. We can see that the prediction set Γ_n^{ϵ} determined by nonconformity measure (11.14) is the interval consisting of z such that

$$|z - \overline{z}_{n-1}| \leq \mathbf{t}_{\epsilon/2,n-2}\sqrt{\frac{n}{n-1}}\hat{\sigma}_{n-1} . \tag{11.16}$$

We obtain the usual prediction set based on the t-test (as in [14, 427], and, implicitly, [105]); now, however, we can see that the errors of this standard procedure (applied in the online fashion) are independent. Some of the facts mentioned in this paragraph will be proved in Sect. 11.5.4.

11.4.2 Gauss Linear Model

We will now consider a rich extension of the Gaussian model. In the repetitive-structure representation $(\Sigma, \mathbf{Z}, t, (P_n))$ of the *Gauss linear model* the example space is of the regression type, $\mathbf{Z} := \mathbf{X} \times \mathbf{Y}$, with the label space being the real line $\mathbf{Y} := \mathbb{R}$ and the object space being the p-dimensional Euclidean space, $\mathbf{X} := \mathbb{R}^p$. The summarizing statistic is

$$t(x_1, y_1, \dots, x_n, y_n) := \left(x_1, \dots, x_n, \sum_{i=1}^{n} y_i x_i, \sum_{i=1}^{n} y_i^2\right) \tag{11.17}$$

(so Σ can be set to $\mathbf{X}^* \times \mathbb{R}^p \times \mathbb{R}$), and $P_n(\cdot \mid \sigma)$, $\sigma \in \Sigma_n$, is the uniform probability distribution on the sphere $t_n^{-1}(\sigma)$ (we consider a point to be a sphere; typically $t_n^{-1}(\sigma)$ will be a point unless $n > p$).

The Gaussian model in the form (11.13) is a special case (using different notation, such as z_i for y_i) corresponding to $p = 1$ and x_i restricted to $x_i = 1$, $i = 1, 2, \dots$. Using $\sum_{i=1}^{n} y_i x_i$ rather than $\sum_{i=1}^{n} y_i$ in (11.17) reflects the possibility that y_i may depend on x_i.

The probability distribution of $z_1, z_2, \dots$ under the linear regression statistical model

$$y_n = w \cdot x_n + \xi_n \tag{11.18}$$

(same as (2.49)), where $w \in \mathbb{R}^p$ is a constant vector and ξ_n are independent random variables with the same zero-mean normal distribution, always agrees with the Gauss linear model.

Our next proposition (and its proof in Sect. 11.5.4) will use the following notation: $\hat{y}_i^n$ is the least squares prediction for the object x_i based on the examples $z_1, \dots, z_n$; $\hat{y}_n$ is a shorthand for $\hat{y}_n^{n-1}$; X_l, $l = 1, 2, \dots$, is the $l \times p$ matrix whose ith row is x_i', $i = 1, \dots, l$ (i.e., X_l is the data matrix, as in (2.38)); and

$$\hat{\sigma}_l^2 := \frac{1}{l-p} \sum_{i=1}^{l} (y_i - \hat{y}_i^l)^2$$

is the standard estimate (as in (7.25)) of the variance of the Gaussian noise ξ_n in (11.18) from the first l examples.

Proposition 11.6 *The conformal predictor determined by the nonconformity measure*

$$A(\sigma, (x, y)) := \left|y - \hat{y}\right| , \tag{11.19}$$

where $\hat{y}$ is the least squares prediction of the x's label y based on the examples summarized by σ, is given, for $n > p+1$ satisfying $\operatorname{rank}(X_{n-1}) = p$, *by the formula*

$$\Gamma_n^\epsilon = \left[\hat{y}_n - \mathbf{t}_{\epsilon/2,n-p-1} V_n, \hat{y}_n + \mathbf{t}_{\epsilon/2,n-p-1} V_n\right] , \tag{11.20}$$

where

$$V_n := \sqrt{1 + x_n'(X_{n-1}'X_{n-1})^{-1}x_n}\hat{\sigma}_{n-1} .$$

Confidence predictor (11.20), generalizing (11.16), will be called the *Student predictor*. The prediction interval (11.20) is standard (see, e.g., [252, (3.54)]), but our results add the usual extra feature: the independence of errors in the online setting.

See Sect. 11.5.4 for a proof of Proposition 11.6. The proof will show that the details of the definition (11.19) are mostly irrelevant. For example, our standard convention is that (x, y) is among the examples summarized by σ, but we will obtain the same prediction interval (11.20) when σ does not include (x, y) (i.e., when in (11.19) we have A^{del} in place of A with a natural definition of A^{del} analogous to that given in (2.35) in the case of the exchangeability model).

Remark 11.7 The methods of this subsection are applicable to time series, although only to the simplest ones: e.g., if

$$y_n = f(n) + \cos\frac{n-a}{T} + \xi_n$$

where $f(n)$ is a polynomial of a known degree p, T is a known constant (the period of the seasonal component), and ξ_n are IID zero-mean normal random variables, we can set

$$x_n := \left(1, n, \ldots, n^p, \cos\frac{n}{T}, \sin\frac{n}{T}\right)$$

and use formula (11.20). Constructing conformal predictors in more interesting cases requires new methods.

11.4.3 Multivariate Gaussian Model

The Gauss linear model is weak in that it does not make any stochastic assumptions whatsoever about the objects x_n (the weakness of this assumption will show in Table 11.1). Now we will make strong Gaussian assumptions about the objects (and in the next subsection we will just assume that the objects are exchangeable).

The summarizing statistic for the *multivariate Gaussian* model is

$$t(z_1, \ldots, z_n) := \left(n, \sum_{i=1}^{n} x_i, \sum_{i=1}^{n} y_i, \sum_{i=1}^{n} x_i x_i', \sum_{i=1}^{n} y_i x_i, \sum_{i=1}^{n} y_i^2 \right) ;$$

equivalently, the statistic can be defined to be the empirical means and covariances of all the variables, i.e., the label and the attributes. Given the value of the summarizing statistic, the distribution of the corresponding examples is, as usual, uniform. The canonical statistical model that agrees with the multivariate Gaussian model is the generalization

$$y_n = w \cdot x_n + c + \xi_n \tag{11.21}$$

of (11.18), where $w \in \mathbb{R}^p$ and $c \in \mathbb{R}$ are parameters, ξ_n are independent random variables with the same zero-mean Gaussian distribution, and, additionally, x_n are generated independently from the same unknown zero-mean multivariate Gaussian distribution on $\mathbb{R}^p$, with the noise random variables $\xi_1, \xi_2, \ldots$ independent of $x_1, x_2, \ldots$. Remember our notation Y_n and X_n for the vector of the first n labels and the data matrix, respectively: see (2.37) and (2.38). Let us extend the data matrix X_n by adding an all-1 column on its left (this corresponds to adding a dummy attribute always equal to 1 to take care of the c in (11.21)).

Suppose the value of the summarizing statistic t on the first n examples output by Reality is known. If we use ridge regression with ridge parameter a (least squares for $a = 0$), the vector of residuals can be written as

$$E := Y_n - X_n \left(X_n' X_n + a I_n \right)^{-1} X_n' Y_n = Y_n - X_n C , \tag{11.22}$$

where $C := (X_n' X_n + a I_n)^{-1} X_n' Y_n$ is a known vector. Since the joint distribution of Y_n and the non-dummy columns of X_n (remember that we extended X_n by adding a dummy column) is invariant with respect to rotations around the vector $(1, \ldots, 1)'$, the distribution of E will also be invariant with respect to such rotations. It might help the reader's intuition to notice that knowing the value of the summarizing statistic t is equivalent to knowing the lengths of and the angles between the following $p + 2$ vectors: Y_n and the $p + 1$ columns of the extended X_n.

In the rest of this section we will assume $n \geq 3$ (with arbitrary conventions for $n = 1, 2$). Let $e_1, \ldots, e_n$ be the components of the vector (11.22) of residuals and $\overline{e}_{n-1}$ be the average of $e_1, \ldots, e_{n-1}$. The standard statistical result mentioned earlier (see (11.15)) allows us to conclude that

$$\sqrt{\frac{n-1}{n}} \frac{e_n - \overline{e}_{n-1}}{\sqrt{\frac{1}{n-2} \sum_{i=1}^{n-1} (e_i - \overline{e}_{n-1})^2}} \tag{11.23}$$

has the t-distribution with $n - 2$ degrees of freedom.

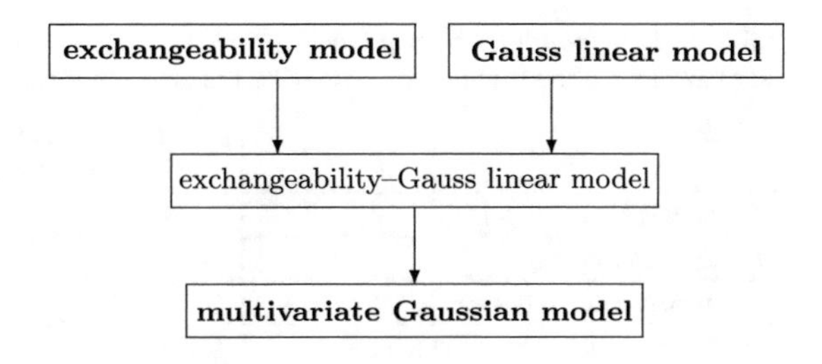

Fig. 11.3 The four models compared in Sect. 11.4.4 (the three main ones are given in boldface)

Let us see how to implement the conformal predictor determined by the nonconformity measure

$$A\left(t(x_1, y_1, \ldots, x_n, y_n), (x_n, y_n)\right) := \frac{|e_n - \overline{e}_{n-1}|}{\sqrt{\sum_{i=1}^{n-1}(e_i - \overline{e}_{n-1})^2}}, \tag{11.24}$$

which is proportional to (11.23) in absolute value; the fact that the right-hand side of (11.24) depends on the first n examples only via $t(x_1, y_1, \ldots, x_n, y_n)$ and (x_n, y_n) can be seen from the representation (11.22), where C is a known vector. (The right-hand side of (11.24) can be replaced by other natural expressions, such as $|e_n - \overline{e}_{n-1}|$ or $|e_n - \overline{e}_n|$, $\overline{e}_n$ being the average of $e_1, \ldots, e_n$, as can be shown using calculations similar to those at the end of Sect. 11.5.4.) First we replace the true value y_n by variable y ranging over $\mathbb{R}$. Each residual e_i becomes a linear (according to (11.22), where C also depends on y) function $e_i(y)$ of y, and the prediction set can be written as

$$\Gamma_n^\epsilon := \left\{ y \in \mathbb{R} : \sqrt{\frac{n-1}{n}} \frac{|e_n(y) - \overline{e}_{n-1}(y)|}{\sqrt{\frac{1}{n-2}\sum_{i=1}^{n-1}(e_i(y) - \overline{e}_{n-1}(y))^2}} < \mathbf{t}_{\epsilon/2,n-2} \right\}.$$

The inequality in this formula is quadratic in y, so Γ_n^ϵ is easy to find. We can see that the prediction set for y_n is an interval (empirically, this is the typical case), the union of two rays, the empty set, or the whole real line. Replacing Γ_n^ϵ by its convex hull $\operatorname{co}\Gamma_n^\epsilon$ in the conformal predictor we have just defined gives us the *multivariate Gaussian predictor*.

11.4.4 Some Comparisons

In this subsection we discuss the four models shown in Fig. 11.3. The three models given in boldface were defined in previous subsections, and the exchangeability–Gauss linear model combines the exchangeability model and the Gauss linear model: its summarizing statistic is

Table 11.1 Steps of the online protocol at which informative prediction becomes possible for the four models; ϵ is the significance level ($\epsilon < 1/2$ is assumed) and p is the number of parameters

Model	The first step at which prediction intervals can become informative
Exchangeability model	$\lceil 1/\epsilon \rceil$
Gauss linear model	$p + 3$
Multivariate Gaussian model	3
Exchangeability–Gauss linear model	$\min(\lceil 1/\epsilon \rceil, p + 3)$

$$t(z_1, \ldots, z_n) := \left(\lbag x_1, \ldots, x_n \rbag, \sum_{i=1}^{n} y_i, \sum_{i=1}^{n} y_i x_i, \sum_{i=1}^{n} y_i^2 \right),$$

and given the summary $t(z_1, \ldots, z_n)$, its Markov kernel generates a random ordering of $\lbag x_1, \ldots, x_n \rbag$ and then generates the labels $(y_1, \ldots, y_n)$ from the uniform distribution on the sphere of the vectors of labels consistent with the summary. The canonical statistical model that agrees with this repetitive structure assumes both that the objects are IID and that the labels are generated by (11.21) with $\xi_1, \xi_2, \ldots$ independent of $x_1, x_2, \ldots$. A natural nonconformity measure is the residual in absolute value, (11.19); this is what we will use in this subsection. We are not aware of efficient implementations of the corresponding conformal predictor in the exchangeability–Gauss linear model, of the kind that we have for the main three models in Fig. 11.3, but one possibility is to use Monte Carlo sampling from the conditional distribution given a summary.

All four models output vacuous prediction sets ($\mathbb{R}$) at any significance level $\epsilon \in (0, 1)$ before seeing a certain number of examples in the online prediction protocol; cf. Table 11.1. It is typical for conformal predictors based on different models to perform similarly soon after this minimal number of examples is reached [405].

For the parametric Gauss linear model, to get a non-vacuous prediction set we need at least p examples, where p is the number of parameters. In point prediction, the number of required examples can be reduced by using feature selection or regularization, but it is not clear how either of these techniques can be applied if we are interested in set prediction and are not willing to sacrifice guaranteed validity. An informative, i.e., bounded, prediction set becomes possible for the nth example in the online prediction protocol only when n reaches $p+3$ (assuming a significance level $\epsilon < 1/2$).

The exchangeability model is nonparametric but, as we know, still admits valid confidence predictors. The smallest possible p-value for the nth example is $1/n$, and so we can start getting informative prediction sets from the $\lceil 1/\epsilon \rceil$th example (when using the nonconformity measure (11.19), we may need to understand $\hat{y}$ in the sense of ridgeless least squares [153] to make sure it always exists). The threshold $1/\epsilon$ can be said to play the role of the number of parameters, and the nonparametric nature of the model is reflected in the fact that $1/\epsilon \to \infty$ as $\epsilon \to 0$. Since $1/\epsilon$ tends to ∞ relatively slowly, such an infinite-dimensional model may be better for the purpose of prediction than a p-dimensional model with a very large p.

Naturally, the exchangeability–Gauss linear model makes it possible to output an informative prediction when either of the two models does it. The corresponding row of Table 11.1 demonstrates, once again, the point we made in Sect. 4.7.1 about the impossibility to condition on everything. If we insist on conditioning on all objects seen so far, we cannot get anything better than (11.20) even under the exchangeability–Gauss linear model. Therefore, we cannot get non-vacuous prediction sets before the number of seen examples exceeds p. On the other hand, this becomes perfectly possible if we sacrifice full conditioning on the objects.

The multivariate Gaussian model makes heaviest assumptions about the data. In the happy situation where those assumptions are justified, we start getting informative predictions almost straight away.

Both Gauss linear and multivariate Gaussian models have been widely discussed in the literature on regression; e.g., Sampson's [301] "two regressions" refers to those models. Fisher [110, Sect. IV.3] emphatically defended the use of the Gauss linear model even in the case where the distribution of the objects is known (with or without parameters).

11.5 Proofs

11.5.1 *Proof of Theorem 11.2*

First we explain the basic idea of the proof. To show that $(p_1, \dots, p_N)$ is distributed as $\mathbf{U}^N$, we use the standard idea of reversing the time (see, e.g., the proof of de Finetti's theorem in [307]). Let P be the distribution on $\mathbf{Z}^N$ generating the examples; it is assumed to agree with the OCM. We can imagine that the data sequence $(z_1, \dots, z_N)$ is generated in two steps: first, the summary σ_N is generated from some probability distribution (namely, the image of the distribution P under the mapping t_N), and then the data sequence $(z_1, \dots, z_N)$ is chosen randomly from $P_N(\cdot|\sigma_N)$. Already the second step ensures that, conditionally on knowing σ_N (and, therefore, unconditionally), the sequence $(p_N, \dots, p_1)$ is distributed as $\mathbf{U}^N$. Indeed, roughly speaking (i.e., ignoring borderline effects), p_N will be the p-value corresponding to the statistic A (the nonconformity measure we are using) and so distributed, at least approximately, as $\mathbf{U}$ (see, e.g., [64, Sect. 3.2]); when the pair (σ_{N-1}, z_N) is disclosed, the value p_N will be settled; conditionally on knowing σ_{N-1} and z_N, p_{N-1} will also be distributed as $\mathbf{U}$, and so on.

We start the formal proof by defining the σ-algebra $\mathcal{G}_n$, $n = 0, 1, \dots, N$, as the one on the sample space $(\mathbf{Z} \times [0,1])^N$ generated by the random elements $\sigma_n := t(z_1, \dots, z_n), z_{n+1}, \tau_{n+1}, z_{n+2}, \tau_{n+2}, \dots, z_N, \tau_N$. In particular, $\mathcal{G}_0$ (the most informative σ-algebra) coincides with the original σ-algebra on $(\mathbf{Z} \times [0,1])^N$; $\mathcal{G}_0 \supseteq \mathcal{G}_1 \supseteq \cdots \supseteq \mathcal{G}_N$.

Fix a smoothed conformal transducer f; it will usually be left implicit in our notation. Let p_n be the random variable $f(z_1, \tau_1, \dots, z_n, \tau_n)$ for each $n = 1, \dots, N$; $\mathbb{P}$ will refer to the probability distribution $P \times \mathbf{U}^N$ (over examples z_n and random numbers τ_n) and $\mathbb{E}$ to the expectation w.r. to $\mathbb{P}$. It will be convenient to write $\mathbb{P}_{\mathcal{G}}(A)$ and $\mathbb{E}_{\mathcal{G}}(\xi)$ for the conditional probability $\mathbb{P}(A \mid \mathcal{G})$ and expectation $\mathbb{E}(\xi \mid \mathcal{G})$, respectively. The proof will be based on the following lemma.

Lemma 11.8 *For any trial $n = 1, \dots, N$ and any $\epsilon \in [0,1]$,*

$$\mathbb{P}_{\mathcal{G}_n}\{p_n \le \epsilon\} = \epsilon .$$

Proof Let us fix a summary σ_n of the first n examples $(z_1, \dots, z_n) \in \mathbf{Z}^n$; we will omit the condition "$\mid \sigma_n$". For every example $\tilde{z}$ define

$$p^+(\tilde{z}) := B_{\mathbf{Z}}\{z : A(\sigma_n, z) \geq A(\sigma_n, \tilde{z})\},$$

$$p^-(\tilde{z}) := B_{\mathbf{Z}}\{z : A(\sigma_n, z) > A(\sigma_n, \tilde{z})\},$$

where $B_{\mathbf{Z}}$ is the marginal distribution of the backward kernel. It is clear that always $p^- \leq p^+$.

Let us say that an example $\tilde{z}$ is

- *strange* if $p^+(\tilde{z}) \leq \epsilon$
- *conforming* if $p^-(\tilde{z}) > \epsilon$
- *borderline* if $p^-(\tilde{z}) \leq \epsilon < p^+(\tilde{z})$.

We will use the notation $p^- := p^-(\tilde{z})$ and $p^+ := p^+(\tilde{z})$ where $\tilde{z}$ is any borderline example (if there are no borderline examples, $p^- = p^+$ can be defined as $\inf p^-(\tilde{z})$ over the conforming $\tilde{z}$ or as $\sup p^+(\tilde{z})$ over the strange $\tilde{z}$). Notice that the $B_{\mathbf{Z}}$-measure of strange examples is p^-, the $B_{\mathbf{Z}}$-measure of conforming examples is $1 - p^+$, and the $B_{\mathbf{Z}}$-measure of borderline examples is $p^+ - p^-$.

By the definition of a smoothed conformal transducer, $p_n \leq \epsilon$ if the example z_n is strange, $p_n > \epsilon$ if the example is conforming, and $p_n \leq \epsilon$ with probability

$$\frac{\epsilon - p^-}{p^+ - p^-}$$

(with, say, $0/0 := 0$) if the example is borderline; indeed, in the latter case, as

$$p_n = p^- + \tau_n(p^+ - p^-),$$

$p_n \leq \epsilon$ is equivalent to

$$\tau_n \leq \frac{\epsilon - p^-}{p^+ - p^-}.$$

Therefore, the overall probability that $p_n \leq \epsilon$ is

$$p^- + (p^+ - p^-)\frac{\epsilon - p^-}{p^+ - p^-} = \epsilon.$$

□

The other basic result that we will need is the following lemma.

Lemma 11.9 *For any trial $n = 1, \ldots, N$, p_n is $\mathcal{G}_{n-1}$-measurable.*

Proof This follows from the definition, (11.5): p_n is defined in terms of $\sigma_n = F(\sigma_{n-1}, z_n)$, z_n and τ_n. The only technicality that might not be immediately obvious is that the function

$$B_{\mathbf{Z}}(\{A(\sigma, \cdot) > c\} \mid \sigma)$$

of $c \in \mathbb{R}$ and $\sigma \in \Sigma$ is measurable. Let $C \in \mathbb{R}$. The set

$$\{(c, \sigma) : B_{\mathbf{Z}}(\{A(\sigma, \cdot) > c\} \mid \sigma) > C\} \tag{11.25}$$

is measurable since it can be represented as $\bigcup_{d \in \mathbb{Q}} (0, d) \times \Sigma_d$, where $\mathbb{Q}$ is the set of rational numbers and Σ_c is the set of σ satisfying the outer inequality in (11.25). □

First we prove that, for any $n = 1, \ldots, N$ and any $\epsilon_1, \ldots, \epsilon_n \in [0, 1]$,

$$\mathbb{P}_{\mathcal{G}_n}\{p_n \leq \epsilon_n, \ldots, p_1 \leq \epsilon_1\} = \epsilon_n \ldots \epsilon_1 \quad \text{a.s.} \tag{11.26}$$

The proof is by induction on n. For $n = 1$, (11.26) is a special case of Lemma 11.8. For $n > 1$ we obtain, making use of Lemmas 11.8 and 11.9, properties 1 and 2 of conditional expectations (see Sect. A.3), and the inductive assumption:

$$\begin{aligned}\mathbb{P}_{\mathcal{G}_n}\{p_n \le \epsilon_n, \dots, p_1 \le \epsilon_1\} &= \mathbb{E}_{\mathcal{G}_n}\left(\mathbb{E}_{\mathcal{G}_{n-1}}\left(1_{p_n \le \epsilon_n} 1_{p_{n-1} \le \epsilon_{n-1}, \dots, p_1 \le \epsilon_1}\right)\right) \\ &= \mathbb{E}_{\mathcal{G}_n}\left(1_{p_n \le \epsilon_n} \mathbb{E}_{\mathcal{G}_{n-1}}\left(1_{p_{n-1} \le \epsilon_{n-1}, \dots, p_1 \le \epsilon_1}\right)\right) \\ &= \mathbb{E}_{\mathcal{G}_n}\left(1_{p_n \le \epsilon_n} \epsilon_{n-1} \dots \epsilon_1\right) = \epsilon_n \epsilon_{n-1} \dots \epsilon_1 \quad \text{a.s.}\end{aligned}$$

By the tower property (property 2 in Sect. A.3), (11.26) immediately implies

$$\mathbb{P}\{p_N \le \epsilon_N, \dots, p_1 \le \epsilon_1\} = \epsilon_N \dots \epsilon_1 .$$

This implies the uniform distribution of $(p_1, \dots, p_N)$ in $[0, 1]^N$ (see, e.g., [329, Lemma 2.2.3]).

Remark 11.10 In our definitions of conformal transducer, conformal predictor, etc., we assumed that the same random number τ_n is used for every potential label y of the new object x_n. In fact, assuming $\mathbf{Y}$ is finite, we can also use a separate random number τ_n^y for each $y \in \mathbf{Y}$, with the random numbers τ_n^y, $n = 1, 2, \dots$, $y \in \mathbf{Y}$, independent. On the other hand, an arbitrary correlation between τ_n^y, $y \in \mathbf{Y}$, can be allowed; Theorems 11.1 and 11.2 will continue to hold as long as the random numbers $\tau_n^{y_n}$, $n = 1, 2, \dots$, are independent.

11.5.2 Proof of Proposition 4.4

Proposition 4.4 is essentially a special case of Theorem 11.1 corresponding to the following *transductive OCM*, which goes slightly beyond the usual notion of OCM:

- The summaries $\Sigma := \mathbf{Z}^{(*)}$ are the bags of elements of $\mathbf{Z}$.
- The forward function $F : \Sigma \times \mathbf{Z}^* \to \Sigma$ is defined similarly to the exchangeability model:

$$F(\sigma, (z_1, \dots, z_k)) := \sigma \cup \lbag z_1, \dots, z_k \rbag$$

  for any sequence $(z_1, \dots, z_k)$, so that F adds all elements of its second argument (a sequence) to its first argument (a bag).
- The backward kernel $B : \Sigma \hookrightarrow \Sigma \times \mathbf{Z}^*$ is defined arbitrarily on $\sigma \in \Sigma$ of size outside $\{l_1, l_2, \dots\}$; and if the size is l_n for some n, $B(\sigma)$ chooses, randomly without replacement, a sequence of k_n elements of σ; that sequence becomes the second element of $B(\sigma)$, and the remaining elements of σ become the first element of $B(\sigma)$.

The proof of Theorem 11.1 still works for such slightly generalized online compression models, where the examples arrive in batches of sizes $k_1, k_2, \dots$. Therefore, the p-values will be independent and uniformly distributed on $[0, 1]$, and the statement of Proposition 4.4 follows.

11.5.3 Proof of Theorem 8.8

It suffices to prove, for a fixed horizon $N \in \{1, 2, \dots\}$, that the random p-values $p_1, p_1', \dots, p_N, p_N'$ are distributed independently and uniformly on $[0, 1]$ (see the derivation of Theorem 11.1 at the end of Sect. 11.2.3). Let us fix such an N.

The rest of this subsection is the modification of the proof of Theorem 11.2 in Sect. 11.5.1. First we give an informal argument and imagine that the data sequence $z_1, \dots, z_n$ is generated in two steps: first a random bag $\lbag z_1, \dots, z_n \rbag$ and then its random ordering. Already the second step ensures that $(p_1, p'_1, \dots, p_N, p'_N)$ are distributed uniformly on $[0,1]^{2N}$ (even conditionally on $\lbag z_1, \dots, z_n \rbag$). This can be demonstrated using the following backward argument. Ignoring borderline effects, p'_N is uniformly distributed on $[0, 1]$ (at least approximately). When y_N is disclosed, p'_N will be settled. Given what we already know, the distribution of p_N will be uniform. When x_N is disclosed, p_N will be settled. Now the distribution of p'_{N-1} given what we already know is uniform, etc.

For the formal proof, we will need the following σ-algebras. Let $\mathcal{G}_n$, $n = 0, \dots, N$, be the σ-algebra

$$\mathcal{G}_n := \sigma\left(\lbag z_1, \dots, z_n \rbag, z_{n+1}, \tau_{n+1}, \tau'_{n+1}, \dots, z_N, \tau_N, \tau'_N\right)$$

generated by the bag $\lbag z_1, \dots, z_n \rbag$ and the other random elements listed in the parentheses. Let $\mathcal{G}'_n$, $n = 1, \dots, N$, be the σ-algebra $\sigma(\mathcal{G}_n, y_n, \tau'_n)$ generated by $\mathcal{G}_n$, the label y_n of the nth example, and the random number τ'_n.

The following two lemmas (analogues of Lemma 11.9) say that

$$\begin{array}{ccccccccccccc} p'_N & & p_N & & p'_{N-1} & & \cdots & & p_2 & & p'_1 & & p_1 \\ \mathcal{G}_N & \subseteq & \mathcal{G}'_N & \subseteq & \mathcal{G}_{N-1} & \subseteq & \mathcal{G}'_{N-1} & \subseteq \cdots \subseteq & \mathcal{G}_1 & \subseteq & \mathcal{G}'_1 & \subseteq & \mathcal{G}_0 \end{array}$$

is a stochastic sequence essentially in the usual sense of probability theory [330, Sect. 7.1.2]: in the second row we have a finite filtration, and the random variables in the first row are measurable w.r.t. to the σ-algebras directly below them.

Lemma 11.11 *For any trial $n = 1, \dots, N$, p'_n is $\mathcal{G}'_n$-measurable.*

Proof The random bag of conformity scores of $z_1, \dots, z_n$ is $\mathcal{G}_n$-measurable, and so, according to the definition (2.67) and the invariance requirement (8.13), p'_n is $\mathcal{G}'_n$-measurable. □

Lemma 11.12 *For any trial $n = 1, \dots, N$, p_n is $\mathcal{G}_{n-1}$-measurable.*

Proof This follows from the definition (8.11) and our requirement that the label-conditional conformal transducer p should be simple. □

We will also need the following results, which can be proved similarly to Lemma 11.8.

Lemma 11.13 *For any trial $n = 1, \dots, N$ and any $\epsilon \in [0, 1]$,*

$$\mathbb{P}_{\mathcal{G}'_n}\{p_n \le \epsilon\} = \epsilon\ .$$

Lemma 11.14 *For any trial $n = 1, \dots, N$ and any $\epsilon \in [0, 1]$,*

$$\mathbb{P}_{\mathcal{G}_n}\left\{p'_n \le \epsilon\right\} = \epsilon\ .$$

Let us now prove the following double sequence of equalities:

$$\mathbb{P}_{\mathcal{G}'_n}\left\{p_n \le \epsilon_n, p'_{n-1} \le \epsilon'_{n-1}, p_{n-1} \le \epsilon_{n-1}, \dots, p'_1 \le \epsilon'_1, p_1 \le \epsilon_1\right\} \\ = \epsilon_n \epsilon'_{n-1} \epsilon_{n-1} \dots \epsilon'_1 \epsilon_1 \qquad (11.27)$$

and

$$\mathbb{P}_{\mathcal{G}_n}\left\{p'_n \le \epsilon'_n, p_n \le \epsilon_n, \ldots, p'_1 \le \epsilon'_1, p_1 \le \epsilon_1\right\} = \epsilon'_n\epsilon_n \ldots \epsilon'_1\epsilon_1 . \tag{11.28}$$

We will use induction arranging these equalities into a single sequence: the equality for $\mathbb{P}_{\mathcal{G}'_1}$, the equality for $\mathbb{P}_{\mathcal{G}_1}$, the equality for $\mathbb{P}_{\mathcal{G}'_2}$, the equality for $\mathbb{P}_{\mathcal{G}_2}$, etc. The first of these equalities is a special case of Lemma 11.13. When proving any other of these equalities, we will assume that all the previous equalities are true.

The equality for $\mathbb{P}_{\mathcal{G}_n}$, $n \in \{1, \ldots, N\}$, follows from

$$\begin{aligned}
&\mathbb{P}_{\mathcal{G}_n}\left\{p'_n \le \epsilon'_n, p_n \le \epsilon_n, \ldots, p'_1 \le \epsilon'_1, p_1 \le \epsilon_1\right\}\\
&\quad = \mathbb{E}_{\mathcal{G}_n}\left(\mathbb{E}_{\mathcal{G}'_n}\left(1_{p'_n\le\epsilon'_n}1_{p_n\le\epsilon_n}\cdots 1_{p'_1\le\epsilon'_1}1_{p_1\le\epsilon_1}\right)\right)\\
&\quad = \mathbb{E}_{\mathcal{G}_n}\left(1_{p'_n\le\epsilon'_n}\mathbb{E}_{\mathcal{G}'_n}\left(1_{p_n\le\epsilon_n}\cdots 1_{p'_1\le\epsilon'_1}1_{p_1\le\epsilon_1}\right)\right)\\
&\quad = \mathbb{E}_{\mathcal{G}_n}\left(1_{p'_n\le\epsilon'_n}\epsilon_n\ldots\epsilon'_1\epsilon_1\right) = \epsilon'_n\epsilon_n\ldots\epsilon'_1\epsilon_1 .
\end{aligned}$$

The first equality is just the tower property (property 2 in Sect. A.3) of conditional expectations. The second equality follows from Lemma 11.11. The third equality follows from the inductive assumption, namely (11.27). The last equality follows from Lemma 11.14.

The equality for $\mathbb{P}_{\mathcal{G}'_n}$, $n \in \{2, \ldots, N\}$, follows from

$$\begin{aligned}
&\mathbb{P}_{\mathcal{G}'_n}\left\{p_n \le \epsilon_n, p'_{n-1} \le \epsilon'_{n-1}, p_{n-1} \le \epsilon_{n-1}, \ldots, p'_1 \le \epsilon'_1, p_1 \le \epsilon_1\right\}\\
&\quad = \mathbb{E}_{\mathcal{G}'_n}\left(\mathbb{E}_{\mathcal{G}_{n-1}}\left(1_{p_n\le\epsilon_n}1_{p'_{n-1}\le\epsilon'_{n-1}}1_{p_{n-1}\le\epsilon_{n-1}}\cdots 1_{p'_1\le\epsilon'_1}1_{p_1\le\epsilon_1}\right)\right)\\
&\quad = \mathbb{E}_{\mathcal{G}'_n}\left(1_{p_n\le\epsilon_n}\mathbb{E}_{\mathcal{G}_{n-1}}\left(1_{p'_{n-1}\le\epsilon'_{n-1}}1_{p_{n-1}\le\epsilon_{n-1}}\cdots 1_{p'_1\le\epsilon'_1}1_{p_1\le\epsilon_1}\right)\right)\\
&\quad = \mathbb{E}_{\mathcal{G}'_n}\left(1_{p_n\le\epsilon_n}\epsilon'_{n-1}\epsilon_{n-1}\ldots\epsilon'_1\epsilon_1\right) = \epsilon_n\epsilon'_{n-1}\epsilon_{n-1}\ldots\epsilon'_1\epsilon_1 .
\end{aligned}$$

Now the second equality follows from Lemma 11.12. The third equality follows from the inductive assumption, namely (11.28) with $n-1$ in place of n. The last equality follows from Lemma 11.13.

Plugging $n := N$ into (11.28), we obtain

$$\mathbb{P}\left\{p_1 \le \epsilon_1, p'_1 \le \epsilon'_1, \ldots, p_N \le \epsilon_N, p'_N \le \epsilon'_N\right\} = \epsilon_1\epsilon'_1\ldots\epsilon_N\epsilon'_N .$$

11.5.4 Proof of Proposition 11.6

It is a standard fact (see, e.g., [349, Sect. 32.10]) that $(y_n - \hat{y}_n)/V_n$ has the t-distribution with $n-p-1$ degrees of freedom; this assumes, however, the standard model (11.18) rather than the uniform conditional distribution of the Gauss linear model. Let us check that $(y_n - \hat{y}_n)/V_n$ will still have the t-distribution with $n-p-1$ degrees of freedom under the uniform conditional distribution.

First note that $(y_n - \hat{y}_n)/V_n$ can be rewritten so that it depends on $y_1, \ldots, y_n$ only through the *n-residuals* $y_i - \hat{y}_i^n$ (i.e., residuals computed from all n examples $z_1, \ldots, z_n$). Indeed, a standard statistical result [252, (4.12)] shows that

$$\hat{\sigma}_{n-1}^2 = \frac{\sum_{i=1}^n (y_i - \hat{y}_i^n)^2 - (y_n - \hat{y}_n^n)^2/(1 - x'_n(X'_nX_n)^{-1}x_n)}{n-p-1} ; \tag{11.29}$$

another standard result ([252, (4.11)], already used in Sect. 2.3.3) shows that

$$y_n - \hat{y}_n = \frac{y_n - \hat{y}_n^n}{1 - x_n'(X_n' X_n)^{-1} x_n} . \tag{11.30}$$

Remember that $Y_n := (y_1, \dots, y_n)'$ is the vector of the first n labels, and let $\hat{Y}_n := (\hat{y}_1^n, \dots, \hat{y}_n^n)'$ be the vector of the first n fitted values. According to the geometric interpretation of the least squares method in the standard model (11.18) (see, e.g., [89, Chaps. 20–21]), the vector of n-residuals is distributed symmetrically around $\hat{Y}_n$ in the space orthogonal to the estimation space $\{X_n w : w \in \mathbb{R}^p\}$. On the other hand, according to (11.17) and the definition of P_n, $P_n(\cdot|\sigma_n)$, where $\sigma_n := t(z_1, \dots, z_n)$, is the uniform distribution on the sphere, of radius equal to the length of the vector of n-residuals, in the hyperplane orthogonal to the estimation space and passing through the projection $\hat{Y}_n$ of Y_n onto the estimation space. Since the ratio $(y_n - \hat{y}_n)/V_n$ (expressed through the n-residuals $y_i - \hat{y}_i^n$) does not change if all n-residuals are multiplied by the same positive constant (and, therefore, its distribution does not change if the random vector of n-residuals is scaled to have a given length), we may replace the normal distribution of (11.18) by our uniform distribution $P_n(\cdot \mid \sigma_n)$.

The proof will be complete if we show that

$$\left| \frac{y_n - \hat{y}_n}{V_n} \right| = \frac{|y_n - \hat{y}_n|}{V_n}$$

is a bona fide nonconformity measure which monotonically increases as $|y_n - \hat{y}_n^n|$ (or $|y_n - \hat{y}_n|$) increases for any fixed $\sigma_n := t_n(z_1, \dots, z_n)$. To see that $|y_n - \hat{y}_n| / V_n$ can be expressed through σ_n and $z_n = (x_n, y_n)$, it suffices to remember that

$$\hat{y}_n = Y_{n-1}' X_{n-1} (X_{n-1}' X_{n-1})^{-1} x_n$$

(see (2.40)) and notice that V_n can also be expressed through σ_n (and z_n):

$$\sum_{i=1}^{n} (y_i - \hat{y}_i^n)^2 = \left\| Y_n - \hat{Y}_n \right\|^2 = \|Y_n\|^2 - \left\| \hat{Y}_n \right\|^2 = \|Y_n\|^2 - \|H_n Y_n\|^2 ,$$

where $H_n = X_n (X_n' X_n)^{-1} X_n'$ is the hat matrix (2.41). Finally, we deduce from (11.29) and (11.30):

$$\begin{aligned} \frac{|y_n - \hat{y}_n|}{V_n} \quad &\uparrow\uparrow \quad \frac{|y_n - \hat{y}_n^n|}{\sqrt{C - c(y_n - \hat{y}_n^n)^2}} \quad \uparrow\uparrow \quad \frac{(y_n - \hat{y}_n^n)^2}{C - c(y_n - \hat{y}_n^n)^2} \\ \uparrow\downarrow \quad \frac{C - c(y_n - \hat{y}_n^n)^2}{(y_n - \hat{y}_n^n)^2} \quad &\uparrow\uparrow \quad \frac{1}{(y_n - \hat{y}_n^n)^2} \quad \uparrow\downarrow \quad |y_n - \hat{y}_n^n| \quad \uparrow\uparrow \quad |y_n - \hat{y}_n| , \end{aligned}$$

where $C > 0$ and c are constants (for a fixed σ_n), $\uparrow\uparrow$ means "changes in the same direction as", and $\uparrow\downarrow$ means "changes in the opposite direction to".

11.6 Context

In the first edition of this book [402, Chap. 8] we had slightly different (but equivalent) definitions of OCMs and repetitive structures. Roughly, now our summaries always include the number of examples summarized.

11.6.1 Kolmogorov's Programme

The general idea of online compression modelling seems to have originated in the work of Andrei Kolmogorov, who is perhaps best known for his axiomatization of probability theory as a branch of measure theory [199]. Kolmogorov, however, never believed that his measure-theoretic axioms per se provide a satisfactory foundation for the applications of probability (as opposed to the mathematical theory of probability). Starting from [202] he embarked on a programme of creating a better foundation, which we started to discuss in Sect. 9.5.1. There is no complete published description of Kolmogorov's programme, but his papers [204, 205] and papers reporting work done by his PhD students [8, 9, 238, 376] provide material for a more or less plausible reconstruction of its main ideas; such an attempt was made in [382, 395].

Andrei N. Kolmogorov (1903–1987).

Used with permission from Albert N. Shiryaev. Elena A. Shiryaeva helped us to obtain it

The standard approach to modelling uncertainty is to choose a family of probability distributions, called a statistical model (see Sect. A.1.1), one of which is believed to be the true distribution generating, or explaining in a satisfactory way, the data. (In some applications of probability theory, the true distribution is assumed to be known, and so the statistical model is a one-element set. In Bayesian statistics, the statistical model is complemented by another element, a prior probability distribution on the elements of the statistical model or their indices.) All modern applications of probability are widely believed to depend on this kind of modelling. (We saw in Sect. 11.4 that, e.g., even such a classical procedure as the confidence predictor (11.16) based on the t-distribution can be directly analyzed in terms of online compression models, but the standard approach would be to do analysis in terms of statistical models.)

In 1965–1970 Kolmogorov suggested a different approach to modelling uncertainty, based on information theory, with the purpose of providing a more direct link between the theory and applications of probability. He started, in [202], from the idea that the object of probability theory is a finitary version of von Mises's notion of collectives, which he called "tables of random numbers", but the development of this idea led him to the general idea of compression modelling. In [204, Sect. 2] he replaced finitary collectives by "Bernoulli sequences" (discussed in detail in Sect. 9.5.1), which provide the first example of what we call "Kolmogorov complexity models". This development became possible only after the introduction of the algorithmic notion of complexity (now called *Kolmogorov complexity*) in [203]. In [205] he defines Markov binary sequences, another example. A third example, Gaussian sequences of real numbers, is described by Asarin [8, 9]; Asarin [8] also describes the Poisson model. We will describe these three examples after a brief general description of Kolmogorov's approach.

The general idea of Kolmogorov's programme is that

> practical conclusions of probability theory can be substantiated as implications of hypotheses of a *limiting*, under given constraints, complexity of phenomena under study

[205, Sect. 4]. In essence, a Kolmogorov complexity model is a way of summarizing information in a data sequence; the summary then provides the constraints under which the complexity of the data sequence is required to be close to the maximum. Using Kolmogorov's algorithmic notion of randomness (a data sequence x is *algorithmically random* in a set A containing x if the Kolmogorov complexity $K(x \mid A)$ is close to the binary logarithm $\log |A|$), we can say that the data sequence is required to be algorithmically random given the summary (i.e., algorithmically random in the set of all data sequences with the same summary).

Each specific Kolmogorov complexity model provides a way of summarizing information in a data sequence. The Bernoulli and Markov models are for binary (consisting of 0 and 1) sequences. The Bernoulli model summarizes a binary sequence by the number of 1s in it. The Markov model summarizes it by the number of 1s after 1s, 1s after 0s, 0s after 1s, and 0s after 0s. Besides, it is always assumed that the length of the data sequence is part of the summary. Accordingly, a finite binary sequence is *Bernoulli* if it has a nearly maximal Kolmogorov complexity in the set of binary sequences of the same length and the same number of 1s; *Markov* binary sequences have a nearly maximal Kolmogorov complexity in the set of binary sequences with the same number of 1s after 1s, 1s after 0s, 0s after 1s, and 0s after 0s. The Gaussian model summarizes a sequence of real numbers by approximate values for its arithmetic mean and variance (11.12), and so *Gaussian* sequences of real numbers are those that are close to being maximally complex in the set of sequences of the same length and with similar mean and variance.

The main features of Kolmogorov's programme appear to be the following (some of these features have not been discussed so far):

- It is based on the idea of *compression*. The compact summary contains, intuitively, all useful information in the data.
- The idea that if the summary is known, the information left in the data is noise, is formalized using the *algorithmic* notion of Kolmogorov complexity: the complexity of the data under the constraint given by the summary should be close to maximal (the requirement of algorithmic randomness).
- Semantically, the requirement of algorithmic randomness means that the conditional distribution of the data given the summary is *uniform*.
- It is preferable to deduce properties of data sequences *directly* from the assumption of limiting complexity, without a detour through standard statistical models (examples of such direct inferences are given in [8] and [9] and hinted at in [205]), especially that Kolmogorov complexity models are not completely equivalent to standard statistical models [376].
- Kolmogorov's programme deals only with finite sets and their elements. This *finitary nature* of Kolmogorov's programme is typical of Kolmogorov's work in general: e.g., in his papers [196, 197] he found it helpful to state even such an apparently asymptotic result as the law of the iterated logarithm in the observable terms, for finite sequences.

11.6.2 Repetitive Structures

The notion of repetitive structure, first introduced by Martin-Löf [239], is a natural outgrowth of Kolmogorov's programme. Different authors used the term "repetitive structure" in different senses (in this book we continue this tradition in that our notion of repetitive structure is somewhat different from those we have seen in literature), and so we will be using "repetitive structure" as a generic term covering several related concepts.

Martin-Löf spent 1964–1965 in Moscow as Kolmogorov's PhD student. His paper [238] is an important contribution to the study of the Bernoulli model, but perhaps his main achievement (published in the same paper) in this area is to restate Kolmogorov's algorithmic notion of randomness in terms of universal statistical tests, thus demonstrating its fundamental character. After 1965 he and Kolmogorov worked on the information-theoretic approach to the applications of probability independently of each other (Martin-Löf having returned to Sweden), but arrived at similar concepts. In his work Martin-Löf was inspired not only by Kolmogorov's programme but also by Gibbs's and Khinchin's ideas in statistical mechanics; this is the origin of names such as "Boltzmann distributions" or "microcanonical distributions" in the theory of repetitive structures.

Martin-Löf [239] gave the definition of repetitive structure, as in Sect. 11.2.4, but with the conditional distributions $P_n(\cdot \mid \sigma)$ being uniform on a finite set and without the condition that t_n should be computable from t_{n-1} and z_n. Martin-Löf's theory of repetitive structures shared the idea of compression and the uniformity of conditional distributions with Kolmogorov's programme. An extra feature of repetitive structures is their *online character*: one consider sequences of all lengths n simultaneously (although the online character of the statistics t_n seems to have entered the theory of repetitive structures through the work of Steffen Lauritzen, who in [214] attributes this notion to Freedman [120, 121]).

Despite being a key contributor to the algorithmic theory of randomness, Martin-Löf did not use the algorithmic notions of complexity and randomness in his theory of repetitive structures. In Chap. 2 we already referred to our observation [395] that these algorithmic notions tend not to lead to mathematical results in their strongest and most elegant form. After having performed their role as a valuable source of intuition, they are often discarded.

The notion of repetitive structure was later studied by Lauritzen; see, especially, his 1982 book [213] and its revised and updated version [214]. Lauritzen's [214, Sect. IV.3] repetitive structures do not involve any probabilities. Dawid [70] was influential in propagating Martin-Löf and Lauritzen's ideas among Bayesian statisticians.

Freedman and Diaconis independently came up with ideas similar to Kolmogorov's (Freedman's first paper [120] in this direction was published in 1962); they were inspired by de Finetti's theorem and the Krylov–Bogolyubov approach to ergodic theory.

The general theory of repetitive structures (as well as its most important special case, the exchangeability model) is now considered to be central to Bayesian model-building. See, e.g., the textbook by Bernardo and Smith [30, Chap. 4], which discusses a wide range of repetitive structures, although without using this name.

The usual approach in the theory of repetitive structures is first to find all probability distributions P on $\mathbf{Z}^\infty$ that agree with the given repetitive structure and then to take the extreme points of the set of all such P as the statistical model. And once we have a statistical model, a wide arsenal of standard methods can be used.

11.6.3 Exchangeability Model

We started our discussion of exchangeability in its historical development in Sect. 2.9.1. As described in Sect. 11.6.1, the binary exchangeability ("Bernoulli") model was the first complexity model considered by Kolmogorov (who arrived at it developing von Mises's ideas).

There exists vast literature (including de Finetti's [77] and Freedman's [120]) on partial exchangeability, a generalization of exchangeability akin to the Mondrian-exchangeability models. Label-conditional OCMs were first described in print by Ryabko [297].

The repetitive structure of Sect. 11.3.7 is a generalization of the one introduced (without using our terminology) by Ryan Tibshirani et al. [355]. Tibshirani et al. only consider the case of covariate shift: the task is to predict the label y_n of x_n, and it is known that $f_1 = \cdots = f_{n-1} = 1$ and f_n depends on z_n only via its object x_n; there are efficient algorithms for estimating the likelihood ratio f_n.

Lauritzen [214, Example 9.1 in Chap. II] characterizes the set of probability measures that agree with the exchangeability OCM with the dataset shift given by $f_1, f_2, \ldots$ in the binary case $\mathbf{Z} = \{0, 1\}$ and under mild restrictions on the sequence $f_1, f_2, \ldots$.

11.6.4 Gaussian Models

The origins of the Gaussian model lie in statistical physics. Especially important is the simplified version of the Gaussian model in which the summarizing statistics are

$$t_n(z_1, \ldots, z_n) = \sqrt{z_1^2 + \cdots + z_n^2},$$

and the conditional distributions $P_n(\cdot \mid \sigma)$ are uniform on the sphere $t_n = \sigma$. Gibbs proposed this as a model of the situation where z_i, $i = 1, \ldots, n$, is the speed of the ith molecule of ideal gas; the value of t_n being known corresponds to the total energy of all molecules being known; the uniform conditional distributions are called microcanonical distributions in statistical physics. Maxwell's law (obtained by Maxwell using the assumption of independence of the components of the molecules' speed along the Cartesian axes) states that the distribution of z_i is normal with zero mean and same variance. According to Bourbaki [34, Chap. 24], Borel (in 1914) appears to be the first to notice that Maxwell's law (representing the standard approach to statistical modelling) is a corollary of Gibbs's model when n is large.

The result that those probability distributions that agree with the simplified Gaussian model are mixtures of power normal distributions $\mathbf{N}_{0,\sigma}^{\infty}$ appears to be due to Freedman [121] (according to Kingman [193]), but it was also independently discovered by Kingman himself [192]. The result about the full Gaussian model is due to Smith [336].

Fisher was the first to prove that (11.15) has the t-distribution with $n - 2$ degrees of freedom (see [104, pp. 610–611], [105, Sect. 5], and [111]). This result was explicitly used for the purpose of prediction by Baker [14]. The name "Gauss liner model" was suggested for the standard linear regression model by Seal [312]. The online compression version of this model was suggested in [384]. Explicit formulas for the multivariate Gaussian predictor are given in Appendix B of the arXiv version of [405]; that appendix also gives details of (essentially a pseudocode for) the conformal predictor for the exchangeability model determined by the nonconformity measure (11.19) (which was used in Sect. 11.4.4, in contrast to the one-sided definition in Sect. 2.3.2).

11.6.5 Markov Model

In the first edition of this book [402, Sect. 8.6] we studied in detail conformal prediction in another online compression model, the Markov model, which uses Kolmogorov's way of summarizing a data sequence produced by a Markov chain, as described in Sect. 11.6.1. An unusual feature of conformal prediction when applied to this model is an apparent efficiency/validity trade-off; see the end of [402, Sect. 8.8] for a detailed discussion.

Chapter 12
Generalized Venn Prediction and Hypergraphical Models

Abstract This chapter has two foci, generalized Venn prediction and hypergraphical models. Generalized Venn prediction extends Venn prediction to general one-off structures and online compression models. An interesting example of one-off structures and online compression models is provided by hypergraphical models. We show that hypergraphical models are a versatile tool and develop both Venn and conformal predictors for them.

Keywords Online compression models · Hypergraphical models · Venn prediction · Conformal prediction

12.1 Introduction

This chapter extends the notion of a Venn predictor (Chap. 6) to the general framework of one-off structures (Sect. 11.2.5). Venn prediction generalized in this way from the exchangeability OOS (as defined in Sect. 11.3.1 and used in Chap. 6 implicitly) still satisfies a similar property of validity.

Another focus of this chapter is the introduction of a new class of one-off structures and online compression models, which we call hypergraphical models; it is analogous to causal networks in machine learning and contingency tables in mathematical statistics. We apply both Venn and conformal prediction to these models. For each hypergraphical model, we define the most refined Venn predictor, which we call the "fully object-conditional Venn predictor" (FOCVP), and we define a very natural conformal predictor based on conditional probability as conformity measure.

We pay a special attention to a subclass of hypergraphical models consisting of what we call "junction-tree models". This subclass is both amenable to straightforward analysis and wide enough to cover many practically interesting models. In particular, we find a simple explicit representation of the FOCVP for junction-tree models. In Sect. 12.4.3 we demonstrate the working of the FOCVP on an artificial dataset randomly generated from a simple hypergraphical model.

V. Vovk et al., *Algorithmic Learning in a Random World*,
https://doi.org/10.1007/978-3-031-06649-8_12

Historically, the main sources of hypergraphical models are statistical physics, path analysis, and contingency tables [215, Sect. 1.1], but the area providing most of the hypergraphical models of interest in our present context is the theory of causal networks; the model of Sect. 12.4.3 is of this type. Transition to junction-tree models is the standard procedure in that area [61, 178, 315].

For our computational experiments with conformal prediction in hypergraphical models (Sect. 12.5) we choose the problem of digit recognition for an artificial dataset and study both unconditional and label-conditional conformal predictors.

In this chapter we only consider the problem of classification: the label space $\mathbf{Y}$ is finite.

12.2 Venn Prediction and Venn-Conditional Conformal Prediction in One-Off Structures

In this section we define and use the notion of a Venn taxonomy in the context of one-off structures in two different ways: first for generalizing Venn predictors, as defined in Chap. 6, and then for generalizing Venn-conditional conformal predictors, as defined in Sect. 4.6.1.

12.2.1 Venn Prediction

The notion of Venn predictor, introduced in Sect. 6.2 for the exchangeability OOS, generalizes easily to arbitrary one-off n-structures $(\Sigma, \mathbf{Z}, t, B)$. A *taxonomy*, or *Venn taxonomy*, is a measurable function $K : \Sigma \times \mathbf{Z} \to \mathbf{K}$, where $\mathbf{K}$ is a countable measurable space (of *categories*) with the discrete σ-algebra. The *Venn predictor determined by* K is the multiprobability predictor which outputs $P := \{P^y : y \in \mathbf{Y}\} \subseteq \mathbf{P}(\mathbf{Y})$ as the prediction for the label of a test object x given a training set $z_1, \dots, z_{n-1}$, where each probability distribution P^y on $\mathbf{Y}$ is defined as follows. Complement the test object x by the "postulated label" y and set

$$\sigma^y := t(z_1, \dots, z_{n-1}, z_n) ,$$

$$\kappa_i^y := K(\sigma^y, z_i), \quad i = 1, \dots, n ,$$

where z_n is understood to be (x, y). Define P^y to be the probability distribution under $B(\cdot \mid \sigma^y)$ of the labels in the test example's category: for all $y' \in \mathbf{Y}$,

$$P^y\{y'\} := \frac{B\left(\left\{z \in \mathbf{Z} : K(\sigma^y, z) = K(\sigma^y, z_n) \ \& \ z^{\text{lab}} = y'\right\} \mid \sigma^y\right)}{B\left(\{z \in \mathbf{Z} : K(\sigma^y, z) = K(\sigma^y, z_n)\} \mid \sigma^y\right)} , \tag{12.1}$$

where z^{lab} stands for the label of the example z (i.e., $(x', y')^{\text{lab}} := y'$). Notice that the denominator in (12.1) is positive almost surely under any probability measure that agrees with the one-off structure. A *Venn predictor* is the Venn predictor determined by some taxonomy.

Now we generalize the notion of validity of Sect. 4.4.3. As before, Y is the random variable that maps any $(z_1, \ldots, z_n) \in \mathbf{Z}^n$ to the label of z_n. A *selector* is a measurable function $S : \mathbf{Z}^n \to \mathbf{Y}$, i.e., a $\mathbf{Y}$-valued random variable. The composition P^S, which is a $\mathbf{P}(\mathbf{Y})$-valued random variable, is also defined as before.

Theorem 12.1 *There exists a selector S such that, under any probability distribution that agrees with the one-off structure, P^S is perfectly calibrated for Y.*

Proof Take $S := Y$ as the selector. It suffices to check that (6.2) holds under $B(\cdot|\sigma)$ for any summary $\sigma \in t(\mathbf{Z}^n)$. Let us fix such a summary σ and let $\mathbb{P}$ stand for $B(\cdot|\sigma)$. We only need, for each $y' \in \mathbf{Y}$, to check the equality

$$\mathbb{P}(Y = y' \mid K(\sigma, Z)) = P\{y'\}, \tag{12.2}$$

where Z is the random example generated by $\mathbb{P}$, Y is its label, and P is the probability distribution of the labels in Z's category $\{z \in \mathbf{Z} : K(\sigma, z) = K(\sigma, Z)\}$ (it is clear that P is measurable w.r.t. to the σ-algebra generated by $K(\sigma, Z)$). And (12.2) follows from the definition (12.1). □

12.2.2 Venn-Conditional Conformal Prediction

Let us now generalize Venn-conditional conformal predictors, introduced in Sect. 4.6.1. In the notation of this section, we give our definitions using the size n of the augmented training set rather than the size $l = n - 1$ of the training set, and we discuss only the smoothed version. The *Venn-conditional conformal predictor* determined by a nonconformity measure A and taxonomy K is defined in terms of p-values as usual,

$$\Gamma^{\epsilon}(z_1, \ldots, z_{n-1}, x) := \{y : p^y > \epsilon\},$$

but now the p-values p^y are

$$\begin{aligned} p^y :=& \frac{B(\{z : K(\sigma^y, z) = K(\sigma^y, (x_n, y)) \,\&\, A(\sigma^y, z) > A(\sigma^y, (x_n, y))\} \mid \sigma^y)}{B(\{z : K(\sigma^y, z) = K(\sigma^y, (x_n, y))\} \mid \sigma^y)} \\ &+ \tau \frac{B(\{z : K(\sigma^y, z) = K(\sigma^y, (x_n, y)) \,\&\, A(\sigma^y, z) = A(\sigma^y, (x_n, y))\} \mid \sigma^y)}{B(\{z : K(\sigma^y, z) = K(\sigma^y, (x_n, y))\} \mid \sigma^y)}, \end{aligned} \tag{12.3}$$

where σ^y is the summary $t(z_1, \ldots, z_{n-1}, (x, y))$ with the postulated test label y, and τ is uniformly distributed on $[0, 1]$, independently of the examples. The denominators in (12.3) are positive a.s. under any probability distribution that agrees with the OOS.

Now we can generalize Proposition 4.5 as follows.

Theorem 12.2 *If examples $z_1, \ldots, z_n$ are generated from a probability distribution on $\mathbf{Z}^n$ that agrees with the OOS, a smoothed Venn-conditional conformal predictor outputs a p-value, p^{y_n}, that is uniformly distributed on* $[0, 1]$, *even conditionally on the category $K(t(z_1, \ldots, z_n), z_n)$ of the test example.*

The proof of Theorem 12.2 is essentially contained in the proof of Theorem 11.2.

12.3 Hypergraphical Models

Starting from this section we assume that the examples are structured, consisting of "variables". Formally, a *hypergraphical structure* is a triple $(V, \mathcal{E}, \Xi)$ where:

- V is a finite set whose elements are called *variables*;
- $\mathcal{E}$ is a family of V's subsets; elements of $\mathcal{E}$ are called *clusters*; the union of all clusters is required to be the whole of V;
- Ξ is a function that maps each variable $v \in V$ into a finite set $\Xi(v)$ of the "values that v can take"; $\Xi(v)$ is called the *frame* of v; to exclude trivial cases, we always assume $\forall v \in V : |\Xi(v)| > 1$.

We will eventually assume that some of the variables are marked as *labels*, but this assumption will not be needed in many of our considerations. A *configuration* on a cluster (or, more generally, V's subset) E is an assignment of an element of $\Xi(v)$ to each $v \in E$. An *example* is a configuration on V; we take $\mathbf{Z}$ to be the set of all examples.

A *table* on a cluster E is an assignment of a nonnegative number to each configuration on E. We will mainly be interested in *natural tables*, which assign only natural (i.e., nonnegative integer) numbers to configurations. (These are known as "contingency tables" in statistics.) The *size* of the table is the sum of values that it assigns to different configurations. A *table set* f assigns to each cluster E a table f_E on this cluster. *Natural table sets* are table sets all of whose tables are natural. We will only be interested in table sets all of whose tables have the same size, which is then called the size of the table set. The number assigned by a natural table set σ to a configuration on a cluster E will sometimes be called the *σ-count* of that configuration.

12.3.1 Hypergraphical Repetitive Structures

Now we are ready to define the hypergraphical repetitive structure and OCM associated with a hypergraphical structure $(V, \mathcal{E}, \Xi)$; as usual, we first define the repetitive structure $(\Sigma, \mathbf{Z}, t, (P_n))$. The table set $t(z_1, \ldots, z_n)$ *generated* by a data sequence $z_1, \ldots, z_n$ assigns to each configuration on each cluster the number of examples among $z_1, \ldots, z_n$ that agree with that configuration (we say that an example z *agrees* with a configuration on a cluster E if that configuration coincides with the restriction $z|_E$ of z to E). In particular, $t()$ is the table set of size 0. The number of data sequences generating a table set σ will be denoted $\#\sigma$ (for $\#\sigma$ to be non-zero the size of σ must exist, and then the length of each sequence generating σ will be equal to its size). The table sets σ with $\#\sigma > 0$ (called *consistent* table sets) are called *summaries*; they form the summary space Σ of the hypergraphical online compression model and repetitive structure associated with $(V, \mathcal{E}, \Xi)$. The conditional probability distribution $P_n(\cdot \mid \sigma)$, where n is the size of σ, is the uniform distribution on the set of all data sequences $z_1, \ldots, z_n$ that generate σ.

The explicit definition of the corresponding hypergraphical OCM $(\Sigma, \Box, \mathbf{Z}, F, B)$ is as follows:

- Σ is the set of all summaries (i.e., consistent table sets); $\Box$ is the *empty* table set, i.e., the one of size 0;
- $\mathbf{Z}$ is the set of all examples (i.e., configurations on V);
- the table set $F(\sigma, z)$ is obtained from σ by adding 1 to the σ-count of each configuration that agrees with z;
- an example z *agrees* with a summary σ if the σ-count of each configuration that agrees with z is positive; if so, we obtain a table set denoted $\sigma \downarrow z$ from σ by subtracting 1 from the σ-count of any configuration that agrees with z; $B(\sigma)$ is defined by

$$B(\{(\sigma \downarrow z, z)\} \mid \sigma) := \frac{\#(\sigma \downarrow z)}{\#\sigma} .$$

Among the probability distributions P that agree with the hypergraphical structure $(V, \mathcal{E}, \Xi)$ (i.e., with the OCM associated with $(V, \mathcal{E}, \Xi)$; we do not always distinguish between hypergraphical structures and the corresponding OCMs and repetitive structures) are power distributions Q^∞ such that each Q (a probability distribution on $\mathbf{Z}$) decomposes into

$$Q\{a\} = \prod_{E \in \mathcal{E}} f_E(a|_E) , \tag{12.4}$$

where a is any configuration on V, f is a fixed table set (not necessarily natural), and $a|_E$ is, as usual, the restriction of a to E.

The exchangeability model with the example space $\mathbf{Z}$ corresponds to the hypergraphical model with only one cluster, $\mathcal{E} = \{V\}$.

12.3.2 Generality of Finitary Repetitive Structures

In this subsection we specialize the discussion of generality and specificity of repetitive structures in Sect. 4.6.4 to the Kolmogorov-type finitary repetitive structures with uniform conditional distributions, including the hypergraphical models. Formally, a repetitive structure $(\Sigma, \mathbf{Z}, t, P_n)$ is *finitary* if the example space $\mathbf{Z}$ is finite, and, for all n and $\sigma \in t(\mathbf{Z}^n)$, $P_n(\cdot \mid \sigma)$ is the uniform probability distribution on the finite set $t^{-1}(\sigma) \subseteq \mathbf{Z}^n$. Let us fix $\mathbf{Z}$ and Σ (the latter can be, e.g., chosen large enough to contain all summaries that we are likely to be interested in). Then a finitary repetitive structure is determined by t, and we will sometimes say that t is the repetitive structure.

The choice of the repetitive structure t reflects the strength of the assumption that we are willing to make about Reality. According to the definition given in Sect. 4.6.4, a finitary repetitive structure t is *more specific* than another finitary repetitive structure t' (denoted as $t' \preceq t$) if $t = f(t')$ for some measurable function $f : \Sigma \to \Sigma$. Intuitively, in this case t performs a greater data compression than t' does and so represents a stronger assumption about Reality. In particular, if $t' \preceq t$, then any probability distribution on $\mathbf{Z}^\infty$ that agrees with t also agrees with t'. The analogous statement is also true for a finite horizon N: any probability distribution on $\mathbf{Z}^N$ that agrees with t also agrees with t'.

12.3.3 Generality of Hypergraphical Models

Fix the set V of variables and the frame $\Xi(v)$ for each variable. The following proposition answers the question when the repetitive structure corresponding to a cluster set $\mathcal{E}_1$ is more specific than the repetitive structure corresponding to a cluster set $\mathcal{E}_2$ (in this case we will say that $\mathcal{E}_1$ is more specific than $\mathcal{E}_2$).

Proposition 12.3 *A cluster set $\mathcal{E}_1$ is more specific than a cluster set $\mathcal{E}_2$, denoted $\mathcal{E}_2 \preceq \mathcal{E}_1$, if and only if for each $E_1 \in \mathcal{E}_1$ there exists $E_2 \in \mathcal{E}_2$ such that $E_1 \subseteq E_2$.*

Proof The part "if" is obvious, so we only prove "only if". Suppose $\mathcal{E}_1$ is more specific than $\mathcal{E}_2$ but there exists an $E \in \mathcal{E}_1$ which is not covered by any element of $\mathcal{E}_2$. Let $k := |E|$; without loss of generality we suppose that the model is binary ($\Xi(v) = \{0, 1\}$ for all v), that $E = V$, and that all subsets of E of size $k - 1$ are in $\mathcal{E}_2$.

Consider the following $k2^{k-1} \times 2^k$ matrix X: the columns of X are indexed by $\{0, 1\}^k$ (they represent the configurations of E); the rows of X are indexed by the sequences in $\{0, 1\}^k$ in which one of the bits is replaced by the symbol "x" (they represent the configurations of the k subsets of E of size $k-1$); the element $X_{i,j}$ of X in row i and column j is 1 if j can be obtained from i by replacing the x with 0 or 1 and is 0 otherwise. If the table corresponding to the cluster E is given by a vector $t \in \mathbb{N}_0^{2^k}$ ($\mathbb{N}_0$ being the set $\{0, 1, \dots\}$ of nonnegative integers), the tables corresponding to the subsets of E of size $k-1$ are given by Xt. Therefore, our goal will be achieved if we show that there are two vectors $t_1, t_2 \in \mathbb{N}_0^{2^k}$ such that $Xt_1 = Xt_2$.

The rank of the matrix X is at most $2^k - 1$, since the sum of all columns is the identical 2. The procedure of Gaussian elimination shows that there exists a zero linear combination of X's columns with rational (and, therefore, with integer) coefficients. Let t_0 be a vector in $\mathbb{Z}^{2^k}$ such that $Xt_0 = 0$. Now we can take as t_1 any vector in $\mathbb{N}^{2^k}$ with sufficiently large elements and set $t_2 := t_1 + t_0$. □

For simplicity we will only consider *reduced* hypergraphical models, i.e., models $(V, \mathcal{E}, \Xi)$ such that no $E_1, E_2 \in \mathcal{E}$ are strictly nested, $E_1 \subset E_2$; in this case we will also say that $\mathcal{E}$ is reduced. This does not limit generality, since we can always replace $\mathcal{E}$ by $\text{red}(\mathcal{E})$, where $\text{red}(\mathcal{E})$ is $\mathcal{E}$ with all $E \in \mathcal{E}$ strictly contained in some $E' \in \mathcal{E}$ removed. (In other words, $\text{red}(\mathcal{E})$ is the only reduced element of $\mathcal{E}$'s equivalence class, where the equivalence of $\mathcal{E}_1$ and $\mathcal{E}_2$ means that $\mathcal{E}_1 \preceq \mathcal{E}_2$ and $\mathcal{E}_2 \preceq \mathcal{E}_1$.)

The set of all reduced hypergraphical structures $\mathcal{E}$ (with V and Ξ fixed) forms a lattice with the join and meet operations

$$\mathcal{E}_1 \vee \mathcal{E}_2 := \text{red}\{E_1 \cap E_2 : E_1 \in \mathcal{E}_1, E_2 \in \mathcal{E}_2\},$$

$$\mathcal{E}_1 \wedge \mathcal{E}_2 := \text{red}(\mathcal{E}_1 \cup \mathcal{E}_2).$$

12.3.4 Junction-Tree Models

An important special case is where we can arrange the clusters of a hypergraphical model into a "junction tree". We will be able to give efficient prediction algorithms only for such junction-tree models; if the hypergraphical model we happen to be interested in is not of this type, it should be replaced by a more general (i.e., less specific) junction-tree model before our prediction algorithms can be applied.

Formally, a *junction tree* for a hypergraphical model $(V, \mathcal{E}, \Xi)$ is an undirected tree (U, S) (with U the set of vertices and S the set of edges) together with a bijective mapping C from the vertices U of the tree to the clusters $\mathcal{E}$ of the hypergraphical model which satisfies the following property: if a vertex v lies on the path from a vertex u to a vertex w in the tree (U, S), then

$$C_u \cap C_w \subseteq C_v$$

(we let C_x stand for $C(x)$). The tree (U, S) will also sometimes be called the junction tree (when the bijection is clear from the context). It is convenient to identify vertices v of the junction tree with the corresponding clusters C_v in $\mathcal{E}$. If $s = \{u, v\} \in S$ is an edge of the junction tree connecting vertices u and v, we will write C_s for $C_u \cap C_v$; C_s will be called the *separator* between C_u and C_v.

We will say "junction-tree structures/models" to mean hypergraphical structures/models in which the clusters are arranged into a junction tree. Fix such a model $(V, \mathcal{E}, \Xi)$ until the end of this section; (U, S) is the corresponding junction tree, F is the forward function, B is the backward kernel, t is the summarizing statistic, and P_n are the Markov kernels, i.e., the conditional probability distributions given a summary (uniform in this case).

12.3.5 Combinatorics of Junction-Tree Models

It is easy to characterize consistent table sets in junction-tree structures. If $E_1 \subseteq E_2 \subseteq V$ and f is a table on E_2, its *marginalization* to E_1 is the table f^* on E_1 such that $f^*(a) = \sum_b f(b)$ for all configurations a on E_1, where b ranges over all configurations on E_2 that agree with a (i.e., such that $b|_{E_1} = a$).

Lemma 12.4 *A natural table set σ on $(V, \mathcal{E}, \Xi)$ is consistent if and only if the following two conditions hold:*

- *each table in σ is of the same size;*
- *if clusters $E_1, E_2 \in \mathcal{E}$ intersect, the marginalizations of their tables to $E_1 \cap E_2$ coincide.*

This lemma is obvious; it, however, ceases to be true if the assumption that $(V, \mathcal{E}, \Xi)$ is a junction-tree structure is dropped.

If σ is a summary and E is a cluster, we earlier defined σ_E as the table that σ assigns to E. If E is a separator, say $E = C_{\{u,v\}}$, σ_E stands for the marginalization of σ_{C_u} (equivalently, by Lemma 12.4, of σ_{C_v}) to E.

The *factorial-product* of a cluster or separator E in a summary σ is, by definition,

$$\mathrm{fp}_\sigma(E) := \prod_{a \in \mathrm{conf}(E)} \sigma_E(a)!$$

(remember that $0! = 1$), where $\mathrm{conf}(E)$ is the set of all configurations on E.

Lemma 12.5 *Consider a summary σ of size n in the junction-tree model. The number of data sequences of length n generating the table set σ equals*

$$\#\sigma = \frac{n! \prod_{s \in S} \mathrm{fp}_\sigma(C_s)}{\prod_{u \in U} \mathrm{fp}_\sigma(C_u)}. \tag{12.5}$$

Proof The proof is by induction on the size of the junction tree. If the junction tree consists of only one vertex u, the right-hand side of (12.5) becomes

$$\frac{n!}{\mathrm{fp}_\sigma(C_u)} = \frac{n!}{\prod_{a \in \mathrm{conf}(C_u)} \sigma_{C_u}(a)!},$$

which is the correct multinomial coefficient.

Now let us assume that (12.5) is true for some tree and prove that it remains true for that tree extended by adding an edge s and a vertex u. (The example space for the new tree will be bigger.) We are required to show that the number of data sequences generating σ is multiplied by

$$\frac{\mathrm{fp}_\sigma(C_s)}{\mathrm{fp}_\sigma(C_u)} = \prod_{a \in \mathrm{conf}(C_s)} \frac{\sigma_{C_s}(a)!}{\prod_{b \in \mathrm{agr}(a)} \sigma_{C_u}(b)!}, \tag{12.6}$$

where $\mathrm{agr}(a)$ is the set of all configurations on C_u that agree with a. It remains to notice that the number of ways in which each sequence of n examples in the old tree can be extended to a sequence of n examples in the new tree is given by the right-hand side of (12.6). □

We will use the shorthand $\sigma_u(z) := \sigma_{C_u}(z|_{C_u})$ for both vertices and edges u of a junction tree; here z is an example or, more generally, a configuration whose domain includes all variables in C_u.

Lemma 12.6 *Given the summary σ of the first n examples $z_1, \ldots, z_n$, the $B(\sigma)$-probability that $z_n = a$ equals*

$$\frac{\prod_{u \in U} \sigma_u(a)}{n \prod_{s \in S} \sigma_s(a)} \tag{12.7}$$

(this ratio is set to 0 if any of the factors in the numerator or denominator is 0; in this case $z_n = a$ does not agree with the summary σ).

Proof Using Lemma 12.5, we obtain for the probability of $z_n = a$:

$$\frac{\#(\sigma \downarrow a)}{\#\sigma} = \frac{(n-1)! \prod_{s \in S} \mathrm{fp}_{\sigma \downarrow a}(C_s) \prod_{u \in U} \mathrm{fp}_\sigma(C_u)}{\prod_{u \in U} \mathrm{fp}_{\sigma \downarrow a}(C_u) n! \prod_{s \in S} \mathrm{fp}_\sigma(C_s)} = \frac{\prod_{u \in U} \sigma_u(a)}{n \prod_{s \in S} \sigma_s(a)}.$$

□

12.3.6 Shuffling Datasets

In this subsection we will see that Lemma 12.6 provides an efficient means of drawing a dataset $z_1, \ldots, z_n$ from the uniform conditional distribution $P_n(\sigma)$, where σ is a summary of size n in the junction-tree model. This can be used for shuffling datasets to make them conform to the given hypergraphical model (see Sect. B.5).

It is convenient first to direct the junction tree, designating an arbitrary vertex as the root $\Box$ and directing all edges from the root (so that the root becomes an ancestor of every vertex). We can then rewrite (12.7) as

$$\frac{\sigma_\Box(a)}{n} \prod_{u \in U \setminus \{\Box\}} \frac{\sigma_u(a)}{\sigma_{u'}(a)}, \tag{12.8}$$

where u' is the separator between u and u's parent. The last formula provides an efficient means of generating a random data sequence $z_1, \dots, z_n$ from a summary σ of size n: to generate z_n, first generate $z_n|_{C_\Box}$ from $\sigma_\Box/n$ (i.e., from the probability distribution that assigns weight $\sigma_\Box(a)/n$ to each configuration a on $C_\Box$), then choose $\Box$'s child u and generate $z_n|_{C_u}$ from $\sigma_u/\sigma_{u'}$ (i.e., from the probability distribution that assigns weight $\sigma_u(a)/\sigma_{u'}(a)$ to each configuration a on C_u that agrees with $z_n|_{u'}$), and so on. After z_n is generated, we can generate z_{n-1} in a similar way from $\sigma \downarrow z_n$, then generate z_{n-2} from $(\sigma \downarrow z_n) \downarrow z_{n-1}$, etc.

12.3.7 *Decomposability in Junction-Tree Models*

Corollary 12.7 *Each power probability distribution Q^∞ on $\mathbf{Z}^\infty$ that agrees with the junction-tree model is decomposable in the sense of* (12.4).

Proof The idea of derivation of this corollary from Lemma 12.6 is standard; see, e.g., [214, Theorem 4.4 in Chap. II]. Let Q be a probability distribution on $\mathbf{Z}$ such that the power distribution Q^∞ on the set of sequences $z_1, z_2, \dots$ agrees with the junction-tree model. It is clear that, for each configuration a on V,

$$Q_n(a) := Q^\infty\left(z_1 = a \mid t(z_1, \dots, z_n), z_{n+1}, z_{n+2}, \dots\right)$$

(the conditional Q^∞-probability that $z_1 = a$) is given by (12.8) with $\sigma := t(z_1, \dots, z_n)$. According to Lévy's "downward" theorem [428, Sect. 14.4] and the Hewitt–Savage zero-one law [330, Theorem 4.1.3], $Q_n(a)$ converge almost surely to $\mathbb{E}_{Q^\infty} Q_n(a) = Q\{a\}$ as $n \to \infty$. Borel's strong law of large numbers shows that all ratios in (12.8) converge almost surely; this completes the proof. $\Box$

12.4 Venn Prediction in Hypergraphical Models

In the case of hypergraphical OCM with one or more vertices marked as labels, there exists a very natural Venn predictor, which will be called the *fully object-conditional Venn predictor* (FOCVP). It is defined to be the Venn predictor determined (for each n, as described in Sect. 11.2.5) by the taxonomy K in which $K(\sigma, z)$ is simply the restriction of z to the non-label variables (cf. our discussion of object-conditional prediction in Sect. 4.7.1). The FOCVP is not only natural but also computationally efficient in the case of junction-tree models. It will be the only Venn predictor considered in this section, but this should not be interpreted as a recommendation to always use it for hypergraphical models: carefully crafted Venn taxonomies will definitely have advantages for small datasets, and conformal predictors (to be discussed in the next section) also remain an attractive option.

12.4.1 Prediction in Junction-Tree Models

In the context of prediction, we consider the situation where the $|V|$ variables of the junction-tree model are divided into the *attributes* V_{obj} and the *label variables* V_{lab}; only a subset of labels, the *target label variables* V_{targ}, have to be predicted.[1] Therefore,

$$V_{\mathrm{targ}} \subseteq V_{\mathrm{lab}} = V \setminus V_{\mathrm{obj}} .$$

The *object space* $\mathbf{X}$ is then the set of all configurations on V_{obj}, the *label space* $\mathbf{Y}$ is the set of all configurations on V_{lab}, and we define the *target label space* $\mathbf{Y}^{\mathrm{targ}}$ to be the set of all configurations on V_{targ}. Each example z has two components: the values $x := z|_{V_{\mathrm{obj}}} \in \mathbf{X}$ (the *object*) taken by the attributes and the values $y := z|_{V_{\mathrm{lab}}} \in \mathbf{Y}$ (the *label*) taken by the label variables; we will write (x, y) to mean z. The values $y^{\mathrm{targ}} := z|_{V_{\mathrm{targ}}} \in \mathbf{Y}^{\mathrm{targ}}$ taken by the target label variables will be called the *target label*.

In this subsection we will give a simple and explicit representation of the FOCVP, for which the category of each example $z = (x, y)$ is defined to be its object x. Suppose we have observed examples $z_1, \dots, z_{n-1}$ and we are given a new object x_n. Consider the "y-completion", in which the object x_n is complemented to the example (x_n, y), where $y \in \mathbf{Y}$ is a label. For each $y'' \in \mathbf{Y}^{\mathrm{targ}}$ we are interested in the conditional $B(\sigma)$-probability, where $\sigma := t(z_1, \dots, z_{n-1}, (x_n, y))$, that the target label y_n^{targ} is y'' given that the values of the attributes are x_n. For each $y' \in \mathbf{Y}$, let $A_{y,y'}$ be the conditional $B(\sigma)$-probability that the label is y' given that the values of the attributes are x_n. The conditional probability $A_{y,y'}$ in the y-completion that $z_n = (x_n, y')$ given that the attributes' values are x_n is proportional to

$$\frac{\prod_{u \in U} \sigma_u((x_n, y'))}{\prod_{s \in S} \sigma_s((x_n, y'))} , \tag{12.9}$$

[1] It might have been more natural to restrict the use of the word "label" only to target labels, and to call non-target labels, for example, nuisance variables; our terminology, however, is more consistent with that of the previous chapters.

Algorithm 12.1 Junction-tree FOCVP in the online protocol

1: $\sigma_0 := \square$
2: **for** $n = 1, 2, \dots$
3: read $x_n \in \mathbf{X}$
4: **for** $y \in \mathbf{Y}$
5: $\sigma := F(\sigma_{n-1}, (x_n, y))$
6: **for** $y' \in \mathbf{Y}$
7: $A_{y,y'} := \frac{\prod_{u \in U} \sigma_u((x_n, y'))}{\prod_{s \in S} \sigma_s((x_n, y'))}$
8: normalize the rows of $A_{y,y'}$
9: set $P_n \subseteq \mathbf{P}(\mathbf{Y}^{\text{targ}})$ to the rows of $A_{y,y'}$ marginalized to V_{targ}
10: read $y_n \in \mathbf{Y}$
11: $\sigma_n := F(\sigma_{n-1}, (x_n, y_n))$.

since, by Lemma 12.6, (12.9) is proportional to the unconditional $B(\sigma)$-probability that $z_n = (x_n, y')$. Overall, A is a $|\mathbf{Y}| \times |\mathbf{Y}|$ matrix with the rows and columns indexed by $\mathbf{Y}$; this matrix determines the Venn predictor's output for the labels (which is the set of the probability distributions represented by the rows of the matrix). To obtain the prediction for the target labels only, the rows of A have to be marginalized to the target labels. Summarizing, we obtain the description of the FOCVP for junction-tree models given as Algorithm 12.1. Venn predictors were described in the one-off mode in Sect. 12.2.1, but in Algorithm 12.1 they are embedded in the online protocol (which we will use in Sect. 12.4.3).

The normalization of the rows of the matrix $A_{y,y'}$ in line 8 of Algorithm 12.1 means that first the sums $S_y := \sum_{y'} A_{y,y'}$ are computed and then each $A_{y,y'}$ is divided by S_y. If $a(y')$, $y' \in \mathbf{Y}$, is a row of the matrix A, its marginalization to V_{targ} (line 9) is defined to be the probability distribution on $\mathbf{Y}^{\text{targ}}$ that gives the weight

$$\sum_{y' : y'|_{V_{\text{targ}}} = y''} a(y')$$

to each $y'' \in \mathbf{Y}^{\text{targ}}$.

12.4.2 *Universality of the Fully Object-Conditional Venn Predictor*

The FOCVP is universal in our usual asymptotic sense: if the examples are generated from a power probability distribution Q^∞ on $\mathbf{Z}^\infty$ that agrees with the given junction-tree model, the maximum distance between the prediction it outputs and the true conditional probability distribution for the next label tends to zero almost surely (the *maximum distance* between a finite set A and a point b being defined as the maximum of the distances between $a \in A$ and b). This follows from, e.g., the proof of Corollary 12.7.

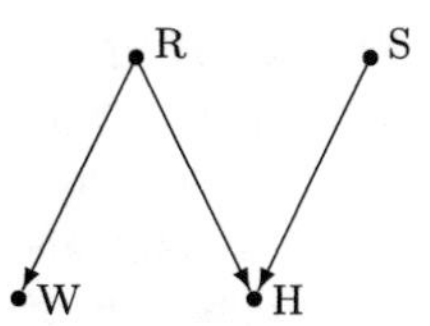

Fig. 12.1 The "wet grass" causal network

12.4.3 An Experiment with a Simple Causal Network

A rich source of hypergraphical models is provided by the theory of causal modelling. We start from a simple example (considered by Pearl [271] and Jensen [178]).

One morning when Mr Holmes leaves his house for work he notices that his grass is wet (H). Was there a rain overnight (R) or did he forget to turn off the sprinkler (S)? Next he checks his neighbour Dr Watson's grass (W). It is wet as well, and so Mr Holmes concludes that wet grass was caused by rain. We can arrange the variables H (Mr Holmes's grass is wet), W (Dr Watson's grass is wet), R (rain), and S (sprinkler) in the causal network shown in Fig. 12.1. The directions of the arrows are intended to reflect the order in which the values of the variables are settled by Reality: first she decides, using a stochastic procedure, on the values of R and S; given the realized value of R she decides on W; and finally, given R and S she decides on H. The probabilities used by Reality are as follows (each variable is assumed binary and takes value 1 if the corresponding event happens and 0 if not):

$$\mathbb{P}\{R = 1\} = 0.2, \quad \mathbb{P}\{S = 1\} = 0.1 \tag{12.10}$$

are the probabilities for R and S,

$$\mathbb{P}\{W = 1 \mid R = 1\} = 1, \quad \mathbb{P}\{W = 1 \mid R = 0\} = 0.2 \tag{12.11}$$

are the conditional probabilities for W given R (Watson may also forget to turn off his sprinkler, but this event is not reflected in the network explicitly), and

$$\begin{aligned} &\mathbb{P}\{H = 1 \mid R = 1, S = 1\} = 1, \quad \mathbb{P}\{H = 1 \mid R = 1, S = 0\} = 1, \\ &\quad \mathbb{P}\{H = 1 \mid R = 0, S = 1\} = 0.9, \quad \mathbb{P}\{H = 1 \mid R = 0, S = 0\} = 0 \end{aligned} \tag{12.12}$$

are the conditional probabilities for H given R and S.

The general approach of causal modelling is to start from a directed acyclic graph (or, more generally, a "chain graph"), such as that in Fig. 12.1, erase the directions of all arrows adding undirected edges between all pairs of vertices that share a common child ("marrying the parents"), and then consider the hypergraphical model whose clusters are the cliques of the resulting undirected graph. If this model is not a junction-tree model (it is in the case of Fig. 12.1), there are computationally efficient ways to find its reasonable junction-tree extension (in the sense of the relation $\preceq$; see

Proposition 12.3). The area of causal modelling thus provides numerous examples of junction-tree models; for details, see [178] or [61].

The preceding paragraph describes only part of the process of causal modelling, which is sometimes called qualitative modelling. Another important ingredient is quantitative modelling [61, pp. 27–29]: the standard approach requires both the structure (such as Fig. 12.1) and the prior probabilities (such as (12.10)–(12.12)). In our approach we do not need the second ingredient: given only the structure, Venn predictors output multiprobability predictions that are automatically valid.

The hypergraph corresponding to the "wet grass" network of Fig. 12.1 has two clusters, $\{W, R\}$ and $\{H, R, S\}$; they form a trivial junction tree with two vertices and one edge between them. The separator is $\{R\}$.

Table 12.1 shows the output of the FOCVP when run on a dataset generated randomly from (12.10)–(12.12). Mr Holmes observes his own and Dr Watson's grass every morning, then checks his sprinkler (if in any doubt), and finally listens to the weather report in the car on his way to work (of course, this pure online protocol can be relaxed, as in Chap. 4). Table 12.1 gives, for selected trials, the trial number n, the observed values H_n, W_n, and S_n of the variables H, W, and S at trial n, the prediction (more precisely, the convex hull of the computed multiprobability prediction) for $S_n = 1$ given the observed value of H_n, and the prediction for $S_n = 1$ given the observed values of H_n and W_n. Naturally, the prediction for $S_n = 1$ given H_n, denoted $P_n(S_n = 1 | H_n)$, is computed using the Junction-tree FOCVP algorithm with W, R, S as the labels and S as the target label, and the prediction for $S_n = 1$ given H_n and W_n, denoted $P_n(S_n = 1 | H_n, W_n)$, is computed using the Junction-tree FOCVP with R, S as the labels and S as the target label.

Trials 3, 8, and 10, where some combination of H and W is observed for the first time, show the full object-conditionality of the predictor used: no information gathered for other combinations is used and the prediction is vacuous, $[0, 1]$. The predictions $P_n(S_n = 1 \mid H_n)$ for both subsequences $\{n : H_n = 0\}$ and $\{n : H_n = 1\}$ are identical to the predictions in the Bernoulli problem (discussed in Sect. 6.2.3). The predictions $P_n(S_n = 1 \mid H_n, W_n)$, however, are not so simple: see trials 14 and 17. The predictions for trials 31–83 clearly show the "explaining away" phenomenon: the conditional probability of $S_n = 1$ drops after Holmes learns that $W_n = 1$. The predictions for trials starting from 1000 (for four different randomly generated datasets) can be compared with the true conditional probabilities: Jensen [178] computes that $S_n = 1$ with probability 0.339 given $H_n = 1$, and $S_n = 1$ with probability 0.161 given $H_n = 1$ and $W_n = 1$. Only one of the eight predictions covers the corresponding true value, but this is not surprising: our current situation (with a simple network and full conditioning on the object) is not so different from the Bernoulli case (Sect. 6.2.3), and so one would expect the order of magnitude n^{-1} for the nth prediction's diameter and $n^{-1/2}$ for the accuracy with which the true probabilities can be estimated. Despite this lack of coverage, the FOCVP's predictions, as we know, are still perfectly calibrated.

Table 12.1 The FOCVP run on a dataset randomly generated from the "wet grass" causal network: the first 17 trials, the trials in the range $n = 18, \ldots, 100$ with $H_n = 1$, and the first trial from $n = 1000$ with $H_n = 1$ and $W_n = 1$; the latter is also given for three other datasets randomly generated from the same causal network (the superscript in square brackets indicates the initial state, if different from 0, of the random number generator)

n	H_n	$P_n(S_n = 1 \mid H_n)$	W_n	$P_n(S_n = 1 \mid H_n, W_n)$	S_n
1	0	[0, 1]	0	[0, 1]	0
2	0	[0, 0.5]	0	[0, 0.5]	0
3	1	[0, 1]	0	[0, 1]	1
4	0	[0, 0.333]	0	[0, 0.333]	0
5	0	[0, 0.25]	0	[0, 0.25]	0
6	0	[0, 0.2]	0	[0, 0.2]	0
7	1	[0.5, 1]	0	[0.5, 1]	1
8	1	[0.667, 1]	1	[0, 1]	0
9	0	[0, 0.167]	0	[0, 0.167]	0
10	0	[0, 0.143]	1	[0, 1]	0
11	0	[0, 0.125]	0	[0, 0.125]	0
12	0	[0, 0.111]	0	[0, 0.111]	0
13	0	[0, 0.1]	0	[0, 0.1]	0
14	1	[0.5, 0.75]	0	[0.647, 1]	1
15	0	[0, 0.091]	0	[0, 0.091]	0
16	0	[0, 0.083]	1	[0, 0.56]	0
17	1	[0.6, 0.8]	1	[0.167, 0.583]	0
31	1	[0.5, 0.667]	1	[0.152, 0.434]	0
32	1	[0.429, 0.571]	1	[0.118, 0.339]	0
34	1	[0.375, 0.5]	1	[0.094, 0.275]	0
41	1	[0.333, 0.444]	0	[0.713, 1]	1
49	1	[0.4, 0.5]	1	[0.085, 0.238]	0
52	1	[0.364, 0.455]	1	[0.082, 0.213]	0
53	1	[0.333, 0.417]	1	[0.072, 0.188]	0
54	1	[0.308, 0.385]	1	[0.065, 0.169]	1
55	1	[0.357, 0.429]	1	[0.088, 0.189]	1
56	1	[0.4, 0.467]	0	[0.829, 1]	1
59	1	[0.438, 0.5]	1	[0.135, 0.231]	0
65	1	[0.412, 0.471]	1	[0.113, 0.202]	1
66	1	[0.444, 0.5]	1	[0.136, 0.222]	0
67	1	[0.421, 0.474]	1	[0.125, 0.205]	0
75	1	[0.4, 0.45]	1	[0.121, 0.194]	0
78	1	[0.381, 0.429]	1	[0.117, 0.185]	0
82	1	[0.364, 0.409]	1	[0.105, 0.169]	0
83	1	[0.348, 0.391]	1	[0.099, 0.159]	0
1001	1	[0.374, 0.378]	1	[0.189, 0.194]	0
$1005^{[1]}$	1	[0.282, 0.286]	1	[0.134, 0.139]	1
$1000^{[2]}$	1	[0.291, 0.295]	1	[0.141, 0.145]	0
$1006^{[3]}$	1	[0.311, 0.315]	1	[0.160, 0.164]	0

12.5 Conformal Prediction in Hypergraphical Models

In this section we still consider a hypergraphical OCM $(\Sigma, \Box, \mathbf{Z}, F, B)$ conforming to a junction tree (U, S); as before, $B_{\mathbf{Z}}$ is the marginal distribution of the backward kernel B on $\mathbf{Z}$ (see (11.3)).

12.5.1 Conditional Probability Conformity Measure

The *conditional probability conformity measure* is defined by

$$A(\sigma, (x, y)) := \frac{B_{\mathbf{Z}}((x, y) \mid \sigma)}{\sum_{y' \in \mathbf{Y}} B_{\mathbf{Z}}((x, y') \mid \sigma)} . \tag{12.13}$$

In other words, $A(\sigma, (x, y))$ is the conditional probability $\mathbb{P}(y \mid x)$ of y given x under $\mathbb{P} := B_{\mathbf{Z}}(\sigma) = B_{\mathbf{Z}}(\cdot \mid \sigma)$. The conditional probability $\mathbb{P}(y \mid x)$ can be easily computed using (12.7).

In this section we are working in the online prediction protocol. At step n we have a training set $z_1, \ldots, z_{n-1}$, and the goal is to predict the label of a test object x_n. The conformal p-values are

$$\begin{aligned} p_n^y := B_{\mathbf{Z}}\left(\left\{z \in \mathbf{Z} : A(\sigma_n^y, z) < A(\sigma_n^y, (x_n, y))\right\} \mid \sigma_n^y\right) \\ + \tau_n B_{\mathbf{Z}}\left(\left\{z \in \mathbf{Z} : A(\sigma_n^y, z) = A(\sigma_n^y, (x_n, y))\right\} \mid \sigma_n^y\right) , \end{aligned} \tag{12.14}$$

where $\sigma_n^y := t(z_1, \ldots, z_{n-1}, (x_n, y))$ (this is essentially (11.5), with $<$ in place of $>$ since now A is a conformity measure), and the conformal prediction sets are

$$\Gamma_n^\epsilon := \{y : p_n^y > \epsilon\} . \tag{12.15}$$

Let us fix a finite range of significance levels ϵ (as in, e.g., Fig. 12.5 below). Computing the prediction sets (12.15) at each step of the online prediction protocol can be carried out in constant time, $O(1)$, for the conditional probability (CP) conformity measure (12.13) (and for a wide range of other conformity measures), assuming a fixed OCM.

Indeed, let σ_n be the summary of the first n examples $z_1, \ldots, z_n$. According to the definition of hypergraphical OCMs, updating σ_n (i.e., computing σ_n from σ_{n-1} and z_n when $n > 0$) can be done in constant time. Given σ_{n-1}, x_n, and a postulated label y (of which there are a fixed finite number), we can compute the probability measure $B(\sigma_n^y)$ defined by (12.7) for the summary $\sigma_n^y := F(\sigma_{n-1}, (x_n, y))$ in constant time. All conformity scores (12.13) can now be computed in constant time. Finally, the p-values p_n^y and prediction sets Γ_n^ϵ can be computed in constant time.

In evaluating the quality of conformal predictors determined by the conformity measure (12.13) we will use the OF (observed fuzziness) and OE (observed excess) criteria. In Sect. 3.1.5 (Theorem 3.1), we saw that these are conditionally proper criteria of efficiency under the exchangeability model; the analogous statement is also true for junction-tree models [102, Sect. 3.4].

12.5.2 LED Datasets

For our experiments we use benchmark LED datasets generated by a program from the UCI repository [91]. The problem is to predict a digit from an image in the seven-segment display shown in Fig. 12.2.

Figure 12.3 shows examples of objects in the dataset (these are the ten "ideal images" of digits; there are also digits corrupted by noise). The seven LEDs (light emitting diodes) can be lit in different combinations to represent a digit from 0 to 9. The program generates examples with noise. There is an ideal image for each digit, as shown in Fig. 12.3. An example has seven binary attributes $s_0, \ldots, s_6$ (s_i is 1 if the ith LED is lit) and a label c, which is a decimal digit. The program randomly chooses a label (0 to 9 with equal probabilities), inverts each of the attributes of its ideal image with probability $p_{\text{noise}} := 1\%$ independently, and adds the noisy image and its label to the dataset. (Originally, p_{noise} was set to 10%, but we always use 1%.)

Let $(S_0, \ldots, S_6, C)$ be the vector of random variables corresponding to the attributes and the label, and let $(s_0, \ldots, s_6, c)$ be an example. According to the data-generating mechanism the probability of the example decomposes as

$$Q\{(s_0, \ldots, s_6, c)\} = Q_7 (C = c) \prod_{i=0}^{6} Q_i (S_i = s_i \mid C = c) \ , \tag{12.16}$$

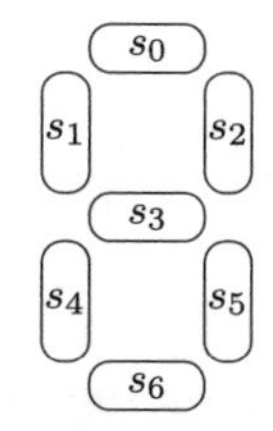

Fig. 12.2 The seven-segment display

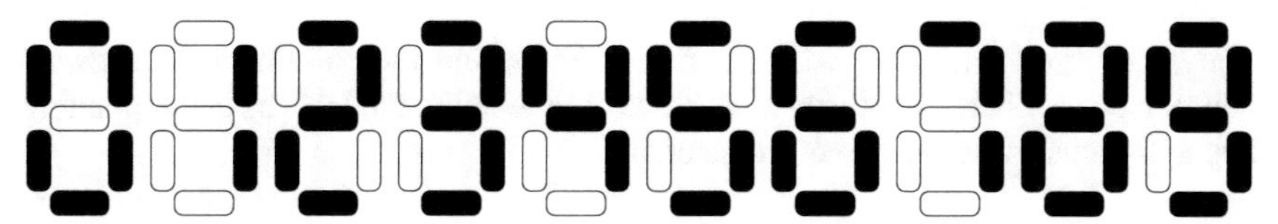

Fig. 12.3 Ideal LED images

where Q_7 is the uniform distribution on the decimal digits and

$$Q_i\left(S_i = s_i \mid C = c\right) := \begin{cases} 1 - p_{\text{noise}} & \text{if } s_i = s_i^c \\ p_{\text{noise}} & \text{otherwise,} \end{cases} \qquad i = 0, \ldots, 6\,, \tag{12.17}$$

$\left(s_0^c, \ldots, s_6^c\right)$ being the attributes of the ideal image for the label c. As usual, examples are generated independently.

12.5.3 Hypergraphical Assumptions for LED Datasets

We consider two hypergraphical models that agree with the decomposition (12.16). These models make different assumptions about the pattern of dependence between the attributes and the label; they do not use the particular noise level p_{noise} or the fact that the same value of p_{noise} is used for all LEDs. For both hypergraphical structures the set of variables is $V := \{s_0, \ldots, s_6, c\}$.

Nontrivial Hypergraphical Model

This is the hypergraphical structure with the clusters

$$\mathcal{E} := \{\{s_i, c\} : i = 0, \ldots, 6\}\ . \tag{12.18}$$

A junction tree for this hypergraphical structure can be defined as a chain with vertices $U := \{u_i : i = 0, \ldots, 6\}$ and the bijection $C_{u_i} := \{s_i, c\}$. By saying that U is a chain we mean that there are edges connecting vertices 0 and 1, 1 and 2, 2 and 3, 3 and 4, 4 and 5, and 5 and 6 (and these are the only edges in the tree). It is clear that this is a junction tree and that $C_s = \{c\}$ for each edge s. It is also clear from (12.16) that the assumption (12.4) is satisfied; e.g., we can set

$$f_{\{s_0,c\}}\left(s_0, c\right) := Q_7\left(C = c\right) \cdot Q_0\left(S_0 = s_0 \mid C = c\right);$$
$$f_{\{s_i,c\}}\left(s_i, c\right) := Q_i\left(S_i = s_i \mid C = c\right), \quad i = 1, \ldots, 6\,.$$

Exchangeability Model

The hypergraphical model with no information about the pattern of dependence between the attributes and the label is the exchangeability model. The corresponding hypergraphical structure has one cluster,

$$\mathcal{E} := \{V\}\ . \tag{12.19}$$

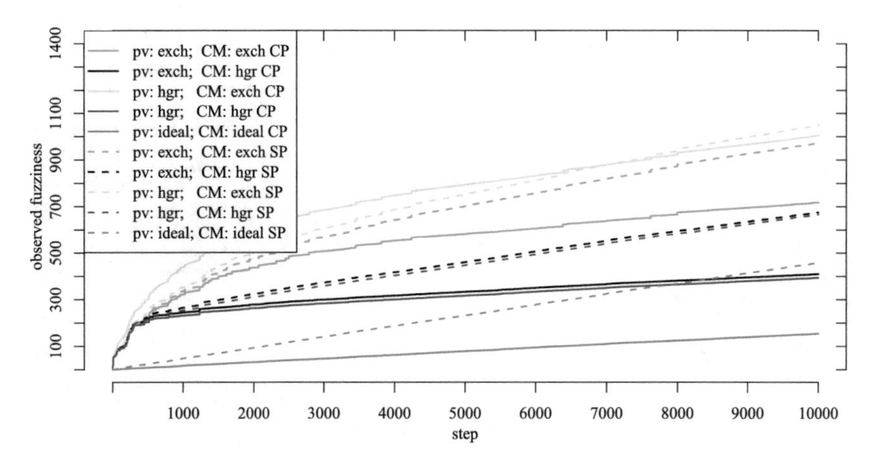

Fig. 12.4 Cumulative observed fuzziness for online predictions. The results are for the LED dataset with 1% noise and 10,000 examples

The junction tree is the one vertex associated with V and no edges.

12.5.4 Results for the LED Datasets

For our computational experiments we create a LED dataset with 10,000 examples. The data are generated according to the model (12.16) with the noise level $p_{\text{noise}} = 1\%$.

Each of Figs. 12.4 and 12.5 corresponds to an efficiency criterion for conformal predictors, OF (observed fuzziness) or OE (observed excess), as defined in Protocol 3.1. Namely, Fig. 12.4 plots OF_n vs $n = 1, \ldots, 10000$ in the online prediction protocol, and Fig. 12.5 plots $\text{OE}^{\epsilon}_{10000}/10000$ (the average excess of predictions) vs $\epsilon \in [0.001, 0.05]$. The conformity measure (CM) for the solid graphs is the conditional probability (CP) conformity measure (12.13). The graphs represented as dashed lines are less important and correspond to the *signed predictability* (SP) conformity measure (CM)

$$A'(\sigma, (x, y)) := \begin{cases} f(x) & \text{if } y = \hat{y}(x) \\ -f(x) & \text{if not} \end{cases}$$

(cf. (3.27)), where the predictability $f : \mathbf{X} \to [0, 1]$ and the choice function $\hat{y} : \mathbf{X} \to \mathbf{Y}$ are defined by the conditions

$$f(x) := \max_{y \in \mathbf{Y}} A(\sigma, (x, y)), \quad A(\sigma, (x, \hat{y}(x))) = f(x) ,$$

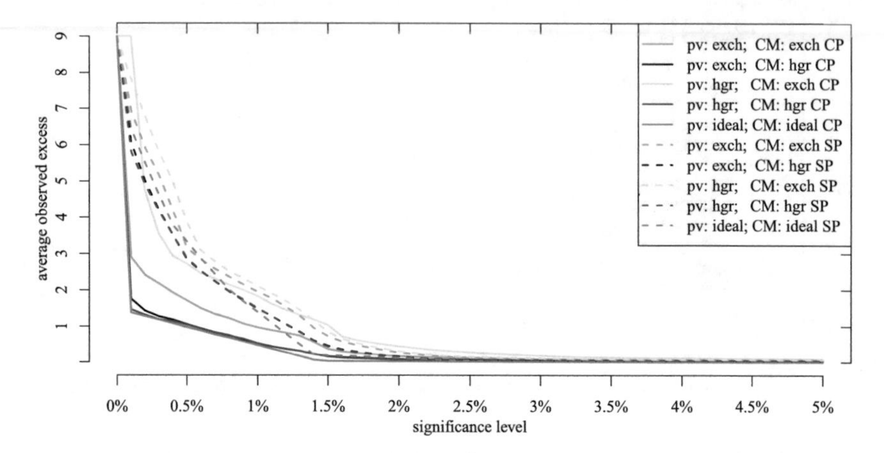

Fig. 12.5 The final average observed excess for significance levels between 0.1% and 5%. The results are for the LED dataset with 1% noise and 10,000 examples

for all $x \in \mathbf{X}$, A being the CP conformity measure (12.13). As discussed in Sect. 3.1.6, this conformity measure optimizes different criteria of efficiency from OF_n and OE_n, and it shows in our plots. In the rest of this section we will only discuss the solid graphs, corresponding to the CP conformity measure.

One of the solid graphs in each figure corresponds to idealised predictors and are drawn only for comparison, representing an unachievable ideal goal. In the idealised case we know the true distribution for the data (given by (12.16), (12.17), and $p_{\text{noise}} = 1\%$). The true distribution is used instead of the backward kernel $B_{\mathbf{Z}}(\sigma_n^y)$ in both (12.14) and (12.13). It gives us the ideal results (the red lines in our plots), leading to the best graphs in each of the figures (remember that for our criteria the lower the better).

We also consider four realistic predictors (which are conformal predictors, unlike the idealised ones). The *pure hypergraphical conformal predictor* (represented by blue lines in our plots) is obtained using the nontrivial hypergraphical model (12.18) both when computing p-values (see (12.14)) and when computing the conformity measure (12.13). Analogously we use the exchangeability model (12.19) to obtain the *pure exchangeability conformal predictor* (green lines in our plots). The two *mixed conformal predictors* (black and yellow lines) are obtained when we use different models to compute the p-values and the conformity scores.

The intuition behind the pure and mixed conformal predictors can be explained using the distinction between hard and soft models that we made earlier in Sect. 2.5.1. The model used when computing the p-values (see (12.14)) is the hard model; the validity of the conformal predictor depends on it. The model used when computing conformity scores (see (12.13)) is the soft model; when it is violated, validity is not affected, although efficiency may suffer. The true data-generating distribution (12.16) conforms to both the exchangeability model and the nontrivial hypergraphical model; therefore, all four conformal predictors are automatically

Table 12.2 The final values of the observed fuzziness in Fig. 12.4 for the black and blue graphs

Seed (10^4)	0	1	...	99	Average	Standard deviation
pv: exch; CM: hgr CP	409.6	422.2	...	402.7	407.0	24.75
pv: hgr; CM: hgr CP	393.6	402.2	...	385.2	384.4	23.99

valid, and we study only their efficiency. (In our current context, it is obvious that the exchangeability model is more general than the nontrivial hypergraphical model, but we can also apply the criterion given in Proposition 12.3.)

In the legends of Figs. 12.4 and 12.5, the hard model used is indicated after "pv" (the way of computing the p-values), and the soft model used is indicated after "CM" (the conformity measure); "exch" refers to the exchangeability model, and "hgr" refers to the nontrivial hypergraphical model.

The most interesting graphs in Figs. 12.4 and 12.5 are the black ones, corresponding to the exchangeability model as the hard model and the nontrivial hypergraphical model as the soft model. The performance of the corresponding conformal predictors is typically better than, or at least close to, the performance of any of the remaining realistic predictors. The fact that the validity of these conformal predictors only depends on the exchangeability assumption makes them particularly valuable. The yellow graphs correspond to the nontrivial hypergraphical model as the hard model and the exchangeability model as the soft model; the performance of the corresponding conformal predictors (rather inane, since it does not make sense for the hard model to be more restrictive than the soft model) is very poor in our experiments.

Now let us discuss each of the two figures, and the corresponding tables, separately. Figure 12.4 shows the cumulative observed fuzziness OF_n vs n. The bottom graph corresponds to the idealised CP predictor. The black and blue graphs, corresponding to the conformal predictors determined by the CP conformity measure using the nontrivial hypergraphical model, are very close; the blue one is slightly lower but the conformal predictor corresponding to the black one still appears preferable as its validity only depends on the weaker exchangeability assumption.

Table 12.2 shows the final values of the observed fuzziness in Fig. 12.4 for the two most important graphs (black and blue) for several seeds of the random number generator. The values of the seed are given in the units of 10,000 (so that 0 stands for 0, 1 for 10,000, 2 for 20,000, etc.); the "10^4" in parentheses serves as a reminder of this. The last two columns of this and next tables give aggregate values: column "average" gives the average of all the 100 values for the different seeds, and column "standard deviation" gives the standard estimate of the standard deviation computed from those 100 values (namely, the square root of the standard unbiased estimate of the variance). The table confirms that the black and blue graphs are close to each other on average (see the penultimate column), although there is a clear tendency for the blue ones to be lower (see the last column; to obtain an estimate of the standard

Table 12.3 The final average observed excess in Fig. 12.5 for the significance level 1% and for the black and blue graphs

Seed (10^4)	0	1	...	99	Average	Standard deviation
pv: exch; CM: hgr CP	0.5218	0.5115	...	0.5197	0.5451	0.1237
pv: hgr; CM: hgr CP	0.5299	0.4868	...	0.5025	0.5228	0.1216

deviation of the average, the value given in the last column should be divided by 10).

Figure 12.5 shows the average observed excess of predictions after observing 10,000 examples as function of the significance level. The black CP graph is again very close to the blue CP graph, corresponding to the pure hypergraphical predictor, except for very low significance levels when the average excess exceeds 1. The closeness at the significance level 1% is confirmed by Table 12.3.

12.5.5 Label-Conditional Conformal Prediction for Hypergraphical OCMs

The usual notion of validity for conformal predictors is unconditional, and in this subsection we will study the label-conditional version, similarly to what we did earlier (Sect. 4.6.7) for the exchangeability model. The categories correspond to labels, which are sometimes called classes, as before.

Formally, *hypergraphical label-conditional conformal predictors* are defined in the same way as hypergraphical conformal predictors in Sect. 12.5.1 except that the definition (12.14) of p-values is modified as

$$\begin{aligned} p_n^y := \; & \frac{B_{\mathbf{Z}}\left(\left\{(x',y') \in \mathbf{Z} : y' = y \;\&\; A(\sigma_n^y,(x',y')) < A(\sigma_n^y,(x_n,y))\right\} \mid \sigma_n^y\right)}{B_{\mathbf{Z}}\left(\{(x',y') \in \mathbf{Z} : y' = y\} \mid \sigma_n^y\right)} \\ & + \tau_n \frac{B_{\mathbf{Z}}\left(\left\{(x',y') \in \mathbf{Z} : y' = y \;\&\; A(\sigma_n^y,(x',y')) = A(\sigma_n^y,(x_n,y))\right\} \mid \sigma_n^y\right)}{B_{\mathbf{Z}}\left(\{(x',y') \in \mathbf{Z} : y' = y\} \mid \sigma_n^y\right)} . \end{aligned} \tag{12.20}$$

Using these p-values instead of (12.14), hypergraphical label-conditional conformal predictors are defined by (12.15). We will sometimes refer to the definition based on (12.14) as "unconditional" conformal predictors.

As in the unconditional case, we use the conditional probability conformity measure (12.13) when computing the p-values (12.20). In the label-conditional case it is still true that the conditional probability conformity measure is optimal in the sense of OE_n^ϵ and in the sense of OF_n: see [406].

12.5.6 Further Results for the LED Datasets

This subsection studies the performance of unconditional and label-conditional conformal predictors under hypergraphical models. First we look at the class-wise validity of these predictors and then we compare their efficiency.

Class-Wise Validity

In our first experiment we assess the final percentage of errors within different classes. For each of the ten classes corresponding to the labels $y \in \{0, \ldots, 9\}$ and at each significance level $\epsilon \in (0, 1)$ the final percentage of errors is calculated by

$$\mathrm{err}^{y,\epsilon} := \frac{\left|\left\{i = 1, \ldots, 10000 : y_i = y \ \& \ y_i \notin \Gamma_i^\epsilon\right\}\right|}{\left|\{i = 1, \ldots, 10000 : y_i = y\}\right|}. \tag{12.21}$$

We study four conformal predictors: the pure exchangeability conformal predictor (the exchangeability model is used for the p-values (12.14) and for the conformity scores (12.13)), the pure hypergraphical conformal predictor (the nontrivial hypergraphical model is used in (12.14) and (12.13)), the *pure exchangeability label-conditional conformal predictor* (the exchangeability model is used in (12.20) and in (12.13)), and the *pure hypergraphical label-conditional conformal predictor* (the nontrivial hypergraphical model is used in (12.20) and (12.13)); the first two are the same predictors as in previous subsections, and the last two are new.

All four predictors make predictions in the online mode, and the final percentage of errors (12.21) is calculated for these predictions and a dense grid of significance levels between 0% and 100%. The percentage of errors $\mathrm{err}^{y,\epsilon}$ plotted against the significance level ϵ will be called the *calibration graph* (for the label y). For valid predictions the calibration graph should be close to the diagonal extending from the bottom left corner (no errors at the significance level 0%, which is achieved when all prediction sets are the whole label set) to the top right corner (errors at each prediction step at the significance level 100%, which is the result of all predictions being the empty predictions).

Figure 12.6 shows the calibration graphs for the two unconditional conformal predictors. In each plot, the ten calibration graphs of different colours correspond to the labels $0, 1, \ldots, 9$. As expected, these unconditional conformal predictions are not class-wise valid. In these plots, calibration graphs that are below the diagonal correspond to easy labels (there are fewer errors than expected), and calibration graphs above the diagonal are for difficult labels (the number of errors is greater than expected). The most difficult digits are 8 and 9: this is not surprising since each of them has 3 other digits at a Hamming distance of 1 from it, more than any other digit (see Fig. 12.3; 0, 6, and 9 are at a Hamming distance of 1 from 8, and 3, 5, and 8 are at a Hamming distance of 1 from 9). The easiest digits are 2 and 4, because these are the only digits that do not have any other digits at a Hamming distance of

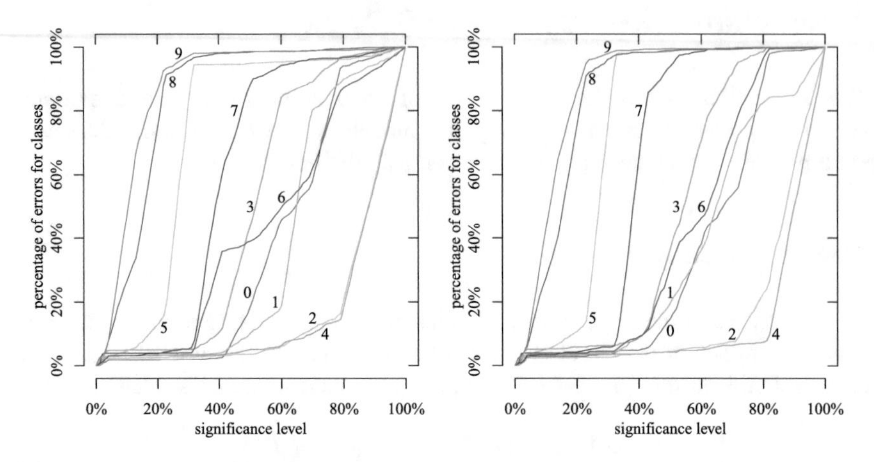

Fig. 12.6 The final percentage of errors for different classes for (unconditional) hypergraphical conformal predictors; different colours correspond to different classes. The predictions are not class-wise valid. The left panel is for the pure exchangeability conformal predictor and the right panel for the pure hypergraphical conformal predictor. The results are for the LED dataset of 10,000 examples with 1% noise

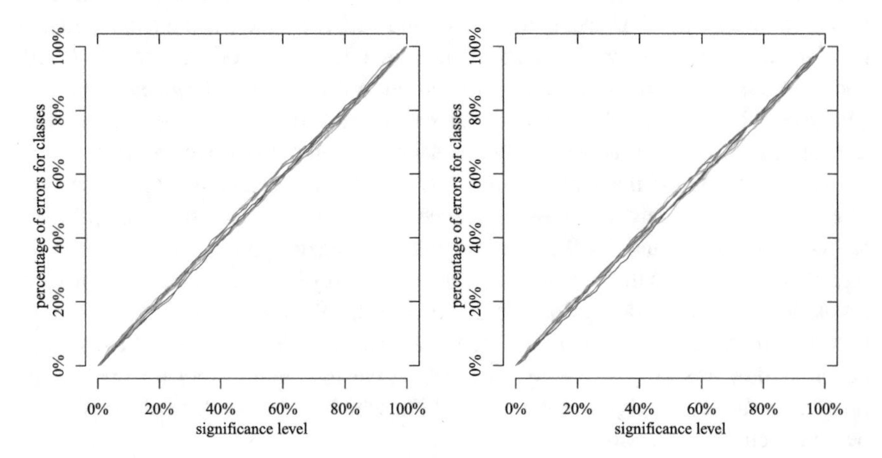

Fig. 12.7 The same as Fig. 12.6 for hypergraphical label-conditional conformal predictors. The predictions are class-wise valid

1 from them (and both have 2 digits at a Hamming distance of 2: 3 and 8 for 2, and 1 and 9 for 4).

Figure 12.7 shows the results for the two label-conditional conformal predictors under hypergraphical models. These predictors are constructed in order to produce class-wise valid predictions, and the experiments just confirm this property.

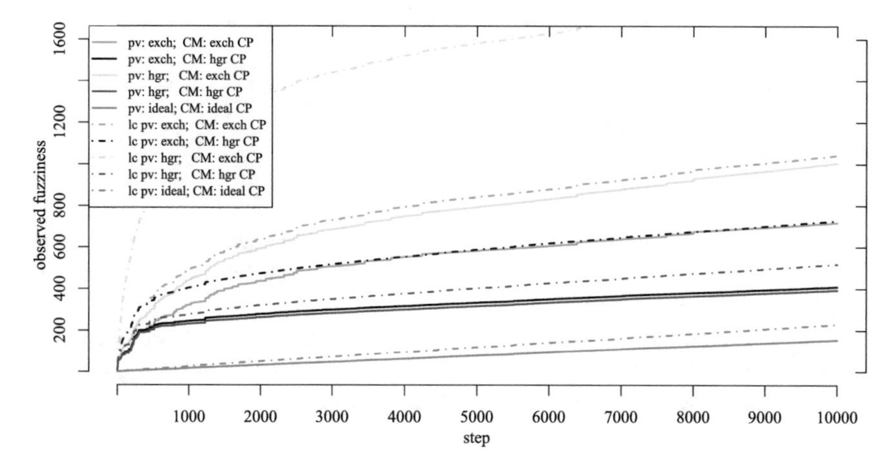

Fig. 12.8 Observed fuzziness for unconditional conformal predictors and label-conditional (lc) conformal predictors. The results are for the LED dataset of 10,000 examples with 1% noise

Efficiency

Let us now compare the efficiency of conformal predictors and label-conditional conformal predictors under hypergraphical models. We do so by calculating the cumulative observed fuzziness OF_n.

Figure 12.8 shows the cumulative observed fuzziness in the online protocol. The solid lines are for unconditional predictors, whose p-values are defined by (12.14), and the dash-dot lines are for label-conditional predictors, whose p-values are defined by (12.20). Again, two models are considered: the exchangeability model and the nontrivial hypergraphical model; each model can be used for calculating the hypergraphical CP conformity scores (12.13) or for p-values ((12.14) and (12.20)). These combinations give four unconditional conformal predictors and four label-conditional conformal predictors. Also, as before, we consider two idealised predictors: the unconditional idealised predictor is obtained using the true distribution for the data instead of the backward kernel $B_{\mathbf{Z}}(\sigma_n^y)$ in both (12.14) and (12.13), and analogously the *label-conditional idealised predictor* is obtained using the true distribution in both (12.20) and (12.13). The notation in the legend is similar to that in the previous set of experiments (see Figs. 12.4 and 12.5) except that "lc" stands for "label-conditional".

As expected, the price to pay for the class-wise validity of label-conditional conformal predictors is that they are less efficient than the corresponding unconditional conformal predictors. But the performance of the pure hypergraphical label-conditional conformal predictor (the second lowest dash-dot line) is not much worse than that for the corresponding unconditional one (the second lowest solid line). The performance of the other label-conditional predictors (unfortunately, including the predictor corresponding to the black line, which we recommended

Table 12.4 The final values of the observed fuzziness in Fig. 12.8 for the black and blue label-conditional graphs

Seed (10^4)	0	1	...	99	Average	Standard deviation
lc pv: exch; CM: hgr CP	726.1	729.7	...	709.6	724.3	31.97
lc pv: hgr; CM: hgr CP	517.5	522.1	...	503.2	512.2	30.96

in the unconditional setting of Sect. 12.5.4) is significantly worse than that of the corresponding unconditional ones. The final values for the black and blue graphs in Fig. 12.8 are given in Table 12.4.

From the computational point of view, the label-conditional way of computing p-values (12.20) is cheaper: for the label-conditional p-values one only needs to look at the conformity scores of configurations with the same label.

12.5.7 Summary of the Experiments

The main finding of this section is that nontrivial hypergraphical models can be useful for conformal prediction when they are true. More surprisingly, in our experiments and the unconditional setting they only need to be used as soft models; the efficiency does not suffer much if the exchangeability model continues to be used as the hard model.

Our experiments with label-conditional conformal predictors under hypergraphical models have demonstrated that they are essential for the class-wise validity. Finally, we have seen that the efficiency of label-conditional conformal predictors is not much worse than that of unconditional ones if the hypergraphical models are used as both the hard and soft models.

12.6 Context

In our exposition of the hypergraphical model we mainly follow [384]. In the first edition of this book [402, Sect. 9.6] we give a natural maximum-likelihood interpretation of Lemma 12.6. For further information about hypergraphs, see [215, Sect. 2.2].

12.6.1 Conformal Prediction in Hypergraphical Models

Here we follow [102], which also discusses in detail signed predictability (the conformity measure that we discussed only briefly in this context) and other criteria of efficiency, namely cumulative unconfidence and the number of multiple predictions (see Sect. 3.1.2). The experimental results in [102] are consistent with our results in Chap. 3: using the conditional probability as conformity measure leads to better performance as measured by observed fuzziness and observed excess, and using the signed predictability leads to better performance as measured by cumulative unconfidence and the number of multiple predictions.

12.6.2 Additive Models

The hypergraphical model, as well as most of the models considered in the previous chapter, are additive in the following abstract sense: Σ is an Abelian semigroup and the statistic t satisfies

$$t(z_1, \ldots, z_n) = t_1(z_1) + \cdots + t_1(z_n)$$

for a function t_1, where "+" is the semigroup operation. The theory of such repetitive structures (with uniform conditional distributions $P_n(\cdot \mid \sigma)$) is especially rich and is treated in [214, Chap. III].

Chapter 13
Contrasts and Perspectives

Abstract This book has emphasized conformal prediction under statistical randomness, the standard assumption in machine learning, and only in this part (starting from Chap. 11) have we extended it to substantially different statistical models. Interestingly, conformal prediction in this wider sense is much more familiar to statisticians. In this concluding chapter, we step back to survey the historical context of conformal prediction, contrasting it with classical inductive, transductive, and Bayesian methods.

In the previous two chapters we generalized conformal and Venn prediction to online compression models. In this chapter we will complement this by discussing conformal predictive distributions in the Gaussian model, which were introduced by Fisher under the name of fiducial predictive distributions.

In conclusion, we review some of the new work on conformal prediction. Relaxing the assumption of randomness and replacing it by other assumptions have been recurring themes in this work.

Keywords Inductive learning · Transductive learning · Fiducial prediction · Bayesian learning · Conformal prediction

13.1 Introduction

We begin this chapter, in Sect. 13.2, by discussing inductive methods. We review the history of inductive learning under statistical randomness, from Jacob Bernoulli's eighteenth-century law of large numbers to relatively recent developments in statistical learning theory.

In Sect. 13.3 we turn to transductive methods, again taking a historical perspective. The vast majority of the authors we are discussing used neither Johnson's *eduction* nor Vapnik's *transduction*. In some of these cases we find instances of conformal prediction. Of particular interest is the prediction interval for a new observation based on Student's t-distribution that Ronald A. Fisher published in 1935 [107]. Our historical review begins with this work, explaining its connections with conformal predictive distributions (Chap. 7), continues with later work on

V. Vovk et al., *Algorithmic Learning in a Random World*,
https://doi.org/10.1007/978-3-031-06649-8_13

tolerance intervals and Dempster–Hill predictive distributions, then moves on to the Vapnik–Chervonenkis approach to transduction, first set forth in their 1974 monograph [370], and ends with a discussion of notions of validity and efficiency in probabilistic prediction.

We discuss Bayesian learning in Sect. 13.4. Viewed from our framework, which allows examples to be governed by any probability distribution Q on $\mathbf{Z}$, Bayesian learning requires us to make additional assumptions. First we assume that Q is one of a small family $(Q_\theta : \theta \in \Theta)$ of probability distributions on $\mathbf{Z}$ (this is the *statistical model*), and then we adopt a probability distribution μ on Θ (this is the *prior distribution*) to express our probabilities for which Q_θ it is. Because of these additional assumptions, Bayesian learning formally lies outside the framework of this book. But it is currently very popular, and the additional assumptions are not always taken for granted by Bayesian statisticians (see, e.g., [30, Chap. 6]), and so we explain briefly how it relates to our methods. Our main conclusion will be unsurprising to anyone familiar with Bayesian learning: although Bayesian predictors are valid on their own terms, this validity depends on the additional assumptions being correct. When these assumptions are violated, Bayesian predictors may lack the kind of validity conformal predictors have. Perhaps more surprising is the fact that a prior distribution can be used to construct a conformal predictor that gives predictions resembling the Bayesian predictions when the Bayesian assumptions are correct but, like any conformal predictor, is valid even if they are incorrect (cf. Sect. 4.3.2).

In Sect. 13.5 we briefly describe some of the new directions in conformal prediction that emerged after the publication of the first edition of this book, namely conformal prediction under dataset shift, causality, and anomaly detection. Inevitably our exposition in that section is very superficial.

13.2 Inductive Learning

Induction means using old observations to formulate a rule for prediction that is then applied to new observations. When we are learning under randomness, the quality of the resulting predictions is affected by the randomness of both sets of observations. Because of the randomness of the new observations, there may be no prediction rule that performs perfectly, and because of the randomness of the old observations, we cannot even expect to find the prediction rule that performs best. As discussed in Chap. 1, a precise mathematical statement of what can be achieved typically involves two positive numbers, often denoted by ϵ and δ; ϵ is a level of imperfection we are willing to tolerate in a prediction rule, and δ bounds the probability we will fail to attain even this level. A typical theoretical result says that a certain method produces, with probability at least $1 - \delta$, a rule whose probability of error is at most ϵ.

The oldest result of this type is Jacob Bernoulli's theorem, which appeared in 1713, in his posthumous *Ars Conjectandi* [31]. Abraham De Moivre [78] improved on this result in 1733, obtaining what we now call the normal approximation to the binomial distribution, and Bernoulli's and De Moivre's results were subsequently generalized in many directions. In this section, we review the generalizations most pertinent to learning under unconstrained randomness, including Vapnik and Chervonenkis's uniform law of large numbers and more recent work on data-dependent bounds.

For historical reasons, we postpone to Sect. 13.3 a discussion of tolerance regions, another important approach to inductive prediction. Similarly to conformal predictors, especially inductive conformal predictors (cf. Sect. 4.7), tolerance regions have both transductive and inductive validity guarantees, and so their theory bears on both this section and Sect. 13.3.

To avoid the possibility of misunderstanding, we should mention that in this section, as in this whole book, we do not address the broad philosophical problem of induction that David Hume introduced in the eighteenth century [173, 284]. In the bulk of this book we are concerned only with learning under randomness, where induction means using randomly chosen observations to find a general rule for making predictions about future randomly chosen observations.

13.2.1 Jacob Bernoulli's Learning Problem

It would be misleading to say that Bernoulli worked with our concept of induction, because his work preceded the concept of a probability distribution. We might even think of his purpose as transductive: he wanted to predict whether an event would happen (or whether something is true or false) by looking at previous observations. But this project led him to formulate what we now recognize as the problem of estimating a probability p from previous randomly chosen observations, and so we can regard him as the founder of the inductive approach to learning under randomness.

Bernoulli likened his observations to pebbles drawn from an urn containing white and black pebbles in the ratio r to s, so that the probability of drawing a white pebble is $p = r/(r+s)$. The problem he considered was that of estimating the ratio r/s. But from the viewpoint of our framework, he was studying the problem of learning under randomness when there are no objects and the labels are binary. As usual, we take the labels to be 1 and 0 rather than white and black. Because there are no objects, we write $\mathbf{Z} = \{0, 1\}$. We write p for the probability of 1; this defines the probability distribution Q on $\mathbf{Z}$. We write $z_1, \ldots, z_l$ for the old observations.

Translated into these terms, Bernoulli's accomplishment was to show how to find, for any positive constants ϵ and δ and any $p \in [0, 1]$, a threshold $N(\epsilon, \delta, p)$ such that

$$Q^l \left\{ (z_1, \ldots, z_l) : \left| \frac{1}{l} \sum_{i=1}^{l} z_i - p \right| \geq \epsilon \right\} \leq \delta \tag{13.1}$$

for any $l \geq N(\epsilon, \delta, p)$. This assumes only that the z_i are independent and equal to 1 with probability p, and it shows that, for large l,

$$\hat{p}_l := \frac{1}{l}\sum_{i=1}^{l} z_i$$

is likely to be an accurate estimate of p. Results of this type, where counts or averages are shown to estimate probabilities or expected values, are now called laws of large numbers.

Bernoulli's threshold $N(\epsilon, \delta, p)$ was remarkably large. In the numerical example he gives at the end of *Ars Conjectandi*, where $p = 0.6$, $\epsilon = 0.02$, and $\delta = 1/1001$, it comes out to 25,550, a huge number in the context of the datasets available in the eighteenth century. As Stigler [343, p. 77] says, it "was more than astronomical; for all practical purposes it was infinite". Bernoulli does not admit to any disappointment, but he ends the book soon after the number 25,550 appears.

Bernoulli's failure to produce a useable error bound has been repeated many times by subsequent authors. The error bounds produced by much of the leading theoretical work are too loose to be of practical value for available datasets, though authors seldom follow Bernoulli's example by providing numerical illustrations that make this shortcoming clear.

Bernoulli's bound can be improved substantially. The essential step in removing the slack was taken by Abraham De Moivre [78] in 1733. De Moivre's theorem, now considered a special case of the central limit theorem, tells us how to calculate approximate probabilities for $z_1 + \cdots + z_l$. In modern terminology, it says that this sum, scaled properly, has an approximately normal distribution. This allows us to approximate the probability in (13.1), not merely bound it. In 1925, Karl Pearson [272] used De Moivre's theorem to show that the lowest valid value for $N(0.02, 1/1001, 0.6)$—i.e., the lowest value of l for which

$$\mathbf{B}_{0.6}^{l}\left\{(z_1, \ldots, z_l) : \left|\frac{1}{l}\sum_{i=1}^{l} z_i - 0.6\right| \geq 0.02\right\} \leq \frac{1}{1001}$$

holds—is approximately 6498. He also found, modifying Bernoulli's argument and using Stirling's formula, which was not known to Bernoulli, what he thought to be a rigorous valid value for $N(0.02, 1/1001, 0.6)$; Sirazhdinov later showed that the value given by Pearson has to be replaced by 6568 (see [285, Sect. 8]).

Even with the improvement brought by De Moivre's theorem, Bernoulli's approach has a drawback that might seem fatal from a rigorously logical point of view: the threshold $N(\epsilon, \delta, p)$ depends on p, which we are supposed not to know. To estimate p we are told to take l to be at least $N(\epsilon, \delta, p)$, but to compute $N(\epsilon, \delta, p)$, we need to know p already. But this difficulty can be solved, and although the solution is awkward (see, e.g., [349, Sects. 19.9–19.11]), it usually produces a value for $N(\epsilon, \delta, p)$ not a great deal different from what we get after the fact for $N(\epsilon, \delta, \hat{p}_l)$.

Although De Moivre's theorem eliminates the slack in Bernoulli's bound, it produces values of $N(\epsilon, \delta, p)$ that are still embarrassingly large. Although we would always like to hope that the probability of error in a serious matter will be less than $1/1001$, most decisions must be based on far fewer than 6568 observations. So statisticians learn to be content with $\delta = 0.01$ or $\delta = 0.05$.

Relatively weak levels of confidence, even when there is little slack in theoretical bounds, remain common in learning under unconstrained randomness, even though we often have two advantages over Bernoulli and De Moivre: much larger databases (at least in some applications) and information about the objects to help us predict the labels. As we have already mentioned, this suggests that it may be too ambitious to try to control both the level of accuracy ϵ and the probability of inaccuracy δ.

Another feature of Bernoulli's and De Moivre's results that we still see in the inductive approach to learning under unconstrained randomness is the logarithmic dependence of the required number of observations on the parameter δ. This is the ubiquitous $\ln \frac{1}{\delta}$ term (see, e.g., (13.4) or [366]). Bernoulli commented on this dependence, pointing out that the same number of additional observations is required every time we multiply the desired odds $(1 - \delta) : \delta$ by 10. When the desired odds are $1000 : 1$ ($\delta = 1/1001$), the number of observations required by Bernoulli's bound is 25,550. When they are increased to 10,000 : 1 ($\delta = 1/10001$), this increases by 5708, to 31,258. When they are increased to 100,000 : 1 ($\delta = 1/100001$), it increases again by 5708, to 36,966. When we take the slack out of Bernoulli's bounds, the dependence is still approximately logarithmic: according to Sirazhdinov, the 6568 observations needed for odds 1000 : 1 goes up by about 2570 every time we multiply these odds by 10—to 9142 for odds 10,000 : 1 and to 11,709 for odds 100,000 : 1.

As a modern example of the logarithmic dependence on δ, we can cite Hoeffding's inequality (see Sect. A.6.3). In the case considered by Bernoulli, it says that

$$\mathbf{B}_p^l \left\{ (z_1, \ldots, z_l) : \left| \frac{1}{l} \sum_{i=1}^{l} z_i - p \right| \geq \epsilon \right\} \leq 2 \exp\left(-2\epsilon^2 l\right) .$$

If we invert this inequality to solve Bernoulli's problem, we obtain

$$N(\epsilon, \delta, p) = \frac{1}{2\epsilon^2} \ln \frac{2}{\delta} .$$

This has the happy feature that it does not depend on p. But it is looser than Sirazhdinov's bound: instead of 6568 for $N(0.02, 1/1001, 0.6)$, it gives 9503.

13.2.2 Statistical Learning Theory

What we now call statistical learning theory was launched by Vapnik and Chervonenkis over 50 years ago, first in the short note [368] published in 1968 and then in the full article [369], with proofs, published in 1971. In this work, they presented a very general and natural inductive method and showed that its performance can be guaranteed, for sufficiently long data sequences, by a uniform law of large numbers.

Let $\mathcal{A}$ be a family of measurable subsets of $\mathbf{Z}$ (in general an arbitrary measurable space). We say that $\mathcal{A}$ *shatters* a finite set $Z \subseteq \mathbf{Z}$ if for any $Z' \subseteq Z$ there exists $A \in \mathcal{A}$ such that $Z' = Z \cap A$. We write $\mathrm{VC}(\mathcal{A})$ for the cardinality of the largest finite set $\mathcal{A}$ shatters, and we call $\mathrm{VC}(\mathcal{A})$ the *VC dimension* of $\mathcal{A}$. When $\mathcal{A}$ shatters arbitrarily large sets, $\mathrm{VC}(\mathcal{A}) := \infty$.

The key result in Vapnik and Chervonenkis's theory is the inequality

$$Q^l \left\{ (z_1, \ldots, z_l) : \sup_{A \in \mathcal{A}} \left| \frac{|\{i = 1, \ldots, l : z_i \in A\}|}{l} - Q(A) \right| > \epsilon \right\}$$
$$< 4 \exp\left(\left(\frac{\mathrm{VC}(\mathcal{A})(1 + \ln(2l/\mathrm{VC}(\mathcal{A})))}{l} - (\epsilon - 1/l)^2 \right) l \right), \qquad (13.2)$$

which holds for every $\mathcal{A}$ satisfying $0 < \mathrm{VC}(\mathcal{A}) < l$ and for every probability distribution Q on $\mathbf{Z}$. (There are many variations on this inequality, none of them canonical. This particular version appears in [366, Theorem 4.4].)

Remark 13.1 To see the potential of inequality (13.2), let us take $\mathbf{Z}$ to be $\{0, 1\}$ and $\mathcal{A}$ to be $\{\emptyset, A\}$, where A is the singleton $\{1\}$. In this case, $\mathrm{VC}(\mathcal{A}) = 1$, we can write p for $Q(\{1\})$, and the inequality becomes

$$Q^l \left\{ (z_1, \ldots, z_l) : \left| \frac{1}{l} \sum_{i=1}^{l} z_i - p \right| > \epsilon \right\} < 4 \exp\left(1 + \ln(2l) - (\epsilon - 1/l)^2 l \right)$$

(we added $\emptyset$ to $\mathcal{A}$ to ensure $\mathrm{VC}(\mathcal{A}) \neq 0$). For fixed ϵ the right-hand side of this inequality can be made arbitrarily small by making l sufficiently large, and so we obtain another proof of Bernoulli's law of large numbers.

The inequality (13.2) is important not simply because it generalizes Bernoulli's theorem but because it does so uniformly in $A \in \mathcal{A}$. The right-hand side does not involve A, and so we can make the probability of a given deviation between $Q(A)$ and $|\{i = 1, \ldots, l : z_i \in A\}| / l$, the empirical frequency of A in the first l examples, small uniformly in A by taking l large enough. This is the Vapnik–Chervonenkis *uniform law of large numbers*: if $\mathrm{VC}(\mathcal{A}) < \infty$, then the empirical frequencies of the sets $A \in \mathcal{A}$ converge to their probabilities $Q(A)$ in probability uniformly.

It is easy to see that this uniform law of large numbers allows us to derive guarantees for inductive classification. Recall that in our framework with $\mathbf{Z} = \mathbf{X} \times \mathbf{Y}$, classification is the case where $|\mathbf{Y}| < \infty$. We extend the concept of VC dimension to a family $\mathcal{F}$ of measurable functions $f : \mathbf{X} \to \mathbf{Y}$ by taking VC($\mathcal{F}$) to be the VC dimension of the family of their graphs

$$\{(x, y) \in \mathbf{X} \times \mathbf{Y} : y = f(x)\} .$$

Choose such a family $\mathcal{F}$, together with parameters $\epsilon > 0$ and $\delta > 0$, and suppose that for at least one function f in $\mathcal{F}$, $f(x_i)$ predicts y_i correctly for most i. If $0 < \mathrm{VC}(\mathcal{F}) < \infty$, then (13.2) suggests the following strategy: choose l such that the right-hand side, with VC($\mathcal{A}$) replaced by VC($\mathcal{F}$), does not exceed δ, and choose a function $\hat{f}$ in $\mathcal{F}$ that minimizes the empirical error

$$E(f) := \frac{|\{i = 1, \ldots, l : y_i \neq f(x_i)\}|}{l} .$$

(This choice of f is often referred to as "empirical risk minimization".) We then know that $\hat{f}$'s probability of error as a prediction rule will not exceed $E(\hat{f}) + \epsilon$ unless an unlikely event—an event of probability at most δ—has happened. The inequality (13.2) further implies that this probability of error will not exceed

$$\inf_{f \in \mathcal{F}} Q\{(x, y) \in \mathbf{X} \times \mathbf{Y} : y \neq f(x)\} + 2\epsilon$$

unless an event of probability 2δ has happened.

There are several useful versions of (13.2), including extensions that are relevant to regression rather than classification. Moreover, it has been shown that the inequality is optimal in several senses. For example, a finite VC dimension is necessary and sufficient for the uniform convergence of frequencies to corresponding probabilities [366, Theorem 4.5].

A number of practical prediction methods recommend using prediction rules having a small empirical error and belonging to a class of finite VC dimension (such as a subclass of neural networks; see also the description of structural risk minimization at the end of this subsection). So the Vapnik–Chervonenkis uniform law of large numbers provides an asymptotic justification for methods actually used. The inequality (13.2) is far too loose, however, to tell us how confident we should be in the accuracy of the specific predictions these methods make. To see the problem, consider the sample size l required in (13.2) to make both ϵ and the probability bound nontrivial—i.e., less than 1. For any result of this kind to have nontrivial implications, l must exceed the VC dimension. It is difficult to assess VC dimension precisely for the classes used in practice, but they are undoubtedly huge. One indication is the bound for the VC dimension of a sigmoid neural network obtained by Karpinski and Macintyre [185, 186]; see also [321, Sect. 20.4]. This bound is roughly $(WN)^2$, where N is the number of computational units and W is the number of independent parameters. For LeNet 1, the first and smallest of the neural networks designed by Yann LeCun's group for recognizing hand-written digits of the type in the USPS dataset, $N = 4635$ and $W = 2578$, and so the bound exceeds 10^{14}. Karpinski and Macintyre's bound can likely be tightened, but it is a very long way to a practical result; even the lower bound of roughly N^2 [321, Sect. 20.4] gives very large values.

The fact that the Vapnik–Chervonenkis uniform law of large numbers and similar results are usually too loose to give guidance about the confidence that we should have in predictions from actual datasets is, of course, well known. See, for example, [257].

Vapnik and Chervonenkis's theory was partially rediscovered by Leslie G. Valiant [359], whose work helped create a large community of computer scientists who have enriched the theory with analyses of computational complexity. Their work came to be known as PAC theory, because they obtained estimates that were Probably Approximately Correct.

Remark 13.2 It is interesting that the Glivenko–Cantelli theorem, which some probabilists consider the fundamental result of mathematical statistics, is also a special case of Vapnik and Chervonenkis's result, (13.2). The Glivenko–Cantelli theorem says that the empirical distribution function

$$F_l(t) := \frac{|\{i = 1, \ldots, l : z_i \leq t\}|}{l}$$

for a random variable ζ ($z_1, z_2, \ldots$ are independent realizations of ζ) converges almost surely to ζ's distribution function F. To derive this result from (13.2), we take $\mathbf{Z}$ to be the real line and $\mathcal{A}$ to be the family of all sets of the form $(-\infty, t]$, $t \in \mathbb{R}$, so that $\mathrm{VC}(\mathcal{A}) = 1$ and thus

$$Q^l \left\{ (z_1, \ldots, z_l) : \sup_{t \in \mathbb{R}} |F_l(t) - F(t)| > \epsilon \right\} < 4 \exp\left(1 + \ln(2l) - (\epsilon - 1/l)^2 l\right). \tag{13.3}$$

This inequality means that the empirical distribution function converges to the true distribution function uniformly in probability, and the convergence almost surely required by Glivenko–Cantelli follows by the Borel–Cantelli lemma. The right-hand side of (13.3) is not too different from the $2\exp(-2\epsilon^2 l)$ obtained specifically for this special case by Dvoretsky et al. [96] and Massart [240]. For further details, see [82, Sect. 12.8].

If we are interested in the case where the "hypothesis space" $\mathcal{F}$ is of infinite VC dimension, it will often be possible to represent $\mathcal{F}$ as the union of a nested sequence of function classes $\mathcal{F}_1 \subseteq \mathcal{F}_2 \subseteq \cdots \subset \mathcal{F}$ of finite VC dimension. For example, if $\mathcal{F}$ is the set of functions computable by neural networks, $\mathcal{F}_k$ may be the set of functions computable by neural networks with no more than k neurons.

In this situation, empirical risk minimization should be replaced by "structural risk minimization"; for details, see [366, Chap. 6].

13.2.3 *The Quest for Data-Dependent Bounds*

The excessive looseness of bounds such as (13.2) is often attributed to their nonconditionality—i.e., the fact that they do not depend on the particular training set $z_1, \ldots, z_l$ at hand. We might hope to identify some training sets that are particularly informative, in the sense that they produce prediction rules with much better accuracy, and we might hope for a distribution Q that produces such favourable training sets with high probability.

One popular and particularly elegant way of obtaining data-dependent bounds, which comes close to being useful, is Littlestone and Warmuth's sample compression approach [114, 235] (see also [321, Chap. 30]). Their theorem (Theorem 4.25 and its corollary Theorem 6.8 in [67]) tells us, as a special case, that the probability of error on a new example by a support vector machine making a prediction with d support vectors is bounded with probability $1 - \delta$ by

$$\epsilon := \frac{1}{l-d}\left(d\left(1 + \ln\frac{l}{d}\right) + \ln\frac{l}{\delta}\right), \tag{13.4}$$

where l is the size of the training set. The full theorem is applicable to any "sample compression scheme"; support vector machines are only one instance, albeit a powerful one, of such a scheme. A similar result holds for regression problems ([67], Theorems 4.26, 4.28, and 4.30).

Though it is one of the tightest data-dependent bounds, Littlestone and Warmuth's bound still falls short of being useful. To see this, we observe that for each of the ten classifiers in the problem of identifying digits in the USPS dataset, (13.4) is approximately

$$\frac{1}{7291 - 274} 274 \left(1 + \ln\frac{7291}{274}\right) \approx 0.17$$

even when we ignore the term $\ln\frac{l}{\delta}$. (There are 7291 training examples, and the average for the 10 classifiers of the number of support vectors for the polynomial kernel of degree 3, the degree that gives the best predictive performance, is 274; see [366, Table 12.2].) Thus the bound on the total probability of a mistake by one or more of the ten classifiers is 1.7, not a useful bound for a probability. There are more sophisticated schemes for multi-label classification using a binary classifier, but they involve separating unnatural classes, such as odd and even digits, and so would lead to large numbers of support vectors.

Another popular way of obtaining data-dependent bounds is the more recent PAC-Bayesian approach [242] (see also [321, Chap. 31]). For reviews of other approaches, see [156, 321].

A whole different tack, which is often advocated but seems to offer little real promise, is to derive bounds that depend on the probability distribution Q, in the hope that the relevant aspects of Q might be estimated from the training set well enough to make the bounds usable. One example is provided by the inequality

$$Q^l \left\{ (z_1, \ldots, z_l) : \sup_{A \in \mathcal{A}} \left| \frac{|\{i = 1, \ldots, l : z_i \in A\}|}{l} - Q(A) \right| > \epsilon \right\} < 4 \exp\left(\left(\frac{H_{\text{ann}}(2l)}{l} - (\epsilon - 1/l)^2 \right) l \right). \quad (13.5)$$

This is a strengthening of (13.2), but the function H_{ann}, called the “annealed entropy”, depends on Q as well as $\mathcal{A}$. Another example is the well-known upper bound

$$\frac{\mathbb{E}\mathcal{K}_{l+1}}{l+1} \quad (13.6)$$

on a support vector machine’s error probability. Here $\mathbb{E}\mathcal{K}_{l+1}$ is the expected number of support vectors among examples $z'_1, \ldots, z'_{l+1}$ randomly generated from Q^{l+1}. (See [366, Theorem 10.5]; Theorems 10.6 and 10.7 are also of this type.) These kinds of bounds leave us with the problem of estimating an aspect of Q and bounding the probable error of this estimate, which seems more daunting than the problem with which we began.

In general, we have several natural levels of data dependence for performance guarantees for learning algorithms:

1. Data-independence: no dependence on data, as in Vapnik and Chervonenkis’s bounds based on (13.2).
2. Inductive data-dependence: dependence on the training set only.
3. Transductive data-dependence: dependence on the training set and the given test objects.
4. *Data-superdependence*: dependence not only on the training set and the test objects but also on unobserved quantities.

Bounds such as (13.6) and (13.5) are data-superdependent in that they depend on the unknown data-generating distribution Q. Another kind of data-superdependence appears when the bounds depend on the labels of the test examples; see [407, Part III] for several examples of such bounds.

13.2.4 The Hold-Out Estimate

Because it has no objects, Bernoulli's problem is of little interest to machine learning, which emphasizes the use of complicated and informative objects. But Bernoulli's theorem nevertheless has an important role in machine learning, because it can be applied when we estimate an error rate from a hold-out sample as discussed in Sect. 1.2.1.

We can use the USPS dataset to illustrate how reasonable the results obtained from the hold-out estimate are. The figures given in Appendix B show that the hold-out estimate gives reasonable results for this dataset (at least producing values much less than one): assuming that the first half (4649 examples) gives a prediction rule whose error rate on the second half is 4%, the generalization error is bounded above by, approximately,

$$4\% + \mathbf{z}_{1\%}\sqrt{\frac{4\% \times 96\%}{4649}} \approx 4.9\%$$

with probability 99%. As we already mentioned, a major disadvantage of this bound is that it does not depend on the object whose label is being predicted, and taking into account the quality of the object will weaken the bound.

Fifty years ago, George Barnard deplored the underuse of the hold-out estimate [346]:

> The simple idea of splitting a sample into two and then developing the hypothesis on the basis of one part and testing it on the remainder may perhaps be said to be one of the most seriously neglected ideas in statistics, if we measure the degree of neglect by the ratio of the number of cases where a method could give help to the number of cases where it is actually used.

His words still ring true today. The accuracies of the methods in machine learning we have been discussing can be assessed much better by looking at their performance on a hold-out sample than by using the loose inequalities that would guarantee their good performance on datasets immensely larger than those available.

The hold-out estimate encounters several difficulties in practice, however:

1. Usually we do not obtain as good a prediction rule using only the training set as we would have obtained using the entire dataset. So far as the development of the prediction rule is concerned, the test set is wasted.
2. The hold-out estimate of prediction accuracy uses only the performance on the test set. So far as the evaluation of the prediction rule is concerned, the training set is wasted.
3. The hold-out estimate gives a single probability of error that applies to all new examples, regardless of how difficult they are. This single probability of error would apply, for example, to all three images in Fig. 1.2.

The last problem can be partly overcome if we find, from the training set, a reasonable division of all objects into a few disjoint classes (perhaps just two, such as "clear images" and "blurred images") and estimate for each class a probability of error from its percentage of wrongly classified test objects. This approach makes our predictions more *conditional* on the information provided by the object. But because it decreases the size of both the training set and the test set for each prediction, it aggravates the first two problems.

These difficulties provide one motivation for the new methods developed in this book.

13.3 Transductive Learning

Transductive prediction (as practised in classical statistics) and testing statistical models are two sides of the same coin—two of the many ways we can use a procedure that tells us whether a dataset agrees with a statistical model.

Testing: If we entertain an a priori possible statistical model M for the data $z_1, \ldots, z_n$, then seeing this data and detecting strong disagreement between it and the statistical model will make us abandon the statistical model.

Prediction: If we strongly believe in the statistical model M and have not yet seen the data, or at least not all of it, then detecting strong disagreement between $z_1, \ldots, z_n$ and M allows us to predict that $z_1, \ldots, z_n$ will not happen. We have been particularly interested in the case where each z_i consists of two components, x_i and y_i, and we have so far seen only $z_1, \ldots, z_{n-1}, x_n$.

We start this historical survey with testing. As we already mentioned, the first test was devised by John Arbuthnott [5] in the early eighteenth century, but Arbuthnott's statistical model was too simple ($\mathbf{B}^n_{1/2}$ with $\mathbf{B}_{1/2}$ the uniform distribution on $\{0, 1\}$) to have interesting implications for prediction. So we start with Student's work and its development by Fisher.

13.3.1 Student and Fisher

In this subsection we take the example space $\mathbf{Z}$ to be $\mathbb{R}$ and use the notation introduced in Sect. 11.4.1. We will assume that $z_1, z_2, \ldots$ are independent $\mathbf{N}_{\mu,\sigma^2}$ random variables.

In 1908 William S. Gosset [138, Sect. III], writing as "Student", correctly guessed the probability distribution of the ratio

$$\frac{\overline{z}_n - \mu}{\hat{\sigma}_n};$$

this ratio does not depend on σ and so can be used for testing, e.g., the hypothesis $\mu = 0$. Fisher derived Student's distribution rigorously by September 1912 (see

[273]), but a demonstration was not published until much later. From the point of view of the general theory it is more natural to consider the ratio

$$\sqrt{n}\frac{\overline{z}_n - \mu}{\hat{\sigma}_n},$$

and the distribution of this ratio is now known as Student's t-distribution (with $n-1$ degrees of freedom).

Student's result is not general enough to lead to an interesting prediction procedure. A more general test was proposed in Fisher's 1925 paper [105], where, after rigorously deriving Student's result, he treated the problem of comparing the means of two independent samples from the same normal distribution. He showed that if $z_1, \dots, z_l, z_{l+1}, \dots, z_{l+k}$ are generated from $\mathbf{N}^{l+k}_{\mu,\sigma^2}$, then

$$\sqrt{\frac{lk(l+k-2)}{l+k}}\frac{\overline{z}_{(1)} - \overline{z}_{(2)}}{\sqrt{S_{(1)} + S_{(2)}}}, \tag{13.7}$$

where

$$\overline{z}_{(1)} := \frac{1}{l}\sum_{i=1}^{l} z_i \qquad S_{(1)} := \sum_{i=1}^{l}\left(z_i - \overline{z}_{(1)}\right)^2$$

$$\overline{z}_{(2)} := \frac{1}{k}\sum_{i=l+1}^{l+k} z_i \qquad S_{(2)} := \sum_{i=l+1}^{l+k}\left(z_i - \overline{z}_{(2)}\right)^2$$

are the means and sums of squared deviations from the means for the two samples, has Student's t-distribution with $l+k-2$ degrees of freedom. This result, as Fisher [107] pointed out in 1935, does have implications for prediction. Setting k to 1 in (13.7), he obtained a result that we mentioned earlier:

$$\sqrt{\frac{l}{l+1}}\frac{z_{l+1} - \overline{z}_l}{\hat{\sigma}_l} \tag{13.8}$$

has Student's t-distribution with $l-1$ degrees of freedom (cf. (11.15)). So z_{l+1} will belong to the interval (11.16) (with $n = l+1$) with probability $1-\epsilon$.[1]

Fisher stated his conclusion concerning z_{l+1} more starkly than we have just done. From (13.8), he concluded that after observing $z_1, \dots, z_l$, we can attribute a fully known probability distribution to z_{l+1}—the distribution of

[1] Another author, George Baker [14], also published this result in 1935, but he was much less influential than Fisher.

Ronald A. Fisher (1890–1962).

Photo by Walter Stoneman. Used with permission of The Royal Society. ©Godfrey Argent Studio

$$\overline{z}_l + \xi\sqrt{\frac{l+1}{l}}\hat{\sigma}_l\ , \tag{13.9}$$

where $\overline{z}_l$ and $\hat{\sigma}_l$ are now known constants and ξ has the t-distribution with $l-1$ degrees of freedom. Fisher called this the *fiducial probability distribution* for z_{l+1}. In earlier articles, starting in 1930 [106], he had similarly derived "fiducial distributions" for the parameters of various statistical models.

Fisher's fiducial argument was the topic of vigorous discussion in the 1930s and 1940s. He defended it until he died in 1962. (See, e.g., Sect. V.4 of his book *Statistical Methods and Scientific Inference*, first published in 1956 [110].) But by the end of the 1930s, most mathematical statisticians had rejected it in favour of Jerzy Neyman's more restrained interpretation of estimation and prediction in the form of intervals, which recognizes that these intervals can have desired unconditional frequency properties but do not have all the properties we expect from probability intervals (see, e.g., Kolmogorov's [201, Sect. 5]). The problem is that after observing $z_1, \ldots, z_l$ we should be interested in the conditional distribution of (13.8) given $z_1, \ldots, z_l$, and this is an unknown normal distribution, not the known t-distribution. Fisher conceded the nonconditionality of the distribution, in his 1935 paper [107] and subsequently, and he saw some force in the objection. In Sect. V.3 of his 1956 book [110], e.g., he wrote this about frequentist verification of fiducial probabilities:

> In carrying out such a verification [...], it is to be supposed that the investigator is not deflected from his purpose by the fact that new data are becoming available from which predictions, better than the one he is testing, could at any time be made. For verification, the original prediction must be held firmly in view. This, of course, is a somewhat unnatural attitude for a worker whose main preoccupation is to improve his ideas. It is perhaps for this reason that some teachers assert that statements of fiducial probability cannot be tested by observations.

In the context of verification of ϵ as the probability of error for (11.16), Fisher appears to suggest repeatedly generating a training set of a fixed size $n-1$ and a new observation, and repeatedly applying the confidence predictor (11.16) trained on each of these training sets (ignoring the previously generated training sets) to predicting the corresponding new observation; it is clear that the limiting frequency of errors will indeed be ϵ. This is unnatural as we keep testing the original prediction (11.16) (z being the new observation) with a constant n, instead of testing better predictions based on all the available data. (This verification protocol is spelled out in detail and discussed further in [405, Sect. 9 of the arXiv report].)

Neyman developed his theory of confidence intervals only for parameters, not for new observations. But Fisher's idea of basing a prediction interval on the t-distribution gained some currency, especially after the appearance of his 1956 book. The idea often appears in books on linear regression and it also appears in the 1974 textbook by Cox and Hinkley [64].

13.3.2 Conformal Predictive Distributions in the Gaussian Model

Fiducial probability distributions for future observations, or *fiducial prediction*, is less prominent in Fisher's oeuvre than fiducial probability distributions for parameters, and it is less controversial. Let us check that the distribution (13.9) can be interpreted as a conformal predictive distribution.

Analogously to the transition from (11.15) to (11.16) in Chap. 11, we can derive from Fisher's statement about the distribution of (13.8) that the conformal p-value corresponding to the value z of the test example is

$$\Pi(z_1, \ldots, z_l, z) = F^t_{l-1}\left(\sqrt{\frac{l}{l+1}}\frac{z - \bar{z}_l}{\hat{\sigma}_l}\right), \tag{13.10}$$

where F^t_{l-1} is the distribution function of Student's t-distribution with $l-1$ degrees of freedom. This is true for a range of natural conformity measures; e.g., the conformity score of z_{l+1} can be defined as (13.8) itself, as the numerator $z_{l+1} - \bar{z}_l$, or simply as z_{l+1}.

As a function of z, (13.10) is both the conformal predictive distribution (of thickness 0, which is not unusual for an online compression model with continuous Markov kernels such as the Gaussian model) and the fiducial predictive distribution function.

13.3.3 Tolerance Regions

Tolerance regions (or, more fully, statistical tolerance regions) were introduced in the 1941 paper [427] by Samuel Wilks as a tool of induction. Wilks was motivated by Walter Shewhart's [325] ideas about industrial mass production.

In this subsection we are only interested in the case where objects are absent, as before, and $\mathbf{Z} = \mathbf{Y}$ is a Euclidean space. For simplicity we will only consider tolerance regions under the randomness model with the additional assumption that the probability distribution generating the individual observations is absolutely continuous (i.e., has a density w.r.t. to the Lebesgue measure), although there has been a lot of work for other statistical models and for discontinuous distributions.

Let $\epsilon, \delta \in (0, 1)$. A measurable function

$$\Gamma : \mathbf{Z}^* \to 2^{\mathbf{Z}} \tag{13.11}$$

(cf. (2.7) and (2.59)) is called an (ϵ, δ)-*tolerance predictor*[2] if

$$\inf_Q Q^l \left\{ (z_1, \ldots, z_l) \in \mathbf{Z}^l : Q(\Gamma(z_1, \ldots, z_l)) \geq 1 - \epsilon \right\} = 1 - \delta ,$$

where $l \in \mathbb{N}$ and Q ranges over the absolutely continuous probability distributions on $\mathbf{Z}$. If ϵ and δ are small, the output $\Gamma(z_1, \ldots, z_l)$ of the tolerance predictor can be used for predicting the future z_i, $i = l + 1, l + 2, \ldots$: the prediction is that $z_i \in \Gamma(z_1, \ldots, z_l)$.

Wilks [427] constructed tolerance predictors in the one-dimensional case and Wald [417] extended Wilks's procedure to the multi-dimensional case. Wald's construction was generalized by Tukey [357] and then further extended by, among others, Fraser [116, 117] and Kemperman [189].

There is a transductive variety of tolerance regions (see [118, 119, 143]; now they are more often referred to as prediction regions or prediction sets), and the ideas for constructing inductive tolerance regions carry over easily to the transductive case.

Let $\epsilon \in (0, 1)$. A measurable function (13.11) is called an ϵ-*tolerance predictor*[3] if

$$Q^{l+1} \left\{ (z_1, \ldots, z_l, z_{l+1}) \in \mathbf{Z}^{l+1} : z_{l+1} \in \Gamma(z_1, \ldots, z_l) \right\} = 1 - \epsilon , \tag{13.12}$$

where $l \in \mathbb{N}$ and Q ranges over the absolutely continuous probability distributions on $\mathbf{Z}$. We will only give a version of Tukey's [357] construction of ϵ-tolerance predictors (although Tukey was interested in (ϵ, δ)-tolerance predictors). Tukey's predictor is essentially the conformal predictor determined by the following nonconformity measure.

For each $n = 1, 2, \ldots$, fix a sequence of measurable functions $\phi_{n,k} : \mathbf{Z} \to \mathbb{R}$, $k = 1, \ldots, n$, and a sequence of real numbers $\alpha_{n,1}, \ldots, \alpha_{n,n}$. We will assume that the Lebesgue measure of $z \in \mathbf{Z}$ satisfying $\phi_{n,k}(z) = c$ is zero for all n, k, and $c \in \mathbb{R}$. For any sequence $z_1, \ldots, z_n$ of n observations, the corresponding nonconformity

[2] More standard terms are, e.g., "$1 - \delta$ tolerance region for a proportion $1 - \epsilon$" [118] or "$(1 - \epsilon)$-content tolerance region at confidence level $1 - \delta$" [119], but we prefer a simpler name.

[3] Or "$1 - \epsilon$ expectation tolerance region" in a more standard terminology [118].

scores (2.19) are defined as follows. Assign the nonconformity score $\alpha_{n,1}$ to all z_i at which $\max \phi_{n,1}(z_i)$ is attained and discard these z_i. Then assign the nonconformity score $\alpha_{n,2}$ to all z_i at which $\max \phi_{n,2}(z_i)$ is attained and discard these z_i. Repeat this procedure, finally assigning the nonconformity score $\alpha_{n,n}$ to all z_i at which $\max \phi_{n,n}(z_i)$ is attained; if there are no z_i left at some stage, do nothing. (With probability one, at each stage $\max \phi_{n,k}$ will be attained at exactly one z_i.)

Tukey proved a general result showing, in particular, that at each significance level ϵ the conformal predictor just described satisfies (13.12), provided ϵ has the form $i/(l+1)$ for an integer i. Fraser [116] noticed that we can allow $\phi_{n,k}$ to depend on the maxima reached by $\phi_{n,1}, \dots, \phi_{n,k-1}$ in the procedure for computing nonconformity scores. Kemperman [189] further noticed that we can allow dependence on the observations where the maxima were reached, not only on the maxima themselves.

Takeuchi (whose idea is published in a rudimentary form in [353] but was explained more fully in his seminars at Stanford University in the late 1970s) arrived at a general notion very similar to that of conformal predictor.

13.3.4 Dempster–Hill Procedure

In this subsection we discuss a classical procedure, briefly mentioned in Chap. 7, that was most clearly articulated by Dempster [79, p. 110] and Hill [159, 160]; therefore, in this book we refer to it as the *Dempster–Hill procedure*.

The Dempster–Hill procedure can be interpreted as the conformal predictive system determined by the conformity measure

$$A(\lbag z_1, \dots, z_n \rbag, z_n) = A(\lbag y_1, \dots, y_n \rbag, y_n) = y_n ; \tag{13.13}$$

it is used when the objects x_i are absent. (Both Dempster and Hill consider this case.) It can be regarded as the special case of the LSPM (Sect. 7.3) for the number of attributes $p = 0$; alternatively, we can take $p = 1$ but assume that all objects are $x_i = 0$. The predictions $\hat{y}$ are always 0 and the hat matrices are $\bar{H} = 0$ and $H = 0$ (although the expressions (7.14) and (7.19) are not formally applicable), which means that (7.11), (7.12), and (7.13) all reduce to (13.13). We will obtain the same predictive distributions for the ordinary LSPM with $p := 1$ and all objects being $x_i := 1$.

Suppose we are given training labels $y_1, \dots, y_{n-1}$ and would like to predict the next label y_n. It is easy to see that the predictive distribution becomes, in the absence of ties (Dempster's and Hill's usual assumption),

$$\Pi_n(y) := \begin{cases} [\frac{i}{n}, \frac{i+1}{n}] & \text{if } y \in (y_{(i)}, y_{(i+1)}) \text{for } i \in \{0, 1, \dots, n-1\} \\ [\frac{i-1}{n}, \frac{i+1}{n}] & \text{if } y = y_{(i)} \text{for } i \in \{1, \dots, n-1\} \end{cases} \tag{13.14}$$

(cf. (7.17)), where $y_{(1)} \le \dots \le y_{(n-1)}$ are the y_i, $i = 1, \dots, n-1$, sorted in the ascending order, $y_{(0)} := -\infty$, and $y_{(n)} := \infty$. This is essentially Hill's assumption

$A_{(n)}$ (which he also denoted A_n); in his words: "A_n asserts that conditional upon the observations $X_1, \ldots, X_n$, the next observation X_{n+1} is equally likely to fall in any of the open intervals between successive order statistics of the given sample" [159, Sect. 1]. The set of all continuous distribution functions F compatible with Hill's $A_{(n)}$ coincides with the set of all continuous distribution functions F satisfying $F(y) \in \Pi_n(y)$ for all $y \in \mathbb{R}$, where Π_n is defined by (13.14).

Notice that the LSPM, as presented in (7.23), is a very natural adaptation of $A_{(n)}$ to the least squares regression.

Since (13.14) is a conformal transducer (provided a point from an interval in (13.14) is chosen randomly from the uniform distribution on that interval), we have our usual guarantees of validity: the distribution of (13.14) is uniform over the interval [0, 1].

It is instructive to modify the conformity measure (13.13) of the Dempster–Hill procedure by making it dependent, in a very simple way, on the objects; it will be monotonic and produce an RPS. We will use the nearest neighbours idea, as in Sect. 7.5, but in a much more limited way. Namely, we set

$$A(\wr (x_1, y_1), \ldots, (x_n, y_n) \int, (x_i, y_i)) := y_i - \hat{y}_i , \tag{13.15}$$

where $\hat{y}_i$ is the label y_j corresponding to the nearest neighbour x_j of x_i: $j \in \arg\min_{k \in \{1,\ldots,n\} \setminus \{i\}} \rho(x_k, x_i)$, ρ being a measurable metric on the object space $\mathbf{X}$. In this example we only consider the case where the pairwise distances $\rho(x_i, x_j), i, j \in \{1, \ldots, n\}$, are all different. (We can easily drop this assumption allowing in (13.15) the index j of the nearest neighbour x_j of x_i, $i \in \{1, \ldots, n\}$, to be chosen randomly from the uniform probability measure on the set $\arg\min_{k \in \{1,\ldots,n\} \setminus \{i\}} \rho(x_k, x_i)$.)

Now let us consider a training set $z_1, \ldots, z_{n-1}$ and a test object x_n. For each $i \in \{1, \ldots, n-1\}$, let $\widehat{y}_i$ be the label y_j corresponding to the nearest neighbour x_j to x_i among $x_1, \ldots, x_{n-1}$: $j \in \arg\min_{k \in \{1,\ldots,n-1\} \setminus \{i\}} \rho(x_k, x_i)$. Define $I \subseteq \{1, \ldots, n-1\}$ to be the set of $i \in \{1, \ldots, n-1\}$ such that x_i is closer to x_n than to any of x_j, $j \in \{1, \ldots, n-1\} \setminus \{i\}$. The conformity scores (7.7) are $\alpha_n^y = y - \hat{y}_n$, $\alpha_i^y = y_i - y$ if $i \in I$, and $\alpha_i^y = y_i - \widehat{y}_i$ if $i \in \{1, \ldots, n-1\} \setminus I$. Solving the equation $\alpha_i^y = \alpha_n^y$ gives $y = C_i := (\hat{y}_n + y_i)/2$ if $i \in I$ and

$$y = C_i := \hat{y}_n + (y_i - \widehat{y}_i) \tag{13.16}$$

if $i \in \{1, \ldots, n-1\} \setminus I$. Assuming, for simplicity, that $C_1, \ldots, C_{n-1}$ are all different, we obtain the conformal predictive distribution

$$\Pi(y_1, \ldots, y_{n-1}, y) = \begin{cases} \left[\frac{i}{n}, \frac{i+1}{n}\right] & \text{if } y \in (C_{(i)}, C_{(i+1)}), i = 0, \ldots, n-1 \\ \left[\frac{i-1}{n}, \frac{i+1}{n}\right] & \text{if } y = C_{(i)}, i = 1, \ldots, n-1 \end{cases} \tag{13.17}$$

(cf. (13.14)), where $C_{(1)}, \ldots, C_{(n-1)}$ is the sequence $C_1, \ldots, C_{n-1}$ sorted in the ascending order, $C_{(0)} := -\infty$, and $C_{(n)} := \infty$.

The naive nearest-neighbour modification of the Dempster–Hill predictive distribution (13.14) would be (13.17) with all C_i defined by (13.16). The conformal predictive distribution is different only for $i \in I$, and I is typically a small set (its expected size is 1). For such i the conformal predictive distribution modifies the residual $y_i - \widehat{y}_i$ in (13.16) by replacing it by $(y_i - \hat{y}_n)/2$. Intuitively, the nearest neighbour to x_i in the augmented set $\{x_1, \ldots, x_n\}$ is x_n, so we would like to use y_n instead of $\widehat{y}_i$; but since we do not know y_n as yet, we have to settle for its estimate $\hat{y}_n$, and the resulting loss of accuracy is counterbalanced by halving the new residual. This seemingly minor further modification ensures the finite-sample property of validity.

13.3.5 Transduction in Statistical Learning Theory

Whereas the vast majority of theoretical results in statistical learning theory are stated in the inductive framework, transduction was an important source of intuition from the very beginning of the theory. For example, the main technical tool of the early theory, the so-called "ghost sample" technique (the technique is described in, e.g., [366, Sect. 4.13], although without using this expression), is of transductive nature. The idea of transduction was described already in the first monograph [370, Chap. VI, Sects. 10–13] devoted to statistical learning theory. In this subsection we will be using a more recent exposition, given in [366, Chap. 8]; we will start from the simplest result and discuss a more interesting case after Proposition 13.3.

As in the case of induction (see the description of inductivist statistical learning theory in Sect. 13.2), we start from a fixed family $\mathcal{F}$ of measurable functions mapping the object space $\mathbf{X}$ to the label space $\mathbf{Y}$; we will assume $|\mathbf{Y}| < \infty$. Let Q be a probability distribution on $\mathbf{Z}$ and l and k be two positive integer numbers (the interesting case is where $l \gg 1$ and $k \gg 1$). Suppose we are given l examples $z_1, \ldots, z_l$ and k unlabelled objects $x_{l+1}, \ldots, x_{l+k}$; our goal is to predict the latters' labels $y_{l+1}, \ldots, y_{l+k}$. We will say that $z_1, \ldots, z_l$ is the *training set* and $x_{l+1}, \ldots, x_{l+k}$ is the *working set*. Vapnik and Chervonenkis (see, e.g., [366, Theorem 8.2 and (8.15)]) found a useful function

$$E = E(\epsilon, l, k, \wr x_1, \ldots, x_{l+k} \int, p) \tag{13.18}$$

such that, for all Q, l, k, and $\epsilon \in (0, 1)$,

$$Q^{l+k}\left\{(z_1, \ldots, z_{l+k}) : \frac{|\{i = l+1, \ldots, l+k : f(x_i) \neq y_i\}|}{k} \leq\right.$$

$$E\left(\epsilon, l, k, \{x_1, \ldots, x_{l+k}\}, \frac{|\{i = 1, \ldots, l : f(x_i) \neq y_i\}|}{l}\right), \forall f \in \mathcal{F}\Bigg\}$$

$$\geq 1 - \epsilon . \qquad (13.19)$$

Equation (13.19) is a transductive analogue of (13.2), and it is applied to the problem of prediction in a similar way. Let ϵ be a small positive constant and suppose that at least one function from $\mathcal{F}$ provides a good prediction rule. If E in (13.18) is reasonably small for the given ϵ, given training and working sets, and a nontrivial range of p, we can apply the following strategy for predicting $y_{l+1}, \ldots, y_{l+k}$. Choose a function $f = \hat{f} \in \mathcal{F}$ with the smallest

$$E\left(\epsilon, l, k, \{x_1, \ldots, x_{l+k}\}, \frac{|\{i = 1, \ldots, l : f(x_i) \neq y_i\}|}{l}\right) \qquad (13.20)$$

(which typically means choosing an f with the best performance

$$\frac{|\{i = 1, \ldots, l : f(x_i) \neq y_i\}|}{l}$$

on the training set, hopefully falling within the nontrivial range of p mentioned earlier); (13.19) implies that, unless an unlikely (of probability at most ϵ) event has occurred, the frequency of errors

$$\frac{\left|\{i = l + 1, \ldots, l + k : \hat{f}(x_i) \neq y_i\}\right|}{k} \qquad (13.21)$$

on the working set will be bounded by (13.20).

Kolmogorov's objection (mentioned in Sect. 13.3.1) to Fisher's fiducial probabilities is also applicable to the Vapnik–Chervonenkis approach to transduction: it appears that we should be interested in the *conditional* probability given $z_1, \ldots, z_l$ (and perhaps also $x_{l+1}, \ldots, x_{l+k}$) that the bound (13.20) on the frequency of errors is satisfied. One important advantage of conditional probabilities, however, does carry over to the bound (13.20). Let ϵ be a small positive constant (the probability of error we are willing to tolerate), $z_1 = (x_1, y_1), z_2 = (x_2, y_2), \ldots$ be the observed sequence of examples, and $l_1, l_2, \ldots$ be a strictly increasing sequence of positive integers. Consider the following scenario of repeated Vapnik–Chervonenkis transduction. First we are given examples $z_1, \ldots, z_{l_1}$ and are asked to predict the labels of new objects $x_{l_1+1}, \ldots, x_{l_2}$. We know that the bound (13.20) on the frequency of errors on the k working examples, where $l := l_1$ and $k := l_2 - l_1$, will hold with probability at least $1-\epsilon$ (even if f is chosen to minimize the frequency of errors on the training set of size l). Next we are told the true labels for $x_{l_1+1}, \ldots, x_{l_2}$, so we now know the full examples $z_1, \ldots, z_{l_2}$. We are asked to predict the labels of new objects $x_{l_2+1}, \ldots, x_{l_3}$. We again know that the bound (13.20) on the frequency of errors on the k working examples, where $l := l_2$ and $k := l_3 - l_2$, holds with probability at least $1 - \epsilon$. We are now told the true labels for $x_{l_2+1}, \ldots, x_{l_3}$, etc. If ϵ were an upper bound on the conditional probability that the bound (13.20) holds, we could deduce from the martingale strong law of large numbers that the limiting (in the sense of lim sup) frequency with which the bound (13.20) is violated does not exceed ϵ. Vapnik and Chervonenkis's result (13.19) by itself does not prevent this limiting frequency (even in the sense of lim inf) from being 1. However, using our standard methods we can deduce the following proposition, which asserts much more, namely, the conservative validity of the procedure based on (13.20) in the online protocol. Let err_n^ϵ be 1 if the bound (13.20) holds for the training set $z_1, \ldots, z_{l_n}$ and the working set $x_{l_n+1}, \ldots, x_{l_{n+1}}$, and let it be 0 otherwise. Remember that the notion of domination is defined in Sect. 2.1.3.

Proposition 13.3 *Let* $\epsilon \in (0, 1)$. *Suppose* (13.19) *holds with* Q^{l+k} *replaced by any exchangeable probability distribution* P *on* $\mathbf{Z}^{l+k}$, *for all* l *and* k. *In the online transduction protocol described above,* err_n^ϵ, $n = 1, 2, \ldots$, *are dominated by independent Bernoulli random variables with parameter* ϵ; *in particular,*

$$\limsup_{n\to\infty} \frac{1}{n} \sum_{i=1}^{n} \mathrm{err}_i^\epsilon \leq \epsilon .$$

The condition of the proposition ((13.19) holding with Q^{l+k} replaced by any exchangeable probability distribution on $\mathbf{Z}^{l+k}$) is satisfied for the usual choices of the function E in (13.18); for details, see Sect. 13.6.

The simplest case of Vapnik–Chervonenkis transduction is not very interesting from the point of view of statistical learning theory: for example, the chosen prediction rule $\hat{f}$ does not usually depend on the working set. Our analysis, however, can be easily extended to more advanced results, such as (8.35) in [366]. In the rest of this section we will briefly describe the procedure of "structural risk minimization" in the case of transduction.

Suppose for each choice of the training set $x_1, \ldots, x_l$ of objects and the working set $x_{l+1}, \ldots, x_{l+k}$ we have a representation of the function class $\mathcal{F}$ as the union of an increasing sequence of function classes

$$\mathcal{F}_1^{\lbag x_1,\ldots,x_{l+k} \rbag} \subseteq \mathcal{F}_2^{\lbag x_1,\ldots,x_{l+k} \rbag} \subseteq \cdots \subseteq \mathcal{F}$$

(depending only on the combined set, or more accurately bag, $\lbag x_1, \ldots, x_{l+k} \rbag$). For example, the classes $\mathcal{F}_r^{\lbag x_1,\ldots,x_{l+k} \rbag}$ for small r may consist of the functions that separate the combined set $\lbag x_1, \ldots, x_{l+k} \rbag$ with a large margin (as in [366, Sect. 8.5] and [370, Sect. VI.10]). Suppose we have a function

$$E = E(\epsilon, l, k, \lbag x_1, \ldots, x_{l+k} \rbag, p, r)$$

such that, for all l and k, all exchangeable distributions P on $\mathbf{Z}^{l+k}$, and all $\epsilon \in (0, 1)$,

$$P\left\{(z_1,\dots,z_{l+k}) : \frac{|\{i=l+1,\dots,l+k : f(x_i)\neq y_i\}|}{k} \leq E\left(\epsilon, l, k, \wr x_1,\dots,x_{l+k}\int, \frac{|\{i=1,\dots,l : f(x_i)\neq y_i\}|}{l}, r\right), \forall r, \forall f \in \mathcal{F}_r^{\wr x_1,\dots,x_{l+k}\int}\right\} \geq 1-\epsilon \quad (13.22)$$

(for examples of such E, see [366, Theorem 8.4 and (8.35)] and [370, Theorem 6.2]).

Equation (13.22) is used for prediction in a more interesting way than (13.19). Let ϵ be a small positive constant. To predict $y_{l+1},\dots,y_{l+k}$, choose r and a function $f = \hat{f} \in \mathcal{F}_r$ with the smallest

$$E\left(\epsilon, l, k, \wr x_1,\dots,x_{l+k}\int, \frac{|\{i=1,\dots,l : f(x_i)\neq y_i\}|}{l}, r\right); \quad (13.23)$$

(13.22) implies that, unless an unlikely (of probability at most ϵ) event has occurred, the frequency of errors (13.21) on the working set will be bounded by (13.23) (with f replaced by $\hat{f}$). It will be easy to check that Proposition 13.3 can be stated and proved for this "structural risk minimization" framework.

13.3.6 Validity and Efficiency in Probabilistic Prediction

The approach of this book is to improve efficiency under the validity constraints (and with conditionality often being an important consideration for both validity and efficiency). In particular, we have been interested (see Chap. 7) in probability distributions for future labels satisfying a natural property of validity. They were introduced independently by Schweder and Hjort [311, Chap. 12] and Shen et al. [324], who also gave several examples of predictive distributions in parametric statistics. Earlier, related notions had been studied extensively in meteorology and statistics; see, e.g., the reviews [72] and [132].

Our approach is in the spirit of Gneiting et al. [134] paradigm (which they trace back to Murphy and Winkler [256]) of *maximizing the sharpness of the predictive distributions subject to calibration*. We refer to calibration as validity and sharpness as efficiency. Martin and Liu [237, Sect. 3.3] state a similar "Efficiency Principle": *Subject to the validity constraint, probabilistic inference should be made as efficient as possible*.

There are several versions and modifications of the notion of calibration. Calibration in probability (Sect. 7.2.2) is regarded as most important, but there is also marginal calibration [132, Definition 3] and the notion of being ideal relative to some information [132, Definition 1], which is very close to the notion of perfect calibration that we used in the context of Venn prediction in Sect. 6.2.1.

13.4 Bayesian Learning

The Bayesian approach to learning was suggested independently by Thomas Bayes, in a paper published posthumously in 1763 [24], and by Pierre Simon Laplace, in a memoir published in 1774 [212] (see also [344]). Since then it has always had its advocates, but it has become particularly popular in the last third of the twentieth century. The main idea of the Bayesian approach is to complement whatever statistical model $(P_\theta : \theta \in \Theta)$ for the data we might have by a new component (the prior $\mu(\mathrm{d}\theta)$ for the parameters) so that a complete probability distribution, $\int P_\theta \mu(\mathrm{d}\theta)$, for the data is obtained. With this probability distribution, the problem of prediction can be solved automatically by computing predictive distributions for new examples; learning from experience is also done automatically by applying Bayes's theorem. Therefore, the Bayesian approach is formally outside the scope of this book, which is about learning under uncertainty and does not assume the knowledge of the probability distribution for the data. It is still informative, however, to look at how the Bayesian approach for different plausible probability distributions compares with our approach; this is what we will do in this section. When the chosen probability distribution is the real one generating the data (realistically, we can know this only for artificial datasets that we ourselves generate), the Bayesian method can be counted on to give valid results; we will see in this section how the validity of those results is affected when the chosen distribution is wrong. By the results of Chaps. 2–4 conformal predictors, even if motivated by Bayesian considerations, are automatically valid under exchangeability.

13.4.1 Bayesian Ridge Regression

We start by giving a standard Bayesian derivation of the ridge regression estimator; this derivation will also provide us with the full conditional distribution for the new label y_{l+1}. As usual, we have training examples $(x_1, y_1), \ldots, (x_l, y_l)$ and a new object x_{l+1}, and our goal is to predict y_{l+1}; the object space is $\mathbf{X} = \mathbb{R}^p$. Let us assume that the objects $x_1, x_2, \ldots$ are fixed (deterministic) and the labels $y_1, y_2, \ldots$ are generated by the rule

$$y_i = w \cdot x_i + \xi_i \tag{13.24}$$

(as in (2.49) and (11.18)), where w is distributed as $\mathbf{N}_{0,(\sigma^2/a)I_p}$, each ξ_i is distributed as $\mathbf{N}_{0,\sigma^2}$, and all these random elements are independent.

The posterior density of w is proportional to

$$\exp\left(-\frac{a}{2\sigma^2}\|w\|^2\right)\prod_{i=1}^{l}\exp\left(-\frac{1}{2\sigma^2}(y_i - w \cdot x_i)^2\right)$$

$$= \exp\left(-\frac{1}{2\sigma^2}\left(a\|w\|^2 + \sum_{i=1}^{l}(y_i - w \cdot x_i)^2\right)\right) . \tag{13.25}$$

It attains its maximum at the w, which we will denote $\hat{w}$, that solves the optimization problem (2.36) (with $n = l$), i.e., at the value given by the ridge regression procedure.

Using the notation X_l and Y_l introduced in Chap. 2 (see (2.37) and (2.38)), we can rewrite the right-hand side of (13.25) as

$$\exp\left(-\frac{1}{2\sigma^2}(w'(X_l'X_l + aI_p)w - 2Y_l'X_l w + Y_l'Y_l)\right)$$
$$\propto \exp\left(-\frac{1}{2\sigma^2}(w - \hat{w})'(X_l'X_l + aI_p)(w - \hat{w})\right) , \tag{13.26}$$

where the expression for $\hat{w}$ is given by the right-hand side (2.40). We can recognize (13.26) as the multivariate normal distribution with mean $\hat{w}$ and variance matrix $V := \sigma^2(X_l'X_l + aI_p)^{-1}$ (see, e.g., [329, Sect. 2.13]); we are primarily interested, however, not in the parameter w but in the next label

$$y_{l+1} = w \cdot x_{l+1} + \xi_{l+1} .$$

The conditional distribution of $w \cdot x_{l+1}$ (given the training examples) is

$$\mathbf{N}\left(\hat{w} \cdot x_{l+1}, x_{l+1}'Vx_{l+1}\right)$$
$$= \mathbf{N}\left(x_{l+1}'(X_l'X_l + aI_p)^{-1}X_l'Y_l, \sigma^2 x_{l+1}'(X_l'X_l + aI_p)^{-1}x_{l+1}\right)$$

(we write $\mathbf{N}(\mu, \sigma^2)$ for $\mathbf{N}_{\mu,\sigma^2}$ if the expression for μ or σ^2 is complicated). The assumption of independence now gives the conditional distribution

$$\mathbf{N}\left(x_{l+1}'(X_l'X_l + aI_p)^{-1}X_l'Y_l, \sigma^2 x_{l+1}'(X_l'X_l + aI_p)^{-1}x_{l+1} + \sigma^2\right) \tag{13.27}$$

for y_{l+1}.

In this book we mainly need (13.27), but for completeness we will also give a kernel version of this procedure. It is possible to rewrite (13.27) in the kernel form using the matrix equation (2.51), similarly to what we did in Chap. 2 to derive the kernel representation of CRR. (This is done in [246, 247].) It is easier, however, to compute the predictive distribution for y_{l+1} directly. Namely, it is easy to see that the covariance between y_i and y_j ($i, j = 1, \dots, l + 1$) under the model

$$y_i = w \cdot F(x_i) + \xi_i$$

(which is (13.24) in the feature space; we are using the same notation as in Sect. 2.3.4 for the mapping $F : \mathbf{X} \to \mathbf{H}$ to the feature space and for the kernel $\mathcal{K}$) is given by

$$\operatorname{cov}(y_i, y_j) = \frac{\sigma^2}{a}\mathcal{K}(x_i, x_j) + \sigma^2 1_{i=j} ;$$

therefore, the theorem on normal correlation (see, e.g., [329, Theorem 2.13.2]) gives the predictive distribution

$$\mathbf{N}\left(Y_l'(K_l + aI_l)^{-1}k_l, \sigma^2 + \frac{\sigma^2}{a}\mathcal{K}(x_{l+1}, x_{l+1}) - \frac{\sigma^2}{a}k_l'(K_l + aI_l)^{-1}k_l\right) \tag{13.28}$$

for y_{l+1}, where K_l and k_l are essentially the same matrix and vector as in (2.53): $(K_l)_{i,j} := \mathcal{K}(x_i, x_j)$, $i, j = 1, \ldots, l$, $(k_l)_i := \mathcal{K}(x_{l+1}, x_i)$, $i = 1, \ldots, l$, and $\mathcal{K}$ is defined by (2.54).

It is clear that at each significance level ϵ the shortest prediction interval is the one symmetrical w.r. to the ridge regression prediction $\hat{y}_{l+1}$. Namely, (13.27) gives the prediction interval

$$\left[\hat{y}_{l+1} - \mathbf{z}_{\epsilon/2}V_l, \hat{y}_{l+1} + \mathbf{z}_{\epsilon/2}V_l\right] \tag{13.29}$$

with

$$\hat{y}_{l+1} := x_{l+1}'(X_l'X_l + aI_p)^{-1}X_l'Y_l ,$$
$$V_l^2 := \sigma^2 x_{l+1}'(X_l'X_l + aI_p)^{-1}x_{l+1} + \sigma^2 ,$$

and (13.28) gives the prediction interval (13.29) with

$$\hat{y}_{l+1} := Y_l'(K_l + aI_l)^{-1}k_l ,$$
$$V_l^2 := \sigma^2 + \frac{\sigma^2}{a}\mathcal{K}(x_{l+1}, x_{l+1}) - \frac{\sigma^2}{a}k_l'(K_l + aI_l)^{-1}k_l .$$

13.4.2 An Illustration

In this subsection we will illustrate the performance of a very special case of Bayesian ridge regression; namely we will set $p := 1$, $x_i := 1$ for all i, and $a := \sigma := 1$. Therefore, our Bayesian model consists of the location family $y_i = w + \xi_i$, where $\xi_i \sim \mathbf{N}_{0,1}$, and the prior distribution $w \sim \mathbf{N}_{0,1}$, all these random variables being independent. The advantage of this toy example is that all our algorithms become very transparent. Of course, the phenomena we will observe are general (see Sect. 13.7.4).

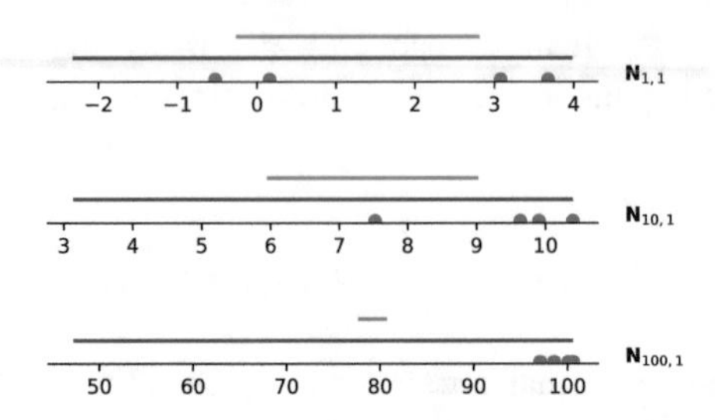

Fig. 13.1 In the top plot, the four observations (shown as big blue dots) are generated from $\mathbf{N}_{1,1}$; in the middle plot, from $\mathbf{N}_{10,1}$; and in the bottom plot, from $\mathbf{N}_{100,1}$. The blue lines are prediction intervals computed by a conformal predictor, and the red lines are Bayes prediction intervals

We convert the predictive distribution (13.27) to a prediction interval as before, by using its $\epsilon/2$ and $1-\epsilon/2$ quantiles. For conformal prediction, we will first use the absolute values (2.30) of the ordinary residuals as nonconformity scores.

The fact that the Bayes predictors are valid only under the postulated model, whereas the conformal prediction intervals are valid under the randomness assumption, is illustrated by Fig. 13.1. The chosen significance level is 20%, and so it is sufficient to have 4 observations in the training set in order to obtain a conformal prediction interval of a finite length. In the top plot, we generate four observations from $\mathbf{N}_{1,1}$; in the middle plot, from $\mathbf{N}_{10,1}$; and in the bottom plot, from $\mathbf{N}_{100,1}$. The blue lines are the conformal prediction intervals, and the red lines are the Bayes prediction intervals.

Remark 13.4 For convenience of the readers who would like to reproduce the plots in Fig. 13.1, the conformal prediction intervals shown there are

$$\left[y_{(1)} \wedge \frac{S-3y_{(4)}}{2},\; y_{(4)} \vee \frac{S-3y_{(1)}}{2}\right],$$

where $y_{(1)}$ and $y_{(4)}$ are the smallest and largest training observations, respectively, and S is the sum of all 4 training observations. This assumes that $\hat{y}$ in (2.30) is computed as $(S+y)/6$, where y is the postulated test observation.

All observations are generated from $\mathbf{N}_{w,1}$ for various constants w. When $w=1$ (and so the Bayesian assumption can be regarded as satisfied), the Bayes prediction intervals are on average only slightly shorter than the conformal prediction interval, and in about 46% of cases they are actually longer. Figure 13.1 shows that for our default seed for the random number generator we have an unusual situation where the Bayes prediction interval is significantly shorter. Since it covers only one out of the 4 training observations, it is clear that the prediction interval uses heavily the prior information about the data-generating distribution. As w grows, the conformal prediction intervals also grow in order to cover the test observation with the required probability, whereas the length of the Bayes prediction intervals remains constant. When $w=100$, and so the Bayesian assumption is clearly violated, the Bayes prediction intervals give very misleading results, and we can be sure that the test observation will not be covered.

In numbers (based on 10^6 computer simulations), the picture is as follows. The length of the Bayes prediction interval is always the same, 3.08. If the 4 observations are generated from $\mathbf{N}_{w,1}$ with $w = 0$, the average length of the conformal prediction intervals is 3.21 (with a significant standard deviation, 1.35, of the lengths). When $w = 1$, the average length of the conformal prediction intervals becomes 3.34 (with standard deviation 1.33). For $w = 10$ and $w = 100$, it becomes 7.58 and 52.57, respectively (with slightly smaller standard deviations).

The conformity measure (2.30) depends heavily on our soft Bayesian model. To make the dependence less heavy, let us proceed as in Sect. 2.3 and use the one-sided conformity measure $\alpha_i = y_i - \hat{y}_i$ for computing the lower prediction bound (i.e., prediction ray that is bounded below) and the opposite one-sided conformity measure $\alpha_i = \hat{y}_i - y_i$ for computing the upper prediction bound. We take the significance level 10% for each of the two prediction rays, which gives us the same upper bound on the probability of error, 20%, as before. Now we need 9 observations for the conformal prediction interval (the intersection of the two prediction rays) to have a finite length; the conformal prediction interval is simply the convex hull $[y_{(1)}, y_{(9)}]$ of the 9 observations (and it does not depend on how we compute $\hat{y}_i$ provided there is no actual dependence of $\hat{y}_i$ on i).

Our numbers are again based on 10^6 computer simulations. Now the distribution of the length of the conformal prediction interval does not depend on w. The average length is 2.97, and the standard deviation of the lengths is 0.81. The conformal prediction interval is shorter than the Bayesian one (whose length shrinks to 2.82, as now we have more training observations) in 45% of the cases. Therefore, the price to pay for the guaranteed coverage does not appear excessive.

In parametric statistics, it is widely believed that the choice of the prior does not matter much: the data will eventually swamp the prior. However, even in parametric statistics the model (such as $\mathbf{N}_{w,1}$) itself may be wrong. In nonparametric statistics, the situation is much worse: "the prior can swamp the data, no matter how much data you have" (Diaconis and Freedman [84, Sect. 4]). In this case, using Bayes prediction intervals becomes particularly problematic.

13.5 Further Development of Conformal Prediction

The first edition of this book [402] could be regarded as a fairly complete account of the theory of conformal prediction at the time of its publication. This is very far from being true for this edition. Now the book only includes those of newer results that are most directly related to the material covered in the first edition. In particular, this means that some of the most interesting and novel ideas have not been included. In this section we briefly review some of the newer literature on conformal prediction that was not mentioned in the previous chapters (mainly because it goes so far beyond the first edition).

13.5.1 Moving Beyond Exchangeability

The most interesting assumptions used in conformal prediction are exchangeability and randomness. In this part (Part IV) of the book, starting from Chap. 11, we have discussed other possible assumptions, but several very interesting options of a different nature have been considered in literature, starting from Ryan Tibshirani et al.'s [355] briefly mentioned in Sect. 11.6.3.

A common theme is dealing with dataset shift, which we discussed in Chap. 8. While we only applied methods of conformal prediction to detecting dataset shift and to making simple predictors (in the terminology of Sect. 2.1.2) more robust (Chap. 10), new research adapts conformal predictors to dataset shift. Gibbs and Candès [130] introduce *adaptive conformal inference*, a modification of conformal prediction that automatically adapts to changes in the distribution of the data and maintains the desired frequency of errors. Barber et al. [20] develop versions of conformal predictors that assign different weights to different training examples; intuitively lower weights mean that those examples are less relevant to the prediction task because, e.g., they are older and so perhaps coming from a significantly different distribution.

Fisch et al. [103] move outside our assumption of exchangeability while remaining inside a wider exchangeable picture. They consider the current prediction task to be an element of an exchangeable family of prediction tasks, and construct a meta-conformal predictor that can learn to learn quickly on the current task.

So far in this subsection connections with exchangeability have been still strong. Time series go beyond exchangeability more radically. Our Remark 11.7 was of this kind, but it was extremely narrow. Even for genuine time series, there may be exchangeability at a higher level; e.g., we can apply methods based on exchangeability to time-series prediction in situations where we have an exchangeable family of time series [339]: each time series in the family may describe a patient, and overall we have a group of patients regarded as exchangeable.

Dashevskiy and Luo [69] proposed a way to adapt conformal prediction to time series (via splitting the data into blocks), but they did not give any theoretical results. Such results were provided by Chernozhukov et al. [48], and the theory was further developed in [429].

13.5.2 Causality

The problem of identifying causal effects is a notoriously difficult one, but there have been important advances over the last decades. A popular approach is based on the Rubin causal model [171, 174, 258], which considers units (such as patients) that may be subjected to different treatments. Suppose the treatments are binary; e.g., $T = 1$ may signify some kind of intervention while $T = 0$ may signify no intervention. With each unit we then associate two numbers, the response $y(0)$ when T for this unit is set to 0, and the response $y(1)$ when T for it is set to 1. A standard task is, given a training set consisting of treated and untreated units and given a test unit, to predict the responses $y(0)$ and $y(1)$ for this unit. For example [174, Sect. 1.3], if I take an aspirin now, will my headache go in an hour's time? To make the task feasible, we assume that for each unit we have its attributes x and that x determines the probability that $T = 1$ (i.e., that the unit is treated) and the probability distribution, independent of T, of $y(0)$ and $y(1)$ (so that x contains all possible "confounders", i.e., factors that affect both T and y). This is known as the *strong ignorability* assumption.

The Rubin causal model, whose simple version goes back to Neyman [258], is somewhat controversial, because for each unit (even one in the training set) we observe either $y(0)$ or $y(1)$, depending on the treatment chosen, and the other element of the pair is counterfactual. There are potential pitfalls in using the model (see, e.g., Dawid's discussion paper [73]), but it remains popular.

A pioneering paper on applying conformal prediction under covariate shift to causal inference is Lei and Candès's [228]; see also [286, Sect. 2.3]. Suppose we would like to predict $y(1)$ for a test unit given training units, all units being IID. The relevant training units are those that received treatment $T = 1$, but the distribution of the attributes for those units (conditional on $T = 1$) is different from the distribution of the attributes for the test unit (which is unconditional). We are in a situation of covariate shift, since the distribution of the label $y(1)$ given the attributes x stays the same [228, Sect. 3.1].

Conformal prediction was also adapted to the Rubin causal model by Chernozhukov et al. [50], who use it to evaluate effects of policy interventions. Their proposed procedure is exactly valid for IID data, and they establish its approximate validity for time series data.

13.5.3 Anomaly Detection

In Part III we discussed what can be called global testing: when can we say that the assumption of randomness has ceased to hold and the distribution of the data changed? A related problem is that of local testing: does a new example look coming from a different distribution as compared with the examples seen so far? Or is it an *outlier*, also known as an *anomaly*? Hawkins's [154] informal definition of an outlier is "an observation which deviates so much from other observations as to arouse suspicions that it was generated by a different mechanism" (cited in [218]). We will refer to the problem of detecting outliers as *anomaly detection*. It is different from global testing considered in Part III in that we are bound to encounter anomalies from time to time even if the data are exchangeable.

The problem of anomaly detection arises both in supervised learning (when we have structured examples containing well-defined labels $y \in \mathbf{Y}$) and in unsupervised learning (when we have unstructured observations). Early work on conformal anomaly detection was done by Laxhammar and Falkman; see, e.g., [218]. A basic way of conformal anomaly detection in an input stream of observations $z_1, z_2, \dots$ is to compute the conformal p-values as the right-hand side of (2.67) and to declare z_n to be an anomaly if $p_n \leq \epsilon$ for some significance level ϵ, which is our target frequency of false alarms. By Proposition 2.4, under exchangeability the probability of an alarm (which we know to be false) will be ϵ and the alarms will be raised independently at different steps; therefore, the long-run frequency of false alarms will be ϵ (or at most ϵ if the p-values are not smoothed). The application that Laxhammar and Falkman concentrated on was detecting anomalous ship trajectories, and an especially interesting nonconformity measure for that application is directed Hausdorff distance [217, Chap. 5].

In supervised learning, we might know that the new example $z_n = (x_n, y_n)$ is an anomaly already after observing x_n. This will be the case when the prediction interval Γ_n^ϵ is empty at the chosen significance level ϵ: whatever y_n is observed, z_n will be anomalous. The uncertainty about the label of x_n resulting from conformal prediction is given by the set of p-values ($p_n^y : y \in \mathbf{Y}$). The nth example z_n is an anomaly if the *credibility* $\max_y p_n^y$ is small (earlier we gave an equivalent definition of credibility as (3.54)). If credibility is low, it looks as if we have a new class (or otherwise, the current object is an unusual representative of an existing class). Such cautious approach to prediction is considered by Guan and Robert Tibshirani [142].

Much more sophisticated procedures of outlier detection built on top of conformal p-values and their modifications are described in [23]. It turns out that properties of validity of conformal p-values can be strengthened and lead to bounds on the false discovery rate for outliers.

13.6 Proof of Proposition 13.3

Vapnik [366, Sect. 8.1] points out that (13.19) (as well as the other results in that chapter) can be strengthened. (Equation (13.19) corresponds to Vapnik's "Setting 2", and we are about to describe the less restrictive "Setting 1"; for a clear discussion, see [80, Sect. 2.1].)

Fix a bag of size $l+k$ of examples, and let P be the uniform probability distribution on the set of the $(l+k)!$ orderings of that bag. Then (13.19) continues to hold if Q^{l+k} is replaced by P.

We can now apply the method used in the proof of Theorem 11.2 in Sect. 11.5.1. It is clear that err_n^ϵ depends on $z_1, \dots, z_{l_{n+1}}$ only through $\Lbag z_1, \dots, z_{l_n} \Rbag$ and $\Lbag z_{l_n+1}, \dots, z_{l_{n+1}} \Rbag$. Let us increase err_n^ϵ (if needed) using randomization to make the probability that $\mathrm{err}_n^\epsilon = 1$ precisely ϵ. It is sufficient to prove that $\mathrm{err}_1^\epsilon, \dots, \mathrm{err}_N^\epsilon$ is a sequence of independent Bernoulli random variables. Therefore, it is sufficient to prove that $\mathrm{err}_N^\epsilon, \dots, \mathrm{err}_1^\epsilon$ is a sequence of independent Bernoulli random variables. This is done as before, using the fact that the conditional probability that $\mathrm{err}_n^\epsilon = 1$ given the bag $\Lbag z_1, \dots, z_{l_{n+1}} \Rbag$ and the sequence $z_{l_{n+1}+1}, z_{l_{n+1}+2}, \dots$ is equal to ϵ, $n = N, \dots, 1$.

13.7 Context

13.7.1 Inductive Prediction

Our historical survey of induction follows, to a large degree, [343, Chap. 2]. Some modern approaches to bounding the Bernoulli parameter p given a data sequence are described by Brown et al. [38]. For reviews of attempts to improve Jacob Bernoulli's $N(0.02, 1/1001, 0.6)$, see [285] and [326]. Barnard's words were quoted second-hand after [276].

Different versions of the Glivenko–Cantelli theorem were proved by Cantelli [41], Glivenko [131], and Kolmogorov [200], whose articles were published in the same volume of the same journal (for details and further developments, see [366], Comments and Bibliographical Remarks).

The study by Langford [211], already mentioned in Chap. 1, found that the hold-out estimate performs much better than the sample compression bound and PAC-Bayes bounds on several datasets. These datasets are low-dimensional, but one can expect the advantage of the hold-out estimate to become even more pronounced as the number of attributes grows. It should be borne in mind that Tables 5.2 and 5.3 in that paper are not based on rigorous bounds on the true error rate of the learned classifier (as the author explains in Sect. 5.3.1).

13.7.2 Transductive Prediction

Fisher imposed numerous restrictions on fiducial distributions that nowadays appear redundant, as discussed in [392, Sect. 2]. One explanation is that he was motivated by a search for fiducial statements that "may claim unique validity" [259, footnote in Fisher's comment]. He was not satisfied with fiducial statements that were merely true (or valid), he wanted them to be the whole truth. In various extensions of fiducial inference, such as by Hannig [149], the requirement of uniqueness has been abandoned.

For a good review of tolerance regions see [143]. The description of Tukey's [357] procedure is given in Sect. 13.3.3 in terms different from Tukey's. For an explanation of the connection between our and Tukey's terminology, see [15, Sect. 2.5].

Dempster referred to the Dempster–Hill procedure as direct probabilities, and Hill as $A_{(n)}/H_{(n)}$. Dempster introduced it in [79, p. 110] and Hill in [159, 160], discussing it further in [161, 162]. Both Dempster and Hill trace their ideas to Fisher's [108, 109] nonparametric version of his fiducial method, but Fisher was interested in confidence distributions for quantiles rather than predictive distributions. Hill [160] also referred to his procedure as Bayesian nonparametric predictive inference, which was abbreviated to nonparametric predictive inference (NPI) by Frank Coolen [10]. We are not using the last term since it seems that all of this book falls under the rubric of "nonparametric predictive inference". An important predecessor of Dempster and Hill was Jeffreys [176], who postulated what Hill later denoted as $A_{(2)}$ (see [210] and [314] for discussions of Jeffreys's paper and Fisher's reaction).

After the discussion of transduction had been published in [370] for the case of classification, it was extended by Vapnik (in the English translation of [364]) to regression. For simplicity, Vapnik and Chervonenkis [370] discuss only the case where the sizes of the training and test sets coincide, $l = k$; this restriction was removed by Vapnik and Sterin [371]. For a further development of the Vapnik–Chervonenkis theory of transduction, see [80].

13.7.3 Online Transduction

In general, an inductive method produces a prediction rule that can be applied to many new examples, and it aims for a high probability that the rule will predict with high accuracy. These goals are attractive but sometimes difficult or impossible to achieve. Conformal methods usually aim for less. Instead of trying to control two parameters—the desired accuracy and the probability of finding a rule with that accuracy—they try to control only the overall frequency of accurate predictions. This means that even when they succeed, they can be criticized for achieving less than an inductive method might achieve. But as we explain in the last subsection of Sect. 10.2 of the first edition of this book [402], the cogency of this criticism is doubtful in the online setting, where we use each rule only once. In the offline setting, where we find a rule from old examples and apply it to many new examples, the repeated use of the rule gives empirical meaning to talk about its probability of predicting accurately. But in the online setting, where a rule for prediction, to the extent that it is even explicitly formulated, is used only once before being replaced by an improved rule, it may be meaningless to talk about the rule's probability of predicting accurately. It makes more sense to emphasize the overall frequency of accurate prediction.

13.7.4 Bayesian Prediction

In Sect. 13.4 we discussed the differences between our approach and the Bayesian approach only in a toy case of regression. In considering the toy example of a location family we follow Wasserman [423, Fig. 1] (and also [310, Chap. 11]). Similar phenomena can be observed in the general case of regression and in the case of classification. The paper by Melluish et al. [248], which we followed in the first edition [402, Sect. 10.3], considers both cases.

The theorem on normal correlation, which we used in the derivation of the kernel representation of Bayesian ridge regression, is known in geostatistics as simple kriging (see, e.g., [66, p. 110]). The name "kriging" was coined by Matheron [241] after the South African mining engineer Krige, but the method itself is not due to Krige [66, p. 106].

13.7.5 Conformal Prediction: Further Developments

See [3] and [431] for recent tutorials on conformal prediction (and [319] for an old one). There have been annual workshops on Conformal Prediction and its Applications (COPA) since 2012, with tutorials on various topics at most of them. ICML 2021 and 2022 hosted Workshops on Distribution-Free Uncertainty Quantification, with conformal prediction as a key topic and several review talks (available on the Internet).

Appendix A
Probability Theory

In this appendix we will give some basic definitions and results of probability theory needed for core results in this book. It is not suitable for a first study; our main goal is to familiarize the reader with our terminology and notation (e.g., we may use simple special cases of notions such as expectation before giving general definitions). The reader who needs an introduction to this material is advised to consult existing excellent textbooks such as [329, 330] and [428] (in our brief review we usually follow Kolmogorov [199], Shiryaev [329, 330], and Devroye et al. [82]). We will rarely give any proofs (and the proofs that we do give will sometimes use notions not defined and results not stated here). This appendix does not treat topics (such as linear regression) that are needed only for applications of this book's core results; references to the relevant literature are given in the main body of the book.

Our exposition is based on Kolmogorov's measure-theoretic axioms of probability. The recent suggestions [317, 320] to base probability theory on the theory of perfect-information games rather than measure theory would have certain advantages, but we preferred the more familiar approach.

A.1 Basics

A.1.1 Kolmogorov's Axioms

A *σ-algebra* $\mathcal{F}$ on a set Ω is a collection of subsets of Ω that contains $\emptyset$ and Ω and is closed under the operations of complementation and taking finite and countable unions and intersections. A *measurable space* is a set Ω equipped with a σ-algebra $\mathcal{F}$ on Ω (so formally the $(\Omega, \mathcal{F})$ is a measurable space; we will, however, often refer to Ω as a measurable space when $\mathcal{F}$ is clear from the context). The elements of $\mathcal{F}$ are called *measurable sets* in Ω. A σ-algebra $\mathcal{F}'$ such that $\mathcal{F}' \subseteq \mathcal{F}$ is called a *sub-σ-algebra* of $\mathcal{F}$.

V. Vovk et al., *Algorithmic Learning in a Random World*,
https://doi.org/10.1007/978-3-031-06649-8

For any family $\mathcal{A}$ of subsets of a set Ω there exists the smallest σ-algebra $\mathcal{F}$ on Ω such that $\mathcal{A} \subseteq \mathcal{F}$ (take as $\mathcal{F}$ the intersection of all σ-algebras that include $\mathcal{A}$). The smallest σ-algebra on the real line $\mathbb{R}$ containing all intervals (a, b) is called the *Borel* σ-algebra. Similarly, the smallest σ-algebra on $[-\infty, \infty]$ containing all intervals (a, b) and the one-element sets $\{-\infty\}$ and $\{\infty\}$ is called *Borel*. (More generally, the Borel σ-algebra on a topological space is defined as the smallest σ-algebra containing all open sets.) When considered as measurable spaces, $\mathbb{R}$ and $[-\infty, \infty]$ will always be assumed equipped with the Borel σ-algebra.

A measurable space is *standard Borel* if it is isomorphic to a measurable subset of the interval $[0, 1]$. The class of standard Borel spaces is very rich: for example, all Polish spaces (such as finite-dimensional Euclidean spaces $\mathbb{R}^n$, $\mathbb{R}^\infty$, functional spaces C and D) are standard Borel; finite and countable products of standard Borel spaces are also standard Borel (see, e.g., [307, Sect. B.3.2]).

A *probability distribution* (or *probability measure*, or simply *distribution*) on a measurable space $(\Omega, \mathcal{F})$ is a function $P : \mathcal{F} \to [0, 1]$ such that: $P(\Omega) = 1$; $P(A \cup B) = P(A) + P(B)$ for all disjoint $A, B \in \mathcal{F}$; and

$$P\left(\bigcup_{i=1}^{\infty} A_i\right) = \lim_{n\to\infty} P(A_n) \tag{A.1}$$

for all nested sequences $A_1 \subseteq A_2 \subseteq \dots$ of sets in $\mathcal{F}$.

The main object studied in probability theory is a *probability space* $(\Omega, \mathcal{F}, P)$, where $(\Omega, \mathcal{F})$ is a measurable space and P is a probability distribution on $(\Omega, \mathcal{F})$. The elements of $\mathcal{F}$ (i.e., the measurable sets in Ω) are also called *events*. A function $\xi : \Omega \to \Xi$, where Ξ is another measurable space, is called *measurable*, or a *random element* in Ξ, if, for every measurable set $A \subseteq \Xi$, the pre-image $f^{-1}(A) \subseteq \Omega$ is measurable; we will say that ξ is $\mathcal{F}$-measurable if the σ-algebra on Ω has to be mentioned explicitly. Two important special cases are: random elements in $\mathbb{R}$ are called *random variables*, and random elements in $[-\infty, \infty]$ are called *extended random variables*. The σ-algebra on Ω *generated* by a random element $\xi : \Omega \to \Xi$ consists of all events of the form $\xi^{-1}(A)$, A ranging over the measurable sets in Ξ. The *distribution* of a random element ξ in Ξ is the probability distribution $P\xi^{-1}$ on Ξ which is the image of P under the mapping ξ:

$$P\xi^{-1}(E) := P(\xi^{-1}(E))$$

for all events $E \subseteq \Xi$. An event E is *almost certain* if $P(E) = 1$; a property of $\omega \in \Omega$ holds *almost surely* (often abbreviated to "a.s.") if the event that this property is satisfied is almost certain.

An element $\omega \in \Omega$ such that $\{\omega\} \in \mathcal{F}$ is said to be an *atom* of the probability space $(\Omega, \mathcal{F}, P)$ (or of the probability measure P) if $P\{\omega\} > 0$ (we usually drop the parentheses in $P(\{\omega\})$); otherwise it is a *non-atom*. We say that $(\Omega, \mathcal{F}, P)$ is *atomless* if $P\{\omega\} = 0$ for all $\omega \in \Omega$.

A *statistical model* is a family of probability distributions $(P_\theta : \theta \in \Theta)$ on the same measurable space (called the *sample space*) indexed by the elements θ of some *parameter space* Θ. Statistical models are the standard way of modelling uncertainty.

The notions of a random variable and event only depend on the underlying measurable space. Therefore, we will sometimes use them even if no probability measure has been fixed (e.g., we may only have a statistical model).

A.1.2 Convergence

There are many senses in which a sequence of random variables $\xi_1, \xi_2, \dots$ can converge to a number $c \in \mathbb{R}$, but in this book we will only be interested in the following two. We say that $\xi_1, \xi_2, \dots$ converges to c *in probability* if

$$\lim_{n\to\infty} P\{\omega \in \Omega : |\xi_n(\omega) - c| > \epsilon\} = 0$$

for any $\epsilon > 0$. The other notion of convergence is where $\xi_1, \xi_2, \dots$ converges to c almost surely (we will sometimes say "with probability one"). Convergence almost surely implies convergence in probability.

Another kind of convergence that we use is convergence in law (or weak convergence) of probability measures on $\mathbb{R}$ (the real line). We say that a sequence of probability measures P_n on $\mathbb{R}$ converges *in law* to a probability measure P on $\mathbb{R}$, $P_n \xrightarrow{\text{law}} P$, if, for any bounded continuous function $f : \mathbb{R} \to \mathbb{R}$, we have $\int f \mathrm{d}P_n \to \int f \mathrm{d}P$ as $n \to \infty$. When talking about convergence in law of random variables, we mean convergence in law of their distributions.

A.2 Independence and Products

Let $(\Omega, \mathcal{F}, P)$ be a probability space. Sub-σ-algebras $\mathcal{F}_1, \dots, \mathcal{F}_n$ of $\mathcal{F}$ are said to be *independent* if, for any choice of events $A_1 \in \mathcal{F}_1, \dots, A_n \in \mathcal{F}_n$,

$$P\left(\bigcap_{i=1}^n A_i\right) = \prod_{i=1}^n P(A_i) .$$

The sub-σ-algebras in an infinite sequence $\mathcal{F}_1, \mathcal{F}_2, \dots$ are *independent* if, for any $n = 1, 2, \dots$, the sub-σ-algebras $\mathcal{F}_1, \dots, \mathcal{F}_n$ are independent. The random elements in a sequence, finite or infinite, $\xi_1, \xi_2, \dots$ are *independent* if the sub-σ-algebras $\mathcal{F}_1, \mathcal{F}_2, \dots$ generated by $\xi_1, \xi_2, \dots$, respectively, are independent.

A.2.1 *Products of Probability Spaces*

If $(Z_1, \mathcal{F}_1, Q_1), \ldots, (Z_n, \mathcal{F}_n, Q_n)$ is a finite sequence of probability spaces, the product

$$(\Omega, \mathcal{F}, P) = \prod_{i=1}^{n} (Z_i, \mathcal{F}_i, Q_i)$$

is defined as follows:

- Ω is the Cartesian product $\prod_{i=1}^{n} Z_i$;
- $\mathcal{F}$ is the smallest σ-algebra on Ω containing all Cartesian products $\prod_{i=1}^{n} A_i$, where $A_i \in \mathcal{F}_i$ for all i;
- P is defined as the only probability distribution on $(\Omega, \mathcal{F})$ such that

$$P\left(\prod_{i=1}^{n} A_i\right) = \prod_{i=1}^{n} Q_i(A_i) ,$$

  for all $A_i \in \mathcal{F}_i$, $i = 1, \ldots, n$.

Analogously, the product

$$(\Omega, \mathcal{F}, P) = \prod_{i=1}^{\infty} (Z_i, \mathcal{F}_i, Q_i)$$

of an infinite sequence of probability spaces $(Z_1, \mathcal{F}_1, Q_1), (Z_2, \mathcal{F}_2, Q_2), \ldots$ is defined as follows:

- Ω is the Cartesian product $\prod_{i=1}^{\infty} Z_i$;
- $\mathcal{F}$ is the smallest σ-algebra on Ω containing the Cartesian product $\prod_{i=1}^{\infty} A_i$ for every sequence $A_1 \in \mathcal{F}_1, A_2 \in \mathcal{F}_2, \ldots$ such that $A_i = Z_i$ from some i on;
- P is defined as the only probability distribution on $(\Omega, \mathcal{F})$ such that

$$P\{(z_1, z_2, \ldots) \in \Omega : z_i \in A_i, \quad i = 1, \ldots, n\} = \prod_{i=1}^{n} Q_i(A_i) ,$$

  for all $n = 1, 2, \ldots$ and all sequences $A_i \in \mathcal{F}_i$, $i = 1, \ldots, n$.

Notice that the random variables

$$\xi_i(z_1, \ldots, z_n) := z_i, \quad i = 1, \ldots, n ,$$

on the product $\prod_{i=1}^{n} (Z_i, \mathcal{F}_i, Q_i)$ are independent; the random variables

$$\xi_i(z_1, z_2, \dots) := z_i, \quad i = 1, 2, \dots ,$$

on the infinite product $\prod_{i=1}^{\infty}(Z_i, \mathcal{F}_i, Q_i)$ are also independent.

Our notation for the elements of the product probability spaces will be

$$\left(\prod_{i=1}^{n} Z_i, \bigotimes_{i=1}^{n} \mathcal{F}_i, \prod_{i=1}^{n} Q_i\right) := \prod_{i=1}^{n}(Z_i, \mathcal{F}_i, Q_i)$$

(where we allow $n = \infty$), or, with the short-hand notation for measurable spaces,

$$\left(\prod_{i=1}^{n} Z_i, \prod_{i=1}^{n} Q_i\right) := \prod_{i=1}^{n}(Z_i, Q_i) .$$

When all the probability spaces (Z_i, Q_i) coincide, $(Z_i, Q_i) = (Z, Q)$ for all i, we will write (Z^n, Q^n) for the product $(\prod_{i=1}^{n} Z, \prod_{i=1}^{n} Q)$ and call it the nth *power* of (Z, Q), with just 'power" meaning "∞th power".

A.2.2 *Randomness Model*

Now can introduce one of the two main statistical models used in this book, the randomness model. The underlying measurable space (i.e., the sample space) is composed of infinite data sequences: it is the product $\mathbf{Z}^{\infty}$, where $\mathbf{Z}$ is the measurable space from which examples are drawn. The *randomness model* is defined to be the set of all power probability distributions Q^{∞}, Q ranging over the probability distributions on $\mathbf{Z}$.

A.3 Expectations and Conditional Expectations

Fix a probability space $(\Omega, \mathcal{F}, P)$. If $\xi : \Omega \to \mathbb{R}$ is a nonnegative random variable and $A \in \mathcal{F}$ is an event, the *Lebesgue integral* $\int_A \xi(\omega) P(\mathrm{d}\omega)$ is defined to be

$$\lim_{\lambda \downarrow 0} \sum_{k=0}^{\infty} k\lambda P\{\omega \in A : k\lambda \le \xi(\omega) < (k+1)\lambda\}$$

(this limit always exists but can be infinite). If $\xi \ge 0$ is an extended random variable, the integral is defined in the same way if $P(A \cap \{\xi = \infty\}) = 0$ and is defined to be ∞ otherwise. For an arbitrary (extended) random variable ξ we set

$$\int_A \xi(\omega) P(\mathrm{d}\omega) := \int_A \xi^+(\omega) P(\mathrm{d}\omega) - \int_A \xi^-(\omega) P(\mathrm{d}\omega) ,$$

provided at least one of $\int_A \xi^+(\omega) P(\mathrm{d}\omega)$, $\int_A \xi^-(\omega) P(\mathrm{d}\omega)$ is finite; if both are infinite, $\int_A \xi(\omega) P(\mathrm{d}\omega)$ does not exist. A shorter alternative notation for $\int_A \xi(\omega) P(\mathrm{d}\omega)$ is $\int_A \xi \mathrm{d}P$, with $\int_\Omega \xi \mathrm{d}P$ abbreviated to $\int \xi \mathrm{d}P$. When the probability space $(\Omega, \mathcal{F}, P)$ is clear from the context, we will write $\mathbb{E}\xi$ for $\int \xi \mathrm{d}P$; $\mathbb{E}_P \xi$ is synonymous with $\int \xi \mathrm{d}P$. (Similarly, we will sometimes write $\mathbb{P}(A)$ or $\mathbb{P}_P(A)$ for $P(A)$.)

Let $\mathcal{G} \subseteq \mathcal{F}$ be a sub-σ-algebra of $\mathcal{F}$ and ξ be a nonnegative extended random variable. The conditional expectation $\mathbb{E}(\xi \mid \mathcal{G})$ of ξ w.r.t. to $\mathcal{G}$ is defined to be a $\mathcal{G}$-measurable extended random variable such that, for any $A \in \mathcal{G}$,

$$\int_A \xi(\omega) P(\mathrm{d}\omega) = \int_A \mathbb{E}(\xi \mid \mathcal{G})(\omega) P(\mathrm{d}\omega)$$

(any two $\mathcal{G}$-measurable extended random variables satisfying this condition coincide almost surely; they will be referred to as *versions* $\mathbb{E}(\xi \mid \mathcal{G})$). For general extended random variables ξ, define

$$\mathbb{E}(\xi \mid \mathcal{G}) := \mathbb{E}(\xi^+ \mid \mathcal{G}) - \mathbb{E}(\xi^- \mid \mathcal{G}) ;$$

this definition will be used only when

$$\min\left(\mathbb{E}(\xi^+ \mid \mathcal{G}), \mathbb{E}(\xi^- \mid \mathcal{G})\right) < \infty \quad \text{a.s.}$$

There may be many *versions* of $\mathbb{E}(\xi \mid \mathcal{G})$, but any two of them coincide almost surely.

In this book we use the following properties of conditional expectations (for proofs, see, e.g., [329, Sect. 2.7.4]):

1. If $\mathcal{G}$ is a sub-σ-algebra of $\mathcal{F}$, ξ and η are bounded random variables, and η is $\mathcal{G}$-measurable,

$$\mathbb{E}(\eta\xi \mid \mathcal{G}) = \eta \mathbb{E}(\xi \mid \mathcal{G})$$

   almost surely.
2. If $\mathcal{G}_1$ and $\mathcal{G}_2$ are sub-σ-algebras of $\mathcal{F}$, $\mathcal{G}_1 \subseteq \mathcal{G}_2 \subseteq \mathcal{F}$, and ξ is a random variable,

$$\mathbb{E}(\mathbb{E}(\xi \mid \mathcal{G}_2) \mid \mathcal{G}_1) = \mathbb{E}(\xi \mid \mathcal{G}_1)$$

   almost surely; in particular, for any sub-σ-algebra $\mathcal{G}$ of $\mathcal{F}$,

$$\mathbb{E}(\mathbb{E}(\xi \mid \mathcal{G})) = \mathbb{E}(\xi) .$$

For an event $A \in \mathcal{F}$ and a sub-σ-algebra $\mathcal{G} \subseteq \mathcal{F}$, the *conditional probability* $\mathbb{P}(A \mid \mathcal{G})$ is defined as the conditional expectation $\mathbb{E}(1_A \mid \mathcal{G})$ of A's indicator function. We sometimes write $\mathbb{E}(\xi \mid \eta)$ (resp. $\mathbb{P}(A \mid \eta)$), where η is a random element, to mean $\mathbb{E}(\xi \mid \mathcal{G})$ (resp. $\mathbb{P}(A \mid \mathcal{G})$) where $\mathcal{G}$ is the σ-algebra generated by η. Both $\mathbb{E}(\xi \mid \eta)$ and $\mathbb{P}(A \mid \eta)$ can be considered functions of η rather than ω: see [428, Sect. A3.2].

A.4 Markov Kernels and Regular Conditional Distributions

Let Ω and Z be two measurable spaces. A function $Q(\omega, A)$, usually written as $Q(A \mid \omega)$, where ω ranges over Ω and A over the measurable sets in Z, is called a *Markov kernel* if:

- as a function of A, $Q(A \mid \omega)$ is a probability distribution on Z, for each $\omega \in \Omega$;
- as a function of ω, $Q(A \mid \omega)$ is measurable, for each measurable $A \subseteq Z$.

We will say that Q is a Markov kernel of the type $\Omega \hookrightarrow Z$, using $\hookrightarrow$ to distinguish Markov kernels from functions of the type $\Omega \to Z$. Unlike functions, Markov kernels map $\omega \in \Omega$ to probability distributions on Z; we will sometimes use the notation $Q(\omega)$ for the probability distribution $A \mapsto Q(A \mid \omega)$ on Z and write $Q(\omega)(\mathrm{d}\zeta)$ for $Q(\mathrm{d}\zeta \mid \omega)$. If $E \subseteq \Omega$ is an event in Ω, the restriction $Q|_E$ is defined to be the same Markov kernel $Q(A \mid \omega)$ but with ω ranging over E.

Lemma A.1 *If $Q : \Omega \hookrightarrow Z$ is a Markov kernel and a bounded function f on $\Omega \times Z$ is measurable, the function $\omega \in \Omega \mapsto \int f(\omega, \zeta) Q(\mathrm{d}\zeta \mid \omega)$ is also measurable.*

Proof The statement of the lemma follows from the standard monotone-class argument (see, e.g., [428, Sect. 3.14]) and the fact that it holds for the indicator functions $f = 1_{A \times B}$ of the rectangles $A \times B$, where $A \subseteq \Omega$ and $B \subseteq Z$ are measurable. □

Let $(\Omega, \mathcal{F}, P)$ be a probability space. If we fix $\omega \in \Omega$, $\mathbb{P}(A \mid \mathcal{G})(\omega)$ will not necessarily be a probability distribution as a function of $A \in \mathcal{F}$. We cannot even guarantee that $\mathbb{P}(A \mid \mathcal{G})(\omega)$ will be a probability distribution for almost all ω. Consider, e.g., the property

$$\mathbb{P}(A \cup B \mid \mathcal{G})(\omega) = \mathbb{P}(A \mid \mathcal{G})(\omega) + \mathbb{P}(B \mid \mathcal{G})(\omega) ,$$

where A and B are disjoint events. For fixed A and B it will hold for almost all ω, but the set of exceptional ω for which it does not hold will depend on the pair (A, B), and the union of the exceptional sets over the potentially uncountable set of all pairs (A, B) is not guaranteed to have probability zero.

Let $\mathcal{G}_1$ and $\mathcal{G}_2$ be two sub-σ-algebras of $\mathcal{F}$. A Markov kernel

$$Q : (\Omega, \mathcal{G}_1) \hookrightarrow (\Omega, \mathcal{G}_2) \tag{A.2}$$

is called a *regular conditional probability* if for each $A \in \mathcal{G}_2$, $Q(A \mid \omega)$ as a function of ω is a version of the conditional probability $\mathbb{P}(A|\mathcal{G}_1)(\omega)$. There may be additional regularity properties that one might like to impose, such as

$$\mathbb{P}\{\omega : \forall A \in \mathcal{G}_1 \cap \mathcal{G}_2 : Q(A \mid \omega) = 1_A(\omega)\} = 1 ,$$

but they do not form part of the definition.

Regular conditional probabilities exist in wide generality. The two most standard results about their existence can be found in [292, Sect. II.89] and [329, Theorem 2.7.5].

The following lemma asserts that, as one would expect, conditional expectations can be computed by averaging over regular conditional probabilities. We will need it only for the case $\mathcal{G}_2 = \mathcal{F}$, but even the more general statement is very easy to prove (and so we will give a proof for this standard result, which, however, can rarely be found in modern textbooks).

Lemma A.2 *Let Q be a Markov kernel from $(\Omega, \mathcal{G}_1)$ to $(\Omega, \mathcal{G}_2)$, where $\mathcal{G}_1$ and $\mathcal{G}_2$ are sub-σ-algebras of $\mathcal{F}$. The Markov kernel Q is a regular conditional probability if and only if, for any bounded $\mathcal{G}_2$-measurable random variable ξ, $\omega_1 \in \Omega \mapsto \int \xi(\omega_2) Q(\mathrm{d}\omega_2 \mid \omega_1)$ is a version of $\mathbb{E}(\xi \mid \mathcal{G}_1)$.*

Proof We are required to prove that Q is a regular conditional probability if and only if

$$\int_E \int \xi(\omega_2) Q(\mathrm{d}\omega_2 \mid \omega_1) P(\mathrm{d}\omega_1) = \int_E \xi(\omega) P(\mathrm{d}\omega) \tag{A.3}$$

for all $\mathcal{G}_1$-measurable E and all bounded $\mathcal{G}_2$-measurable ξ.

If we take $\xi = 1_A$ for $A \in \mathcal{G}_2$, (A.3) will become

$$\int_E Q(A \mid \omega_1) P(\mathrm{d}\omega_1) = P(E \cap A) , \tag{A.4}$$

which means that $Q(A \mid \omega)$ is a version of the conditional probability of A given $\mathcal{G}_1$ and, therefore, that Q is a regular conditional probability.

Let us now assume that Q is a regular conditional probability; we are required to prove that (A.3) holds for all $\mathcal{G}_1$-measurable E and bounded $\mathcal{G}_2$-measurable ξ; fix such an E. We know that (A.3) holds for $\xi = 1_A$ with $A \in \mathcal{G}_2$ (see (A.4)); it remains to apply the standard monotone-class argument [428, Sect. 3.14], already used in the proof of Lemma A.1. □

If $\xi_1 : \Omega \to \Xi_1$ and $\xi_2 : \Omega \to \Xi_2$ are random elements, a *regular conditional distribution* of ξ_2 given ξ_1 is the Markov kernel $R : (\Omega, \mathcal{G}_1) \hookrightarrow \Xi_2$ defined by $R(\omega) := Q(\omega)\xi_2^{-1}$, $\omega \in \Omega$, where Q is a regular conditional probability (A.2), $\mathcal{G}_1$ is the σ-algebra on Ω generated by ξ_1, and $\mathcal{G}_2$ is the σ-algebra generated by ξ_2; a regular conditional distribution R exists if and only if a regular conditional probability (A.2) exists. We can again consider R to be a function of ξ_1 rather than of ω [428, Sect. A3.2].

A.5 Exchangeability

Let $\mathbf{Z}$ be a measurable space. We say that a probability distribution P on the measurable space $\mathbf{Z}^n$ of sequences of length n, where $n \in \{1, 2, \dots\}$, is *exchangeable* if

$$P(E) = P\left\{z_1, \dots, z_n : z_{\pi(1)} \dots z_{\pi(n)} \in E\right\}$$

for any measurable $E \subseteq \mathbf{Z}^n$ and any permutation π of the set $\{1, \dots, n\}$ (in words: if the distribution of the sequence $z_1 \dots z_n$ is invariant under any permutation of the indices). We say that a probability distribution P on the power measurable space $\mathbf{Z}^\infty$ is *exchangeable* if the marginal distribution P_n of P on $\mathbf{Z}^n$ (defined by

$$P_n(E) := P\left\{(z_1, z_2, \dots) \in \mathbf{Z}^\infty : z_1 \dots z_n \in E\right\}$$

for all events $E \subseteq \mathbf{Z}^n$) is exchangeable for each $n = 1, 2, \dots$ (in words: if the distribution of the sequence $z_1 z_2 \dots$ is invariant under any permutation of a finite number of the indices). This is equivalent to the definition given in Chap. 2 (see (2.2)). The *exchangeability model* is defined to be the set of all exchangeable probability distributions on $\mathbf{Z}^\infty$.

A.5.1 De Finetti's Representation Theorem

The question of relation between randomness and exchangeability models is a popular topic in the foundations of Bayesian statistics (see, e.g., [307, Chap. 1]). Every power probability distribution Q^∞ on $\mathbf{Z}^\infty$ is exchangeable, and under a weak regularity condition every exchangeable probability distribution on $\mathbf{Z}^\infty$ is a mixture of power distributions; this is de Finetti's representation theorem (see, e.g., [307, Theorem 1.49]).

De Finetti's Theorem *Suppose* $\mathbf{Z}$ *is a standard Borel space. A probability distribution* P *on* $\mathbf{Z}^\infty$ *is exchangeable if and only if* P *is a mixture of power distributions:*

$$P = \int Q^\infty \mu(\mathrm{d}Q)$$

for some probability distribution μ *on the space* $\mathbf{P}(\mathbf{Z})$ *of all probability distributions on* $\mathbf{Z}$ *(equipped with the smallest* σ*-algebra such that all evaluation functions* $Q \mapsto Q(E)$ *are measurable,* E *ranging over the events in* $\mathbf{Z}$*).*

A.5.2 Conditional Probabilities Given a Bag

We will need the following simple result about the existence of a regular conditional probability for exchangeable distributions. We define the *bag σ-algebra* on $\mathbf{Z}^n$ as the family of events $E \subseteq \mathbf{Z}^n$ such that

$$(z_1, \ldots, z_n) \in E \Longrightarrow (z_{\pi(1)}, \ldots, z_{\pi(n)}) \in E$$

for all permutations π of the set $\{1, \ldots, n\}$.

Lemma A.3 *Let P be an exchangeable distribution on $\mathbf{Z}^n$ for some $n \in \mathbb{N}$ and let $\mathcal{G}$ be the bag σ-algebra on $\mathbf{Z}^n$. The Markov kernel C which maps each $\omega = (z_1, \ldots, z_n) \in \mathbf{Z}^n$ to the probability distribution $C(\omega)$ on $\mathbf{Z}^n$ concentrated on the set of all permutations $(z_{\pi(1)}, \ldots, z_{\pi(n)})$ and assigning the same probability $1/n!$ to each of these permutations will be a regular conditional probability w.r. to $\mathcal{G}$ in the probability space $(\mathbf{Z}^n, P)$.*

Proof For any bounded random variable ξ on $\mathbf{Z}^n$ set

$$\overline{\xi}(z_1, \ldots, z_n) := \frac{1}{n!} \sum_{\pi} \xi(z_{\pi(1)}, \ldots, z_{\pi(n)}) ,$$

the sum being over all $n!$ permutations π of $\{1, \ldots, n\}$. By Lemma A.2, we are required to prove that

$$\int_E \left(\overline{\xi}(z_1, \ldots, z_n) - \xi(z_1, \ldots, z_n) \right) P(\mathrm{d}z_1, \ldots, \mathrm{d}z_n) = 0 \tag{A.5}$$

for any set $E \subseteq \mathbf{Z}^n$ in the bag σ-algebra.

First we notice that, if Ω is a measurable space and $G : \Omega \to \Omega$ is a bijection measurable in both directions, then for every measurable function $f : \Omega \to \mathbb{R}$, every measurable set $E \subseteq \Omega$, and every probability distribution P on Ω,

$$\int_E f dP = \int_{E'} f' \mathrm{d}P' ,$$

where the set E', function f', and probability distribution P' are defined by

$$E' := G^{-1}(E), \quad f'(\omega) := f(G(\omega)), \quad P'(A) := P(G(A)) .$$

Applying this to $\Omega := \mathbf{Z}^n$, $G(z_1, \ldots, z_n) := (z_{\pi(1)}, \ldots, z_{\pi(n)})$, where π is a permutation, and $f := \xi$, we obtain

$$\int_E \left(\xi(z_{\pi(1)}, \ldots, z_{\pi(n)}) - \xi(z_1, \ldots, z_n)\right) P(\mathrm{d}z_1, \ldots, \mathrm{d}z_n) = 0$$

(remember that $E' = E$ and $P' = P$). Finally, averaging over all π gives (A.5). $\square$

A.6 Theory of Martingales

Let $(\Omega, \mathcal{F}, P)$ be a probability space. It will be convenient to use the adjectives "increasing" and "decreasing" in the extended sense (as we do throughout the book); in particular, a sequence of σ-algebras $\mathcal{F}_n$ is *increasing* if $\mathcal{F}_1 \subseteq \mathcal{F}_2 \subseteq \ldots$ and a sequence of random variables ξ_n is *increasing* if $\xi_1 \leq \xi_2 \leq \ldots$ ($\xi \leq \eta$ can be defined as "$\xi(\omega) \leq \eta(\omega)$ for all ω" or as "$\xi \leq \eta$ almost surely"; it does not matter which definition is used for the mathematical results stated in this appendix).

A *filtration* is an increasing sequence of sub-σ-algebras $\mathcal{F}_0 \subseteq \mathcal{F}_1 \subseteq \mathcal{F}_2 \subseteq \cdots \subseteq \mathcal{F}$; when we say that $\mathcal{F}_1 \subseteq \mathcal{F}_2 \subseteq \cdots \subseteq \mathcal{F}$ is a filtration, we always mean that it is complemented by $\mathcal{F}_0 := \{\emptyset, \Omega\}$. Let $\mathcal{F}_\infty$ be the smallest σ-algebra containing all $\mathcal{F}_n$. We say that a sequence of random elements ξ_n, where $n = 1, 2, \ldots$ or $n = 0, 1, 2, \ldots$, is *adapted* if each ξ_n is $\mathcal{F}_n$-measurable. A sequence of random elements $\xi_1, \xi_2, \ldots$ is *predictable* if each ξ_n is $\mathcal{F}_{n-1}$-measurable.

We say that an adapted sequence of random variables $\xi_0, \xi_1, \ldots$ is a *martingale* if $\mathbb{E}(\xi_n \mid \mathcal{F}_{n-1}) = \xi_{n-1}$ for all $n = 1, 2, \ldots$, and we say that an adapted sequence of random variables $\xi_1, \xi_2, \ldots$ is a *martingale difference* if $\mathbb{E}(\xi_n \mid \mathcal{F}_{n-1}) = 0$ for all $n = 1, 2, \ldots$. A very useful generalization of the notion of a martingale is that of a *supermartingale*: this is an adapted sequence of random variables $\xi_0, \xi_1, \ldots$ such that $\mathbb{E}(\xi_n \mid \mathcal{F}_{n-1}) \leq \xi_{n-1}$ for all $n = 1, 2, \ldots$.

We occasionally use the notion of a *submartingale*, i.e., an adapted sequence of random variables $\xi_0, \xi_1, \ldots$ such that $\mathbb{E}(\xi_n \mid \mathcal{F}_{n-1}) \geq \xi_{n-1}$ for all $n = 1, 2, \ldots$. A *compensator* of such a submartingale is a nonnegative increasing predictable sequence $\eta_1, \eta_2, \ldots$ such that $\xi_n - \eta_n$, $n = 0, 1, \ldots$, is a martingale, where $\eta_0 := 0$. (Any two compensators coincide almost surely.)

If there is no a priori fixed filtration on the given probability space, we say that a sequence of random variables $\xi_0, \xi_1, \ldots$ is a *martingale* (resp. *supermartingale*) if it is a martingale (resp. supermartingale) w.r. to the filtration $\mathcal{F}_n$, $n = 0, 1, \ldots$, such that each $\mathcal{F}_n$ is generated by the random variables $\xi_1, \ldots, \xi_n$ (in particular, $\mathcal{F}_0 := \{\emptyset, \Omega\}$ and ξ_0 is a constant). This rather old-fashioned notion of a martingale is used in Chap. 8.

Analogous definitions can be given for a finite filtration, $\mathcal{F}_1, \ldots, \mathcal{F}_N$; in this case, martingales, supermartingales, and submartingales are finite sequences $\xi_0, \xi_1, \ldots, \xi_N$, and predictable sequences and martingale differences are finite sequences $\xi_1, \ldots, \xi_N$.

A.6.1 Basic Results

The following is a simple version of Ville's inequality; it holds for both finite and infinite filtrations.

Ville's Inequality *If (ξ_n) is a nonnegative supermartingale w.r. to a filtration $(\mathcal{F}_n)$ with $\mathcal{F}_0 = \{\emptyset, \Omega\}$ and c is a positive constant, then*

$$\mathbb{P}\left\{\sup_n \xi_n \geq c\right\} \leq \frac{\xi_0}{c}.$$

In stating the following simple but useful result we will use the logical notation $A \Leftrightarrow B$ for the symmetric difference of events A and B.

Borel–Cantelli–Lévy Lemma *If $\mathcal{F}_0, \mathcal{F}_1, \ldots$ is a filtration and $A_n \in \mathcal{F}_n$ for $n = 1, 2, \ldots$, then*

$$\left(\sum_{n=1}^{\infty} \mathbb{P}(A_n \mid \mathcal{F}_{n-1}) < \infty\right) \Longleftrightarrow \left(\sum_{n=1}^{\infty} 1_{A_n} < \infty\right)$$

almost surely.

The Borel–Cantelli–Lévy lemma generalizes the part of the classical Borel–Cantelli lemma that deals with sequences of independent events A_n (the independence of the events A_n means that the σ-algebras in the sequence

$$\mathcal{F}_n := \{\emptyset, A_n, \Omega \setminus A_n, \Omega\}$$

are independent).

Borel–Cantelli Lemma *Let $A_1, A_2, \ldots$ be a sequence of events.*

- *If $\sum_n \mathbb{P}(A_n) < \infty$,*

$$\sum_n 1_{A_n} < \infty \quad a.s.$$

- *If the events $A_1, A_2, \ldots$ are independent, and $\sum_n \mathbb{P}(A_n) = \infty$,*

$$\sum_n 1_{A_n} = \infty \quad a.s.$$

A.6.2 Limit Theorems

In this subsection we will state some fundamental limit theorems of the theory of martingales (for proofs, see [330], [317], and [320]).

Martingale Strong Law of Large Numbers *Let* (ξ_n) *be a martingale difference w.r. to a filtration* $\mathcal{F}_0, \mathcal{F}_1, \ldots$ *and let* (A_n) *be an increasing predictable sequence w.r. to the same filtration with* $A_1 > 0$ *and* $A_\infty = \infty$ *a.s. If*

$$\sum_{i=1}^{\infty} \frac{\mathbb{E}(\xi_i^2 \mid \mathcal{F}_{i-1})}{A_i^2} < \infty \quad a.s. \,,$$

then

$$\frac{1}{A_n} \sum_{i=1}^{n} \xi_i \to 0 \quad (n \to \infty) \quad a.s.$$

These are important special cases.

Kolmogorov's Strong Law of Large Numbers *Suppose* $\xi_1, \xi_2, \ldots$ *is a sequence of independent zero-mean random variables and* $A_1, A_2, \ldots$ *is an increasing sequence of positive numbers such that* $A_n \to \infty$ $(n \to \infty)$. *If*

$$\sum_{i=1}^{\infty} \frac{\mathbb{E}(\xi_i^2)}{A_i^2} < \infty \,, \tag{A.6}$$

then

$$\frac{1}{A_n} \sum_{i=1}^{n} \xi_i \to 0 \quad (n \to \infty) \quad a.s.$$

Borel's Strong Law of Large Numbers *If* $\xi_1, \xi_2, \ldots$ *is a sequence of independent binary (i.e., taking values in* $\{0, 1\}$*) random variables with* $\mathbb{E}(\xi_n) = p$, $n = 1, 2, \ldots$, *then*

$$\frac{1}{n} \sum_{i=1}^{n} \xi_i \to p \quad (n \to \infty) \quad a.s.$$

The following result is a martingale version (due to Stout [348]) of Kolmogorov's law of the iterated logarithm (it uses the usual logical convention that the event $A \Rightarrow B$ is the union of B and the complement of A).

Martingale Law of the Iterated Logarithm *Let (ξ_n) be a martingale difference w.r. to a filtration $\mathcal{F}_0, \mathcal{F}_1, \ldots$ and let (A_n), (c_n) be increasing positive predictable sequences w.r. to this filtration. If $|\xi_n| \le c_n$ for all n, then, almost surely,*

$$\left(A_n \to \infty \;\&\; c_n = o\left(\sqrt{\frac{A_n}{\ln \ln A_n}}\right)\right) \Longrightarrow \limsup_{n\to\infty} \frac{\sum_{i=1}^n \xi_i}{\sqrt{2A_n \ln \ln A_n}} = 1 .$$

A.6.3 Hoeffding's Inequality

Hoeffding's Inequality *Let $\mathcal{F}_0, \ldots, \mathcal{F}_n$ be a filtration. For any deterministic sequence $c_1, \ldots, c_n$ of positive numbers, any predictable sequence $v_1, \ldots, v_n$ w.r. to $(\mathcal{F}_i)$, any martingale difference $\xi_1, \ldots, \xi_n$ w.r. to $(\mathcal{F}_i)$ such that $|\xi_i - v_i| \le c_i$, $i = 1, \ldots, n$, and any $\epsilon > 0$,*

$$\mathbb{P}\left\{\frac{1}{n}\sum_{i=1}^n \xi_i \ge \epsilon\right\} \le \exp\left(-\frac{\epsilon^2 n^2}{2\sum_{i=1}^n c_i^2}\right)$$

and

$$\mathbb{P}\left\{\frac{1}{n}\sum_{i=1}^n \xi_i \le -\epsilon\right\} \le \exp\left(-\frac{\epsilon^2 n^2}{2\sum_{i=1}^n c_i^2}\right) .$$

The proof of this result will easily follow from the following one-step inequality.

Lemma A.4 *Let ξ be a random variable such that $\mathbb{E}\xi = 0$ and $a \le \xi \le b$ for constants a and b and let $s > 0$. Then*

$$\mathbb{E}\left(\mathrm{e}^{s\xi}\right) \le \mathrm{e}^{s^2(b-a)^2/8} .$$

Proof Assume, without loss of generality, that $s = 1$ (the general case follows from this special case by replacing ξ, a, b with $s\xi, sa, sb$, respectively). Since the function $x \mapsto \mathrm{e}^x$ is convex, we can assume, without loss of generality, that the distribution of ξ is concentrated on $\{a, b\}$. Since $\mathbb{E}\xi = 0$, the mass at a is $b/(b-a)$ and the mass at b is $-a/(b-a)$ (remember that $a \le 0$ and $b \ge 0$; we only consider the nontrivial case $a \ne b$); therefore, we are only required to prove that

$$\frac{b}{b-a}\mathrm{e}^a - \frac{a}{b-a}\mathrm{e}^b \le \mathrm{e}^{(b-a)^2/8} . \tag{A.7}$$

(The formal version of this argument is that, by the convexity of the exponential function,

$$e^x \le \frac{x-a}{b-a}e^b + \frac{b-x}{b-a}e^a$$

for $x \in [a, b]$; it remains to find the expectations of the two sides.)

Introducing the notation

$$u := b - a, \quad p := -\frac{a}{b-a}, \quad 1 - p := \frac{b}{b-a},$$

we can rewrite (A.7) as

$$pe^{(1-p)u} + (1-p)e^{-pu} \le e^{u^2/8}$$

or, equivalently,

$$\phi(u) := -pu + \ln(1 - p + pe^u) \le u^2/8 .$$

It remains to notice that $\phi(0) = 0$, $\phi'(0) = 0$ and $\phi''(u) \le 1/4$; the last inequality follows from the fact that the geometric mean never exceeds the arithmetic mean:

$$\phi''(u) = \left(-p + \frac{p}{p + (1-p)e^{-u}}\right)' = \frac{p(1-p)e^{-u}}{(p + (1-p)e^{-u})^2} \le \frac{1}{4} .$$

□

Lemma A.4 shows that

$$\exp\left(s\sum_{i=1}^{n}\xi_i - \frac{1}{8}s^2\sum_{i=1}^{n}(b_i - a_i)^2\right), \quad n = 0, 1, 2, \ldots ,$$

where a_i and b_i are predictable sequences and $\xi_i \in [a_i, b_i]$ is a martingale difference, is a supermartingale. In conjunction with Ville's inequality it implies Hoeffding's inequality (s should be chosen optimally for the given ϵ and $c_1, \ldots, c_n$).

A.7 Bibliographical Remarks

Kolmogorov's axioms are proposed in his book [199]. The origins and legacy of this book are discussed in [318].

A.7.1 Conditional Probabilities

The classical definition of the conditional probability of an event A given another event B is $\mathbb{P}(A \mid B) := \mathbb{P}(A \cap B)/\mathbb{P}(B)$, but it only works if $\mathbb{P}(B) > 0$. One of the main contributions of Kolmogorov's *Grundbegriffe* [199] was to extend this definition to the case where $\mathbb{P}(B) = 0$ is allowed; the price was that the definition had to be given for all B in a partition of Ω simultaneously, and the definition made sense not always but only almost surely. From the modern point of view, presented in Sect. A.3, Kolmogorov defined $\mathbb{P}(A \mid \mathcal{G})$ only for $\mathcal{G}$ obtained from the original σ-algebra $\mathcal{F}$ and a partition: an event $A \in \mathcal{F}$ is included in $\mathcal{G}$ if and only if it is the union of some elements of the partition; his definition, however, extends trivially to the standard definition (given above) applicable to an arbitrary σ-algebra $\mathcal{G} \subseteq \mathcal{F}$. (For a detailed discussion, see [318].)

One of the difficulties of Kolmogorov's definition is demonstrated by Dieudonné's [86] famous example, in which the conditional probabilities $\mathbb{P}(A \mid \mathcal{G})(\omega)$ do not have versions that would form a probability distribution as a function of A for almost all ω. This prompted the development of the theory of regular conditional probabilities, in which conditional probabilities $\mathbb{P}(A \mid B)$ are defined simultaneously for all events $A \in \mathcal{G}_2$ and $B \in \mathcal{G}_1$ ranging over sub-σ-algebras $\mathcal{G}_2$ and $\mathcal{G}_1$ of $\mathcal{F}$.

A.7.2 Martingales

A mathematical notion of a martingale was introduced explicitly and used for the purposes of the foundations of probability by Ville [373]; earlier, it had been used by several people, including Lévy and Kolmogorov, without an explicit definition. What we call martingales and supermartingales are usually called generalized martingales and supermartingales, respectively; see [330, Theorem 7.1.1] for equivalent definitions. The simple version of Ville's inequality given in this appendix was proved by Ville [373, p. 100]; more sophisticated versions appeared in Doob's [87]. For further historical details and references, see [317, 320].

A.7.3 Hoeffding's Inequality

Hoeffding's inequality was first published in [164]; its martingale version is sometimes referred to as the Hoeffding–Azuma inequality (after Azuma's paper [12]), although already [164, p. 18] contains the martingale extension. Our exposition follows [82].

Appendix B
Datasets and Computations

In this appendix we will describe three of the datasets used in the book. A separate section, Sect. B.4, is devoted to the important problem of "normalization" of the objects; some other technical issues, including details of randomization, are discussed in Sect. B.5.

B.1 USPS Dataset

The USPS (US Postal Service) dataset is a standard benchmark for testing classification algorithms. It consists of 7291 training examples and 2007 test examples collected from real-life US zip codes (mail passing through the Buffalo, NY, post office). In our experiments, we always merge the training set and the test set, in this order, obtaining what we call the *full USPS dataset* (or just USPS dataset). After that we often apply a random permutation.

Each example consists of an image (16×16 matrix with entries in the interval $(-1, 1)$ that describe the brightness of individual pixels) and its label (0 to 9).

It is well known that the USPS dataset is heterogeneous; in particular, the training and test sets seem to have different distributions. For example, the 1-nearest neighbour algorithm makes 5.7% errors on the USPS test set but only 2.3% on a test set of the same size randomly chosen from the full dataset. (See, e.g., [122].)

The USPS dataset has been used in hundreds of papers and books; see, e.g., [219] and [366]. Among the results reported in the literature for the error rate on the test set are: 2.5% for humans, 5.1% for the five-layer neural network LeNet 1, 4.0% for a polynomial support vector machine (with the polynomial kernel of degree 3). A very good result of 2.7% was obtained by Simard et al. [334] without using any learning methods. Since we are usually interested in results for a randomly permuted USPS dataset, the error rates that we obtain are not comparable with the error rates on the test set alone.

V. Vovk et al., *Algorithmic Learning in a Random World*,
https://doi.org/10.1007/978-3-031-06649-8

In our experiments we used MATLAB for the first edition, but for the new figures produced for the second edition, we used Python with the `matplotlib` library.

B.2 Wine Quality Dataset

In Sect. 8.4 we use the version of the Wine Quality dataset available in the UCI Machine Learning repository [91]. There are eleven attributes (such as residual sugar and alcohol) that may be useful for predicting the label. The label is mostly between 4 and 7. In our experiments described in Sect. 8.4 we use `scikit-learn` [274] with the default values of all parameters, and for drawing pictures we use `matplotlib` (with the default `matplotlib` boxplots).

B.3 Bank Marketing Dataset

In our experiments in Chap. 10 we use the full version of the dataset as given in the `openml.org` repository, which is easier to process in `scikit-learn` than the UCI version. The dataset consists of 45,211 examples representing telemarketing calls for selling long-term deposits offered by a Portuguese retail bank, with data collected from 2008 to 2013.

The labels are 1 and 2, which we encode as 0 and 1, respectively (and in Chap. 10 we never mention the original labels). Label 1 (after the transformation) indicates a successful sale, and such examples comprise only 12% of all labels. Examples of attributes are the client's age, the code for the job, and the marital status.

In our experiments reported in Sects. 10.3.3 and 10.5.2, we use random forest and other popular prediction algorithms as implemented in `scikit-learn` [274] with the default values of all parameters. We chose random forest for representing in our figures since it consistently produces good results in our experiments.

B.4 Normalization

The performance of many machine-learning algorithms improves greatly if the objects are pre-processed. Suppose we are given a training set $(x_1, y_1), \ldots, (x_{n-1}, y_{n-1})$, a test object x_n, and our goal is to predict the label y_n; each object x_i is a vector in $\mathbb{R}^K$ and its components are denoted $x_{i,1}, \ldots, x_{i,K}$. (In the case of the USPS dataset, $K = 256$, in the case of the Wine Quality dataset, $K = 11$, and in the case of the Bank Marketing dataset, $K = 20$.)

The attributes of the Wine Quality and Bank Marketing datasets have different orders of magnitude simply because of their nature: e.g., for the latter, some attributes are binary (such as marital status) and some are numeric (such as age).

Therefore, it often leads to better results if we somehow normalize the columns of the data matrix $x_{i,k}$. In our experiments with the Wine Quality and Bank Marketing datasets, we apply the linear transformation

$$x_{i,k} \mapsto x'_{i,k} := a_k + b_k x_{i,k}$$

that makes the mean and standard deviation of each of the attributes equal to 0 and 1, respectively, over the training set.

The kind of normalization required for the USPS dataset is different: the attributes, being the intensities of individual pixels, are directly comparable, but the brightness and contrast of different hand-written images seems to be irrelevant to their classification. Therefore, it appears useful to normalize the rows of the data matrix $x_{i,k}$. In many of our experiments we apply the linear transformation

$$x_{i,k} \mapsto x'_{i,k} := a_i + b_i x_{i,k} ,$$

where a_i and b_i are chosen to make the mean and standard deviation of each row equal to 0 and 1, respectively. Since the pre-processing of each example is done independently of the other examples, even in online experiments we can do all pre-processing in advance. No pre-processing is done in Chap. 8 on change detection: all experiments are run on the original USPS dataset. In the experiments reported in the other chapters this dataset is randomly permuted and all the objects are normalized.

The normalization procedure that we have just described may fail if the standard deviation of a column or row is zero (because of that column or row being constant), but such cases never happen in our experiments.

B.5 Randomization and Reshuffling

Most of our experimental results involve randomization: our algorithms may be randomized (e.g., smoothed conformal predictors are), or we might "reshuffle" a dataset to make sure it conforms to an interesting assumption, such as exchangeability (when reshuffling is done, this is always mentioned explicitly). The main features of the plots in this book are not significantly affected by the details of randomization.

For reproducibility, we follow a simple policy of the choice of the initial seed for the pseudorandom number generator (which we abbreviate to "random number generator"). By default, all reported results in the first edition were obtained by setting the initial seed of the random number generator (in MATLAB) to 0. For the results added in the second edition (except for Sect. 12.5, still using MATLAB), the default initial seed is 2022 (in NumPy). Since the main purpose of computational experiments in this book is to illustrate theoretical results, we rarely use other initial seeds. (When we do need several seeds, as in the left panel of Fig. 8.13, we choose consecutive seeds starting from 2022.)

The most basic case of reshuffling is a random permutation of the dataset performed to make sure that the assumption of exchangeability is satisfied. We often do it for the USPS dataset.

We have been saying "reshuffle" because the examples in the original dataset are already in a somewhat random order, but Reality's attempt at shuffling is often only half-hearted (see Chap. 8 for results about the USPS dataset), and in the main body of the book we sometimes say "shuffle", especially when we are trying to achieve conformity with an assumption that is stronger than exchangeability.

After the general notion of an online compression model is introduced in Sect. 11.2, the idea of shuffling also becomes more general: given a dataset $z_1, \dots, z_n$, we can find its summary $\sigma := t(z_1, \dots, z_n)$ and then draw another data sequence $z'_1, \dots, z'_n$ from the conditional distribution $P_n(\mathrm{d}z_1, \dots, \mathrm{d}z_n \mid \sigma)$. If the model is very specific, shuffling is much more intrusive than a random permutation is, and might be better described as generation of a new dataset sharing some characteristics with the original dataset. In Sect. 12.3.6 we explicitly describe an efficient procedure of shuffling for junction-tree models, an important class of online compression models. However, we never use shuffling in this more general sense in our computational experiments in this book.

B.6 Bibliographical Remarks

In our description of the USPS dataset we partly followed [219]. The original papers about the Wine Quality and Bank Marketing datasets are [56] and [253], respectively.

Appendix C
FAQ

In this appendix we give our answers to several questions we have been asked by our colleagues and students. We will discuss prediction under unconstrained randomness unless a different model is explicitly mentioned.

C.1 Unusual Features of Conformal Prediction

Isn't your Proposition 2.4 too strong to be true? It is generally believed that to make categorical assertions about error probabilities some Bayes-type assumptions are needed and that the assumption of unconstrained randomness is not sufficient. For example, in the theory of PAC learning an error probability ϵ *is only asserted with some probability* $1 - \delta$.

It should be remembered that Proposition 2.4 does not assert that the probability of error, $\mathrm{err}_n = 1$, is ϵ conditionally on knowing the whole past (2.5); it is only asserted that it is ϵ unconditionally and conditionally on knowing $\mathrm{err}_1, \ldots, \mathrm{err}_{n-1}$. (Actually, it is quite obvious that the probability of error is often not equal to ϵ if the whole past is known: if the prediction set is empty, the conditional probability of error is 1; to balance this, the conditional probability that a nonempty prediction set is wrong will tend to be less than ϵ.)

How is it possible to achieve probability of error exactly ϵ *in the problem of classification? For example, in the binary case there are only two possible labels, and you cannot expect that one of these labels will have probability exactly* ϵ.

It is impossible to achieve the conditional probability of error equal to ϵ given the observed examples, but it is the unconditional probability of error that equals ϵ. Therefore, it implicitly involves averaging over different data sequences, and this gives us the leeway needed to obtain a probability precisely equal to ϵ.

V. Vovk et al., *Algorithmic Learning in a Random World*,
https://doi.org/10.1007/978-3-031-06649-8

Suppose the prediction set is empty at the chosen significance level. Does it mean that the result of conformal prediction is useless in this case?

Even if you are interested in only one significance level, say ϵ, the empty prediction still carries some information: you know that the object whose label you are predicting is unusual (in the long run the frequency of seeing such unusual objects is at most ϵ). Are you sure there was no mistake in recording the object? Do you still believe in the exchangeability assumption? If the answer to these questions is "yes" and you would still like to have a nonempty prediction, you have no choice but to look at what happens at the other significance levels. (As clear from Chap. 1, we share the standard view that it is never wise to concentrate on just one significance level.) Look at the smallest significance level ϵ'' at which the prediction set is empty and at the smallest significance level ϵ' at which the prediction set is a singleton. (Cf. the definitions of confidence and credibility in (3.53) and (3.54).) If ϵ' is small and the difference between ϵ' and ϵ'' is significant, you can be fairly sure that the singular prediction set at the significance level $(\epsilon' + \epsilon'')/2$ will be correct.

Don't Theorem 2.2 and Proposition 2.4 contradict each other? As you explain in Sect. 2.1.4, if we extend each object x_n by adding to it the random number τ_n, smoothed conformal predictors become confidence predictors.

Theorem 2.2 does not make any assumptions on the distribution of each object, whereas Proposition 2.4 assumes that the component τ_n of the nth extended object is distributed uniformly on [0, 1].

C.2 Independence of Errors

You say that, in the offline mode of prediction, errors made by a conformal predictor on different test examples are not independent, since predictions are computed from the same training set. However, given the randomness assumption, how is it possible for the errors to be dependent? Intuitively it seems the errors should still be independent (although not necessarily committed with a probability exactly equal to the significance level ϵ).

When we say that the errors are not independent, we mean their unconditional independence. Conditionally on the training set, the errors are indeed independent. Here the situation is very different from the online mode of prediction, where errors are unconditionally independent but the notion of conditional independence given the training set becomes meaningless. A possible intuitive explanation why errors on different test examples are not independent unconditionally may go as follows: when you observe a lot of errors on part of the test set, your expectation of an error on unseen test examples increases (perhaps you are observing so many errors because you were unlucky with your training set).

As far as I understand, one of the most important results in conformal prediction is that the p-values for a random sequence of examples $z_1, z_2, \dots$ are mutually independent in the online framework. This property follows from Theorem 11.1 in the case of smoothed transducers. But what about the case of the deterministic transducer producing conservative p-values? Does it follow from Theorem 11.1 that the conservative p-values are also independent?

Conservative p-values are no longer independent; however, this is not so important in practice since their conservative validity still shows in empirical frequencies of error: this follows from their being dominated by independent random variables, as discussed in Chap. 2 (see the discussion following (2.11)). Let us give an example of a situation where there is no independence under the randomness assumption. Suppose the example space is $\mathbf{Z} = \{0, 1\}$. Therefore, the data-generating distribution produces examples independently from a Bernoulli distribution; the probability of 1 will be assumed to be different from 0 and 1. The nonconformity score of an example is defined to be that example itself (i.e., the nonconformity score of 1 is always 1 and the nonconformity score of 0 is always 0). Let $n \geq 2$ be an integer. Conditionally on knowing that the $(2n + 1)$th p-value is $(n + 1)/(2n + 1)$ (which implies that there are exactly n 1s among the first $2n$ examples), we have:

- The $(2n)$th p-value is $1/2$ with probability $1/2$ (this happens when the $(2n)$th example is 1) and 1 with probability $1/2$ (this happens when the $(2n)$th example is 0).
- The $(2n - 1)$th p-value given the $(2n)$th and $(2n + 1)$th p-values is

$$\begin{cases} n/(2n-1) & \text{with probability } n/(2n-1) \\ 1 & \text{with probability } (n-1)/(2n-1) \end{cases}$$

  if the $(2n)$th p-value is 1 and is

$$\begin{cases} (n-1)/(2n-1) & \text{with probability } (n-1)/(2n-1) \\ 1 & \text{with probability } n/(2n-1) \end{cases}$$

  if the $(2n)$th p-value is $1/2$.

We can see that the distribution of the $(2n - 1)$th p-value given the $(2n)$th and $(2n + 1)$th p-values depends on the $(2n)$th p-value. So there is no independence.

C.3 Ways of Applying Conformal-Type Predictors

What should be the size of the calibration set in inductive conformal prediction?

This obviously depends on the size of the training set (you do not want to have a tiny proper training set). But if your training set is large enough, the rule of thumb is to include at least $10/\epsilon$ training examples in the calibration set, where ϵ is your significance level (or your smallest significance level if you have several values of ϵ in mind, which is usually a good idea). This rule of thumb ensures that the boundaries of the prediction set are determined by a group of examples of size at least 10 (and so are not unduly affected by randomness).

C.4 Conformal Prediction vs Standard Methods

In your approach to classification, the simplest conditionally proper criterion of efficiency (see Table 3.1) is N (the average size of the prediction sets, given by (3.2)). You prove that Err_n^ϵ *grows as* $n\epsilon$ *(where* ϵ *is your chosen significance level) plus random noise and observe that in experiments the average size of the prediction sets is usually small. In the standard approach (e.g., in the PAC theory) one trivially has the average size of the prediction sets of 1 and observes that in experiments* Err_n^ϵ *is usually small. There is a complete symmetry and you cannot claim that your approach is better.*

This symmetry is superficial. Imagine that we are given a new object x_n having observed examples (x_i, y_i), $i = 1, \dots, n-1$. Suppose that for a small significance level ϵ a conformal predictor outputs a one-element prediction set $\{y\}$ and the standard approach outputs the simple prediction y. We can see that the prediction set $\{y\}$ is a singleton, and we know that the predictor has a small (equal to ϵ) probability of error. This gives us more information than the simple prediction y does: we knew in advance a simple prediction was going to be a singleton, and no reliable inference about the probability of error can be drawn from the smallness of the number of errors so far. To use the language of the theory of martingales (see Sect. A.6), the main asymmetry between Err_n^ϵ and the average size of the prediction sets is that the size $\left|\Gamma_n^\epsilon\right|$ of the prediction set at step n is a predictable quantity, whereas making an error err_n^ϵ at step n is not.

Could you summarize the main differences between conformal prediction and the standard approach to prediction?

Conformal predictors implement transductive rather than inductive learning. The basic validity result about conformal predictors is proved in the online rather than offline learning protocol. To state our results in the simplest and most general form we use online compression rather than traditional statistical modelling. Table C.1

Table C.1 Three dichotomies for hedged prediction

Inductive	Transductive
Offline	Online
Statistical modelling	Online compression modelling

shows these three differences; conformal prediction is mostly concerned with the right-hand column and traditional machine learning is mostly concerned with the left-hand column.

Cross-validation and conformal prediction have similar goals: both answer the question "How confident can we be in our prediction?" What's the difference?

These are some differences:

- Cross-validation evaluates a prediction algorithm *en masse*, whereas conformal prediction tries to say something about a given test object.
- Unlike conformal prediction, cross-validation does not have any simple properties of validity (its properties of validity are complicated and/or asymptotic).
- Cross-validation might be easier to interpret; it is still the standard method.

In the main body of the book we discuss a related method, the hold-out estimate.

Suppose I have a plausible online compression model M but know little about the set $\mathcal{P}$ of probability distributions on $\mathbf{Z}^\infty$ that agree with M; in particular, $\mathcal{P}$ may turn out to be empty or a singleton. Should I be worried about this?

In our opinion, in practical applications you can safely ignore the foundational questions such as whether $\mathcal{P}$ is rich enough. You know that for each finite horizon N there are plenty of probability distributions on $\mathbf{Z}^N$ that agree with M, and you are never going to reach infinity.

In Chaps. 2–4 you show how one can use standard machine-learning methods to devise nonconformity measures. Are there any formal connections between those standard methods and conformal predictors based on those methods?

We are not aware of any formal connections that always hold; it is often true, however, that the simple prediction produced by a machine-learning method will belong to the prediction set produced by the corresponding conformal predictor, unless that prediction set is empty.

In Fig. 13.1 the Bayesian model would become comparable with conformal prediction if you allowed the noise variance σ^2 to vary and put a prior on it. Isn't your comparison unfair?

Our point is that, no matter how general your Bayesian model is, it will sometimes be wrong. When you know how exactly it is violated, in hindsight you can always explain how your model could be improved to make this violation impossible. But after that the new model can be violated in new ways (e.g., the noise can be heavy-tailed). There are no universal priors (see, e.g., [267]).

References

For the authors who used different forms of their name in different publications, we use their later choices. In the case of Russian authors, we use a modern transliteration and give the patronymics as initials. Sometimes we use the following abbreviations for conferences (especially when their proceedings are published by the *Proceedings of Machine Learning Research*): ALT stands for "International Conference on Algorithmic Learning Theory" (annual conference on theoretical machine learning), COLT stands for "Conference on Learning Theory" (another annual conference on theoretical machine learning), COPA stands for "Conformal and Probabilistic Prediction and Applications" (annual workshop on conformal prediction), and ICML stands for "International Conference on Machine Learning" (annual conference on machine learning).

1. Aizerman, M.A., Braverman, E.M., Rozonoer, L.I.: Метод потенциальных функций в теории обучения машин (The Method of Potential Functions in the Theory of Machine Learning). Nauka, Moscow (1970)
2. Aldrich, J., Johnson, W.E.: Cambridge thought on probability. Int. J. Approx. Reason. **141**, 146–158 (2022)
3. Angelopoulos, A.N., Bates, S.: A gentle introduction to conformal prediction and distribution-free uncertainty quantification. Tech. Rep. arXiv:2107.07511 [cs.LG], arXiv.org e-Print archive (2022)
4. Anscombe, F.J.: Rejection of outliers. Technometrics **2**, 123–147 (1960)
5. Arbuthnott, J.: An argument for divine Providence, taken from the constant regularity observ'd in the births of both sexes. Philos. Trans. R. Soc. Lond. **27**, 186–190 (1710–1712)
6. Archibald, R.C.: A rare pamphlet of Moivre and some of his discoveries. Isis **8**, 671–683 (1926). A facsimile copy of De Moivre's pamphlet [78] starts on p. 677
7. Aronszajn. N.: Theory of reproducing kernels. Trans. Am. Math. Soc. **68**, 337–404 (1950)
8. Asarin, E.A.: Some properties of Kolmogorov δ-random finite sequences. Theory Probab. Appl. **32**, 507–508 (1987)
9. Asarin, E.A.: On some properties of finite objects random in the algorithmic sense. Soviet Math. Doklady **36**, 109–112 (1988). The Russian original published in 1987
10. Augustin, T., Coolen, F.P.A.: Nonparametric predictive inference and interval probability. J. Stat. Plan. Inference **124**, 251–272 (2004)

V. Vovk et al., *Algorithmic Learning in a Random World*,
https://doi.org/10.1007/978-3-031-06649-8

11. Ayer, M., Brunk, H.D., Ewing, G.M., Reid, W.T., Silverman, E.: An empirical distribution function for sampling with incomplete information. Ann. Math. Stat. **26**, 641–647 (1955)
12. Azuma, K.: Weighted sums of certain dependent random variables. Tohoku Math. J. **19**, 357–367 (1967)
13. Bahadur, R.R.: A note on quantiles in large samples. Ann. Math. Stat. **37**, 577–580 (1966)
14. Baker, G.A.: The probability that the mean of a second sample will differ from the mean of a first sample by less than a certain multiple of the standard deviation of the first sample. Ann. Math. Stat. **6**, 197–201 (1935)
15. Balasubramanian, V.N., Ho, S.-S., Vovk, V. (Eds.): Conformal Prediction for Reliable Machine Learning: Theory, Adaptations, and Applications. Elsevier, Amsterdam (2014)
16. Banachiewicz, T.: Sur l'inverse d'un cracovien et une solution générale d'un système d'équations linéares. Comptes rendus mensuels des Séances de la Classe des Sciences Mathématiques et Naturelles de l'Académie Polonaise des Sciences et des Lettres **4**, 3–4 (1937)
17. Banachiewicz, T.: Zur Berechnung der Determinanten, wie auch der Inversen, und zur darauf basierten Auflösung der Systeme linearer Gleichungen. Acta Astron. C **3**, 41–67 (1937)
18. Barber, R.F.: Is distribution-free inference possible for binary regression? Electron. J. Stat. **14**, 3487–3524 (2020)
19. Barber, R.F., Candès, E.J., Ramdas, A., Tibshirani, R.J.: Predictive inference with the jackknife+. Ann. Stat. **49**, 486–507 (2021)
20. Barber, R.F., Candès, E.J., Ramdas, A., Tibshirani, R.J.: Conformal prediction beyond exchangeability. Tech. Rep. arXiv:2202.13415 [stat.ME], arXiv.org e-Print archive (2022)
21. Barlow, R.E., Bartholomew, D.J., Bremner, J.M., Brunk, H.D.: Statistical Inference Under Order Restrictions: The Theory and Application of Isotonic Regression. Wiley, London (1972)
22. Barnard, G.A.: Pivotal inference and the Bayesian controversy. Bullet. Int. Stat. Inst. **47**, 543–551 (1977)
23. Bates, S., Candès, E., Lei, L., Romano, Y., Sesia, M.: Testing for outliers with conformal p-values. Tech. Rep. arXiv:2104.08279 [stat.ME], arXiv.org e-Print archive (2021)
24. Bayes, T.: An essay towards solving a problem in the doctrine of chances. Philos. Trans. R. Soc. Lond. **53**, 370–418 (1763)
25. Bellotti, A.: Constructing normalized nonconformity measures based on maximizing predictive efficiency. Proc. Mach. Learn. Res. **128**, 41–54 (2020). COPA 2020
26. Bellotti, A.: Optimized conformal classification using gradient descent approximation. Tech. Rep. arXiv:2105.11255 [cs.LG], arXiv.org e-Print archive (2021)
27. Belyaev, Y.: Bootstrap, resampling, and Mallows metric. Tech. Rep., Department of Mathematical Statistics, Umeå University, Sweden (1995)
28. Belyaev, Y., Sjöstedt–de Luna, S.: Weakly approaching sequences of random distributions. J. Appl. Probab. **37**, 807–822 (2000)
29. Berger, J.O.: Statistical Decision Theory and Bayesian Analysis, 2nd edn. Springer, New York (1993)
30. Bernardo, J.M., Smith, A.F.M.: Bayesian Theory. Wiley, Chichester (1994)
31. Bernoulli, J.: Ars Conjectandi. Thurnisius, Basel (1713). English translation, with an introduction and notes, by Edith Dudley Sylla: The Art of Conjecturing, together with Letter to a Friend on Sets in Court Tennis. Johns Hopkins University Press, Baltimore (2006). Russian translation (second edition, with commentaries by Oscar B. Sheynin and Yurii V. Prokhorov): О законе больших чисел, Nauka, Moscow (1986)
32. Bhattacharya, P.K.: Some aspects of change-point analysis. In: Change-Point Problems. IMS Lecture Notes—Monograph Series, vol. 23. Institute of Mathematical Statistics, Hayward (1994)
33. Billingsley, P.: Convergence of Probability Measures, 2nd edn. Wiley, New York (1999)
34. Bourbaki, N.: Elements of the History of Mathematics. Springer, Berlin (1994)
35. Breiman, L.: Random forests. Mach. Learn. **45**, 5–32 (2001)
36. Brent, R.P., Zimmermann, P.: Modern Computer Arithmetic. Cambridge University Press, Cambridge (2011)

37. Brier, G.W.: Verification of forecasts expressed in terms of probability. Month. Weather Rev. **78**, 1–3 (1950)
38. Brown, L.D., Cai, T.T., DasGupta, A.: Interval estimation for a binomial proportion (with discussion). Stat. Sci. **16**, 101–133 (2001)
39. Brunk, H.D.: Maximum likelihood estimates of monotone parameters. Ann. Math. Stat. **26**, 607–616 (1955)
40. Burnaev, E., Vovk, V.: Efficiency of conformalized ridge regression. JMLR Workshop Conf. Proc. **35**, 605–622 (2014). COLT 2014
41. Cantelli, F.P.: Sulla determinazione empirica della leggi di probabilità. Giornale dell'Istituto Italiano degli Attuari **4**, 421–424 (1933)
42. Carlsson, L., Eklund, M., Norinder, U.: Aggregated conformal prediction. In: Iliadis, L., Maglogiannis, I., Papadopoulos, H., Sioutas, S., Makris, C. (eds.) AIAI Workshops, COPA 2014. IFIP Advances in Information and Communication Technology, vol. 437, pp. 231–240. Springer, Berlin (2014)
43. Carroll, R.J.: On the distribution of quantiles of residuals in a linear model. Tech. Rep. Mimeo Series No. 1161, Department of Statistics, University of North Carolina at Chapel Hill (1978). Available on the web (accessed in March 2022)
44. Cauchois, M., Gupta, S., Duchi, J.C.: Knowing what you know: valid and validated confidence sets in multiclass and multilabel prediction. J. Mach. Learn. Res. **22**(81), 1–42 (2021)
45. Cesa-Bianchi, N., Lugosi, G.: Prediction, Learning, and Games. Cambridge University Press, Cambridge (2006)
46. Chatterjee, S., Hadi, A.S.: Sensitivity Analysis in Linear Regression. Wiley, New York (1988)
47. Chen, G.: Empirical processes based on regression residuals: theory and applications. Ph.D. Thesis, Department of Mathematics and Statistics, Simon Fraser University (1991)
48. Chernozhukov, V., Wüthrich, K., Zhu, Y.: Exact and robust conformal inference methods for predictive machine learning with dependent data. Proc. Mach. Learn. Res. **75**, 732–749 (2018). COLT 2018
49. Chernozhukov, V., Wüthrich, K., Zhu, Y.: Distributional conformal prediction. Proc. Natl. Acad. Sci. USA **118**(48), e2107794118 (2021)
50. Chernozhukov, V., Wüthrich, K., Zhu, Y.: An exact and robust conformal inference method for counterfactual and synthetic controls. J. Am. Stat. Assoc. **116**, 1849–1864 (2021)
51. Cherubin, G., Chatzikokolakis, K., Jaggi, M.: Exact optimization of conformal predictors via incremental and decremental learning. Proc. Mach. Learn. Res. **139**, 1836–1845 (2021). ICML 2021
52. Clopper, C.J., Pearson, E.S.: The use of confidence or fiducial limits illustrated in the case of the binomial. Biometrika **26**, 404–413 (1934)
53. Colombo, N., Vovk, V.: Training conformal predictors. Proc. Mach. Learn. Res. **128**, 55–64 (2020). COPA 2020
54. Cook, R.D., Weisberg, S.: Residuals and Influence in Regression. Chapman and Hall, New York (1982)
55. Cormen, T.H., Leiserson, C.E., Rivest, R.L., Stein, C.: Introduction to Algorithms, 3rd edn. MIT Press, Cambridge (2009)
56. Cortez, P., Cerdeira, A., Almeida, F., Matos, T., Reis, J.: Modeling wine preferences by data mining from physicochemical properties. Decis. Support Syst. **47**, 547–553 (2009)
57. Cournot, A.-A.: Exposition de la théorie des chances et des probabilités. Hachette, Paris (1843)
58. Cover, T.M.: This week's citation classic: Cover T M & Hart P E. Nearest neighbor pattern classification. IEEE Trans. Inform. Theory **IT-13**:21–7 (1967). Current Contents **13**:20 (1982)
59. Cover, T.M., Hart, P.E.: Nearest neighbor pattern classification. IEEE Trans. Inf. Theory **13**, 21–27 (1967)
60. Cover, T.M., Thomas, J.A.: Elements of Information Theory, 2nd edn. Wiley, Hoboken (2006)
61. Cowell, R.G., Dawid, A.P., Lauritzen, S.L., Spiegelhalter, D.J.: Probabilistic Networks and Expert Systems. Springer, New York (1999)

62. Cox, D.R.: Some problems connected with statistical inference. Ann. Math. Stat. **29**, 357–372 (1958)
63. Cox, D.R.: Two further applications of a model for binary regression. Biometrika **45**, 562–565 (1958)
64. Cox, D.R., Hinkley, D.V.: Theoretical Statistics. Chapman and Hall, London (1974)
65. Cramér, H.: Mathematical Methods of Statistics. Princeton University Press, Princeton (1946)
66. Cressie, N.A.C.: Statistics for Spatial Data, rev. edn. Wiley, New York (1993)
67. Cristianini, N., Shawe-Taylor, J.: An Introduction to Support Vector Machines and Other Kernel-Based Methods. Cambridge University Press, Cambridge (2000)
68. Dale, A.I.: A study of some early investigations into exchangeability. Historia Math. **12**, 323–336 (1985)
69. Dashevskiy, M., Luo, Z.: Time series prediction with performance guarantee. IET Commun. **5**, 1044–1051 (2011)
70. Dawid, A.P.: Intersubjective statistical models. In: Koch, G.S., Spizzichino, F. (eds.) Exchangeability in Probability and Statistics, pp. 217–232. North-Holland, Amsterdam (1982)
71. Dawid, A.P.: Calibration-based empirical probability (with discussion). Ann. Stat. **13**, 1251–1285 (1985)
72. Dawid, A.P.: Probability forecasting. In: Kotz, S., Johnson, N.L., Read, C.B. (eds.) Encyclopedia of Statistical Sciences, vol. 7, pp. 210–218. Wiley, New York (1986). Reprinted in the second edition (2006) on pp. 6445–6452 (Volume 10)
73. Dawid, A.P.: Causal inference without counterfactuals. J. Am. Stat. Assoc. **95**, 407–424 (2000)
74. Dawid, A.P., Vovk, V.: Prequential probability: principles and properties. Bernoulli **5**, 125–162 (1999)
75. de Finetti, B.: Funzione caratteristica di un fenomeno aleatorio. Memorie della Reale Accademia Nazionale dei Lincei: Classe di scienze fisiche, matematiche, e naturali, Serie 6 **4**, 86–133 (1930)
76. de Finetti, B.: La prévision, ses lois logiques, ses sources subjectives. Annales de l'Institut Henri Poincaré **7**, 1–68 (1937). An English translation of this article is included in [209]
77. de Finetti, B.: Sur la condition d'équivalence partielle. In: Actualités Scientifiques et Industrielles, vol. 739. Hermann, Paris (1938). An English translation, under the title "On the condition of partial exchangeability", is included in [175], pp. 193–206
78. De Moivre, A.: Approximatio ad summam terminorum binomii $\overline{a+b}|^n$ in seriem expansi (1733). Included in [6]
79. Dempster, A.P.: On direct probabilities. J. R. Stat. Soc. B **25**, 100–110 (1963)
80. Derbeko, P., El-Yaniv, R., Meir, R.: Explicit learning curves for transduction and application to clustering and compression algorithms. J. Artif. Intell. Res. **22**, 117–142 (2004)
81. Devroye, L., Lugosi, G.: Combinatorial Methods in Density Estimation. Springer, New York (2001)
82. Devroye, L., Györfi, L., Lugosi, G.: A Probabilistic Theory of Pattern Recognition. Springer, New York (1996)
83. Diaconis, P., Freedman, D.A.: Finite exchangeable sequences. Ann. Probab. **8**, 745–764 (1980)
84. Diaconis, P., Freedman, D.: On the consistency of Bayes estimates (with discussion). Ann. Stat. **14**, 1–67 (1986)
85. Dietterich, T.G., Bakiri, G.: Solving multiclass learning problems via error-correcting output codes. J. Artif. Intell. Res. **2**, 263–286 (1995)
86. Dieudonné, J.: Sur le théorème de Lebesgue–Nikodym. III. Annales de l'Institut Fourier **23**, 25–53 (1948)
87. Doob, J.L.: Stochastic Processes. Wiley, New York (1953)
88. Dorai-Raj, S.: Binomial confidence intervals for several parameterizations (2014). Version 1.1-1
89. Draper, N.R., Smith, H.: Applied Regression Analysis, 3rd edn. Wiley, New York (1998)

90. Du, W., Polunchenko, A.S., Sokolov, G.: On robustness of the Shiryaev–Roberts change-point detection procedure under parameter misspecification in the post-change distribution. Commun. Stat. Simul. Comput. **46**, 2185–2206 (2017)
91. Dua, D., Graff, C.: UCI machine learning repository (2019). URL http://archive.ics.uci.edu/ml
92. Dudley, R.M.: Weak convergence of probabilities on nonseparable metric spaces and empirical measures on Euclidean spaces. Ill. J. Math. **10**, 109–126 (1966)
93. Dudley, R.M.: Measures on non-separable metric spaces. Ill. J. Math. **11**, 449–453 (1967)
94. Dudley, R.M.: Real Analysis and Probability. Cambridge University Press, Cambridge (2002). Original edition published in 1989 by Wadsworth
95. Durbin, J.: Weak convergence of the sample distribution function when parameters are estimated. Ann. Stat. **1**, 279–290 (1973)
96. Dvoretzky, A., Kiefer, J.C., Wolfowitz, J.: Asymptotic minimax character of a sample distribution function and of the classical multinomial estimator. Ann. Math. Stat. **27**, 642–669 (1956)
97. Eckel, F.A., Walters, M.K.: Calibrated probabilistic quantitative precipitation forecasts based on the MRF ensemble. Weather Forecast. **13**, 1132–1147 (1998)
98. Efron, B., Tibshirani, R.J.: An Introduction to the Bootstrap. Chapman and Hall, New York (1993)
99. Eklund, M., Norinder, U., Boyer, S., Carlsson, L.: The application of conformal prediction to the drug discovery process. Ann. Math. Artif. Intell. **74**, 117–132 (2015)
100. Fawcett, T., Flach, P.A.: A response to Webb and Ting's on the application of ROC analysis to predict classification performance under varying class distributions. Mach. Learn. **58**, 33–38 (2005)
101. Fedorova, V., Nouretdinov, I., Gammerman, A., Vovk, V.: Plug-in martingales for testing exchangeability on-line. In: Langford, J., Pineau, J. (eds.) Proceedings of the Twenty Ninth International Conference on Machine Learning, pp. 1639–1646. Omnipress, Madison (2012)
102. Fedorova, V., Gammerman, A., Nouretdinov, I., Vovk, V.: Hypergraphical conformal predictors. Int. J. Artif. Intell. Tools **24**(6), 1560003 (2015). COPA 2013 Special Issue
103. Fisch, A., Schuster, T., Jaakkola, T., Barzilay, R.: Few-shot conformal prediction with auxiliary tasks. Proc. Mach. Learn. Res. **139**, 3329–3339 (2021). ICML 2021
104. Fisher, R.A.: The goodness of fit of regression formulae and the distribution of regression coefficients. J. R. Stat. Soc. **85**, 597–612 (1922)
105. Fisher, R.A.: Applications of "Student's" distribution. Metron **5**, 90–104 (1925)
106. Fisher, R.A.: Inverse probability. Proc. Cambridge Philos. Soc. **26**, 528–535 (1930)
107. Fisher, R.A.: The fiducial argument in statistical inference. Ann. Eugenics **6**, 391–398 (1935)
108. Fisher, R.A.: Student. Ann. Eugenics **9**, 1–9 (1939)
109. Fisher, R.A.: Conclusions fiduciaires. Annales de l'Institut Henri Poincaré **10**, 191–213 (1948)
110. Fisher, R.A.: Statistical Methods and Scientific Inference, 3rd edn. Hafner, New York (1973). Included in [112]. First edition: 1956
111. Fisher, R.A.: Statistical Methods for Research Workers, 14th (revised and enlarged) edn. Hafner, New York (1973). Included in [112]. First edition: 1925
112. Fisher, R.A.: Statistical Methods, Experimental Design, and Scientific Inference. Oxford University Press, Oxford (1991). Edited by Bennett, J.H.
113. Fix, E., Hodges, J.L., Jr: Discriminatory analysis. Nonparametric discrimination: small sample performance. Tech. Rep. Report Number 11, Project Number 21-49-004, USAF School of Aviation Medicine, Randolph Field, TX (1951). Reprinted in International Statistical Review **57**, 238–247 (1989)
114. Floyd, S., Warmuth, M.K.: Sample compression, learnability, and the Vapnik–Chervonenkis dimension. Mach. Learn. **21**, 269–304 (1995)
115. Fong, E., Holmes, C.C.: Conformal Bayesian computation. In: Advances in Neural Information Processing Systems 34 (NeurIPS 2021) (2021)

116. Fraser, D.A.S.: Sequentially determined statistically equivalent blocks. Ann. Math. Stat. **22**, 372–381 (1951)
117. Fraser, D.A.S.: Nonparametric tolerance regions. Ann. Math. Stat. **24**, 44–55 (1953)
118. Fraser, D.A.S.: Nonparametric Methods in Statistics. Wiley, New York (1957)
119. Fraser, D.A.S, Guttman, I.: Tolerance regions. Ann. Math. Stat. **27**, 16–32 (1956)
120. Freedman, D.A.: Invariants under mixing which generalise de Finetti's theorem. Ann. Math. Stat. **33**, 916–923 (1962)
121. Freedman, D.A.: Invariants under mixing which generalise de Finetti's theorem: continuous time parameter. Ann. Math. Stat. **34**, 1194–1216 (1963)
122. Freund, Y., Schapire, R.E.: Experiments with a new boosting algorithm. In: Machine Learning: Proceedings of the Thirteenth International Conference, pp. 148–156 (1996)
123. Gács, P.: Exact expressions for some randomness tests. Zeitschrift für Mathematische Logik und Grundlagen der Mathematik **26**, 385–394 (1980)
124. Gama, J., Medas, P., Castillo, G., Rodrigues, P.: Learning with drift detection. In: Bazzan, A.L.C., Labidi, S. (eds.) Advances in Artificial Intelligence: SBIA 2004, pp. 286–295. Springer, Berlin (2004)
125. Gammerman, A.: Machine learning: progress and prospects. Royal Holloway, University of London, Egham, Surrey, TW20 0EX (1997). This booklet is based on an inaugural lecture delivered on 11 December 1996
126. Gammerman, A., Vapnik, V., Vovk, V.: Transduction in pattern recognition (1997). Manuscript submitted to the Fifteenth International Joint Conference on Artificial Intelligence in January 1997. Extended version published as [127]. The algorithm proposed in this paper was described in a 1996 public lecture [125]
127. Gammerman, A., Vovk, V., Vapnik, V.: Learning by transduction. In: Cooper, G.F., Moral, S. (eds.) Proceedings of the Fourteenth Conference on Uncertainty in Artificial Intelligence, pp. 148–155. Morgan Kaufmann, San Francisco (1998)
128. Gauss, C.F.: Theoria Combinationis Observationum Erroribus Minimis Obnoxiae, Pars Posterior. Dieterich, Göttingen (1823)
129. Getoor, R.K., Sharpe, M.J.: Conformal martingales. Invent. Math. **16**, 271–308 (1972)
130. Gibbs, I., Candès, E.J.: Adaptive conformal inference under distribution shift. In: Advances in Neural Information Processing Systems 34 (NeurIPS 2021) (2021)
131. Glivenko, V.I.: Sulla determinazione empirica di probabilità. Giornale dell'Istituto Italiano degli Attuari **4**, 92–99 (1933)
132. Gneiting, T., Katzfuss, M.: Probabilistic forecasting. Annu. Rev. Stat. Appl. **1**, 125–151 (2014)
133. Gneiting, T., Raftery, A.E.: Strictly proper scoring rules, prediction, and estimation. J. Am. Stat. Assoc. **102**, 359–378 (2007)
134. Gneiting, T., Balabdaoui, F., Raftery, A.E.: Probabilistic forecasts, calibration and sharpness. J. R. Stat. Soc. B **69**, 243–268 (2007)
135. Good, I.J.: Rational decisions. J. R. Stat. Soc. B **14**, 107–114 (1952)
136. Good, I.J.: The Estimation of Probabilities: An Essay in Modern Bayesian Methods. MIT Press, London (1965)
137. Goodfellow, I., Bengio, Y., Courville, A.: Deep Learning. MIT Press, Cambridge (2016)
138. Gossett, W.S. (Student): On the probable error of a mean. Biometrika **6**, 1–25 (1908)
139. Graham, R.L.: An efficient algorithm for determining the convex hull of a finite planar set. Inf. Process. Lett. **1**, 132–133 (1972)
140. Greenland, S.: Valid P-values behave exactly as they should: some misleading criticisms of P-values and their resolution with S-values. Am. Stat. **73**(1), 106–114 (2019)
141. Grünwald, P.D.: The Minimum Description Length Principle. MIT Press, Cambridge (2007)
142. Guan, L., Tibshirani, R.: Prediction and outlier detection in classification problems. J. R. Stat. Soc. B **84**, 524–546 (2022)
143. Guttman, I.: Statistical Tolerance Regions: Classical and Bayesian. Griffin, London (1970)
144. Györfi, L., Kohler, M., Krzyżak, A., Walk, H.: A Distribution-Free Theory of Nonparametric Regression. Springer, New York (2002)

145. Haag, J.: Sur une problème général de probabilités et ses diverses applications. In: Proceedings of the International Congress of Mathematicians, Toronto, 1924, pp. 659–674. Toronto University Press, Toronto (1928)
146. Hájek, A.: The reference class problem is your problem too. Synthese **156**, 563–585 (2007)
147. Hamill, T.M., Colucci, S.J.: Verification of Eta-RSM short-range ensemble forecasts. Month. Weather Rev. **125**, 1312–1327 (1997)
148. Hamill, T.M., Colucci, S.J.: Evaluation of Eta-RSM ensemble probabilistic precipitation forecasts. Month. Weather Rev. **126**, 711–724 (1998)
149. Hannig, J.: On generalized fiducial inference. Stat. Sin. **19**, 491–544 (2009)
150. Harries, M.: Splice-2 comparative evaluation: electricity pricing. Tech. Rep. UNSW-CSE-TR-9905, Artificial Intelligence Group, School of Computer Science and Engineering, University of New South Wales (1999)
151. Hastie, T., Tibshirani, R., Friedman, J.: The Elements of Statistical Learning, 2nd edn. Springer, New York (2009)
152. Hastie, T., Tibshirani, R., Wainwright, M.: Statistical Learning with Sparsity: The Lasso and Generalizations. CRC Press, Boca Raton (2016)
153. Hastie, T., Montanari, A., Rosset, S., Tibshirani, R.: Surprises in high-dimensional ridgeless least squares interpolation. Ann. Stat. **50**, 949–986 (2022)
154. Hawkins, D.: Identification of Outliers. Chapman and Hall, London (1980)
155. Henderson, H.V., Searle, S.R.: On deriving the inverse of a sum of matrices. SIAM Rev. **23**, 53–60 (1981)
156. Herbrich, R., Williamson, R.C.: Learning and generalization: theoretical bounds. In: Arbib, M.A. (ed.) Handbook of Brain Theory and Neural Networks, 2nd edn., pp. 3140–3150. MIT Press, Cambridge (2002)
157. Herbster, M., Warmuth, M.K.: Tracking the best expert. Mach. Learn. **32**, 151–178 (1998)
158. Hewitt, E., Savage, L.J.: Symmetric measures on Cartesian products. Trans. Am. Math. Soc. **80**, 470–501 (1955)
159. Hill, B.M.: Posterior distribution of percentiles: Bayes' theorem for sampling from a population. J. Am. Stat. Assoc. **63**, 677–691 (1968)
160. Hill, B.M.: De Finetti's theorem, induction, and $A_{(n)}$ or Bayesian nonparametric predictive inference (with discussion). In: Lindley, D.V., Bernardo, J.M., DeGroot, M.H., Smith, A.F.M. (eds.) Bayesian Statistics, vol. 3, pp. 211–241. Oxford University Press, Oxford (1988)
161. Hill, B.M.: Bayesian nonparametric prediction and statistical inference. In: Goel, P.K., Iyengar, N.S. (eds.) Bayesian Analysis in Statistics and Econometrics. Lecture Notes in Statistics, vol. 75, chap. 4, pp. 43–94. Springer, New York (1992)
162. Hill, B.M.: Parametric models for A_n: splitting processes and mixtures. J. R. Stat. Soc. B **55**, 423–433 (1993)
163. Hoaglin, D.C., Welsch, R.E.: The hat matrix in regression and ANOVA. Am. Stat. **32**, 17–22 (1978)
164. Hoeffding, W.: Probability inequalities for sums of bounded random variables. J. Am. Stat. Assoc. **58**, 13–30 (1963)
165. Hoerl, A.E.: Optimum solution of many variables equations. Chem. Eng. Prog. **55**, 69–78 (1959)
166. Hoerl, A.E.: Applications of ridge analysis to regression problems. Chem. Eng. Prog. **58**, 54–59 (1962)
167. Hoerl, R.W.: Ridge analysis 25 years later. Am. Stat. **39**, 186–192 (1985)
168. Hoerl, A.E., Kennard, R.W.: Ridge regression: biased estimation for nonorthogonal problems. Technometrics **12**, 55–67 (1970)
169. Hoerl, A.E., Kennard, R.W.: Ridge regression: applications to nonorthogonal problems. Technometrics **12**, 69–82 (1970). Erratum: **12**, 723
170. Hoff, P.: Bayes-optimal prediction with frequentist coverage control. Tech. Rep. arXiv:2105.14045 [math.ST], arXiv.org e-Print archive (2021). To appear in *Bernoulli*
171. Holland, P.W.: Statistics and causal inference. J. Am. Stat. Assoc. **81**, 945–960 (1986)

172. Huber, P.J., Ronchetti, E.M.: Robust Statistics, 2nd edn. Wiley, Hoboken (2009). First edition (by Huber): 1981
173. Hume, D.: A Treatise of Human Nature. Noon (vols. 1–2, 1739) and Longman (vol. 3, 1740), London (1739–1740)
174. Imbens, G.W., Rubin, D.B.: Causal Inference for Statistics, Social, and Biomedical Sciences: An Introduction. Cambridge University Press, New York (2015)
175. Jeffrey, R.C. (ed.): Studies in Inductive Logic and Probability, vol. 2. University of California Press, Berkley (1981)
176. Jeffreys, H.: On the theory of errors and least squares. Proc. R. Soc. Lond. A **138**, 48–55 (1932)
177. Jeffreys, H.: Theory of Probability, 3rd edn. Oxford University Press, Oxford (1961)
178. Jensen, F.V.: An Introduction to Bayesian Networks. UCL Press, London (1996)
179. Johansson, U., König, R., Löfström, T., Boström, H.: Evolved decision trees as conformal predictors. In: de la Fraga, L.G. (ed.) Proceedings of the 2013 IEEE Conference on Evolutionary Computation, vol. 1, pp. 1794–1801. Cancun, Mexico (2013)
180. Johnson, W.E.: Logic, Part I. Cambridge University Press, Cambridge (1921)
181. Johnson, W.E.: Logic, Part II: Demonstrative Inference: Deductive and Inductive. Cambridge University Press, Cambridge (1921)
182. Johnson, W.E.: Logic, Part III: The Logical Foundations of Science. Cambridge University Press, Cambridge (1921)
183. Kac, M., Kiefer, J.C., Wolfowitz, J.: On test of normality and other tests of goodness-of-fit, based on distance method. Ann. Math. Stat. **26**, 189–211 (1955)
184. Kahneman, D., Slovic, P., Tversky, A. (eds.): Judgment under Uncertainty: Heuristics and Biases. Cambridge University Press, Cambridge, England (1982)
185. Karpinski, M., Macintyre, A.J.: Polynomial bounds for VC dimension of sigmoidal neural networks. In: Proceeding of the Twenty-Seventh Annual ACM Symposium on the Theory of Computing, pp. 200–208. ACM Press, New York (1995)
186. Karpinski, M., Macintyre, A.J.: Polynomial bounds for VC dimension of sigmoidal and general Pfaffian neural networks. J. Comput. Syst. Sci. **54**, 169–176 (1997)
187. Kechris, A.S.: Classical Descriptive Set Theory. Springer, New York (1995)
188. Kelly, J.L.: A new interpretation of information rate. Bell Syst. Tech. J. **35**, 917–926 (1956)
189. Kemperman, J.H.B.: Generalized tolerance limits. Ann. Math. Stat. **27**, 180–186 (1956)
190. Khinchin, A.Y.: Sur les classes d'événements équivalents. Математический сборник **39**, 40–43 (1932)
191. Kılınç, B.E.: The reception of John Venn's philosophy of probability. In: Hendricks, V.F., Pedersen, S.A., Jørgensen, K.F. (eds.) Probability Theory: Philosophy, Recent History and Relations to Science, pp. 97–121. Kluwer, Dordrecht (2001)
192. Kingman, J.F.C.: On random sequences with spherical symmetry. Biometrika **59**, 492–493 (1972)
193. Kingman, J.F.C.: Uses of exchangeability. Ann. Probab. **6**, 183–197 (1978)
194. Knight, F.H.: Risk, Uncertainty, and Profit. Houghton Mifflin, Boston (1921)
195. Köhn, J.: Uncertainty in Economics: A New Approach. Springer, Cham (2017)
196. Kolmogorov, A.N.: Sur une formule limite de M. A. Khintchine. Comptes rendus des Séances de l'Académie des Sciences **186**, 824–825 (1928)
197. Kolmogorov, A.N.: Über das Gesetz des iterierten Logarithmus. Math. Ann. **101**, 126–135 (1929)
198. Kolmogorov, A.N.: Sur la notion de la moyenne. Atti della Reale Accademia Nazionale dei Lincei. Classe di scienze fisiche, matematiche, e naturali. Rendiconti Serie VI **12**(9), 388–391 (1930)
199. Kolmogorov, A.N.: Grundbegriffe der Wahrscheinlichkeitsrechnung. Springer, Berlin (1933). Published in English as *Foundations of the Theory of Probability* (Chelsea, New York). First edition, 1950; second edition, 1956
200. Kolmogorov, A.N.: Sulla determinazione empirica di unna legge di distribuzione. Giornale dell'Istituto Italiano degli Attuari **4**, 83–91 (1933)

201. Kolmogorov, A.N.: Определение центра рассеивания и меры точности по ограниченному числу наблюдений (The estimation of the mean and precision from a finite sample of observations). Известия АН СССР, *серия математическая* **6**, 3–32 (1942)
202. Kolmogorov, A.N.: On tables of random numbers. Sankhya Ind. J. Stat. A **25**, 369–376 (1963)
203. Kolmogorov, A.N.: Three approaches to the quantitative definition of information. Problems Inf. Transm. **1**, 1–7 (1965)
204. Kolmogorov, A.N.: Logical basis for information theory and probability theory. IEEE Trans. Inf. Theory **IT-14**, 662–664 (1968)
205. Kolmogorov, A.N.: Combinatorial foundations of information theory and the calculus of probabilities. Russian Math. Surv. **38**, 29–40 (1983). This article was written in 1970 in connection with Kolmogorov's talk at the International Mathematical Congress in Nice
206. Kolmogorov, A.N., Shiryaev, A.N.: Применение марковских процессов к обнаружению разладок производственного процесса (1960). Доклад на VI Совещании по теории вероятностей и математической статистике, Vilnius
207. Kuhn, T.S.: The Structure of Scientific Revolutions, 3rd edn. University of Chicago Press, Chicago (1996). First published in 1962
208. Kuleshov, V., Fenner, N., Ermon, S.: Accurate uncertainties for deep learning using calibrated regression. Proc. Mach. Learn. Res. **80**, 2796–2804 (2018). ICML 2018
209. Kyburg, H., Smokler, H. (eds.): Studies in Subjective Probability, 2nd edn. Krieger, New York (1980)
210. Lane, D.A.: Fisher, Jeffreys, and the nature of probability. In: Fienberg, S.E., Hinkley, D.V. (eds.) R. A. Fisher: An Appreciation. Lecture Notes in Statistics, vol. 1, pp. 148–160. Springer, Berlin (1980)
211. Langford, J.: Tutorial on practical prediction theory for classification. J. Mach. Learn. Res. **6**, 273–306 (2005)
212. Laplace, P.S.: Mémoire sur la probabilité des causes par les événements. Mémoires de mathématique et de physique, presentés à l'Académie royale des sciences, par divers savans & lûs dans ses assemblées **6**, 621–656 (1774). English translation: Statistical Science **1**, 364–378 (1986)
213. Lauritzen, S.L.: Statistical Models as Extremal Families. Aalborg University Press, Aalborg (1982)
214. Lauritzen, S.L.: Extremal Families and Systems of Sufficient Statistics. Lecture Notes in Statistics, vol. 49. Springer, New York (1988)
215. Lauritzen, S.L.: Graphical Models. Clarendon Press, Oxford (1996)
216. Lawless, J.F., Fredette, M.: Frequentist prediction intervals and predictive distributions. Biometrika **92**, 529–542 (2005)
217. Laxhammar, R.: Conformal anomaly detection: detecting abnormal trajectories in surveillance applications. Ph.D. thesis, University of Skövde, Sweden (2014)
218. Laxhammar, R., Falkman, G.: Online learning and sequential anomaly detection in trajectories. IEEE Trans. Pattern Anal. Mach. Intell. **36**, 1158–1173 (2014)
219. LeCun, Y., Boser, B.E., Denker, J.S., Henderson, D., Howard, R.E., Hubbard, W.E., Jackel, L.D.: Handwritten digit recognition with a back-propagation network. In: Touretzky, D.S. (ed.) Advances in Neural Information Processing Systems, vol. 2, pp. 396–404. Morgan Kaufmann, San Mateo (1990)
220. Lee, Y., Barber, R.F.: Distribution-free inference for regression: discrete, continuous, and in between. In: Advances in Neural Information Processing Systems 34 (NeurIPS 2021) (2021)
221. Legendre, A.M.: Nouvelles méthodes pour la détermination des orbites des comètes. Courcier, Paris (1805)
222. Lehmann, E.L., Romano, J.P.: Testing Statistical Hypotheses, 4th edn. Springer, Cham (2022). First edition (by Lehmann): 1959
223. Lei, J.: Classification with confidence. Biometrika **101**, 755–769 (2014)
224. Lei, J.: Fast exact conformalization of lasso using piecewise linear homotopy. Biometrika **106**, 749–764 (2019)

225. Lei, J., Wasserman, L.: Distribution-free prediction bands for nonparametric regression. J. R. Stat. Soc. B **76**, 71–96 (2014)
226. Lei, J., Rinaldo, A., Wasserman, L.: A conformal prediction approach to explore functional data. Ann. Math. Artif. Intell. **74**, 29–43 (2015)
227. Lei, J., Robins, J., Wasserman, L.: Distribution free prediction sets. J. Am. Stat. Assoc. **108**, 278–287 (2013)
228. Lei, L., Candés, E.J.: Conformal inference of counterfactuals and individual treatment effects. J. R. Stat. Soc. B **83**, 911–938 (2021)
229. Levin, L.A.: Uniform tests of randomness. Sov. Math. Dokl. **17**, 337–340 (1976)
230. Levin, L.A.: Randomness conservation inequalities; information and independence in mathematical theories. Inf. Control **61**, 15–37 (1984)
231. Lévy, P.: Théorie de l'addition des variables aléatoires. Gauthier-Villars, Paris (1937). Second edition: 1954
232. Li, M., Vitányi, P.: An Introduction to Kolmogorov Complexity and Its Applications, 4th edn. Springer, Berlin (2019)
233. Linusson, H., Norinder, U., Boström, H., Johansson, U., Löfström, T.: On the calibration of aggregated conformal predictors. Proc. Mach. Learn. Res. **60**, 154–173 (2017). COPA 2017
234. Lipton, Z., Wang, Y.-X., Smola, A.: Detecting and correcting for label shift with black box predictors. Proc. Mach. Learn. Res. **80**, 3122–3130 (2018). ICML 2018
235. Littlestone, N., Warmuth, M.K.: Relating data compression and learnability. Tech. Rep., University of California, Santa Cruz (1986)
236. Littlestone, N., Warmuth, M.K.: The weighted majority algorithm. Inf. Comput. **108**, 212–261 (1994)
237. Martin, R., Liu, C.: Inferential Models: Reasoning with Uncertainty. CRC Press, Boca Raton (2016)
238. Martin-Löf, P.: The definition of random sequences. Inf. Control **9**, 602–619 (1966)
239. Martin-Löf, P.: Repetitive structures (with discussion). In: Barndorff-Nielsen, O.E., Blæsild, P., Schou, G. (eds.) Proceedings of Conference on Foundational Questions in Statistical Inference, pp. 271–294. Aarhus, Denmark (1974)
240. Massart, P.: The tight constant in the Dvoretzky–Kiefer–Wolfowitz inequality. Ann. Probab. **18**, 1269–1283 (1990)
241. Matheron, G.: Principles of geostatistics. Econ. Geol. **58**, 1246–1266 (1963)
242. McAllester, D.A.: Some PAC-Bayesian theorems. In: Proceedings of the Eleventh Annual Conference on Computational Learning Theory, pp. 230–234. ACM Press, New York (1998). Journal version: [243]
243. McAllester, D.A.: Some PAC-Bayesian theorems. Mach. Learn. **37**, 355–363 (1999)
244. McDonald, D.: A cusum procedure based on sequential ranks. Naval Res. Logist. **37**, 627–646 (1990)
245. Medarametla, D., Candés, E.: Distribution-free conditional median inference. Electron. J. Stat. **15**, 4625–4658 (2021)
246. Melluish, T.: Transductive algorithms for finding confidence information for regression estimation in the typicalness framework. Ph.D. Thesis, Royal Holloway, University of London (2005)
247. Melluish, T., Saunders, C., Nouretdinov, I., Vovk, V.: Comparing the Bayes and typicalness frameworks. Tech. Rep. CLRC-TR-01-05, Computer Learning Research Centre, Royal Holloway, University of London (2001)
248. Melluish, T., Saunders, C., Nouretdinov, I., Vovk, V.: Comparing the Bayes and typicalness frameworks. In: De Raedt, L., Flach, P.A. (eds.) Machine Learning: ECML'2001. Proceedings of the Twelfth European Conference on Machine Learning. Lecture Notes in Computer Science, vol. 2167, pp. 360–371. Springer, Heidelberg (2001)
249. Mercer, J.: Functions of positive and negative type, and their connection with the theory of integral equations. Philos. Trans. R. Soc. Lond. A **209**, 415–446 (1909)
250. Micchelli, C.A., Xu, Y., Zhang, H.: Universal kernels. J. Mach. Learn. Res. **7**, 2651–2667 (2006)

251. Mohammadi, M.: On the bounds for diagonal and off-diagonal elements of the hat matrix in the linear regression model. REVSTAT Stat. J. **14**, 75–87 (2016)
252. Montgomery, D.C., Peck, E.A., Vining, G.G.: Introduction to Linear Regression Analysis, 6th edn. Wiley, Hoboken (2021)
253. Moro, S., Cortez, P., Rita, P.: A data-driven approach to predict the success of bank telemarketing. Decis. Supp. Syst. **62**, 22–31 (2014)
254. Mugantseva, L.A.: Testing normality in one-dimensional and multi-dimensional linear regression. Theory Probab. Appl. **22**, 591–602 (1977)
255. Murphy, A.H., Epstein, E.S.: Verification of probabilistic predictions: a brief review. J. Appl. Meteorol. **6**, 748–755 (1967)
256. Murphy, A.H., Winkler, R.L.: A general framework for forecast verification. Month. Weather Rev. **115**, 1330–1338 (1987)
257. NeuroCOLT: Generalisation bounds less than 0.5 (2002). NeuroCOLT Workshop, Windsor, England
258. Neyman, J.: On the application of probability theory to agricultural experiments. Essay on principles. Section 9. Stat. Sci. **5**, 465–480 (1990). Master thesis. Originally published in *Roczniki Nauk Rolniczych* **10**, 1–51 (1923)
259. Neyman, J.: On the two different aspects of the representative method: the method of stratified sampling and the method of purposive selection (with discussion). J. R. Stat. Soc. **97**, 558–625 (1934). Reprinted in [260, pp. 98–141]. Fisher's comment: 614–619
260. Neyman, J.: A Selection of Early Statistical Papers of J. Neyman. Cambridge University Press, Cambridge (1967)
261. Norris, J.R.: Markov Chains. Cambridge University Press, Cambridge (1997)
262. Nouretdinov, I., Vovk, V.: Criterion of calibration for transductive confidence machine with limited feedback. Theor. Comput. Sci. **364**, 3–9 (2006). ALT 2003 Special Issue
263. Nouretdinov, I., Vovk, V., Vyugin, M., Gammerman, A.: Pattern recognition and density estimation under the general i.i.d. assumption. In: Helmbold, D., Williamson, B. (eds.) Proceedings of the Fourteenth Annual Conference on Computational Learning Theory and Fifth European Conference on Computational Learning Theory. Lecture Notes in Artificial Intelligence, vol. 2111, pp. 337–353. Springer, Berlin (2001)
264. Nouretdinov, I., Melluish, T., Vovk, V.: Ridge regression confidence machine. In: Proceedings of the Eighteenth International Conference on Machine Learning, pp. 385–392. Morgan Kaufmann, San Francisco (2001)
265. Nouretdinov, I., V'yugin, V.V., Gammerman, A.: Transductive confidence machine is universal. In: Gavaldà, R., Jantke, K.P., Takimoto, E. (eds.) Proceedings of the Fourteenth International Conference on Algorithmic Learning Theory. Lecture Notes in Artificial Intelligence, vol. 2842, pp. 283–297. Springer, Berlin (2003)
266. Nouretdinov, I., Volkhonskiy, D., Lim, P., Toccaceli, P., Gammerman, A: Inductive Venn–Abers predictive distribution. Proc. Mach. Learn. Res. **91**, 15–36 (2018). COPA 2018
267. Oakes, D.: Self-calibrating priors do not exist (with discussion). J. Am. Stat. Assoc. **80**, 339–342 (1985)
268. Page, E.S.: Continuous inspection schemes. Biometrika **41**, 100–115 (1954)
269. Papadopoulos, H., Proedrou, K., Vovk, V., Gammerman, A.: Inductive confidence machines for regression. In: Elomaa, T., Mannila, H., Toivonen, H. (eds.) Proceedings of the Thirteenth European Conference on Machine Learning. Lecture Notes in Computer Science, vol. 2430, pp. 345–356. Springer, Berlin (2002)
270. Papadopoulos, H., Vovk, V., Gammerman, A.: Qualified predictions for large data sets in the case of pattern recognition. In: Proceedings of the International Conference on Machine Learning and Applications (ICMLA 2002), pp. 159–163. CSREA Press, Las Vegas (2002)
271. Pearl, J.: Probabilistic Reasoning in Intelligent Systems: Networks of Plausible Inference. Morgan Kaufmann, San Francisco (1988)
272. Pearson, K.: James Bernoulli's theorem. Biometrika **17**, 201–210 (1925)
273. Pearson, E.S.: Studies in the history of probability and statistics. XX: Some early correspondence between W. S. Gosset, R. A. Fisher and Karl Pearson, with notes and comments. Biometrika **55**, 445–457 (1968)

274. Pedregosa, F., Varoquaux, G., Gramfort, A., Michel, V., Thirion, B., Grisel, O., Blondel, M., Prettenhofer, P., Weiss, R., Dubourg, V., Vanderplas, J., Passos, A., Cournapeau, D., Brucher, M., Perrot, M., Duchesnay, É.: Scikit-learn: machine learning in Python. J. Mach. Learn. Res. **12**, 2825–2830 (2011)
275. Pelillo, M.: Alhazen and the nearest neighbor rule. Pattern Recognit. Lett. **38**, 34–37 (2014)
276. Picard, R.R., Berk, K.N.: Data splitting. Am. Stat. **44**, 140–147 (1990)
277. Pierce, D.A., Kopecky, K.J.: Testing goodness of fit for the distribution of errors in regression models. Biometrika **66**, 1–5 (1979)
278. Pinelis, I.: Exact bounds on the closeness between the Student and standard normal distributions. ESAIM Probab. Stat. **19**, 24–27 (2015)
279. Plackett, R.L.: Studies in the history of probability and statistics. XXIX: The discovery of the method of least squares. Biometrika **59**, 239–251 (1972)
280. Pollak, M.: Average run lengths of an optimal method of detecting a change in distribution. Ann. Stat. **15**, 749–779 (1987)
281. Poor, H.V., Hadjiliadis, O.: Quickest Detection. Cambridge University Press, Cambridge (2009)
282. Popper, K.R.: Objective Knowledge: An Evolutionary Approach, rev. edn. Clarendon Press, Oxford (1979). First edition: 1972
283. Popper, K.R.: The Open Universe. Hutchinson, London (1982)
284. Popper, K.R.: The Logic of Scientific Discovery. Routledge, London (1999). First published in German in 1934; first English edition 1959
285. Prokhorov, Y.V.: Закон больших чисел и оценки вероятностей больших уклонений (1986). This is Commentary II to the second Russian edition of Jacob Bernoulli's *Ars Conjectandi* [31]
286. Qiu, H., Dobriban, E., Tchetgen, E.T.: Distribution-free prediction sets adaptive to unknown covariate shift. Tech. Rep. arXiv:2203.06126v1 [stat.ME], arXiv.org e-Print archive (2022)
287. Ramdas, A., Ruf, J., Larsson, M., Koolen, W.: Testing exchangeability: fork-convexity, supermartingales and e-processes. Int. J. Approx. Reason. **141**, 83–109 (2022)
288. Reichenbach, H.: The Theory of Probability. University of California Press, Berkeley (1949). This is an English translation (by Ernest H. Hutten and Maria Reichenbach) of the 1935 German original *Wahrscheinlichkeitslehre*. Because of the numerous changes, the English version was designated as the second edition of the book
289. Revuz, D., Yor, M.: Continuous Martingales and Brownian Motion, 3rd edn. Springer, Berlin (1999)
290. Robbins, H.: A remark on Stirling's formula. Am. Math. Month. **62**, 26–29 (1955)
291. Roberts, S.W.: A comparison of some control chart procedures. Technometrics **8**, 411–430 (1966)
292. Rogers, L.C.G., Williams, D.: Diffusions, Markov Processes, and Martingales, Vol. 1: Foundations, 2nd edn. Wiley, Chichester (1994). Reissued by the Cambridge University Press in *Cambridge Mathematical Library* (2000)
293. Romano, Y., Patterson, E., Candès, E.J.: Conformalized quantile regression. In: Advances in Neural Information Processing Systems, vol. 32. Curran Associates, Red Hook (2019)
294. Romano, Y., Sesia, M., Candès, E.J.: Classification with valid and adaptive coverage. In: Advances in Neural Information Processing Systems, vol. 33. Curran Associates, Red Hook (2020)
295. Rosenblatt, M.: Remarks on a multivariate transformation. Ann. Math. Stat. **23**, 470–472 (1952)
296. Rüschendorf, L.: Random variables with maximum sums. Adv. Appl. Probab. **14**, 623–632 (1982)
297. Ryabko, D.: Relaxing i.i.d. assumption in online pattern recognition. Tech. Rep. CS-TR-03-11, Department of Computer Science, Royal Holloway, University of London (2003)
298. Ryabko, D.: Pattern recognition for conditionally independent data. J. Mach. Learn. Res. **7**, 645–664 (2006)

299. Ryabko, D., Vovk, V., Gammerman, A.: Online region prediction with real teachers, On-line compression modelling project, http://www.alrw.net/old/old.html, Working Paper #7 (2003)
300. Sadinle, M., Lei, J., Wasserman, L.: Least ambiguous set-valued classifiers with bounded error levels. J. Am. Stat. Assoc. **114**, 223–234 (2019)
301. Sampson, A.R.: A tale of two regressions. J. Am. Stat. Assoc. **69**, 682–689 (1974)
302. Saunders, C., Gammerman, A., Vovk, V.: Ridge regression learning algorithm in dual variables. In: Shavlik, J.W. (ed.) Proceedings of the Fifteenth International Conference on Machine Learning, pp. 515–521. Morgan Kaufmann, San Francisco (1998)
303. Saunders, C., Gammerman, A., Vovk, V.: Transduction with confidence and credibility. In: Dean, T. (ed.) Proceedings of the Sixteenth International Joint Conference on Artificial Intelligence, vol. 2, pp. 722–726. Morgan Kaufmann, San Francisco (1999)
304. Savage, L.J.: The Foundations of Statistics. Wiley, New York (1954). Second edition: Dover, New York (1972)
305. Schafer, D., Pierce, D.: Hugh Daniel Brunk: 1919–2009. IMS Bullet. **39**(10), 6–7 (2010)
306. Schapire, R.E., Freund, Y.: Boosting: Foundations and Algorithms. MIT Press, Cambridge (2012)
307. Schervish, M.J.: Theory of Statistics. Springer, New York (1995)
308. Schölkopf, B., Smola, A.J.: Learning with Kernels. MIT Press, Cambridge (2002)
309. Schölkopf, B., Janzing, D., Peters, J., Sgouritsa, E., Zhang, K., Mooij, J.: On causal and anticausal learning. In: Langford, J., Pineau, J. (eds.) Proceedings of the Twenty Ninth International Conference on Machine Learning, pp. 459–466. Omnipress, Madison (2012)
310. Schölkopf, B., Luo, Z., Vovk, V. (eds.): Empirical Inference: Festschrift in Honor of Vladimir N. Vapnik. Springer, Berlin (2013)
311. Schweder, T., Hjort, N.L.: Confidence, Likelihood, Probability: Statistical Inference with Confidence Distributions. Cambridge University Press, Cambridge (2016)
312. Seal, H.L.: Studies in the history of probability and statistics. XV: The historical development of the Gauss linear model. Biometrika **54**, 1–24 (1967)
313. Seber, G.A.F., Lee, A.J.: Linear Regression Analysis, 2nd edn. Wiley, Hoboken (2003)
314. Seidenfeld, T.: Jeffreys, Fisher, and Keynes: predicting the third observation, given the first two. In: Cottrell, A.F., Lawlor, M.S. (eds.) New Perspectives on Keynes, pp. 39–52. Duke University Press, Durham (1995)
315. Shafer, G.: Probabilistic Expert Systems. SIAM, Philadelphia (1996)
316. Shafer, G.: The language of betting as a strategy for statistical and scientific communication (with discussion). J. R. Stat. Soc. A **184**, 407–478 (2021)
317. Shafer, G., Vovk, V.: Probability and Finance: It's Only a Game! Wiley, New York (2001)
318. Shafer, G., Vovk, V.: The sources of Kolmogorov's *Grundbegriffe*. Stat. Sci. **21**, 70–98 (2006). A greatly expanded version is published as arXiv report 1802.06071 under the title "The origins and legacy of Kolmogorov's *Grundbegriffe*"
319. Shafer, G., Vovk, V.: A tutorial on conformal prediction. J. Mach. Learn. Res. **9**, 371–421 (2008)
320. Shafer, G., Vovk, V.: Game-Theoretic Foundations for Probability and Finance. Wiley, Hoboken (2019)
321. Shalev-Shwartz, S., Ben-David, S.: Understanding Machine Learning: From Theory to Algorithms. Cambridge University Press, New York (2014)
322. Shawe-Taylor, J., Cristianini, N.: Kernel Methods for Pattern Analysis. Cambridge University Press, Cambridge (2004)
323. Shen, A., Uspensky, V.A., Vereshchagin, N.: Kolmogorov Complexity and Algorithmic Randomness. American Mathematical Society, Providence (2017)
324. Shen, J., Liu, R., Xie, M.: Prediction with confidence—a general framework for predictive inference. J. Stat. Plann. Infer. **195**, 126–140 (2018)
325. Shewhart, W.A.: Economic Control of Quality of Manufactured Product. Van Nostrand, New York (1931)
326. Sheynin, O.B.: История теории вероятностей до XX века (The History of Probability Theory before the Twentieth Century). Северо-Западный заочный государственный технический университет, St. Petersburg (2003)

327. Shiryaev, A.N.: On optimum methods in quickest detection problems. Theory Probab. Appl. **8**, 22–46 (1963)
328. Shiryaev, A.N.: Quickest detection problems: fifty years later. Sequential Anal. **29**, 345–385 (2010). Editor's special invited paper
329. Shiryaev, A.N.: Probability-1, 3rd edn. Springer, New York (2016)
330. Shiryaev, A.N.: Probability-2, 3rd edn. Springer, New York (2019)
331. Shiryaev, A.N.: Stochastic Disorder Problems. Springer, Cham (2019)
332. Siegmund, D., Yakir, B., Zhang, N.R.: Detecting simultaneous variant intervals in aligned sequences. Ann. Appl. Stat. **5**, 645–668 (2011)
333. Silverman, B.W., Jones, M.C.: E. Fix and J. L. Hodges (1951): an important contribution to nonparametric discriminant analysis and density estimation: commentary on Fix and Hodges (1951). Int. Stat. Rev. **57**, 233–238 (1989)
334. Simard, P., LeCun, Y., Denker, J.: Efficient pattern recognition using a new transformation distance. In: Hanson, S.J., Cowan, J.D., Giles, C.L. (eds.) Advances in Neural Information Processing Systems, vol. 5, pp. 50–58. Morgan Kaufmann, San Mateo (1993)
335. Sjöstedt–de Luna, S.: Some properties of weakly approaching sequences of distributions. Stat. Probab. Lett. **75**, 119–126 (2005)
336. Smith, A.F.M.: On random sequences with centred spherical symmetry. J. R. Stat. Soc. B **43**, 208–209 (1981)
337. Snell, J.L.: Gambling, probability, and martingales. Math. Intell. **4**(3), 118–124 (1982)
338. Snell, J.L.: A conversation with Joe Doob. Stat. Sci. **12**, 301–311 (1997)
339. Stankevičiūtė, K., Alaa, A.M., van der Schaar, M.: Conformal time-series forecasting. In: Advances in Neural Information Processing Systems 34 (NeurIPS 2021) (2021)
340. Steinwart, I.: On the influence of the kernel on the consistency of support vector machines. J. Mach. Learn. Res. **2**, 67–93 (2001)
341. Steinwart, I., Christmann, A.: Support Vector Machines. Springer, New York (2008)
342. Stigler, S.M.: Gauss and the invention of least squares. Ann. Stat. **9**, 465–474 (1981). Revised version published as Chap. 17 of [345]
343. Stigler, S.M.: The History of Statistics: The Measurement of Uncertainty Before 1900. Harvard University Press, Cambridge (1986)
344. Stigler, S.M.: Laplace's 1774 memoir on inverse probability. Stat. Sci. **1**, 359–378 (1986)
345. Stigler, S.M.: Statistics on the Table: The History of Statistical Concepts and Methods. Harvard University Press, Cambridge (1999)
346. Stone, M.: Cross-validatory choice and assessment of statistical predictions (with discussion). J. R. Stat. Soc. B **36**, 111–147 (1974). Barnard's comment (proposing the vote of thanks): 133–135
347. Stone, C.J.: Consistent nonparametric regression (with discussion). Ann. Stat. **5**, 595–645 (1977)
348. Stout, W.F.: A martingale analogue of Kolmogorov's law of the iterated logarithm. Zeitschrift für Wahrscheinlichkeitstheorie und verwandte Gebiete **15**, 279–290 (1970)
349. Stuart, A., Ord, K.J., Arnold, S.: Kendall's Advanced Theory of Statistics, Vol. 2a: Classical Inference and the Linear Model, 6th edn. Arnold, London (1999)
350. Stutz, D., Dvijotham, K., Cemgil, A.T., Doucet, A.: Learning optimal conformal classifiers. In: Tenth International Conference on Learning Representations (2022)
351. Sutton, R.S., Barto, A.G.: Reinforcement Learning: An Introduction. MIT Press, Cambridge (1998)
352. Taillardat, M., Mestre, O., Zamo, M., Naveau, P.: Calibrated ensemble forecasts using quantile regression forests and ensemble model output statistics. Month. Weather Rev. **144**, 2375–2393 (2016)
353. Takeuchi, K.: Statistical Prediction Theory (in Japanese). Baifukan, Tokyo (1975)
354. Takeuchi, J., Kawabata, T., Barron, A.R.: Properties of Jeffreys mixture for Markov sources. IEEE Trans. Inf. Theory **59**, 438–457 (2013)
355. Tibshirani, R.J., Barber, R.F., Candès, E.J., Ramdas, A.: Conformal prediction under covariate shift. In: Advances in Neural Information Processing Systems, vol. 32, pp. 2530–2540. Curran Associates, Red Hook (2019)

356. Torres-Sospedra, J., Montoliu, R., Martínez-Usó, A., Avariento, J.P, Arnau, T.J., Benedito-Bordonau, M., Huerta, J: `UJIIndoorLoc`: a new multi-building and multi-floor database for WLAN fingerprint-based indoor localization problems. In: 2014 International Conference on Indoor Positioning and Indoor Navigation (IPIN 2014), pp. 261–270. Institute of Electrical and Electronics Engineers (2014)
357. Tukey, J.W.: Nonparametric estimation II: statistically equivalent blocks and tolerance regions – the continuous case. Ann. Math. Stat. **18**, 529–539 (1947)
358. Turing, A.M.: Computing machinery and intelligence. Mind **59**, 433–460 (1950)
359. Valiant, L.G.: A theory of the learnable. Commun. ACM **27**, 1134–1142 (1984)
360. van den Burg, G.J.J., Groenen, P.J.F.: GenSVM: a generalized multiclass support vector machine. J. Mach. Learn. Res. **17**(224), 1–42 (2016)
361. van der Vaart, A.W.: Asymptotic Statistics. Cambridge University Press, Cambridge (1998)
362. van der Vaart, A.W., Wellner, J.A.: Weak Convergence and Empirical Processes: With Applications to Statistics. Springer, New York (1996)
363. Vapnik, V.: Вапник, В.Н.: Задача обучения распознаванию образов. Знание, Moscow (1971)
364. Vapnik, V.: Estimation of Dependences Based on Empirical Data. Springer, New York (1982). This is the English translation of: Вапник, В.Н., Восстановление зависимостей по эмпирическим данным, Nauka, Moscow (1979)
365. Vapnik, V.: The Nature of Statistical Learning Theory. Springer, New York (1995). Second edition: 2000
366. Vapnik, V.: Statistical Learning Theory. Wiley, New York (1998)
367. Vapnik, V.: Empirical inference science: Afterword of 2006. In: Estimation of Dependences Based on Empirical Data. Springer, New York (2006). This book is a reprint of [365] with an added afterword
368. Vapnik, V., Chervonenkis, A.: On the uniform convergence of relative frequencies of events to their probabilities. Soviet Math. Dokl. **9**, 915–918 (1968)
369. Vapnik, V., Chervonenkis, A.: On the uniform convergence of relative frequencies of events to their probabilities. Theory Probab. Appl. **16**, 264–280 (1971)
370. Vapnik, V., Chervonenkis, A.: Вапник, В.Н. and Червоненкис, А.Я.: Теория распознавания образов (Theory of Pattern Recognition). Nauka, Moscow (1974). German translation: Wapnik, W., and Tscherwonenkis, A., *Theorie der Zeichenerkennung*, Akademie-Verlag, Berlin (1979)
371. Vapnik, V., Sterin, A.: Ordered minimization of total risk in a pattern-recognition problem. Autom. Remote Conrol **10**, 1495–1503 (1977). Russian original in: *Автоматика и телемеханика* **10**, 83–92
372. Venn, J.: The Logic of Chance. Macmillan, London (1866). Our references are to the third edition (1888)
373. Ville, J.: Étude critique de la notion de collectif. Gauthier-Villars, Paris (1939)
374. Volkhonskiy, D., Burnaev, E., Nouretdinov, I., Gammerman, A., Vovk, V.: Inductive conformal martingales for change-point detection. Proc. Mach. Learn. Res. **60**, 132–153 (2017). COPA 2017
375. von Neumann, J., Morgenstern, O.: Theory of Games and Economic Behavior, 3rd edn. Princeton University Press, Princeton (1953). First edition: 1944
376. Vovk, V.: On the concept of the Bernoulli property. Russian Math. Surv. **41**, 247–248 (1986). Russian original: О понятии бернуллиевости. Another English translation with proofs: arXiv report 1612.08859
377. Vovk, V.: Aggregating strategies. In: Fulk, M., Case, J. (eds.) Proceedings of the Third Annual Workshop on Computational Learning Theory, pp. 371–383. Morgan Kaufmann, San Mateo (1990)
378. Vovk, V.: Universal forecasting algorithms. Inf. Comput. **96**, 245–277 (1992)
379. Vovk, V.: A logic of probability, with application to the foundations of statistics (with discussion). J. R. Stat. Soc. B **55**, 317–351 (1993)
380. Vovk, V.: Derandomizing stochastic prediction strategies. Mach. Learn. **35**, 247–282 (1999)
381. Vovk, V.: Competitive on-line statistics. Int. Stat. Rev. **69**, 213–248 (2001)

382. Vovk, V.: Kolmogorov's complexity conception of probability. In: Hendricks, V.F., Pedersen, S.A., Jørgensen, K.F. (eds.) Probability Theory: Philosophy, Recent History and Relations to Science, pp. 51–69. Kluwer, Dordrecht (2001)
383. Vovk, V.: On-line confidence machines are well-calibrated. In: Proceedings of the Forty-Third Annual Symposium on Foundations of Computer Science, pp. 187–196. IEEE Computer Society, Los Alamitos (2002)
384. Vovk, V.: Well-calibrated predictions from on-line compression models. Theor. Comput. Sci. **364**, 10–26 (2006). ALT 2003 Special Issue
385. Vovk, V.: Conditional validity of inductive conformal predictors. Mach. Learn. **92**, 349–376 (2013)
386. Vovk, V.: Cross-conformal predictors. Ann. Math. Artif. Intell. **74**, 9–28 (2015)
387. Vovk, V.: The fundamental nature of the log loss function. In: Beklemishev, L.D., Blass, A., Dershowitz, N., Finkbeiner, B., Schulte, W. (eds.) Fields of Logic and Computation II: Essays Dedicated to Yuri Gurevich on the Occasion of His 75th Birthday. Lecture Notes in Computer Science, vol. 9300, pp. 307–318. Springer, Cham (2015)
388. Vovk, V.: Transductive conformal prediction. Int. J. Artif. Intell. Tools **24**(6), 1560001 (2015). COPA 2013 Special Issue
389. Vovk, V.: Conformal testing in a binary model situation. Proc. Mach. Learn. Res. **152**, 131–150 (2021). COPA 2021
390. Vovk, V.: Protected probabilistic regression. Tech. Rep. arXiv:2105.08669 [cs.LG], arXiv.org e-Print archive (2021)
391. Vovk, V.: Testing randomness online. Stat. Sci. **36**, 595–611 (2021)
392. Vovk, V.: Conformal predictive distributions: an approach to nonparametric fiducial prediction. In: Berger, J., Meng, X.-L., Reid, N., Xie, M. (eds.) Handbook of Bayesian, Fiducial, and Frequentist Inference. Chapman and Hall, London (2022, to appear)
393. Vovk, V.: Universal predictive systems. Pattern Recognit. **126**, 108536 (2022). COPA 2019 Special Issue
394. Vovk, V., Petej, I.: Venn–Abers predictors. In: Zhang, N.L., Tian, J. (eds.) Proceedings of the Thirtieth Conference on Uncertainty in Artificial Intelligence, pp. 829–838. AUAI Press, Corvallis (2014)
395. Vovk, V., Shafer, G.: Kolmogorov's contributions to the foundations of probability. Problems Inf. Trans. **39**, 21–31 (2003)
396. Vovk, V., V'yugin, V.V.: On the empirical validity of the Bayesian method. J. R. Stat. Soc. B **55**, 253–266 (1993)
397. Vovk, V., Wang, R.: Combining p-values via averaging. Biometrika **107**, 791–808 (2020)
398. Vovk, V., Wang, R.: E-values: calibration, combination, and applications. Ann. Stat. **49**, 1736–1754 (2021)
399. Vovk, V., Zhdanov, F.: Prediction with expert advice for the Brier game. J. Mach. Learn. Res. **10**, 2445–2471 (2009)
400. Vovk, V., Gammerman, A., Saunders, C.: Machine-learning applications of algorithmic randomness. In: Proceedings of the Sixteenth International Conference on Machine Learning, pp. 444–453. Morgan Kaufmann, San Francisco (1999)
401. Vovk, V., Fedorova, V., Nouretdinov, I., Gammerman, A.: Criteria of efficiency for set-valued classification. Ann. Math. Artif. Intell. **81**, 21–46 (2017). Special Issue on Conformal and Probabilistic Prediction with Applications
402. Vovk, V., Gammerman, A., Shafer, G.: Algorithmic Learning in a Random World. Springer, New York (2005). This is the first edition of this book
403. Vovk, V., Nouretdinov, I., Gammerman, A.: Testing exchangeability on-line. In: Fawcett, T., Mishra, N. (eds.) Proceedings of the Twentieth International Conference on Machine Learning, pp. 768–775. AAAI Press, Menlo Park (2003)
404. Vovk, V., Shafer, G., Nouretdinov, I.: Self-calibrating probability forecasting. In: Thrun, S., Saul, L.K., Schölkopf, B. (eds.) Advances in Neural Information Processing Systems, vol. 16, pp. 1133–1140. MIT Press, Cambridge (2004)
405. Vovk, V., Nouretdinov, I., Gammerman, A.: On-line predictive linear regression. Ann. Stat. **37**, 1566–1590 (2009). See also arXiv:math/0511522 [math.ST] (November 2011)

406. Vovk, V., Petej, I., Fedorova, V.: From conformal to probabilistic prediction. In: Iliadis, L., Maglogiannis, I., Papadopoulos, H., Sioutas, S., Makris, C. (eds.) AIAI Workshops, COPA 2014. IFIP Advances in Information and Communication Technology, vol. 437, pp. 221–230 (2014)
407. Vovk, V., Papadopoulos, H., Gammerman, A. (eds.): Measures of Complexity: Festschrift for Alexey Chervonenkis. Springer, Cham (2015)
408. Vovk, V., Petej, I., Fedorova, V.: Large-scale probabilistic predictors with and without guarantees of validity. In: Advances in Neural Information Processing Systems, vol. 28, pp. 892–900. Curran Associates, Red Hook (2015)
409. Vovk, V., Nouretdinov, I., Manokhin, V., Gammerman, A.: Conformal predictive distributions with kernels. In: Rozonoer, L., Mirkin, B., Muchnik, I. (eds.) Braverman's Readings in Machine Learning: Key Ideas from Inception to Current State. Lecture Notes in Artificial Intelligence, vol. 11100, pp. 103–121. Springer, Cham (2018)
410. Vovk, V., Shen, J., Manokhin, V., Xie, M.: Nonparametric predictive distributions based on conformal prediction. Mach. Learn. **108**, 445–474 (2019). COPA 2017 Special Issue
411. Vovk, V., Petej, I., Toccaceli, P., Gammerman, A., Ahlberg, E., Carlsson, L.: Conformal calibration. Proc. Mach. Learn. Res. **128**, 84–99 (2020). COPA 2020
412. Vovk, V., Nouretdinov, I., Gammerman, A.: Conformal testing: binary case with Markov alternatives. Proc. Mach. Learn. Res. **179**, 207–218 (2022). COPA 2022
413. Vovk, V., Petej, I., Gammerman, A.: Protected probabilistic classification. Tech. Rep. arXiv:2107.01726 [cs.LG], arXiv.org e-Print archive (2021). Short version published as poster extended abstract: Proceedings of Machine Learning Research **152**, 297–299 (2021). COPA 2021
414. Vovk, V., Petej, I., Nouretdinov, I., Ahlberg, E., Carlsson, L., Gammerman, A.: Retrain or not retrain: conformal test martingales for change-point detection. Proc. Mach. Learn. Res. **152**, 191–210 (2021). COPA 2021
415. V'yugin, V.V.: Algorithmic entropy (complexity) of finite objects and its applications to defining randomness and amount of information. Sel. Math. Soviet. **13**, 357–389 (1994)
416. Wahba, G.: Spline Models for Observational Data. SIAM, Philadelphia (1990)
417. Wald, A.: An extension of Wilks' method for setting tolerance limits. Ann. Math. Stat. **14**, 45–55 (1943)
418. Wald, A.: Sequential tests of statistical hypotheses. Ann. Math. Stat. **16**, 117–186 (1945)
419. Wald, A.: Sequential Analysis. Wiley, New York (1947)
420. Wald, A.: Statistical Decision Functions. Wiley, New York (1950)
421. Wald, A., Wolfowitz, J.: Optimum character of the sequential probability ratio test. Ann. Math. Stat. **19**, 326–339 (1948)
422. Walsh, J.B.: A property of conformal martingales. Séminaire de probabilités (Strasbourg) **11**, 490–492 (1977)
423. Wasserman, L.: Frasian inference. Stat. Sci. **26**, 322–325 (2011)
424. Watkins, C., Weston, J.: Support vector machines for multiclass pattern recognition. In: Proceedings of the Seventh European Symposium on Artificial Neural Networks, pp. 219–224 (1999)
425. Webb, G.I., Ting, K.M.: On the application of ROC analysis to predict classification performance under varying class distributions. Mach. Learn. **58**, 25–32 (2005)
426. Werner, H., Carlsson, L., Ahlberg, E., Boström, H.: Evaluation of updating strategies for conformal predictive systems in the presence of extreme events. Proc. Mach. Learn. Res. **152**, 229–242 (2021). COPA 2021
427. Wilks, S.S.: Determination of sample sizes for setting tolerance limits. Ann. Math. Stat. **12**, 91–96 (1941)
428. Williams, D.: Probability with Martingales. Cambridge University Press, Cambridge (1991)
429. Xu, C., Xie, Y.: Conformal prediction interval for dynamic time-series. Proc. Mach. Learn. Res. **139**, 11559–11569 (2021). ICML 2021

430. Zadrozny, B., Elkan, C.: Transforming classifier scores into accurate multiclass probability estimates. In: Hand, D., Keim, D., Ng, R. (eds.) Proceedings of the Eighth ACM SIGKDD International Conference on Knowledge Discovery and Data Mining, pp. 694–699. ACM Press, New York (2002)
431. Zeni, G., Fontana, M., Vantini, S.: Conformal prediction: a unified review of theory and new challenges. Bernoulli **29**, 1–23 (2023)

Index

V. Vovk et al., *Algorithmic Learning in a Random World*,
https://doi.org/10.1007/978-3-031-06649-8

D

Printed in the United States
by Baker & Taylor Publisher Services